Professional Practice in Earth Sciences

Series Editor

James W. LaMoreaux, Tuscaloosa, AL, USA

Books in Springer's Professional Practice in Earth Sciences Series present state-of-the-art guidelines to be applied in multiple disciplines of the earth system sciences. The series portfolio contains practical training guidebooks and supporting material for academic courses, laboratory manuals, work procedures and protocols for environmental sciences and engineering. Items published in the series are directed at researchers, students, and anyone interested in the practical application of science. Books in the series cover the applied components of selected fields in the earth sciences and enable practitioners to better plan, optimize and interpret their results. The series is subdivided into the different fields of applied earth system sciences: Laboratory Manuals and work procedures, Environmental methods and protocols and training guidebooks.

More information about this series at http://www.springer.com/series/11926

Shuning Dong · Wanfang Zhou ·
Qisheng Liu · Hao Wang · Yadong Ji

Methods and Techniques for Preventing and Mitigating Water Hazards in Mines

Springer

Shuning Dong
China Coal Technology & Engineering Group Corp
Xi'an Research Institute
Xian, Shaanxi, China

Qisheng Liu
China Coal Technology & Engineering Group Corp
Xi'an Research Institute
Xian, Shaanxi, China

Yadong Ji
China Coal Technology & Engineering Group Corp
Xi'an Research Institute
Xian, Shaanxi, China

Wanfang Zhou
Zeo Environmental, LLC
Knoxville, TN, USA

Hao Wang
China Coal Technology & Engineering Group Corp
Xi'an Research Institute
Xian, Shaanxi, China

ISSN 2364-0073 ISSN 2364-0081 (electronic)
Professional Practice in Earth Sciences
ISBN 978-3-030-67058-0 ISBN 978-3-030-67059-7 (eBook)
https://doi.org/10.1007/978-3-030-67059-7

This Springer imprint is published by the registered company Springer Nature Switzerland AG
The registered company address is: Gewerbestrasse 11, 6330 Cham, Switzerland

Preface

Mine water inflows have a variety of origins. Access shafts or open pits often intercept aquifers before reaching the mining horizons. These aquifers may be shallow phreatic or deeper confined, and their nature may range from porous alluvial deposits or sandstones to intensely fractured and karstified limestones. Underground tunnels, galleries, and working faces may encounter faults or karst collapse columns that can transmit groundwater through relatively impermeable zones from aquifers into mine workings. Similarly, the presence of faults and other discontinuities in the rock mass may present water problems for developments under bodies of water such as lakes, reservoirs, rivers, and the sea, or to developments in close proximity to abandoned mine pools, unsealed or poorly sealed shafts and boreholes. To these pre-existing water pathways must be added the induced or modified water pathways due to mine development and mineral extraction. Hydraulically conductive channels may be created by mining through roof or floor strata to connect with the sources of water. Water hazards in the form of water inrush often occur during mining and account for many of mine disasters and casualties in many countries of the world. According to statistics by Chinese State Administration of Production Safety Supervision and Management, the water inrush disasters are second only to the gas explosion disasters in coal mines in the serious and extraordinarily serious accident categories. Mine water inrushes not only cause casualties, but also are the most serious of mine accidents in terms of economic losses, accident emergency rescue, and mine restoration effort.

Major accomplishments have been made in predicting, preventing, detecting, and mitigating the water hazards in mines during the last three decades. Practices have demonstrated that conceptual site models that describe the mine water inflow pathways from water sources to working areas are essential to address both the safety issues and environmental impacts posed by mining activities. The traditional dewatering method must be carefully evaluated during environmental impact studies to comply with the current regulatory policy of green and water-conservation mining. Sophisticated multi-porosity and multi-permeability numerical models are available to predict mine water inflows more accurately. Advanced geophysical techniques and early warning monitoring devices are being implemented to detect and capture anomalies that are indicative of water inrushes. Water inrush risk assessment is supplemented with new methods including the vulnerability index approach that addresses the multi-factor, nonlinear, and mathematically non-amenable water inrush processes. Development of directional drilling technique makes it possible to implement targeted grouting proactively in specific stratum on a regional scale or to seal off localized water pathways in response to water inrush incidents.

This book presents the research results and practices of the above-mentioned topics in ten chapters. Chapter 1 gives the definition of water inrush in mines and provides an overview of the hazards associated with water inrushes. Chapter 2 presents the water inrush mechanisms and methods to evaluate water inrush risks. Chapter 3 describes various methods in predicting water inflow into mines from heterogeneous karst aquifers. Chapters 4 through 7 document eleven case studies in which the state-of-the-art methods and technologies are applied to solving water hazard problems under different conditions. The case studies in Chaps. 4 and 5 deal with water hazards posed from aquifers underlying and overlying coal seams, respectively. The case study in Chap. 6 deals with water hazards posed from an abandoned mine pool, whereas the case study in Chap. 7 discusses water hazards associated with a surface river in a quarry. Chapters 8 and 9 describe environmental impact assessments for an underground mine and an open pit quarry, respectively. Both cases emphasize the importance of determining the water sources and pathways. Chapter 10 provides the best management practices in mine water prevention and control.

The greatest part of this practical reference book is the technology applications in the case studies. The book would not be complete without support of those who participated in these projects. We are in debt to all of them. We would like to thank Xi'an Research Institute of China Coal Technology & Engineering Group Corp for providing most of the case studies and drafting most of the graphics. In particular, the following professionals are acknowledged for their significant contributions to the projects: Jin Dewu, Huang Xuanming, Cheng Jiayuan, Wang Xinwen, Li Yunchao, Nan Shenghui, Zheng Shitian, Zhu Mingcheng, Ji Zhongkui, Wang Shidong, Cao Haidong, Li Gongyu, Liu Yingfeng, Liu Yang, Zhao Baofeng, Xu Feng, Wang Qiangmin, and Zhou Zhenfang. We would also like to express our gratitude to Drs. James W. LaMoreaux, Annet Buettner, and Prasanna Kumar Narayanasamy of Springer Nature for their support and encouragement in preparation of the book.

Xian, China — Shuning Dong
Xian, China — Qisheng Liu
Xian, China — Hao Wang
Xian, China — Yadong Ji
Knoxville, USA — Wanfang Zhou
November 2020

Contents

Chapter 1
Water Hazards in Coal Mines and Their Classifications

1.1 Introduction

Mine water hazards are often caused by mine water inrushes. Mine water inrush is a phenomenon in which a large volume of water unexpectedly gushes into underground workings or open pit mines when tunneling or mining exposes water-bearing media, such as high-pressure confined aquifers, surface water bodies, or underground mine pools. Mine water inrush generally occurs dramatically and can flood underground workings in a short period of time, jeopardizing mine production and causing casualties (Gui et al. 2017; Zhang et al. 2017; LaMoreaux et al. 2014). China is currently the largest coal producing country in the world; its coal resources encompass a large geographical area with various environments (Sun et al. 2015). Furthermore, China consists of multiple tectonic plates that were spliced through numerous tectonic movements, resulting in complex hydrogeological conditions in many coal mines (Yin et al. 2018; Zhao et al. 2017). Therefore, China is one of the countries in which coal mining is most seriously impacted by mine water inrushes (Yang et al. 2017; Lu et al. 2017; Yin et al. 2018).

Statistics on serious and major coal mine accidents by China's State Administration of Work Safety (Sun et al. 2015; Wang and Meng 2018) indicate that mine water inrush is the second most serious disaster type after gas explosions. More than 2.5 $\times$ 1010 tons of coal reserves are currently threatened with water inrush, mainly in developed industrial areas including north China, east China, and south China where coal reserves account for approximately 70% of the national coal resource. In the last 18 years between 2000 and 2017, the water-related hazards and casualties in coal mines of China have been greatly reduced (LaMoreaux et al. 2014; Wei et al. 2016; Qiao et al. 2017; Wang et al. 2017; Wu et al. 2017; Zhang et al. 2017; Xu et al. 2018). The improvement is primarily attributed to three factors: (1) progress of science and technology in mine water control and management, (2) updated mining equipment and mining technology, and (3) implementation of more stringent regulations on mine safety.

S. Dong et al., *Methods and Techniques for Preventing and Mitigating Water Hazards in Mines*, Professional Practice in Earth Sciences,
https://doi.org/10.1007/978-3-030-67059-7_1

Although the number of accidents and casualties has declined, the casualties and property losses induced by these accidents are still concerns. It is imperative to recognize the current situation of mine water inrush, study countermeasures for preventing water inrush and take effective engineering controls (Zhang et al. 2014, 2018a, b; Yang et al. 2017). Mine water inrush disaster control and water resource utilization will still be one of the important research topics to achieve accident-free and scientific mining in China (Zhang 2005; Qiu et al. 2017; Li et al. 2018).

1.2 Water Inrush Conceptual Site Models for Coal Mines of China

Development of conceptual site models (CSMs) will provide necessary guidance for further improving the safety records in coal mining. A CSM is a description of relationships between sources and receptors under both natural and man-made environments based on existing knowledge (Zhou and Lei 2017). It describes sources of water recharging the mine as well as complete, potentially complete, or incomplete exposure pathways from the sources to the mining area. The CSM serves as a planning tool, a modeling and data interpretation aid, and a device that assists the project team in communicating with the public, integrating information and making informed decisions. These decisions can range from water inrush risk assessment to engineering measures for risk reduction. A well-thought CSM also provides a structure to summarize and display information about a site and identify additional information needed to develop technically sound decisions. Table 1.1 presents a general CSM for coal mines of China. The CSMs include the ecological environment as another essential receptor in evaluating engineering measures for mine water control.

1.2.1 Development of Water Inrush Conceptual Site Models

The CSM includes three types of water sources: surface water, water in underground mine pools, and groundwater. Each type of water source is divided into different categories based on their origins. The threats posed by water inrushes consist of direct threats and indirect threats, and for the purposes of this document are differentiated by the terms hazard and risk. Water inrush presents a hazard of direct physical injury to miners and damage to working spaces in tunnels and working panels. Water inrush also presents a risk to the environment through indirect exposures. Surface subsidence, water resource reduction and contamination, and adverse impacts on biodiversity are examples of these indirect consequences of water inrushes. Both hazards and risks must be considered in water inrush assessment. The degree of hazard and risk posed by water inrush is usually proportional to the strength of water

Table 1.1 Water inrush CSMs for coal mines of China

Water Sources	Pathways	Water Inrushes						Applicability to Regions of China (Figure 1.1)
		Hazard or Direct Injury			Risk to Ecological Environment			
		Worker Safety & Health	Flooding of Tunnel	Flooding of Working Panel	Surface Subsidence	Water Resource	Crops and Biodiversity	
Surface water: Precipitation Runoff	Shaft	●	○	○	◎	●	◎	1-6
	Mining-induced fracture	●	○	○	◎	●	◎	
	Sinkhole	●	○	○	◎	●	◎	
	Poorly sealed borehole	●	○	○	◎	●	◎	
Surface water: River/Lake/Reservoir	Shaft	○	○	○	◎	●	◎	
	Mining-induced fracture	●	○	○	◎	●	◎	
	Sinkhole	○	○	○	◎	●	◎	
	Poorly sealed borehole	●	○	○	◎	●	◎	
Surface water: Mining-induced Subsidence Pond	Shaft	●	○	○	◎	●	◎	
	Mining-induced fracture	●	○	○	◎	●	◎	
	Sinkhole	●	○	○	◎	●	◎	
	Poorly sealed borehole	●	○	○	◎	●	◎	
Water in abandoned mines: Accumulated water in old mining area of same mine	Direct exposure by tunneling	●	○	○	◎	●	◎	1-6
	Mining-induced fracture	●	○	○	◎	●	◎	
	Hydraulically conductive fault/fracture	●	○	○	◎	●	◎	
Water in abandoned mines: Accumulated water in neighboring mine	Direct exposure by tunneling	●	○	○	◎	●	◎	
	Mining-induced fracture	●	○	○	◎	●	◎	
	Hydraulically conductive fault/fracture	●	○	○	◎	●	◎	

(continued)

Table 1.1 (continued)

Water Sources		Pathways	Water Inrushes						Applicability to Regions of China (Figure 1.1)
			Hazard or Direct Injury			Risk to Ecological Environment			
			Worker Safety & Health	Flooding of Tunnel	Flooding of Working Panel	Surface Subsidence	Water Resource	Crops and Biodiversity	
Groundwater: Groundwater directly intercepted by mining activities	Fractured sandstone aquifer	Direct exposure by tunnel	●	○	○	◎	●	◎	1–6
		Direct exposure by mining panel	●	○	○	◎	●	◎	
	Thin-bedded limestone aquifer	Direct exposure by tunnel	●	○	○	◎	●	◎	
		Direct exposure by mining panel	●	○	○	◎	●	◎	
Groundwater: Groundwater overlying working area while separated by an aquitard under natural conditions	Porous medium aquifer	Mining-induced fracture in overlying formation	●	○	○	◎	●	◎	1–6
		Hydraulically conductive fault/fracture	●	○	○	◎	●	◎	
		Hydraulically conductive karst collapse column	○	○	○	◎	●	◎	
	Fractured sandstone aquifer	Mining-induced fracture in overlying formation	●	○	○	◎	●	◎	
		Hydraulically conductive fault/fracture	●	○	○	◎	●	◎	
		Hydraulically conductive karst collapse column	○	○	○	◎	●	◎	
	Thin-bedded limestone aquifer	Mining-induced fracture in overlying formation	●	○	○	◎	●	◎	
		Hydraulically conductive fault/fracture	●	○	○	◎	●	◎	
		Hydraulically conductive karst collapse column	○	○	○	◎	●	◎	

(continued)

Table 1.1 (continued)

Water Sources		Pathways	Water Inrushes						Applicability to Regions of China (Figure 1.1)
			Hazard or Direct Injury			Risk to Ecological Environment			
			Worker Safety & Health	Flooding of Tunnel	Flooding of Working Panel	Surface Subsidence	Water Resource	Crops and Biodiversity	
Groundwater: Groundwater underlying working areas while separated by an aquitard under natural conditions	Thin-bedded limestone aquifer	Mining-induced fracture in underlying formation	●	○	○	◎	●	◎	1–3
		Hydraulically conductive fault/fracture	●	○	○	◎	●	◎	
		Hydraulically conductive karst collapse column	○	○	○	◎	●	◎	
	Thick-bedded limestone aquifer	Mining-induced fracture in underlying formation	●	○	○	◎	●	◎	
		Hydraulically conductive fault/fracture	●	○	○	◎	●	◎	
		Hydraulically conductive karst collapse column	●	○	○	◎	●	◎	

sources, the pumping capacity of the mine, and the ecological sensitivity to water regime changes. The CSM also describes the pathways from each source to various receptors. Direct exposure to water-bearing formations, mining-induced fractures in overlying or underlying aquitards, hydraulically conductive faults and karst collapse columns, and poorly sealed boreholes are potential pathways, making water inrush possible (Zhou and Li 2001; Tan et al. 2010; Yin et al. 2018).

CSM development is an iterative process that reflects the progressive understanding of a study site from initial hydrogeological investigation to water control during coal mining. Potential sources of water and receptors should be documented in the initial CSM. Complete or potentially complete exposure pathways are focuses on hydrogeological investigation and data collection. For example, analysis of the groundwater pathway will usually entail some hypotheses on groundwater flow velocity or direction relative to potential receptors. If these parameters are not known, they can be measured through hydrogeological investigation or interpreted through modeling or professional judgment. If the results from data collection confirm the interpretations, the CSM is updated to show that the hypothesis is correct. However, if results do not support the predicted outcome, the hypothesis should be restated, leading to a revision to the existing CSM.

Geographically, China can be divided into six regions (Fig. 1.1). Applicability of the CSM to different regions is presented in Table 1.1 as well. Surface water and

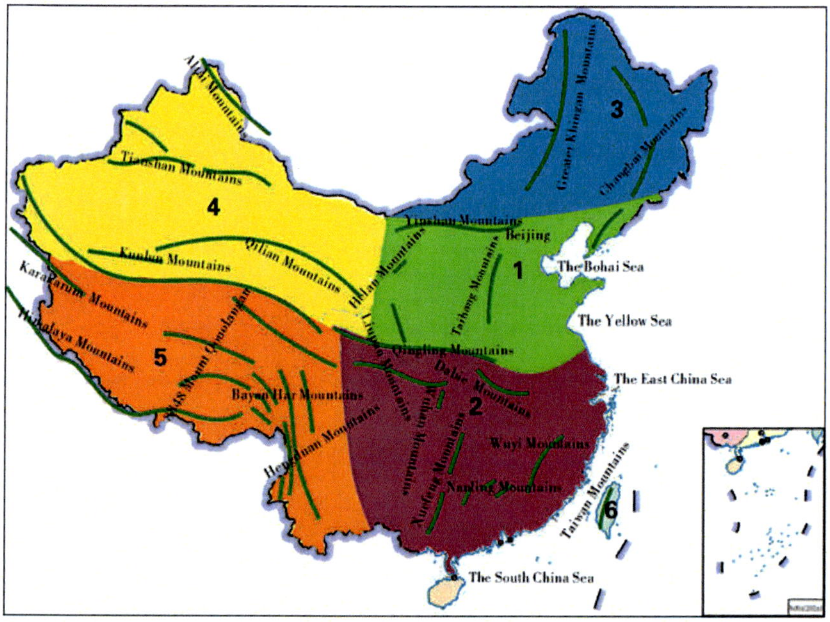

Fig. 1.1 CSM-based water inrush regions in China

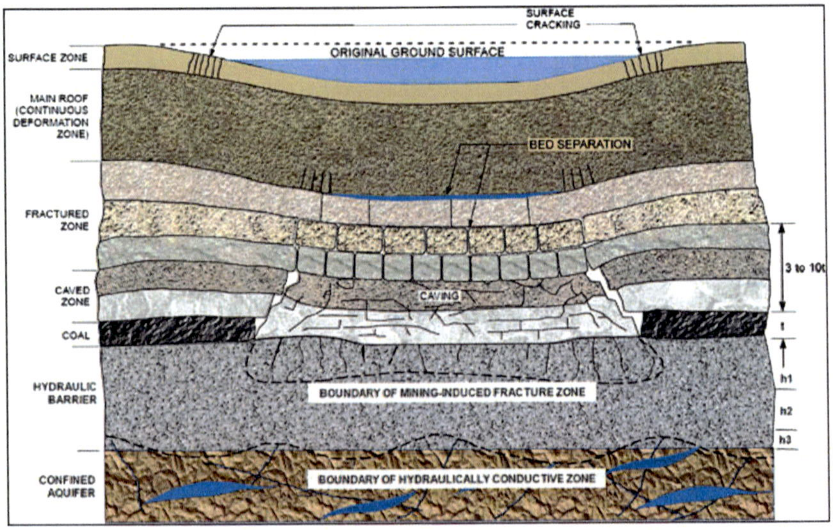

Fig. 1.2 CSM showing water threats from water bodies both underlying and overlying the coal seam (t—coal seam thickness; h1—height of mining-induced fracture zone in the top of aquitard immediately underlying the coal seam; h2—thickness of intact aquitard; h3—height of hydraulically conductive zone in the bottom of aquitard)

water in underground mine pools occur in all the regions, whereas water inrushes from aquifers both overlying and underlying coal seams occur only in Regions 1 through 3.

The water inrush CSM can also be shown in diagrams to illustrate the pathways from water sources to receptors. Figure 1.2 shows a CSM for water inrush threats from both the underlying and overlying aquifers. Various factors that may affect occurrence of water inrushes are displayed in the CSM. Natural water-conductive faults or fractures as well as mining-induced fractures can be parts of complete pathways for water bodies either overlying or underlying the coal seam (Tan et al. 2013; Li et al. 2015; Wang et al. 2017).

1.2.2 *Benefits of Water Inrush Conceptual Site Models*

Water inrushes in coal mines of China pose threats to workers' safety, mine property, and the ecological environment. Representative water inrush CSMs provide an innovative tool to better understand water inrush mechanisms, help with pathway analysis in water inrush risk assessment and identify data gaps for further investigations, promote integration of water hazard control and ecological environment risk reduction. They also drive proactive engineering measures for water inrush prevention and mine water control to eliminate or reduce water inrush hazards and ecological deterioration risks. Therefore, the authors recommend that a CSM be developed for each mine and each mining area and updated on a regular basis. Development of water inrush CSMs for coal mines of China can serve the following purposes:

- Provide an innovative tool to better understand the water inrush mechanism, and design large-scale analog simulation models. With the increase in both mining depth and mining scale, additional problems may be caused by high stress, high temperature, high water pressure and high gas pressure. The water inrush mechanisms can be complex with diversified characteristics. Delayed water inrush or mining-induced bed separation water inrush in the shallow zone occurs more frequently. The CSM approach will help identify the complete, potentially complete, and incomplete pathways for either mathematical simulations or physical analog models.
- Help with pathway analysis in water inrush risk assessment and identify data gaps for further hydrogeological investigations. Water inrush risks are typically performed for complete pathways, whereas water inrush probability is provided for potentially complete pathways. Potential complete pathways between sources and receptors are often the data gaps, which form the basis for additional data collection efforts. New data on sources, interactions and receptors are compared to the current CSM to refine it as necessary. This in turn may result in additional data gaps that may impact the design of site characterization. The CSM may also help identify modeling that may be required to determine whether there is an unacceptable risk to receptors.
- Promote both hazard control of water and risk reduction of the ecological environment. The early Middle Jurassic coal fields in the northwest and western side of north China are in arid and semi-arid environments where the annual precipitation ranges from 250 to 450 mm. The natural ecological environment is fragile and sensitive to any changes of the water table. The CSMs call for a strategy that will conserve the water resource and protect the ecological system while keeping the water under control for a safe working environment in the mines. In east China, mining is severely threatened by the confined karst water from the Ordovician limestone and water in underground mine pools. Representative CSMs in this area can help resolve conflicting issues concerning coal mining safety, water resource management, and ecological protection and lay a foundation to achieve not only safe mining but also green mining.
- Drive proactive engineering measures for water inrush prevention and mine water control to eliminate or reduce water inrush hazards and ecological deterioration risks. Any engineering measures will be targeted to either water source removal or pathway elimination or both in the CSMs. Water inrush risks are often high in north China's coalfields where the aquitard or hydraulic barrier between the lower group coal seams and the underlying confined aquifer is relatively small. Large-scale grouting operations to reinforce the hydraulic barrier are considered to be effective engineering measures to eliminate the pathways. Grouting reinforcement technology injects a large quantity of grout directly into the targeted formation

according to rational design. Thin-bedded aquifers can also be the targeted formation for grouting, depending on their spatial relationship with the coal seams. The grout will fill the karst-fissures and cracks of the aquifer, transforming the aquifer into a hydraulic barrier.

1.3 Classification of Water Inrush for Coal Mines of China

1.3.1 Principles for Classification of Mine Water Inrush

Mine water inrush studies suggest that different types of water inrush generally call for different prevention and control technologies. Therefore, it is necessary to classify water inrushes according to their distinct characteristics. Due to diversity and complexity of mine water inrush, the systematic, consensus and comprehensive classifications of mine water inrush are rarely reported at present. In fact, scientific classification of mine water inrush is a huge systematic classification project and needs to develop qualitative, quantitative, and combined classification methods. Furthermore, it requires a large amount of field data and cases of water inrush as well as supports of relevant scientific theories. Therefore, the classification of mine water inrush has important theoretical and practical value for basic theory research, investigation and exploration, evaluation and prediction, design of detection equipment, prevention and control technologies, and comprehensive utilization of mine water.

The division bases need to be established prior to classification of mine water inrush. According to case histories of water inrush and specific characteristics, water recharge source, water flow channel, water recharge strength, harm forms, economic loss and casualties, and time-varying characteristics are used as the bases for classifying water inrush. Water recharge source refers to all of the water sources that exist in and show hydraulic connections with ore bodies and surrounding rock strata that can cause continuous mine water inrush during the mining process. Water recharge channel is the path for these water sources entering into mines. Harm forms indicate pit water inrush showing characteristics, such as abnormal temperature and corrosivity. In addition, economic losses and casualties measure the magnitude of economic losses and the number of casualties directly resulting from mine water inrush, respectively. Time-varying characteristics show the temporal relationship between mine water inrush and the progress of mining engineering.

1.3.2 Types of Mine Water Inrush

Classifying mine water inrush in accordance with water recharge source. Based on the nature of water recharge source, water inrush can be divided into water inrush of natural and artificial sources (Fig. 1.3). The natural water recharge source includes the mine water inrush directly recharged by atmospheric precipitation, mine water

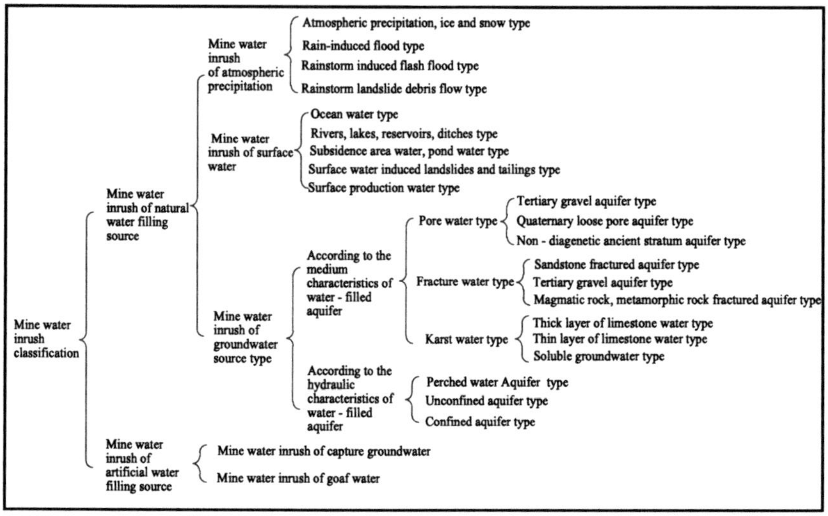

Fig. 1.3 Classification of water inrush based on recharge sources

inrush of surface water recharge sources (large-scale surface water bodies, such as seas, lakes, rivers, pools, bogs and reservoirs) and water inrush of groundwater source. Among them, the groundwater source type water inrush can be divided into unconsolidated pore water recharge source type, bedrock fissure water recharge source type and karst water recharge source type in soluble karst rocks through medium characteristics of water recharge aquifers. According to hydraulic characteristics of water recharge aquifers, the groundwater source type water inrush can be divided into perched water source type, phreatic water recharge source type and confined water recharge source type. In addition, the artificial source water inrush consists of mine water disasters of groundwater captured source type and goaf water source type.

The classification can also be carried out according to the location and contact relationship between minable seams and water recharge aquifers. In accordance with the relative locations of minable seams and water recharge aquifers, water inrush can be classified water inrush from coal roof, coal floor and periphery water recharge source. Furthermore, based on the contact relationship between minable seams and water recharge strata, water inrush can be further divided into six types with direct and indirect roof water recharge source, direct and indirect floor water recharge source, and direct and indirect surrounding water recharge source.

Water disasters can also be classified in light of the water recharge channel. Water recharge channel can be divided into natural and artificial water recharge source passages. The natural water recharge source passages include water inrush with point karst collapse column passage, linear fracture (fissure) zone passage, narrow strip concealed outcrop passage, plain fracture network (thinning area of partial plain aquifuge) and earthquake-induced passage. The artificial water recharge source passages are divided into those with passages in roof caving fractured zone, roof cut caving fractured zone and roof pump caving zone, floor rock pressure failure zone, floor draft zone of confined water, ground karst collapse zone and poorly-sealed boreholes.

Mine water inrush can be classified according to the harm forms into normal temperature, moderate to high temperature or corrosive water disasters. One can divide mine water inrush in accordance with economic losses and causalities. In accordance with causalities or direct economic losses, mine water disasters can be divided into extremely large, very large, large, and minor ones. Classification can be carried out according to time-varying characteristics as well. Based on time-varying characteristics, water inrush is divided into instant, hysteresis, or gradually varying ones.

1.3.3 Characteristics of Mine Water Inrushes

Mine water inrush directly recharged by atmospheric precipitation: Atmospheric precipitation is the main supply source of groundwater, and all ore deposits filled with water are directly or indirectly related to atmospheric precipitation. The source of atmospheric precipitation described here indicates the only water source for direct water recharge of ore deposits. There is a synchronous correlation or delayed correlation between the disaster time and the precipitation time. Moreover, catastrophic risk is related to precipitation and rainfalls and is generally proportional to rainfalls.

Water inrush of surface water: For ore deposits close to large-scale surface water bodies, such as seas, lakes, rivers, reservoirs and pools, it is critical to clarify the influences of surface water under natural conditions and after mining on ore deposit mining. This is a key process of hydrogeological exploration in mining areas and hydrogeological work in mines. Surface water is generally large in volumes. Once surface water forms hydraulic connection with mining activities and influencing ranges, the catastrophic risk is likely to rise.

Water inrush of groundwater: This type of disasters is complex. According to medium characteristics of water recharge aquifers, the water inrush can be divided into those with sources of pore water recharge unconsolidated sediments, bedrock fissure water recharge and karstic water recharge in soluble rocks. Furthermore, perched water, phreatic water and confined aquifer source water disasters are included in this type in accordance with hydraulic characteristics of water recharge aquifers. From the perspectives of water-bearing media and hydraulic characteristics, confined aquifers of karstic water show strong water abundance in general. Therefore, once such aquifers are connected due to mining activities and influencing ranges, the water inrush of karstic water source generally demonstrates the greatest catastrophic risk.

Water inrush of captured water: Because of mining, the groundwater cone of depression constantly extends, and thereby mining activities strongly transform natural groundwater flow fields in mining areas. The new supply water source obtained in artificial groundwater flow fields is known as captured water source. The captured water source includes spring water in groundwater drainage areas, surface water (seas, lakes, and rivers), the neighboring aquifers in one side of drainage areas in groundwater flow zone in mining areas and groundwater in the adjacent hydrogeological units. Therefore, the catastrophic risk is generally proportional to water abundance of the supply source.

Water inrush of goaf water: Because part of goaf remained open after mining, the goaf becomes filled with water in the late stage, becoming mine pools. If the edge of water bodies is mined, water in the goaf can suddenly gush into underground mines, causing mine water inrush accidents. According to statistics, this type of inrush has the largest number and strongest catastrophic risk in serious mine water disasters. Such a type of mine water inrush is unexpected with large amounts of water inflow, causing great damages. The goaf water is often acid and shows high concentration of hydrogen sulfide gas. However, due to the limited size of the water-storing space, the water flow may last for a short duration, and the water can be easily drained.

Mine water inrush of roof water: Mine water inrush of roof source occurs when mining activities and influencing ranges (caving and fissured zone and water-flowing structure) affect the aquifers overlying the ore body. The catastrophic risk is directly related to the water abundance and connectivity of the overlying water recharge aquifers. Greater catastrophic risk results from stronger water abundance and connectivity of aquifers in the influencing ranges of mining activities.

Mine water inrush of floor water: Mine water inrush of floor source is triggered when the mining activities and influencing ranges (zone destroyed by mine pressure and water-flowing structures) affect aquifers underlying the ore body. Similar to the water inrush of roof source, the catastrophic risk of floor water inrush is directly correlated with water abundance in and connectivity to the underlying aquifers. If aquifers in the influencing ranges of mining activities show remarkable water abundance and connectivity, the catastrophic risk of mine inrush is strong.

Mine water inrush of periphery water: Such disasters result from mining activities and influencing ranges affecting the aquifers around the ore bodies. The water recharge sources can be direct or indirect. The catastrophic risk has a proportional relation with water abundance of surrounding water recharge aquifers and connectivity of fissures. In general, direct water recharge sources in above three types of mine water inrush refer to the water source directly contacted with mined ore or water source which can be affected by roof water flowing caving zone or floor rock pressure failure zone and thereby contacted with mined ore bodies. The indirect water recharge source indicates the water recharge source that enters mines by passing through water-resisting rocks through certain water flowing structures or via leakage. It mainly distributes around mined ore body but does not directly contact with ore bodies or locates outside normal caving zone or zone destroyed by mine pressure.

Mine water inrush through natural water recharge passage: While mining ore bodies, various paths for water recharge source entering pits are referred to as water-flowing passages. Moreover, the mine water disaster caused by water gushing into pits through non-artificial water-flowing passages is called water inrush of natural water flowing passage. The characteristics are described as follows:

- In case of water inrush through karst collapse column passages, the groundwater in coal series strata and different water recharge aquifers can be hydraulically connected by karst collapse column passages, thus increasing catastrophic risk of such mine water disasters.
- Water inrush through passages in linear fracture (fissure concentrated) zone mainly takes place in fault concentrated zone, fault intersection point, fault convergence or fault tip. The passages link the close hydraulic connection between water recharge rock strata, thus causing mine water disasters.
- In view of water inrush of narrow strip concealed outcrop passage, according to the practical experience in China, the Quaternary pore aquifer group is very likely hydraulically connected to the coal series and water-recharge aquifer group of the thick carbonate formations at the narrow strip concealed outcrops. As a result, water disaster happens through narrow strip concealed outcrop passages, resulting in high catastrophic risk.
- Water inrush through passage in plain fracture networks (thinning area of local plain aquifuge). In the northern area of north China type coalfield, stresses have been released through rock fracturing in the brittle water-resisting strata under multi-stage tectonic stresses in the geological history. Therefore, concentrated cracks and joints in different directions are present in the water-resisting strata. These fractures and joints form plain extended fracture networks with a planar distribution. With the increases of groundwater head difference in the upper and lower water recharge aquifer groups, such fracture networks form vertical water exchange in a plain leaky form and cause water disasters of plain fracture network passages.
- Water inrush through earthquake-induced passage. When strong earthquakes occurred, fractures in different scales were formed near the epicenter by coupling of cyclic tension and compression of seismic forces with shear. Mine water inrush disaster occurs when fractures near the coal seams develop and connect with surrounding aquifers.

Mine water inrush through artificial water flowing passages: The mine water inrush caused by water gushing into mines through artificial water flowing passages is known as inrush of artificial water flowing passage. Such type includes water inrush with passages in the following media:

- Roof caved-in zone
- Roof fractured zone
- Roof caved-in collapse zone
- Zone destroyed by mine pressure of floor
- Floor penetration zone by confined water

- Ground karst collapse
- Poorly sealed borehole.

The first three sub-types are similar and caused by the fact that upper aquifers are connected due to roof rock damages trigged by mining activities. The difference is that caved-in zone is mainly developed in horizontal or gently tilted strata, while caved-in collapse zone is mainly developed in steep dip strata. Furthermore, water-conductive fracture zone forms in thick and extremely thick strata of sandstone or coarse sandstone with a large modulus of elasticity on the roof of coal seams. Caving does not happen in limited mining ranges. When caving occurs, it takes place in a large range, damaging roofs, or floors. Similarly, sub-types for floor strata are similar. They lead to mine water disasters because the lower aquifers are connected to the mining area as a result rock damages induced by mining activities. The differences lie in that zone damaged by mine pressure in zone destroyed by mine pressure of floor is formed in strata closely neighboring the lower ore bodies, while draft zone of confined groundwater in floor penetration zone develops in the top of lower aquifers of ore bodies. Large-scale water pumping and dewatering practices of karst water-recharge deposits, surface karst collapse is well developed in mining areas and surrounding areas. These collapses allow surface water and atmospheric precipitation to be filled into mines. When roadways or working face is advanced to intercept poorly sealed boreholes, groundwater in roof and floor water recharge aquifers of coal seams gush into tunneling face via these boreholes.

Normal temperature, moderate to high temperature and corrosive water inrush: The normal temperature water disaster refers to water inrush in the normal temperature range of local groundwater. Under the effects of abnormal geothermal, the water disaster in which the temperature of water inrush is higher than the normal water temperature is known as moderate to high temperature water disaster. Corrosive water disaster means that the source of water inrush is corrosive to mining machinery equipment, drainage equipment and roadways.

Instant, hysteresis, skipping and gradually varied water inrush: Instant water disaster refers to the water inrush occurring in the working face of mines while hysteretic water disaster refers to that appearing in the goaf behind the working face. With the gradual growth of mining depth and large-scale mining of deep coal seams, the crustal stress and water pressure of the mining environment also increase. Hysteresis water disasters induced by different passages including faults, fracture concentrated zones or karst collapse columns occur more frequently in recent years. Skipping water disaster refers to water inrush during which the inrush amount constantly changes with time while gradually varied water disaster is water inrushes during which the inrush amount gradually increases or decreases.

1.4 Hydrogeological Classification for Mine Water Hazard Control

1.4.1 Criteria of Hydrogeological Classification

Methods and techniques for water control and management depend on the CSM of each mine. Based on their hydrogeological complexity, coal mines in China are divided into four classes: simple, moderately complex, complex, and extremely complex. The hydrogeological classification helps with design of the supplemental exploration, understanding of the mine hydrogeological condition, and the prevention and control of water inrush event (Wu 2014). Table 1.2 presents the hydrogeological classification criteria, which are based on the following factors:

- The degree to which the aquifers or water bodies are affected or destroyed by mining
- The distribution of old mine pooling within the mine and its vicinity
- The amount of water inflow
- The amount of water inrush
- The risk posed by water hazard
- The difficulty in water prevention and control.

The hydrogeological classification can be an iterative process. The classification system is updated as more data become available and better understanding of the mine conditions is obtained. The amount of water inflow is determined by the representative of the main aquifer in the mine. When there are more coal seams in the same mine field and the hydrogeological conditions vary greatly, the coal mines should be evaluated based on their hydrogeological complexity. The hydrogeological type of the mine is typically biased to the complex end if uncertainty in the data is present.

1.4.2 Hydrogeological Classification of Coal Mines in China

Based on the data collected by China's State Administration of Coal Mine Safety in 2012 (Sun et al. 2015), approximately 1% coal mines were extremely complex in hydrogeological conditions, 7% coal mines were complex, 36% coal mines were moderate complexity, and 56% coal mines were simple. Figure 1.4 shows the distribution of hydrogeological classifications for the surveyed coal mines. Approximately 70% of the hydrogeologically complex and extremely complex mines are in Shanxi, Heilongjiang, Anhui, Shandong, Henan, Hunan, Chongqing, Sichuan, and Guizhou provinces. Preventing and controlling water inrushes are extremely important in these mines.

Table 1.2 Hydrogeological classifications of mines

Classification basis			Category			
			Simple	Moderate	Complex	Extremely complex
Aquifers and bodies of water damaged or affected by mining	Aquifer properties and recharge conditions		Pores, fissures, karst aquifers damaged or affected by mining, poor recharge conditions, or few sources of recharge	Pores, fissures, karst aquifers damaged or affected by mining, general recharge conditions, certain supply water sources	The main factors that are damaged or affected by mining are karst aquifer, thick layer of gravel aquifer, old mine pools, and surface water. The recharge conditions are good, and the recharge water source is abundant	The karst aquifer, the old mine pools and the surface water are damaged or affected by the mining. The recharge conditions are very good, the supply sources are extremely abundant, and the surface drainage conditions are poor
	Specific water yield q (L/s/m)		$q \leq 0.1$	$0.1 < q \leq 1.0$	$1.0 < q \leq 5.0$	$q > 5.0$
Characteristics of old mine pooling within the mine and its vicinity			No old mine pooling	There is a small amount of old mine pooling, and the location, range, and amount of water are clear	There is a small amount of old mine pooling, and the location, range, and amount of water are unclear	There is a large amount of old mine pooling, and the location, range, and amount of water are unclear
Mine water inflow (m^3/h)		Normal Q1 Maximum Q2	$Q1 \leq 180$ $Q2 \leq 300$	$180 < Q1 \leq 600$ $300 < Q2 \leq 1,200$	$600 < Q1 \leq 2,100$ $1,200 < Q2 \leq 3,000$	$Q1 > 2,100$ $Q2 > 3,000$

(continued)

Table 1.2 (continued)

Classification basis	Category			
	Simple	Moderate	Complex	Extremely complex
Water inrush Q3 (m^3/h)	None	$Q3 \leq 600$	$600 < Q3 \leq 1,800$	$Q3 > 1,800$
Extent to which mining is affected by water	Mining works are not affected by water	There is occasional water inrush in the mine, and the mining project is affected by water hazard, but it does not threaten the safety of the mine	There is water inrush in the mine, and the mining activities and mine safety are threatened by water hazard	The mine has frequent water inrush, and the mining activities and mine safety are seriously threatened by water hazard
Difficulty in water prevention and control	Simple water control	Water prevention work is simple or easy to implement	The amount of water prevention and control works is extensive and difficult to implement	The amount of water control engineering is extremely expensive and difficult to implement

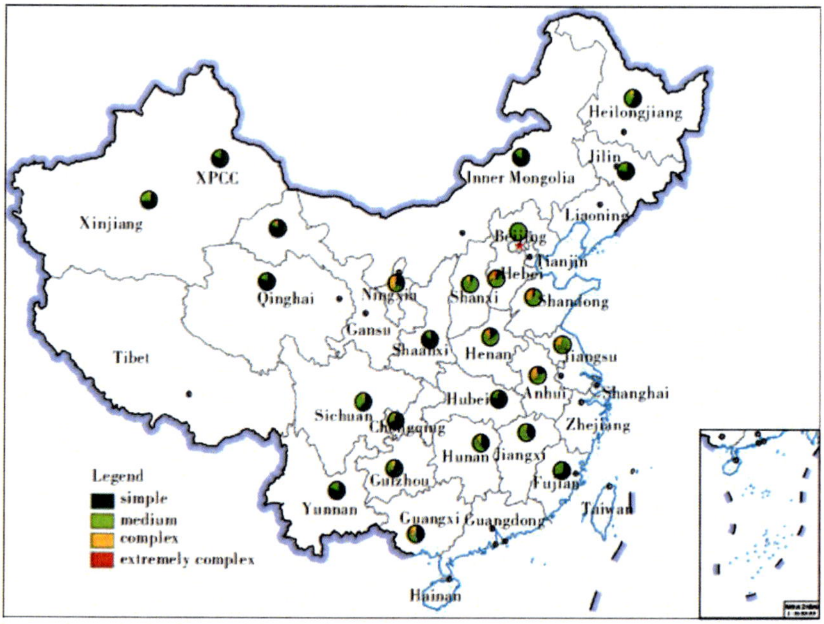

Fig. 1.4 Hydrogeological classification of coal mines in China

1.4.3 Hydrogeological Characteristics of Mines

Table 1.3 summarizes statistics on mine water inflow in select coal mines in China (Sun et al. 2015). More than 0.2 billion m^3 of water was discharged annually from coal mines in 13 provinces/regions: Hebei, Shanxi, Inner Mongolia, Heilongjiang, Jiangxi, Shandong, Henan, Hunan, Chongqing, Sichuan, Guizhou, Yunnan, and Shaanxi. The discharges from these areas accounted for approximately 83% of China's annual mine water discharge, which totaled 7.17 billion m^3. There were 61 mines dealing with average inflows that exceeded 1,000 m^3/h. There were 18 such mines in Henan, 11 in Hebei, 7 in Shandong, 6 in Inner Mongolia, 5 in Heilongjiang, 3 in Shaanxi, 2 in Shanxi, Guangxi, and Chongqing, and 1 in: Liaoning, Jiangsu, Anhui, Jiangxi, and Sichuan. The highest inflow of water into a single underground mine was at the Shendongjinjie mine in Shaanxi province, which had an average inflow of 4,900 m^3/h and a maximum inflow of 5,499 m^3/h. The second was the Yanmazhuang mine in Henan province, with an average inflow of 4,500 m^3/h and a maximum inflow of 5,400 m^3/h. Among the open pits, the maximum discharge was at the Yuanbaoshan mine in Inner Mongolia, in which the average inflow was 11,250 m^3/h and the maximum inflow was 12,500 m^3/h.

Data from the reported mines indicates that the ratio of the total maximum inflow to the total average inflow was 1.9. In nine provinces and regions, Guangxi, Hubei, Jiangxi, Chongqing, Guizhou, Hunan, Sichuan, Fujian, and Yunnan, the total

Table 1.3 Summary of mine water inflow in select coal mines in China

Region	Normal mine water inflow (10,000 m^3/h)	Maximum mine water inflow (10,000 m^3/h)	Ratio of maximum/normal inflows
Guangxi	0.944	2.854	3.02
Hubei	1.456	4.245	2.92
Jiangxi	1.537	4.351	2.83
Chongqing	3.357	8.779	2.62
Guizhou	5.357	13.442	2.51
Hunan	4.764	11.632	2.44
Sichuan	2.951	6.597	2.24
Fujian	1.43	2.878	2.01
Yunnan	1.949	3.906	2.00
Xinjiang Production and Construction	0.234	0.45	1.92
Qinghai	0.039	0.066	1.69
Jiangsu	0.433	0.731	1.69
Henan	8.83	14.655	1.66
Jilin	0.84	1.377	1.64
Liaoning	1.192	1.925	1.61
Shanxi	5.659	9.105	1.61
Shandong	4.948	7.84	1.58
Shaanxi	2.865	4.418	1.54
Ningxia	0.773	1.178	1.52
Gansu	0.633	0.955	1.51
Heilongjiang	3.093	4.605	1.49
Anhui	1.68	2.471	1.47
Beijing	0.128	0.188	1.47
Inner Mongolia	3.938	5.74	1.46
Hebei	3.23	4.455	1.38
Xinjiang	0.99	1.135	1.15

maximum inflow was at least twice the total average inflow. These areas are all located in south China. Seasonal precipitation affects their discharges greatly. Coal mines in Guangxi province, where the average inflow was 9,440 m^3/h and the maximum inflow was 28,540 m^3/h, had the highest ratio of 3.02. A higher ratio of maximum inflow to average inflow suggests a closer hydraulic connection between precipitation, surface water, and groundwater. A wide variation in the two numbers indicates that a large amount of water flows into the mine during the wet season. It is normal for underground mine inflow to increase dramatically during the rainy season in south China, so the coal mines there should have enough backup drainage capacity to be able to respond to a flooding emergency to prevent inundation. When very large storms are expected, it may be best to stop mining and withdraw the miners. The mine should restart production only when the threat of imminent danger has passed.

1.5 Advances in Prevention and Control Technologies of Mine Water Hazards

Prevention and control of mine water hazards are a challenging task and need to respond to various hydrogeological conditions, complex structure and dynamic but concealed water inrush factors, change in mining methods, and ever evolving regulations. The challenges are more pronounced in large-scale mining of the deep-level resources. China has made great progress in recent years in the field of water inrush prevention and control and resource utilization. In battling against the water hazards, the Chinese mining engineers and hydrogeologists have advanced both theory and practice of mining hydrogeology.

1.5.1 Updated Mining Principles

The two-legged principles, "investigation must be done if any water hazard is suspected and excavation occurs only after investigation" have been used in China for mine water inrush prevention and control for many years. Research results and practices have updated the basic principle and its corresponding comprehensive measures as follows:

- Assess water hazard risk first;
- Investigation must be done if any water hazard is suspected;
- Excavation occurs only after investigation; and
- Engineering measures should be in place before mining.

Two additional principles, risk assessment and engineering controls, are added to the previous principles. The potential position or concealed danger of water inrushes must be identified before mining. Identification of the potential position and water disasters risk assessment in advance are the important issues for mine enterprise and field engineering technician. Commonly used engineering measures include installation of early warning system, plugging, dewatering, drainage, and interception.

1.5.2 Evolution of Water Inrush Coefficient

Decades' practices in mine water hazard control and prevention helped evolution of the water inrush coefficient equation and confirmed the concept of water inrush coefficient and its hydrogeological significance. The water inrush coefficient equation, T = p/M where T is defined as the water inrush coefficient; p is the potentiometric pressure in megapascal [MPa] of the underlying aquifers; and M is the aquiclude thickness in meter [m] between the aquifer and coal seam, was proposed by Chinese scholars during the Jiaozuo hydrogeology consortium in 1964. The water inrush coefficient is similar to the concept of relative water-resisting layer thickness (1/T) which was proposed by Hungarian engineers. Using a large number of water inrush cases and basic data in Jiaozuo, Fengfeng and Zibo, Jingxing, Handan, Feicheng mining areas, the Chinese scholars calculated the water inrush coefficient threshold values of 0.1 and 0.06 MPa/m in the tectonically inactive areas and the tectonically active areas, respectively. Obviously, the water inrush coefficient formula is reciprocal of the relative water-resisting layer thickness. The mathematical meaning of reciprocal seems simple, but the hydrogeology meaning is of great significance. The water inrush coefficient reflects the most basic rule of groundwater seepage in hydrogeology, i.e., the core of the Darcy's law. The equation reflects the energy dissipations of the floor water permeating through water-resisting layer between coal seam and aquifers under the water pressure as the driver, and finally into the working and development areas. It describes the whole dynamic process of the coal floor water-inrush. The water pressure p is defined as the floor water-resisting layer bottom pressure rather than the coal seam floor water pressure.

With the improved understanding of floor failure, water resisting layer lithology combination and the original confined groundwater rise development induced by mining and excavating, different water inrush coefficient equations have been proposed, and these formulas described more of the actual hydrogeological concept model of water-inrush from coal floor in China.

1.5.3 Supplemental Investigation and Water Inrush Prediction

Innovations were made of mine hydrogeological supplemental investigation and exploration concept and technology (Dong 2007; Peng 2008). The exploration requirements have increased significantly. Due to limited scope in hydrogeological exploration and uncertain water recharge hydrogeological condition in many mines or in the deep extending development zone in China, mine hydrogeological exploration degree can rarely satisfy the high strength development requirements, especially in the northwest, northeast and the west of north China. The 2011 Coal Mine Water Prevention and Control Regulation stipulated mine hydrogeological exploration must be added under seven scenarios. In the northeast, northwest and the deep extending development zone, the supplemental hydrogeology exploration degree had improved; exploration concept had been accepted; exploration techniques and methods had innovated; and made certain progresses in detection of geological structure, water source, water channel, and gob area and mine water inflow forecast. In recent years, the following techniques have been applied to construction of hydrogeological concept site model and hydrogeological parameters determination:

- Surface high resolution 3-dimensional (3D) seismic exploration technique for tectonic exploration
- Ground transient electromagnetic (TEM) exploration
- High density resistivity method
- Frequency sounding method
- Controlled source audio-frequency magnetotellurics technique (CSAMT)
- Magnetic dipole source frequency sounding method
- Borehole radio waves perspective method, particularly for investigation of aquifer water-abundance, water recharge channel structure, permeable structure, or water-bearing structure exploration
- Underground high-resolution 3D seismic exploration technique
- Underground electric resistivity imaging for old mine pools
- Single-well, multi-wells, group wells steady and unsteady flow pumping (dewatering) test
- Tracer test for hydraulic connections
- Pulse interference test.

The achievements stipulated the analysis procedures of the abnormal water inrush symptom information as the important content of water disasters prediction. Water inrush symptoms may appear with varying degrees such as change of dripping water quantity and quality, coal or rock deformation and stress, water and rock temperature, or gas physical and chemical properties before approaching or in the limit equilibrium state. The emergence of the above information does not suggest occurrence of a water inrush. The key is to observe closely the dynamic change characteristics of symptom information. If the initial symptom information is not obvious but its dynamic change is quick, then the risk becomes higher. On the contrary, even if the initial value is obvious but there is no change or the dynamic change is very slow, then the water inrush risk is not high while the dynamic monitoring continues. Once abnormal symptoms appear, all underground work should immediately stop and any precautionary measures should be taken including reporting to the mine control center, sounding an alarm, withdrawing all the worker threatened by water disasters, organizing professionals to perform the root cause analysis for the water inrush, carrying out necessary detection, and making scientific judgment. No mining should be conducted prior to finding out the cause and eliminating the hidden danger.

Mine water hazard prediction theory has improved continuously, and the integrated technique system of disaster evaluation has matured. Supplemental hydrogeology exploration provides a very important geological information for mine structure, the relations of main coal seam and sedimentary structure, the recharge source, water flow pathways and old mine pools. When the advanced scientific evaluation method and model are used to analyze, simulate and process the available information, the results can be evaluated and predicted in zoning characteristics and water hazards risk levels from the overlying aquifer (roof), underlying aquifer (floor), mined-out space, and geologic structure in the process of mine production. The prediction results can be effectively used to make prevention measures for different types of mine water disaster. For floor water inrush prediction, more sophisticated evaluation methods such as the vulnerability index and the five maps + double coefficient method consider more factors than those based on the water inrush coefficient method. For roof water disaster, the three maps + double prediction method can solve the key technical problems of water recharge source, pathway, and flow intensity.

1.5.4 Advanced Detection and Dewatering Technologies

1.5.4.1 Pre-mining Water Hazard Detection and Mitigation

Advanced water exploration and drainage are a traditional technology in mining. They play an important role in forecasting and avoiding serious water inrush accidents when the mine hydrogeological condition and water recharge factors are not determined with certainty. The general detection and dewatering means include geophysical, geochemical, and drilling investigations. The more economic and non-destructive geophysics are used first to delineate the anomalous areas. Drilling is then used to verify the geophysical results and carry out dewatering underground. It is mandatory that the advanced detection and dewatering are carried out using dedicated drilling machines by properly trained professionals.

1.5.4.2 Underground Geophysical Advanced Exploration Technique

Years' research, field test and analysis have resulted in great progress in underground geophysical technology and can help with many aspects of the underground mining engineering. Many technologies are now available for advanced detection of working face such as direct current electrical resistivity imaging method, TEM method, Rayleigh wave method, induced polarization method, 3D seismic high-resolution method, ground penetrating radar, and sonar method. The most common methods currently used are electrical resistivity imaging method and TEM method. The advantage of the resistivity imaging method is that it is contact detection without detection blind spots. The limitation is to carry out single direction detection in front of excavating face with the detection distance less than 60–80 m. The volume effect is relatively large, and the interpretation is not unique. The advantage of the TEM method is more direction detection in underground roadway, and the detection range is relatively greater (80–120 m). The volume effect is relatively small. However, its limitation is that the detecting results consist of approximately 20 m blind area, and it is non-contact detection. In terms of working face advanced detection, the main detection methods include audio-frequency electric imaging, TEM method and radio imaging, channel wave seismic method and elastic wave tomography. Audio-frequency electric imaging and TEM method are used mainly to detect water-abundance of coal seam roof and floor aquifers and burnt rock, distribution of old mine pools, and water-bearing structure, and permeable structure. Radio imaging, channel wave seismic method, and elastic wave tomography method are mainly used to detect coal structure, coal thickness change, magmatic rocks and the burned area. The new geophysical detection technology also includes borehole televiewer, borehole geophysical logging and borehole hydrogeophysical logging.

1.5.4.3 Underground Directional Drilling

Underground directional drilling technique includes branch borehole-drilling technique and directional drilling for advanced water detection and dewatering in working face. Advanced drilling exploration and dewatering was traditional technology, but still widely used as an effective way to prevent and control water disasters. Because the advanced underground drilling for water exploration and discharge are all straight boreholes, each excavating face needs at least three straight boreholes if only floor water is investigated. More straight boreholes are needed if detection is for both the roof and floor water. A lot of advanced drillings seriously affect the advance rate. The advanced directional drilling of water exploration and dewatering can be oriented with directional settings, and the desired trajectory can be designed in accordance with the requirements. One directional drilling can substitute multiple straight drillings, greatly reducing the amount of drilling efforts. In addition, the position of directional drilling can be placed on the back of the development face so it can solve the interference between the advanced drilling and mining advance rate. The directional drilling revolutionizes the underground water exploration and dewatering. Other advantages of directional drilling include small borehole density, high drilling trajectory controllability, multidirectional, less invalid work, parallel operation with excavating, accurate detection of target layer, and high discharge efficiency. In addition, the directional drilling and branch borehole-drilling technique offer excellent application prospects in directional grouting for reinforcement of floor water-resisting layer and transformation of aquifers to aquicludes, advanced exploration and dewatering of roof and floor aquifers and old mine pools, detection of adverse geologic abnormal body and control of geological structure and coal thickness change.

Rapid recognition of water inrush results from statistical analysis of multi-source information such as hydrological geochemistry, water pressure and water temperature. Many water inrushes receive recharges from different types of water supply such as atmospheric precipitation, surface water, groundwater, and old mine pools. Databases are developed for various types of water source to document their hydrogeochemical characteristics such as water temperature and water level (pressure). Mathematical models are established for rapid recognition of the water sources for the underground mine water inrush. Inputs of water pressure (level), temperature and water chemistry analysis results will help quickly and accurately recognize the water supply sources of the water inrush point.

1.5.5 Early Warning Technique

Where water disasters result from mining induced fractures transmitting water into mining areas there are signals indicative of their occurrence (Wang et al. 2005). The formation and occurrence of mine water disasters have an evolution process from conception, development to occurrence, and in different stages, the stress and strain in the geological tectonic position, water pressure (level), water inflow, water chemistry and water temperature can release the corresponding water inrush symptoms. Timely, accurately, effectively monitoring of these symptoms requires development of a full set of technology system for monitoring, recognition and early warning. In-situ water hazard monitoring and early warning technology have been applied in many mines in China. The early warning system developed by Xi'an Research Institute of China Coal Technology and Engineering Group Corp. uses water pressure (level), water temperature and three-component strain grating sensors and fiber grating communication technology. The high precision microseismic monitoring system provides real-time, dynamic and planar diagnosis and stereo display of coal seam roof and floor deformation and failure process, and geological structures such as faults and collapse columns activation intensity.

1.5.6 Innovated Grouting Technique

This grouting technique started applications in the floor water-inrush prevention in the late 1980's. Due to strong recharge aquifer, high potentiometric pressure, thin coal floor thickness, or presence of a tectonic fracture zone and water-conducting fractured zone, use of dewatering depressurization method led to high dewatering and discharge cost and groundwater resource loss. In addition, there is no guarantee of mining safety. Under such circumstances, the grouting technique is the preferred method including floor water-resisting layer reinforcement, aquifer transformation and plugging of local water channels. The technique mainly aims at floor water inrush disasters prevention and involves the following: use of the existing ventilation gateway and haulage gateway to conduct geophysical exploration and drilling; detection of floor water resisting layer fracture development pattern and aquifer water-abundance zone; and grouting to reinforce the floor water-resisting layer and/or transform aquifer to aquiclude. This technique uses the grouting method to solves the common floor water inrush problem under natural conditions or when dewatering is not recommended. The current technique is mainly applied in thin-bedded limestone formation and the top of the Ordovician limestone for reinforcement and reconstruction before mining.

With mining depth increasing and the lower group coal seam mining, frequent water inrush accidents, serious water disaster occur. The regional or local grouting techniques on the ground and underground show obvious advantages for the rapid sealing mine water bursting and eliminating potential water inrush disasters. The grouting can be performed with curtain grouting technique in the concealed vulnerable position of the concentrated water recharge zone and the preferential flow zone, partial pre-grouting for aquifer reconstruction and plugging of the water channel. The directional grouting can be applied on the top of karst collapse column to build water plugs to cut off the recharge channel of the Ordovician limestone and build water-resistance wall in the water recharge roadway. The rapid directional drilling and branch hole-building technique, innovative grouting technology on the ground and in the underground, performance evaluation methods, and standards of grouting effectiveness provide a powerful technical support for mine water hazard prevention and control.

References

Dong SN (2007) Current situation and prospect of coal mine geological guarantee technologies to improve safety and efficiency. Coal Sci Technol 35(3):1–5

Gui HR, Song XM, Lin ML (2017) Water-inrush mechanism research mining above karst confined aquifer and applications in North China coalmines Arab. J Geosci 10:180. https://doi.org/10.1007/s12517-017-2965-5

LaMoreaux JW, Wu Q, Zhou WF (2014) New development in theory and practice in mine water control in China. Carbonates Evaporites 29:141–145. https://doi.org/10.1007/s13146-014-0204-7

Li GY, Mou L, Zhou WF (2015) Paleokarst crust of Ordovician limestone and its utilization in evaluating water inrushes in coalmines of North China. Carbonates Evaporites 30:365–371. https://doi.org/10.1007/s13146-015-0231-z

Li WP, Liu Y, Qiao W, Zhao CX, Yang DD, Guo QC (2018) An improved vulnerability assessment model for floor water bursting from a confined aquifer based on the water inrush coefficient method. Mine Water Environ 37:196–204. https://doi.org/10.1007/s10230-017-0463-3

Lu T, Liu SD, Wu RX, Wang B, Hu XW (2017) A review of geophysical exploration technology for mine water disaster in China: applications and trends. Mine Water Environ 36:331–340. https://doi.org/10.1007/s10230-017-0467-z

Peng SP (2008) Present study and development trend of the deepen coal resource distribution and mining geologic evaluation. Coal 17(2):1–11

Qiao W, Li WP, Li T, Chang JY, Wang QQ (2017) Effects of coal mining on shallow water resources in semiarid regions: a case study in the Shennan mining area, Shaanxi, China. Mine Water Environ 36(1):104–113. https://doi.org/10.1007/s10230-016-0414-4

Qiu M, Han J, Zhou Y, Shi LQ (2017) Prediction reliability of water inrush through the coal mine floor. Mine Water Environ 36:217–225. https://doi.org/10.1007/s10230-017-0431-y

Sun WJ, Zhou WF, Jiao J (2015) Hydrogeological classification and water inrush accidents in China's coal mines. Mine Water Environ 35:214–220. https://doi.org/10.1007/s10230-015-0363-3

Tan Y, Yu FH, Chen L (2013) A new approach for predicting bedding separation of roof strata in underground coalmines. Int J Rock Mech Min Sci 61(7):183–188. https://doi.org/10.1016/j.ijrmms.2013.02.005

Tan YL, Zhao TB, Xiao YX (2010) In situ investigations of failure zone of floor strata in mining close distance coal seams. Int J Rock Mech Min Sci 47(5):865–870. https://doi.org/10.1016/j.ijrmms.2009.12.016

Wang JM, Dong SN, Liu QS (2005) The principal of early warning for groundwater hazards in coal mine and its application. Coal Geol Explor 33(8):1–4

Wang X, Meng FB (2018) Statistical analysis of large accidents in China's coalmines in 2016. Nat Hazards 92:311–325. https://doi.org/10.1007/s11069-018-3211-5

Wang G, Wu MM, Wang R, Xu H, Song X (2017) Height of the mining-induced fractured zone above a coal face. Eng Geol 216:140–152. https://doi.org/10.1016/j.enggeo.2016.11.024

Wei JC, Wu FZ, Yin HY, Guo JB, Xie DL, Xiao LL, Zhi HF, Liliana L (2016) Formation and height of the interconnected fractures zone after extraction of thick coal seams with weak overburden in western China. Mine Water Environ 36(1):59–66. https://doi.org/10.1007/s10230-016-0396-2

Wu Q (2014) Progress, problems and prospects of prevention and control technology of mine water and reutilization in China. J China Coal Soc 39(5):795–805

Wu Q, Guo XM, Shen JJ, Xu S, Liu SQ, Zeng YF (2017) Risk assessment of water inrush from aquifers underlying the Gushuyuan coal mine, China. Mine Water Environ 36:96–103. https://doi.org/10.1007/s10230-016-0410-8

Xu ZM, Sun YJ, Gao S, Zhao XM, Duan RQ, Yao MH, Liu Q (2018) Groundwater source discrimination and proportion determination of mine inflow using ion analyses: a case study from the Longmen coal mine, Henan Province, China. Mine Water Environ 37:385–392. https://doi.org/10.1007/s10230-018-0512-6

Yang BB, Sui WH, Duan LH (2017) Risk assessment of water inrush in an underground coal mine based on GIS and fuzzy set theory. Mine Water Environ 36:617–627. https://doi.org/10.1007/s10230-017-0457-1

Yin HY, Zhou W, LaMoreaux JW (2018) Water inrush conceptual site models for coal mines of China. Environ Earth Sci 77(746):1–7. https://doi.org/10.1007/s12665-018-7920-6

Zhang JC (2005) Investigations of water inrushes from aquifers under coal seams. Int J Rock Mech Min Sci 42:350–360. https://doi.org/10.1016/j.ijrmms.2004.11.010

Zhang JX, Jiang HQ, Deng XJ, Ju F (2014) Prediction of the height of the water-conducting zone above the mined panel in solid backfill mining. Mine Water Environ 33:317–326. https://doi.org/10.1007/s10230-014-0310-8

Zhang SZ, Guo WJ, Li YY (2017) Experimental simulation of water-inrush disaster from the floor of mine and its mechanism investigation. Arab J Geosci 10(22):503–513. https://doi.org/10.1007/s12517-017-3287-3

Zhang DY, Sui WH, Liu JW (2018a) Overburden failure associated with mining coal seams in close proximity in ascending and descending sequences under a large water body. Mine Water Environ 37:322–335. https://doi.org/10.1007/s10230-017-0502-0

Zhang WJ, Li SC, Wei JC, Zhang QS, Liu RT, Zhang X, Yin HY (2018b) Grouting rock fractures with cement and sodium silicate grout conclusions. Carbonates Evaporites 33:211–222. https://doi.org/10.1007/s13146-016-0332-3

Zhao TB, Guo WY, Tan YL, Lu CP, Wang CW (2017) Case histories of rock bursts under complicated geological conditions. Eng Geol Environ (2):1–17. https://doi.org/10.1007/s10064-017-1014-7

Zhou WF, Lei MT (2017) Conceptual site models for sinkhole formation and remediation. Environ Earth Sci 76:818. https://doi.org/10.1007/s12665-017-7129-0

Zhou WF, Li GY (2001) Geological barrier—a natural rock stratum for preventing confined karst water from flowing into mines in North China. Environ Geol 40:1003–1009. https://doi.org/10.1007/s002540100289

Chapter 2
Mine Water Inrush Mechanisms and Prediction Methods

2.1 Overview of Water Inrush Studies

The study of mine water inrush can be traced to the 1940s when a Hungarian engineer, Weg Frence, proposed the concept of relative water-resisting layer aiming at the karst water inrush from coal floors (Qian et al. 2000). The researcher suggested that water inrush is correlated with the thickness of water-resisting aquiclude and the groundwater pressure of the underlying aquifer, and the water inrush process is controlled by the ratio of the thickness of equivalent water-resisting aquiclude to the value of the groundwater pressure. The ratio was called relative water-resisting layer thickness. Approximately 88% of the water inrush occurred where the thickness of relative water-resisting aquifuge was less than 1.5 m/atm in the mining process. Thus, many countries in which mining occurred above confined aquifers followed this concept and suggested that water inrush will not occur if the relative thickness is greater than a more conservative threshold of 2 m/atm (Wu et al. 2004; He et al. 2005). From the 1960s to 1970s, the Hungary Mining Technology Committee included the thickness of relative water-resisting aquiclude into their mining safety codes and provided explanations on different mining conditions (Wu et al. 2013a, b). Scholars in many countries such as former Soviet Union and Yugoslavia also began to study the effect of the relative water-resisting aquiclude during this period involving the influence of stress change caused by mining on the thickness of relative water-resisting layer and relationship between groundwater flows and rock structures (Kuznetsov and Trofimov 2002; Wolkersdorfer and Bowell 2005). Xi'an Exploration Group of China Coal Research Institute began to investigate mine water inrush in 1960s and proposed applying the water inrush coefficient as the criterion to predict occurrence of water inrushes from the underlying aquifers. Details of the water inrush coefficient concept is discussed in Sect. 2.4.

From the 1970s to the late 1980s, scholars specialized in rock mechanics investigated the failure mechanism of floors while studying the stability of coal and rock pillars. For example, Santos and Bieniawski (1967) analyzed the load-bearing

S. Dong et al., *Methods and Techniques for Preventing and Mitigating Water Hazards in Mines*, Professional Practice in Earth Sciences,
https://doi.org/10.1007/978-3-030-67059-7_2

capacity of floors by introducing the critical energy release point based on the improved Hoek–Brown's rock strength criterion. Li (1999) proposed three-zone theory, which indicates that there are three zones in the underlying coal seam floors just like the mining induced three zones in overlying strata. The three zones consist of floor failure zone, intact rock stratum, and the confined water rising zone. Wang and Liu (1993) put forward the theory of in-situ tension and failure at zero position, which points out that the mining and groundwater pressures exert a combination effect on working face, and the influence range on coal seams can be divided into three sections: advanced pressure compressing, pressure-releasing expansion, and post-mining pressure compressing. Since 1990s, the methods for studying mine water inrush have been diversified with the development of science and technology. For instance, the United States Bureau of Mines analyzed the stability of roadways under multiple environments by using finite element model. Qian established the key strata theory of rocks in mining floors according to the stratified structure characteristic of coal floors (Qian et al. 1996). The theory indicates that a stratum with the largest load-bearing capacity in the coal floor below the mining failure zone and above the aquifer, which is referred to as the key strata, plays a key role in controlling a water inrush. Institute of Geology of Chinese Academy of Sciences proposed the theory of high-permeability passage, which states that the occurrence of water inrush from floors depends on the existence of water inrush passages (Wang and Liu 1993). It is divided into two cases, i.e., presence of water inrush passages connecting with the water source through hydrogeological structures of the floor and absence of natural water inrush passages in the coal floors but passages with high permeability are formed due to the deformation and damages to weak sections under the effects of engineering stress, crustal stress and groundwater. In the 1990s, Xi'an Exploration Group of China Coal Research Institute proposed the theory of coupling of rock mechanics and hydraulics. The theory holds that floor water inrush results from interactions between rock, water, and stress (Zhou and Li 2001). The mining pressure makes the floor aquiclude produce a certain depth of the permeable fracture, reducing the strength of the rock, weakening the aquiclude performance, resulting in the re-distribution of the seepage field. When the confined water along the water rupture zone to further rise, rocks are softened due to water and continue to expand the cracks, until the result of the interaction between the two is enhanced. When the minimum principal stress of the rock mass is less than the confined water pressure, fracturing expansion occurs, which leads to the water inrush (Xie et al. 2007).

The development history of mine water inrush theory reflects the continuous improvement in understanding of mine water inrush disaster, which is an iterative process from practice to theory and vice versa. Because the water inrush is affected by geological and hydrogeological conditions, mining method, and mining practices, there is no universal mechanism that fits all situations. Although the studies by others provide useful references to understanding the water inrush processes, the water inrush mechanism is site-specific.

2.2 Water Inrush Mechanisms in North China's Coalfields

Coalfields in North China encompassing more than ten provinces contain six to seven coal seams in the Permo-Carboniferous strata. The lower three seams, accounting for 37% of the total reserves, are threatened with karst water from the underlain Ordovician limestone. Hundreds of water inrush incidence have occurred in which a large amount of water suddenly flows into tunnels or working faces under high potentiometric pressure.

2.2.1 Hydrogeological Background

Figure 2.1 shows the typical lithology on north China's coalfields. The coal seams in the coalfields of North China lie in the Permo-Carboniferous strata. The Taiyuan Formation of Carboniferous system has a thickness of 95–163 m, consisting of argillaceous shale and sandstone. From the top to the bottom, the coal seams are Xia-jia, Da-xing, Xiao-qing, Shan-qing, Ye-qing and Yi-zuo and their total average thickness is 9 m. Their roofs consist of mainly thin-bedded limestone with varying thickness from 2 to 7 m. Except for the lowest layer of limestone that may have hydraulic connection with the underlain Ordovician limestone, water in the rest of the thin bedded limestone is generally static. It is relatively easy to dewater or drain. The Shanxi Formation of the Permian system has a thickness of 90 m. It includes one coal seam Da-mei with the thickness of 6–7 m. Beneath the Taiyuan Formation is the Benxi Formation, which is 18–53 m thick and consists of arenaceous shale, bauxite shale and iron ores. The Ordovician limestone is a highly permeable confined aquifer. Its average thickness is 650 m. Due to the potential impacts of the confined water in the Ordovician limestone on the mining activities, the three lower coal seams, accounting for 37% of the total reserve, are listed as prospective reserves.

One of the major impacts of the groundwater on the mining activities is the unpredictable occurrence of water inrush, in which a significant amount of water suddenly invades the underground working areas from the underlying aquifer under potentiometric pressure. Hundreds of water inrushes have occurred. Figure 2.2 shows the schematic modes of typical flow paths of the karst water to the underground workings.

The position of water inrush is often related to geologic structures. Faults, fracture zones, karst collapse columns, anticline and synclinal axes are more susceptible to water inrush and constitute the weak zones and preferential flow paths for the confined karst water. Because of fault displacements, in some locations the coal seams are in direct contact with the limestone aquifers. Faults can also connect different aquifers. In the coalfields of north China, over 78% of water inrushes are related to faults, fractures, and karst collapse columns. Figure 2.3 shows schematically water inrushes through a fault and a karst collapse column.

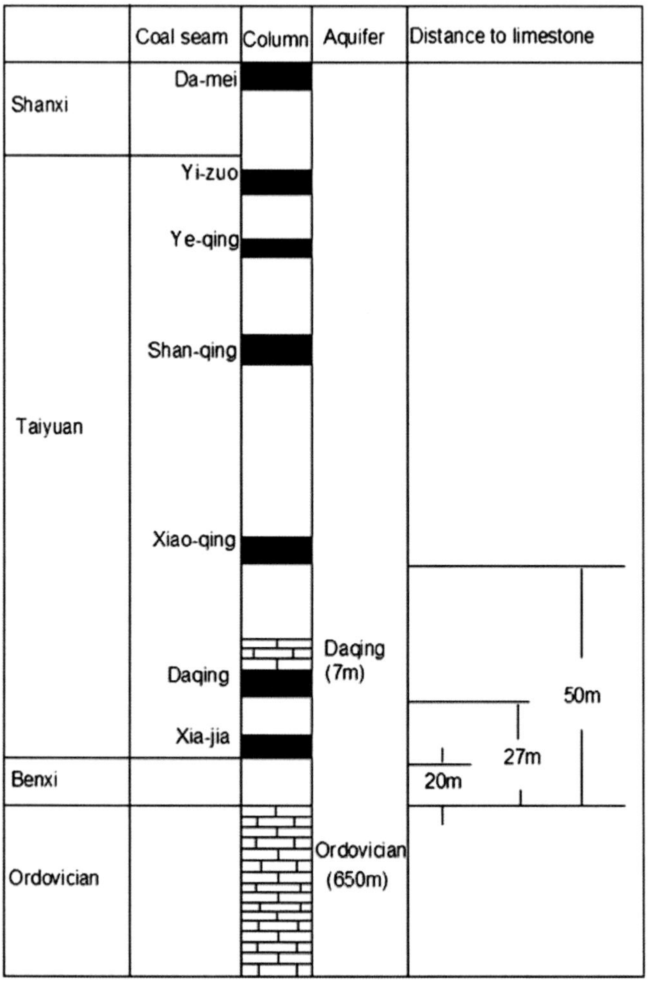

Fig. 2.1 Geologic column in coalfields of north China

2.2.2 *Relationship Between Aquiclude Thickness and Groundwater Pressure*

Water inrush is the result of interaction between water and rocks in the geologic stratum between the coal seams and the Ordovician limestone. Because this stratum provides water resistance against water inrush, it is sometimes referred to as the geologic barrier. It occurs when the strength of the stratum is not strong enough to resist the water pressure. The position of the water inrush is often related to geologic structures. Adjacency, intersection and pinch of faults, anticline and synclinal axes

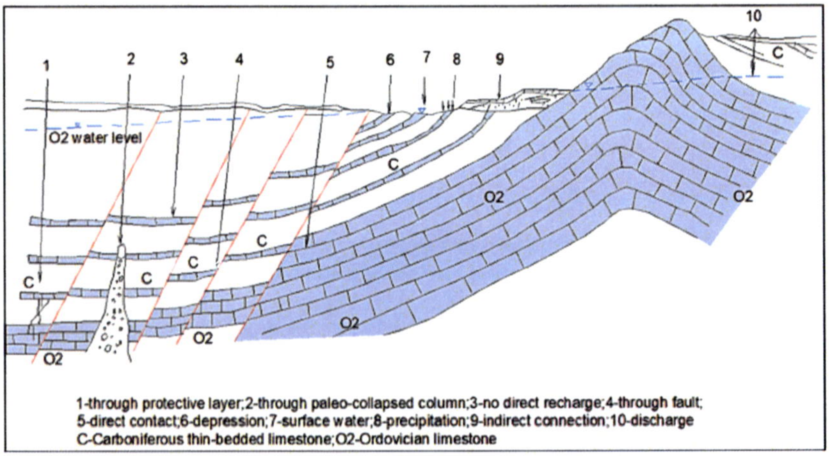

Fig. 2.2 Schematic models of water inrush in north China's coalfields

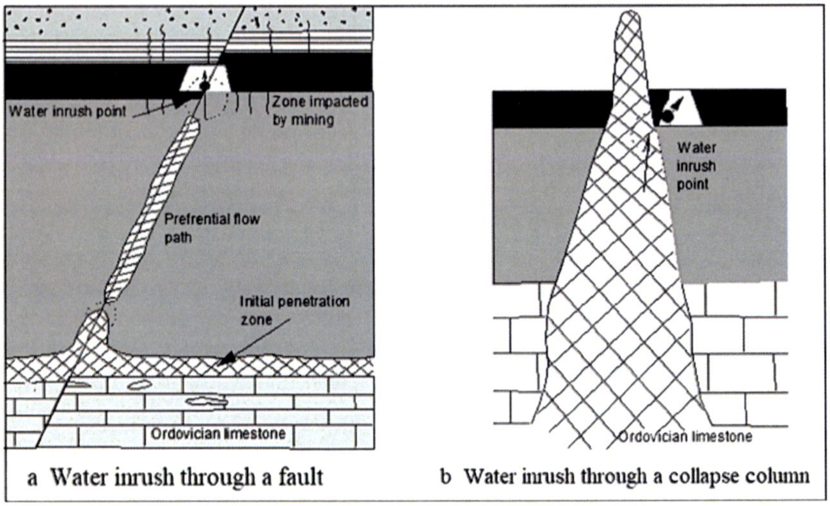

Fig. 2.3 Schematic models of water inrushes through **a** a fault and **b** a karst collapse column

are more susceptible to water inrush. In the studied area, over 78% of water inrushes are related to faults and the northeast fracture group controls 62% of them.

Figure 2.4 shows the data points of water inrushes collected at Fengfeng coalfield. The vertical coordinate represents the thickness of the rock layer between the coal seams and the Ordovician limestone, and the horizontal coordinate represents the potentiometric pressure of the karst water. The invasion points are concentrated mostly on the up right, while the safety points on the down left. Between them, there is a natural limit approaching to a straight line. Clearly, the rock layer between the

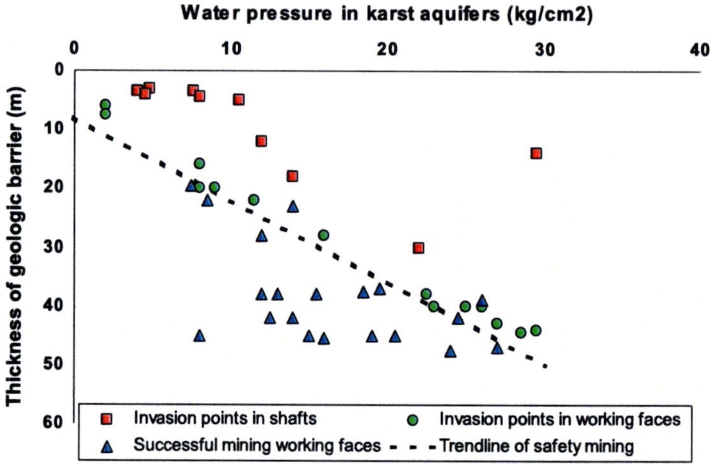

Fig. 2.4 Thickness of geologic barrier versus water pressure in Fengfeng coalfield

coal seam and the threatening aquifer acts as a geologic barrier or a hydraulic barrier that prevents the water in the Ordovician limestone from invading and flowing into the underground cavities.

The effectiveness of the geologic barrier depends on its thickness, lithology, and integrity. Water inrushes are unlikely to occur when the geologic barrier is thick. The four upper coal seams are free from water invasion because the geologic barrier is over 100 m. Hard rocks such as limestone and sandstone, have high intensity, for example, a layer of medium-grained sandstone of 2 m can bear 7 kg/cm^2 of water pressure. Flexible rocks such as shale do not have intensity as high as hard rocks. However, they may have higher capacity of water resistance, as found in another north China's coalfield (Fig. 2.5). The interception for shale seems to be larger than that for the hard rock, but the slope is relatively sharp, indicating water inrush could occur under higher water pressure for the same thickness of the geologic barrier. When the geologic barrier consists of inter-bedding layers of flexible and hard rocks, water invasion can hardly take place. As shown in Fig. 2.6, the interception for such an arrangement of rocks is very small, only 3 m. Coals were mined successfully under a water pressure of 8 kg/cm^2 when the geologic barrier was 13 m thick.

Obviously, water invasion occurs more easily under higher water pressure. When Shaqing coal seam was extracted at −90 m below sea level in one coalfield, water pressure in the Ordovician limestone was 22 kg/cm^2 and the protective layer was 40–45 m thick. No water inrushes took place. However, when the mining level was extended to −170 m, the water pressure increased to 30 kg/cm^2, and six water inrushes took place already although the protective layer remained the same.

The non-zero interception to the vertical axis implies that water inrush could take place even when the groundwater is not under pressure. We envisioned that this is the effect of mining activities. Part of the protective layer might be destroyed by mining

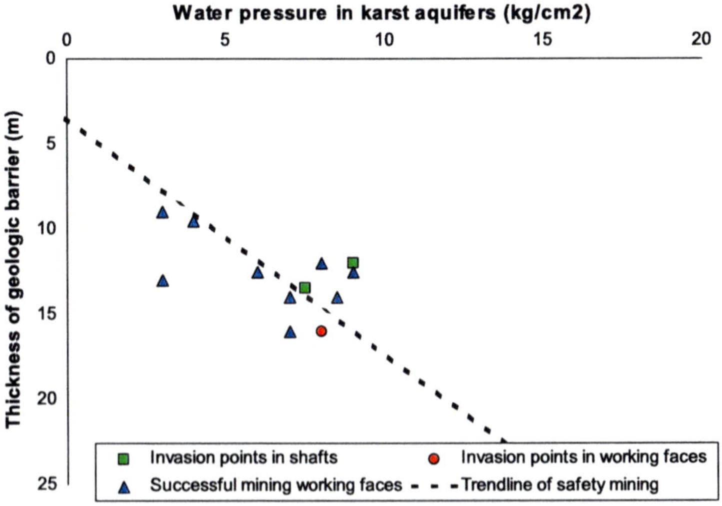

Fig. 2.5 Thickness of geologic barrier versus water pressure in Jiaozuo coalfield

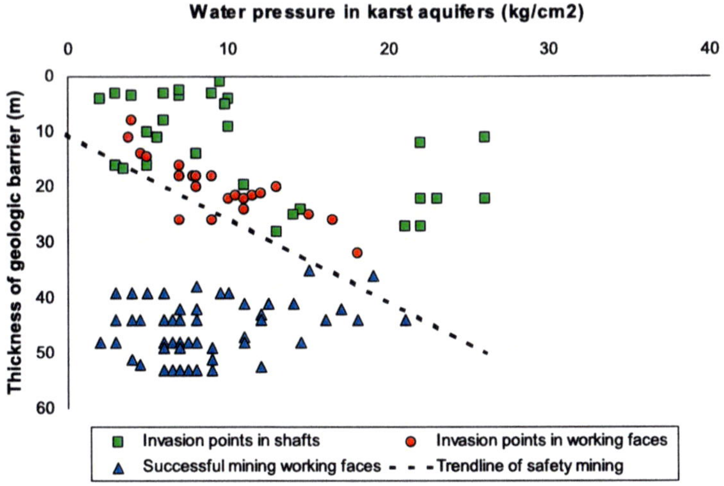

Fig. 2.6 Thickness of geologic barrier versus water pressure in Handan coalfield

operation. This is clearly illustrated by the different distribution of the invasion points at the working faces and in the shafts. For the same geologic barrier, water inrushes are more likely to occur at the working face than in the shaft. The space and span of the working faces has a significant influence on water-resisting capacity of the geologic layer. In addition, water invasions in shafts might happen with a delay of one to two years after excavation of the shaft because of the long-term effect of shaft excavation on the floor.

2.2.3 Impact of Mining Activities on Geologic Barrier

Regarding the influence of mining activities on the protective layer, the data from gas-discharge in coalmines could be used as a reference to the destroyed thickness. Gas was liberated 20–80 m below the layer after mining. Due to different properties of gas and water, fractures caused by mining can conduct gas but may not be conducive to water. The thickness of the rock through which both gas and water can flow is the parameter of concern. Two water injection tests (#1 and #2) were conducted in one coalfield to investigate the destroyed thickness. In each water injection test, 5 angled boreholes were drilled into the protective layer of a mining slope (Fig. 2.7). The vertical distance between the boreholes was 1 m and the shallowest borehole (borehole 1) was 2.5 m below the slope floor. The geologic barrier was composed of sandy shale, argillaceous limestone, and coal. A small fault with 0.5 m displacement was observed in the barrier.

A long-wall extraction approach was employed for the tests. Figure 2.8 shows the influence of mining activities on the volume of water injected for water injection test #1. The distance to the mining face is expressed by the horizontal coordinate with zero at the face. To the left and right are the distances to the mining direction and to

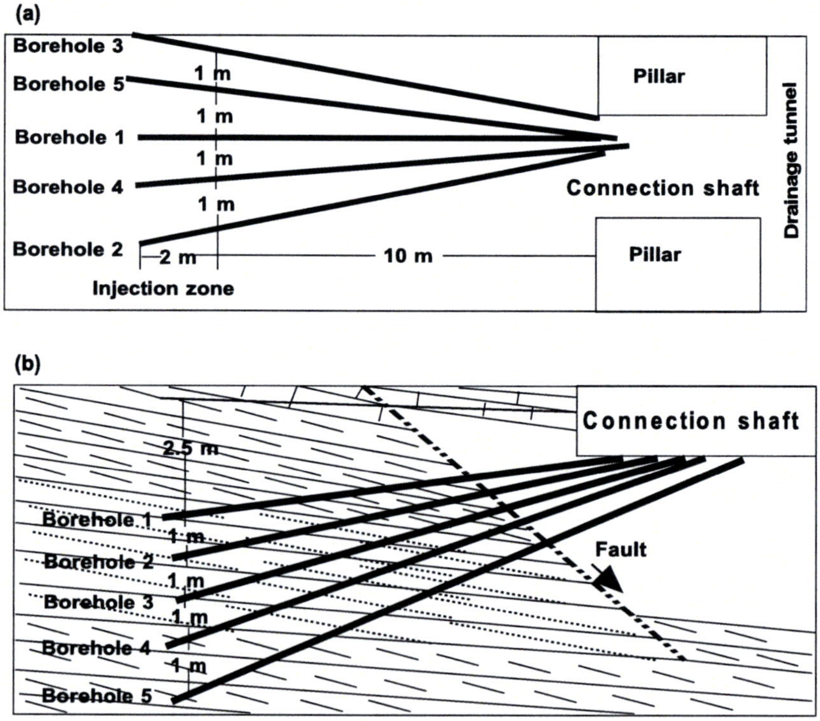

Fig. 2.7 Water injection tests. a Plane view; b Profile view

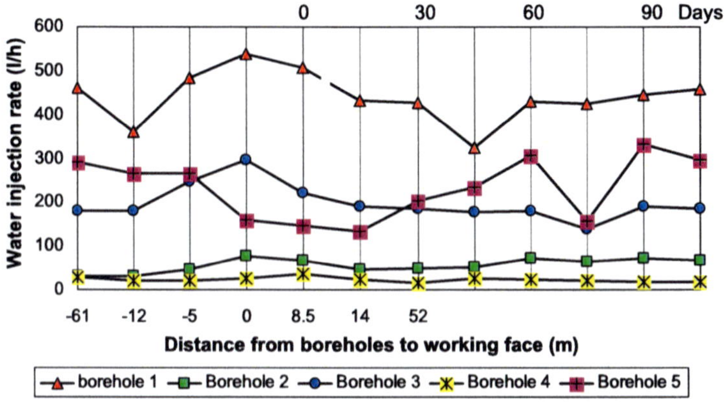

Fig. 2.8 Water injection rate during mining for Test #1

the extracted zone, respectively. The time (days) calculated according to the average speed of face advance during the test is shown in the figure as well. The volume of water (liter/hour) flowing through the boreholes of various depths and horizontal distance is shown on the vertical coordinate.

Borehole 5 was discharging water before the test with a water pressure of 0.68–0.85 kg/cm^2. The water injection pressure applied to this borehole was 1.5 kg/cm^2. The pressure applied to the other boreholes was 1 kg/cm^2. The water flow in borehole 5 increased when the central borehole was 14 m away from the working face after mining, while water flow in boreholes 2 and 3 seemed to decrease. The water flow in boreholes 2 and 3 had a slight increase at 52 m and the water flow decreased in borehole 4. Figure 2.9 shows the influence of mining activities on volume of water injected for test #2. Water pressure of 1 kg/cm^2 was applied to all the boreholes. In

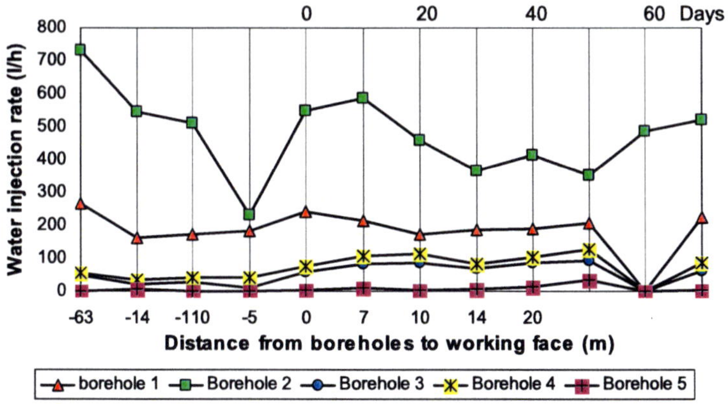

Fig. 2.9 Water injection rate during mining for Test #2

general, water flow started after water injection reached its maximum volume at the working face.

Change of the amount of water injected through the boreholes implies the effect of mining activities on the geologic barrier. The thickness of the geologic barrier destroyed by the mining could be analyzed by the water overflow from the abandoned galleries through the boreholes. Fluctuations of the water volume injected before and after mining decreased as the depth of the tested segment of the borehole increased. When the depth reached 7–8 m, the fluctuation became very small. In addition, the geologic barrier that was initially impermeable began to conduct water through holes at 6–7 m below the slope after the mining. On average, the thickness of the geologic barrier destroyed by mining activity was approximately 8 m in north China.

2.2.4 Laboratory Experiments on Failure of Geologic Barrier

Experiments were conducted in laboratory to test the failure mechanism of the geologic barrier under high water pressure in a tri-axial filter (Fig. 2.10). The rock sample is clayey limestone with two fractures perpendicular to each other. It is 150 mm in diameter and 400 mm long. A water pressure (P_w) of 35 kg/cm^2 was applied to the sample. Stresses (σ) simulating the lateral earth stress of 200 m below the ground were applied to the cylinder of the sample. The applied stress varied to simulate three mining-related stages—compression, dilation, and recovery.

In initial phase (0–1 min), the water flow rate is greater than 300 m^3/min, where $P_w \gg \sigma$. With the increase of the stress, the rock underwent the compression process (1–23 min) where $P_w < \sigma$. The fractures were gradually forced to close, and the water flow decreased to zero. At the dilation period, the earth stress decreased due to the pressure release caused by excavation. When $P_w > \sigma$ (28–29 min), the fractures opened, and water began to flow again. The maximum water flow was 120 m^3/min. During the recovery period, the roof would collapse, and the stress gradually increased to its normal. When $P_w < \sigma$ (>29 min), the fractures closed again, and water flow receded.

The experimental results indicate that the failure of the fractured rock sample as a hydraulic barrier depends on the relationship between the water pressure and the lateral stress. The water flow was observed when the hydraulic pressure exceeded the lateral stress. This process is very much similar to the spontaneous hydrofracturing in which the wedging effect takes place when the hydraulic pressure exceeds the hydrofracturing pressure (P_{hf}). P_{hf} is approximately the same as the minimum earth stress in value and it is a regular practice to measure the minimum earth stress by hydrofracturing tests in boreholes (Kesseru 1997). Therefore, the effectiveness of the geologic barrier can be quantitatively described by measuring its hydrofracturing pressure or the minimum stress. The presence of fractures is pre-requisite to the spontaneous hydrofracturing.

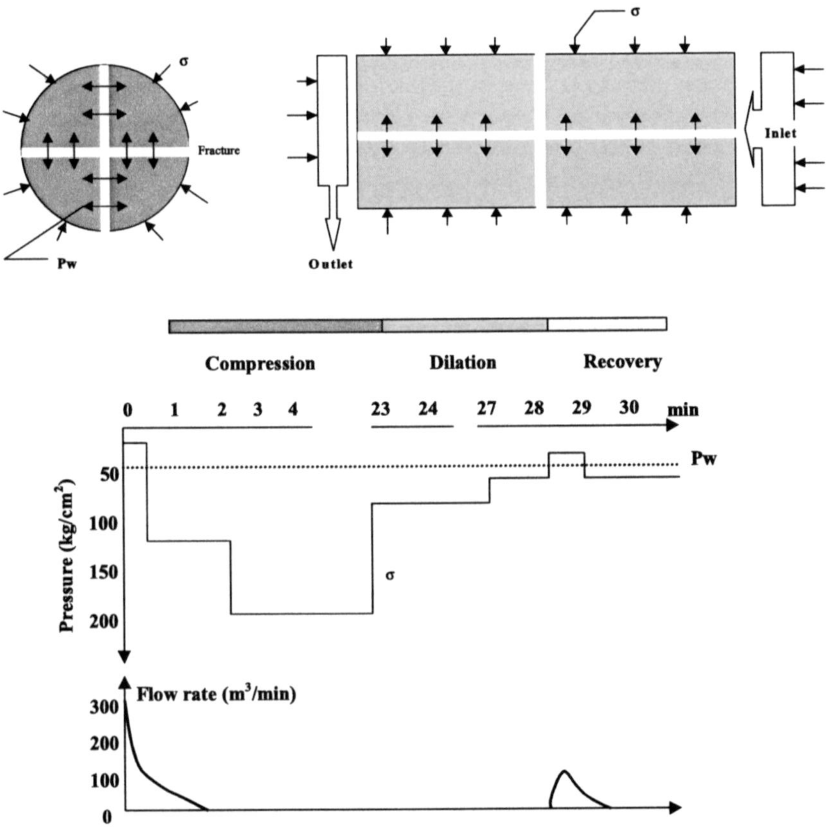

Fig. 2.10 Experiments of water inrush through geologic barrier

2.2.5 Initial Conductive Zone in Geologic Barrier

The permeability of the geologic barrier is in general low; however, there is no distinctive contact plane between the barrier and the Ordovician limestone. The Carboniferous barrier is unconformably overlain the Ordovician limestone. There is a long period during which the limestone underwent various weathering processes. The weathering processes resulted in an irregular limestone surface. In addition, the geologic barrier has itself undergone numerous tectonic movements. Fractures exist in the barrier. When the fractures in the barrier are connected with the underlying limestone, the karst water in the limestone penetrates upward into the barrier under its potentiometric pressure. The area that has already been invaded by the karst water prior to mining is defined as the initial conductive zone. This zone cannot play an effective role in preventing karst water from flowing into the mines. However, the existence of this zone provides the essential condition for the wedging effect. At locations where this zone does not exist, the barrier would remain its integrity even the water pressure exceeds the hydrofracturing pressure. The barrier breaks only when the water pressure exceeds its shearing strength by bending of the barrier.

Continuous monitoring of the potentiometric pressure while drilling into the geologic barrier is necessary to detect the height of this zone. Table 2.1 lists the exploratory results at one mine. Clearly, the initial conductive zone is closely associated with the lithology and structure of the barrier. It more likely develops in harder rocks with better-developed fractures. High conductive zone may develop along fractures and around collapsed zones, which is schematically illustrated in Fig. 2.11. Therefore, fractures and karst collapse columns are likely to become avenues for groundwater flow or contaminant transport (Zhou 1997).

Table 2.1 Measured height of the initial conductive zone

Thickness of geologic barrier (m)	Potentiometric pressure of Ordovician limestone (kg/cm^2)	Height of initial conductive zone (m)	Characteristics of geologic barrier
21.4	11.0	0.92	Fine sandstone, bauxite, clay stone, very few fractures
19.7	10.5	0.0	
23.6	12.6	0.0	Fine sandstone, coal seam, bauxite, few fractures
21.6	12.6	<7.6 (inferred value)	
22.4	12.0	8.5	Fine to medium-grained sandstone, fractures well developed
20.1	12.0	6.67	
17.8	12.0	7.07	
20.8		<6.3	

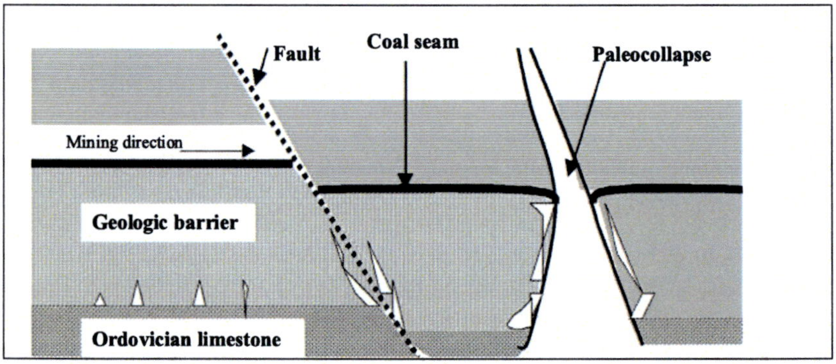

Fig. 2.11 Schematic illustration of initial conductive zone

2.2.6 In-Situ Hydrofracturing Tests

In-situ hydrofracturing tests were conducted in two geologic media—shalestone and fine sandstone, and bauxite. The test holes were drilled into the test rock from the working faces. The diameter of the test holes is 100 mm except for the last 1 m where the diameter was changed to 58 mm. Packers were used to isolate the last 1 m. Water injection rate is 0.1 m^3/min, which led to the gradual built-up of the pressure. The occurrence of hydrofracturing was observed in the working face, as shown in Fig. 2.12.

Table 2.2 shows the test results. In the shalestone and fine sandstone, two sets of joints develop in the rock with their densities of 14/m and 8/m, respectively. Similar joint sets develop in the bauxite. Their densities are 5/m and 3.7/m, respectively. The in-situ hydrofracturing tests indicated:

- The water inrush (seepage) occurred preferentially through joints or fractures.
- The hydrofracturing pressure of the shalestone and fine sandstone is larger than that of the bauxite.
- The water pressure remained at a certain level after the water inrush occurred, which is of a typical property of hard rocks.
- The hydrofracturing pressure decreases with depth.

Figure 2.13 shows the linear relationships established at shallow depths for both barriers, from which the hydrofracturing pressure at one point could be approximately calculated. The calculated pressure is then used to compare with the local potentiometric pressure in the Ordovician limestone to determine the risk of water inrush during mining activities. However, cautions must be taken when the equations are extrapolated to deep areas.

Geologic barrier is a natural rock stratum that prevents groundwater from flowing into active mines. It can also be used to contain contaminants when similar geologic conditions exist. The effectiveness of a geologic barrier can be quantitatively evaluated by measuring its hydrofracturing pressure. The hydrofracturing pressure

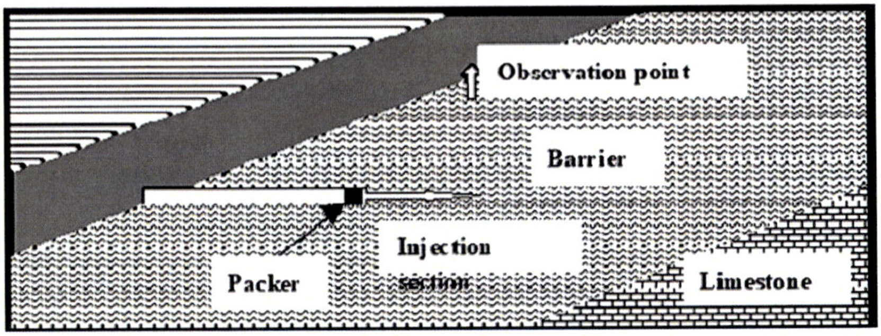

Fig. 2.12 In-situ hydrofracturing tests

Table 2.2 Results of in-situ hydrofracturing tests

Geologic barrier	Vertical depth (m)	Water pressure (kg/cm^2)	Duration of water injection (min)	Characteristics of hydrofracturing	Hydrofracturing pressure (kg/cm^2)
Shalestone and fine sandstone	0.5	3		Instantaneous seepage, while the water pressure decreased to 2 kg/cm^2	3
	1.1	10	20	Water flow through two sets of joints	10
	1.5	5	25	Seepage through one set of joints, while the water pressure decreased to 8 kg/cm^2	15
		10	15		
		15	5		
	3	3	10	Water flow through joints	20
		5	20		
		8	15		
		10	15		
		15	25		
		20	1		
Bauxite	0.5	3	20	Seepage through one set of joints	1.5
	0.7	5	1	Seepage through two sets of joints	4
	1	5	15	Seepage through one set of joints	5

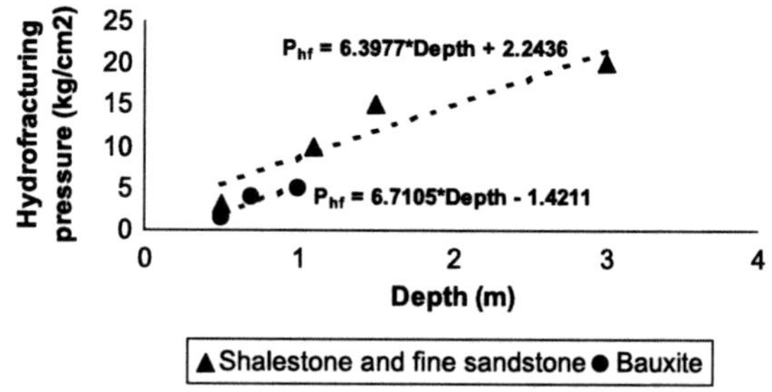

Fig. 2.13 Hydrofracturing pressure versus depth

decreases with depth due to the impact of mining activities. The initial conductive zone provides fractures for the wedging effect to take place. Water inrush occurs when the water pressure exceeds the hydrofracturing pressure in the initial conductive zone.

2.3 Water Inrush Mechanism Through Karst Collapse Columns

2.3.1 Karst Collapse Columns and Their Relationship with Mining

Karst collapse columns are another geologic feature through which water inrushes occurred. Karst collapse columns are widely distributed in northern China, especially in the provinces of Shanxi, Hena, and Hebei, as listed in Table 2.3. They have been found in over 50 coalfields and their total number exceeds 3000 with an intensity of up to 70 collapses/km^2. In some mining stopes karst collapse structures comprise 30% of the total mined areas. They are recognizable in plan view as patches of breccia with miscellaneous lithological composition, generally derived from overlying strata and completely enclosed in lower bedrock. Diameters range from tens to hundreds of meters with the largest measuring 1050 m. In profile, they take the form of vertical cylinders several hundred meters deep. No bedding is apparent inside these structures and the different rocks are intermixed and poorly sorted. They generally contain higher proportions of displaced blocks and the adjoining strata are offset as a result of dissolution-collapse. Fragments tend to be sharply angular, typically rotated, show little sign of wear and appear to have dropped from their original stratigraphic position. Figure 2.14 shows some common profiles of karst collapses reported in northern China. While the majority of the karst collapses are buried underground, some are exposed to surface and expressed as depressions.

Usually the infill materials in the karst collapse structures are tabular 5–40 cm angular fragments which display random orientation. Sides are subparallel and contacts between host and fills are sharp and irregular. In most cases, the matrix consists of clastic sediments without cement or mineralization. These structures are generally perpendicular to the ground surface. However, they can become inclined as a result of tectonic movements but remain perpendicular to the surrounding strata. Voids may be present at the top of the structures and drill bits can drop noticeably during borehole drilling. Closed depressions sometimes form in the surficial sediments without any apparent fluctuation in water level or any construction works taking place.

Table 2.3 Karst collapse column reported in northern China's coalfields

Province	Location	Number of Karst collapses	Max. size of collapses (m)			Water inrush
			Length	Width	Height	
Shaanxi	Xishan Mine, Taiyuan	1300	40			
	Yangquan Mine	960	300		600	
	Huoxian Mine	405	60			2
	Zhuangjia zhuang Mine, Fenxi	360	30			1
	Shenlin Mine	75	30			
Hebei	Jingxing Mine	112	300	23	300	1
	Fenfeng Mine No. 1	6	200	40		
	Fengfeng Mine No. 4	20	73	150		
	Yangquhe Mine	6	204	57		
	Fangezhuang Mine, Kailuan	9	95	74	280	4
	Tangjia zhuang Mine, Kailuan	5	116		50	
Henan	Hebi Mine	22	200	60	398	
	Tongyie Mine, Anyang	22				1
	Lifeng Mine, Jiaozuo	1				1
Jiangsu	Dahuangsha Shaft, Xuzhou	17	200	146		1
	Qingshan zong Shaft No. 1	1				1
	Qingzhuan Mine	1	38	19		
	Liuxing Mine	1		20		
	Jincheng Mine	27	125	62		
	Xinhe Mine	3	136	100		
	Jiahe Mine	1	135	100		1
	Wangzhuang Mine	1	122	93		

(continued)

Table 2.3 (continued)

Province	Location	Number of Karst collapses	Max. size of collapses (m)			Water inrush
			Length	Width	Height	
	Quanzhao Mine	7	60	40		
	Hanqiao Mine	4	170	100		
	Dongzhuang Mine	1	280	200		
	Yian Mine	1				

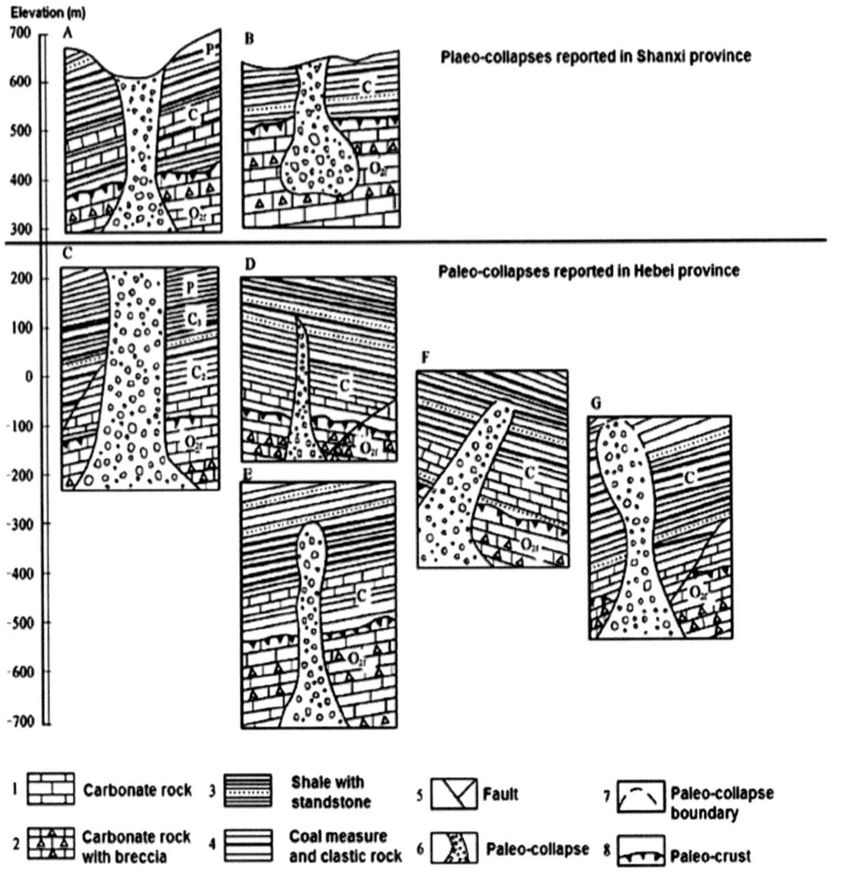

Fig. 2.14 Select karst collapse profiles reported in northern China. Elevation in meters is shown on the left side of the figure

2.3.2 *Hydrogeological Characteristics of Karst Collapse Columns*

The karst collapses found in northern China are of different hydrogeological types depending primarily upon the lithology of their internal rock blocks, extents of weathering and cementation, and the secondary structures associated with the collapses. Based on the exposed collapses from drillings and excavations, the karst collapses can be permeable, impermeable and poorly permeable. In different karst collapses or different locations of the same karst collapse, the rock blocks may have different hydrogeological properties. The permeable karst collapses consist of weathered rock blocks, but they are typically unconsolidated and not cemented. The impermeable karst collapses consist of weathered rock blocks that are cemented by weathered shale and mudstone. The poorly permeable karst collapses consist of partially cemented rock blocks with secondary fractures around the border of the collapses. Table 2.4 summarizes the hydrogeological properties of some of the karst collapses.

Figure 2.15 shows profiles of the three types of karst collapses. All of the three karst collapses are exposed in the Fangezhuang mine. Karst collapse No. 9 is permeable; karst collapse No. 2 is impermeable; and karst collapse No. 1 poorly permeable.

2.3.3 *Water Inrushes Through Karst Collapse Columns*

The absence of coal seams and the sudden inrush of karst water from the Ordovician limestone have been encountered in the mines of the Permo-Carboniferous coalfields

Table 2.4 Summary of hydrogeological properties of karst collapses

Coalfield, Province	Number of reported Karst collapses	Number of impermeable collapses	Percent of impermeable collapses
Yangquan, Shanxi	450	450	100
Xishan, Shanxi	1300	400	30.8
Fengfeng, Hebei	65	30	42.6
Jingxing, Hebei	120	80	66.7
Fengezhuang, Kailuan, Hebei	13	3	23
Tangjiazhuang, Kailuan, Hebei	6	6	100
Xinwen, Shandong	14	7	50
Feicheng, Shandong	10	9	90
Tengxian, Shandong	5	1; the rest are poorly permeable	20
Yanzhou, Shandong	2	2	100

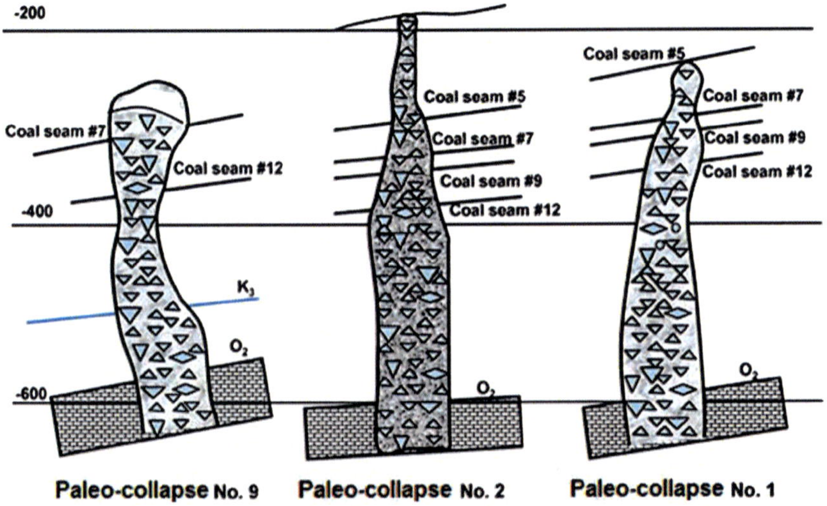

Fig. 2.15 Longitudinal cross sections of karst collapses No. 9 (highly permeable), No. 2 (impermeable) and No. 1 (poorly permeable) in Fangezhuang Coal Mine. Elevation in meters is shown on the left side of the figure

of northern China. These events are due to the presence of karst collapse structures. Table 2.5 lists 15 water inrushes reported from the mines, including the largest inrush in the world, which occurred in Fangezhuang Mine in 1984. Karst water gushed into the mine at a flow rate of 34 m^3/s at a depth of 313 m below mean sea level (bmsl). The surface level is 27 m above mean sea level (amsl). The whole mine was flooded within 21 h and as a result, the regional water table in the Ordovician limestone dropped from 5.94 to 111.09 m bmsl. The cone of water depression covered 84 km^2 with a north–south axis of 25 km. The fall in the level of the water table in the Ordovician limestone caused serious problems for local residents. These included the drying up of their water supply wells, contamination of the groundwater and the formation of new sinkholes. The water inrush led to the development of 17 cover collapses, with resulting sinkholes ranging in diameter from 2.5 to 3 m and with depths of 3–12 m.

Three basic conditions are required to cause a water inrush. These are the presence of:

- Water-bearing karst conduits or a nearby water body;
- Water-permeable internal structures in the karst collapses; and
- Water pressure differences between the karst aquifer and the working area.

The relative location of a mine to the active flow zone or karst conduits in karst aquifers determines the amount of water that can flow into the mine. In the presence of a large water-pressure difference, hydrofracturing will facilitate the upward flow of Ordovician karst water into mines. Apparently inactive karst collapse structures or those which have been cemented can be reactivated by activities such as mining,

Table 2.5 Case histories of water inrushes through karst collapses in northern China

Mine	Date	Flow rate (m^3/s)	Description	Hazard
Tongyie Mine, Anyuang	1965	0.39	A water inspection borehole drilled into a karst collapse structure from a drift. The initial water flow rate was 0.5 m^3/min and water flow rate increased to 23.3 m^3/min. An exploration borehole revealed 17 cavities within 50 m of the collapse with the maximum bit-drop of 2.59 m	The whole mine was flooded
Lifeng Mine, Jiaozuo	1967	2	Karst developed very well in the area. Intensive mine water drainage reactivated the karst collapse structure	A working stope was flooded
Fagezhuang Mine, Kailuan	1978	1	Water flowed into the mine from the sandstone, which is 160 m above the underlying Ordovician limestone. A sluice gate was constructed to isolate the water inflow area and a 0.2-m fracture was revealed, connected with a karst collapse structure	Part of a drift and a working stope (70,188 m^3) was flooded
Fagezhuang Mine, Kailuan	1983	0.23	A small fault with displacement of 0.2–0.5 m was intercepted by a working stope. Karst water flowed through a karst collapse structure into the fault and then to the working stope	The working stope was flooded
Fagezhuang Mine, Kailuan	1984	34	This is the biggest water inflow incident in the world. The mining coal seam was 180 m above the Ordovician limestone, but they are connected by a karst collapse structure. The reactivation of the karst collapse may be associated with a recent earthquake in this area. Grouting boreholes revealed that the top of the sinkhole was unfilled with sediments. Due to the water inrush, 17 cover sinkholes were induced on the surface	The whole mine was flooded, and the adjacent three mines were threatened

(continued)

Table 2.5 (continued)

Mine	Date	Flow rate (m^3/s)	Description	Hazard
Huoxian Mine	1967	0.13	Karst water flowed into an excavating drift through a karst collapse structure and a connecting fault	The drift was abandoned
Huaxian Mine	1984	0.06	Karst water from a karst collapse structure flowed into surrounding fractures and then into the horizontal drift in the Ordovician limestone	The drift was abandoned
Wucun, Huixian	1999	0.67	Karst water from a karst collapse structure flowed into the mining area	The panel was flooded
Dongpan, Xingtai	2003	19.5	Karst water from a karst collapse structure flowed into the mining area	The mine was flooded

pumping water, dam construction, and landfill development. They may also be triggered by natural events such as neo-tectonic movements and earthquakes. Table 2.5 provides some examples, and Fig. 2.16 shows three scenarios where karst collapse structures can lead to water inrushes. Mining drifts do not have to intercept karst collapse structures directly to cause a geohazards but may instead intercept faults

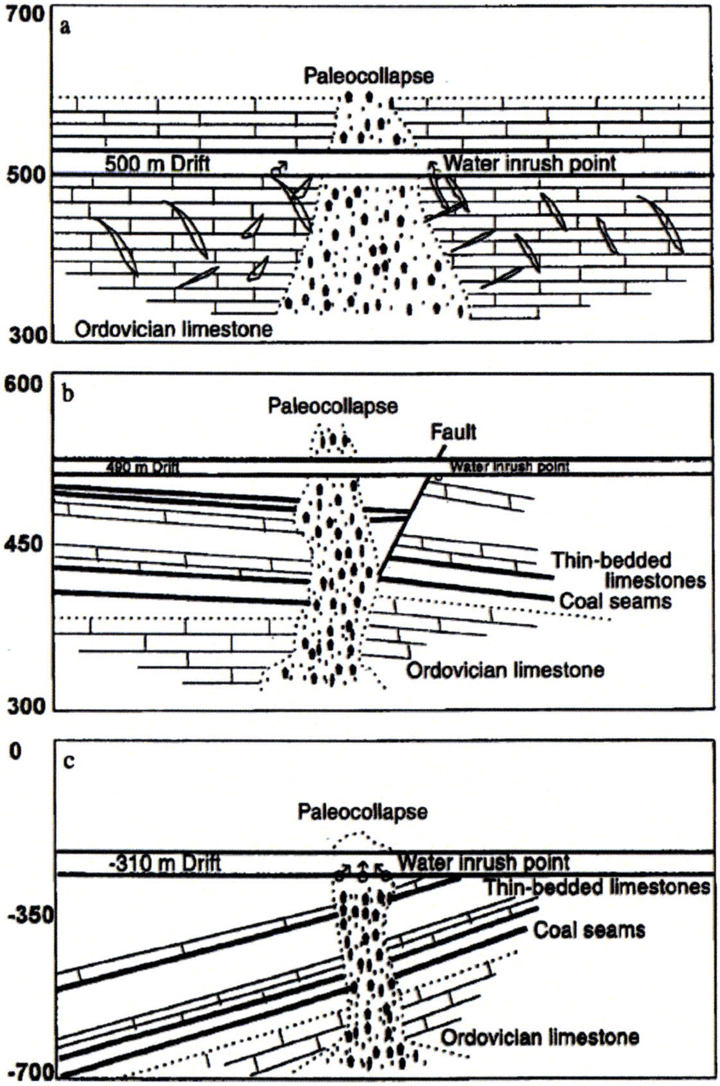

Fig. 2.16 Scenarios where karst water from Ordovician limestone flows into workings through karst collapse structures. **a** through collapse-connected fissures; **b** through collapse-connected faults; **c** through collapses

or fractures connected to them. However, once a water inrush occurs and significant water flows into the workings, the whole mine or quarry may become flooded. In cases where different aquifers, several hundred meters apart, become hydraulically connected, the impacts on safety, economy and the environment can be alarming.

2.4 Prediction Methods

2.4.1 *Water Inrush Coefficient Method*

The water inrush coefficient method has been a viable method to predict water inrushes since it was proposed in the 1960s. The water inrush coefficient refers to the water pressure that can be undertaken by unit water-resisting aquiclude, which is expressed by the following equation (Guan 2012; Dong 2010):

$$T_s = P/M$$

Where T_s is the water inrush coefficient, in unit of MPa/m; M is the thickness of floor aquiclude between the mining coal seam and the underlying aquifer, in unit of m; and P is the water pressure exerted on to the floor aquiclude, in unit of MPa.

When the water pressure exceeds the critical water inrush coefficient, water inrush possibly occurs. Based statistics of actual water inrushes, the critical water inrush coefficients are 0.10 MPa/m for normal segments and 0.06 MPa/m for complicated tectonic segments. The hydrogeological model of the water inrush coefficient method is simple and easy to use. Although the basic form remains to be the core in the calculation of the water inrush coefficient, variations in the equation expression have been developed over the decades, as summarized in Table 2.6.

2.4.2 *Vulnerability Index Method*

Mining practices indicate that there are multiple factors that cause a water inrush from the coal floor, and the water inrush process exhibits a complex non-linear dynamic characteristic. Figure 2.17 list the factors that potentially affect the water inrush occurrence. This list is not exhaustive, and other factors may have to be considered for a specific mine.

Because the traditional water inrush coefficient method only considers groundwater pressure and aquiclude thickness, it fails to fully describe the water inrush from coal floors (Wu et al. 2004). The vulnerability index method was proposed in the 1990s (Wu et al. 2009; Wu et al. 2013a, b) for evaluating the water inrush risk from coal floors. The vulnerability index method is based the multi-source information integration theory. Basic model of the vulnerability index method is:

Table 2.6 Water inrush coefficient and its variations

Time	1960s	1970s	1980s	2005	2009
Equation	$T_s = \frac{P}{M}$	$T_s = \frac{P}{M-C_p}$	$T_s = \frac{P}{\sum_{i=1}^{n} M_i m_i - C_p}$	$T_s = \frac{P_c}{M \cdot \sum_{i=1}^{n} \xi_i M_i - M_l - M_m}$	$T_s = \frac{P}{M}$
Symbol description	P: water pressure, MPa M: thickness of water resisting layer, m	C_P: depth of coal seam destroyed by mine pressure, m	M_i: thickness of the ith water resisting layer, m m_i: conversion coefficient of equivalent thickness of the ith water resisting layer	P_c: Residual water pressure in intrusion point on water head intrusion belt, MPa ξ_i: Equivalent anti-hydro-pressure intensity factor of different strata M_l: Destruction zone of seam floor, m M_m: Thickness of water head original intrusion belt, m	M: thickness of effective water resisting layer of coal seam floor, m
Rationales for modification	Original form proposed at water prevention consortium at Jiaozuo, China	Impact of mining was considered	Stratified aquifuge and mechanical property were considered	Floor failure, separation group of confining beds, confined water intrusions were considered	Effective water resisting layer was used

$$VI = \sum_{i=1}^{N} S_i i x I_i$$

where:

VI Vulnerability index of coal floor water inrush;
S_i Contribution of main controlling factor i on the inrush or relative weight;
I_i Dimensionless value of main control factor i after normalization; and
N Number of main control factors.

After the investigation area is divided into grids, geographic information system (GIS) is used to calculate the VI in each grid. The method can reasonably reflect the water inrush process from coal floors and forecast water inrush from floors by considering the interaction among various influence factors of water inrush from floors and the relative weights of factors. Vulnerability index method is used for predicting water inrush from coal floors by coupling the mathematic model with the GIS with functions in analyzing and processing spatial information. The mathematic model determines the weight coefficients of multiple control factors of water inrush. The vulnerability index method reflects the complicated mechanisms in water inrush

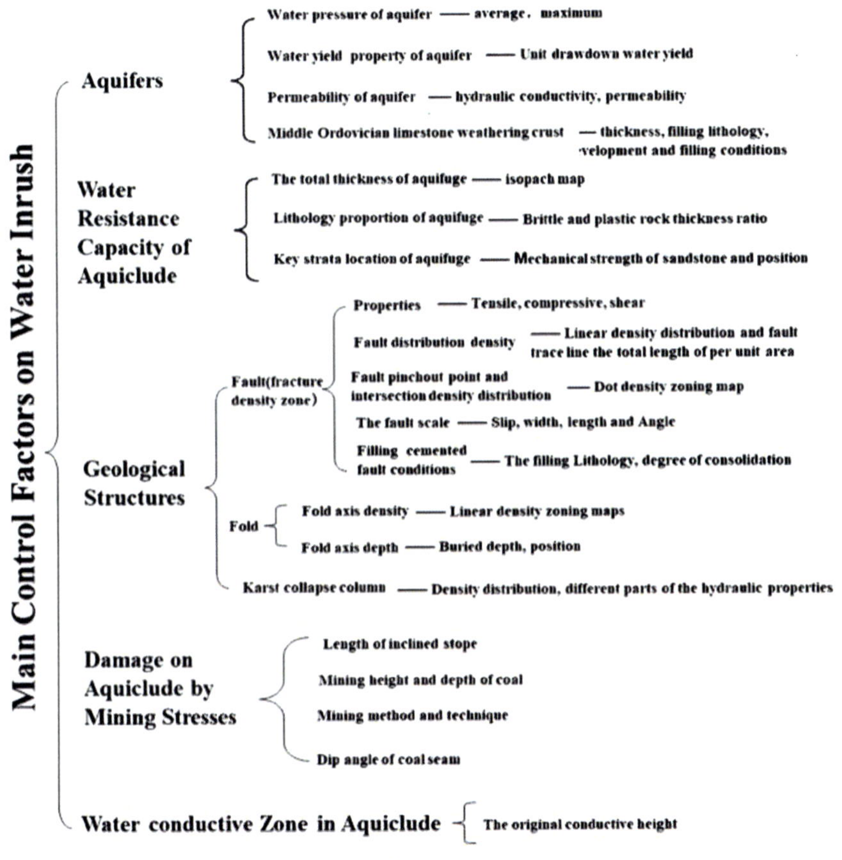

Fig. 2.17 Factors that potentially affect water inrush occurrence

processes and can predict the water inrushes reasonably well. Details of this method can be found in published literature (Wu et al. 2009; Wu et al. 2013a, b). The specific steps are described as follows:

- Collect data about the mine under study. Based on the analysis of mine water inrush mechanism in the mine, the main controlling factors of water inrush from coal floors are determined through three aspects—water abundance of aquifers, water-resisting layer of floor and geological structure. Typical factors include equivalent thickness of effective aquiclude, thickness of brittle rock under broken zone by underground pressure, distribution of faults and folds, water abundance of aquifer and water pressure of aquifer.
- The geological and hydrogeological data are used to establish the attribute database and develop the thematic map for each factor. The collected data including drilling data, coordinates of water monitoring points and quantitative values are input into GIS to develop the thematic maps of the main controlling

factors by using functions of GIS including data storage, spatial data analysis and processing, and result display. The maps reflect the spatial characteristics of the factors. Figure 2.18 shows an example.

- The weights of main controlling factors can be calculated by using either constant weight mode or variable weight model. The variable weight model considers the relative importance of controlling factors and the influence of changes of index values of controlling factors on the comprehensive evaluation value in different evaluation units. The role is realized by adjusting weights in response to the change of index values. Evaluation results are more in line with the actual situations. The relevant theory of the variable weight model can be obtained from references (Wu et al. 2013a, b).
- The superposition process is conducted to compose information storage layers of relevant factors into one information storage layer, so that the produced information storage layer contains the information of all relevant factors. The evaluation

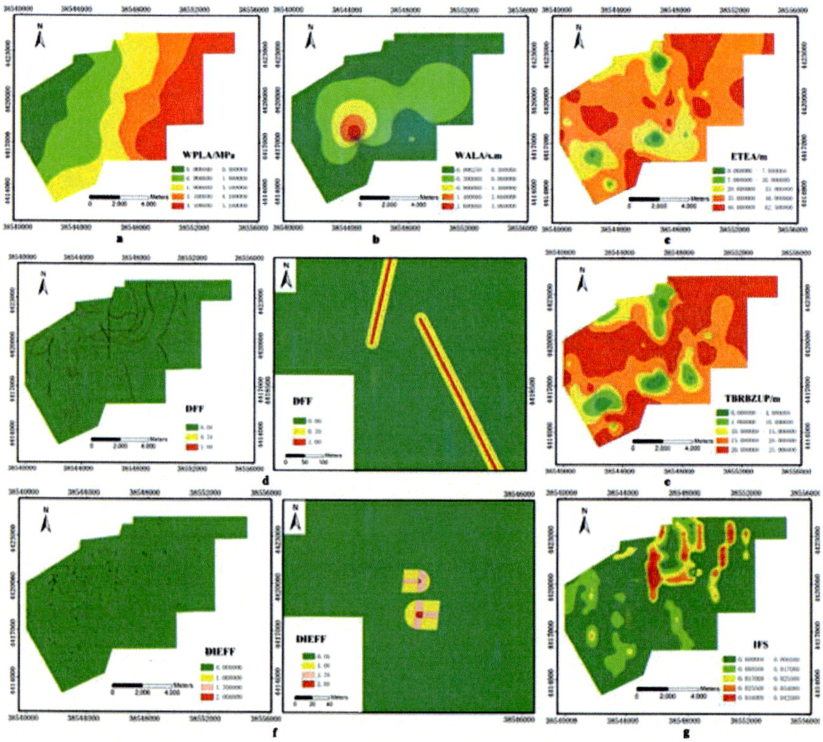

Fig. 2.18 Thematic maps of main factors influencing the floor water bursting (**a** water pressure of limestone aquifer; **b** water abundance of limestone aquifer; **c** equivalent thickness of effective aquiclude; **d** distribution of faults and folds; **e** thickness of brittle rock under broken zone by underground pressure; **f** distribution of intersection and endpoint of faults and folds; **g** index of fault scale)

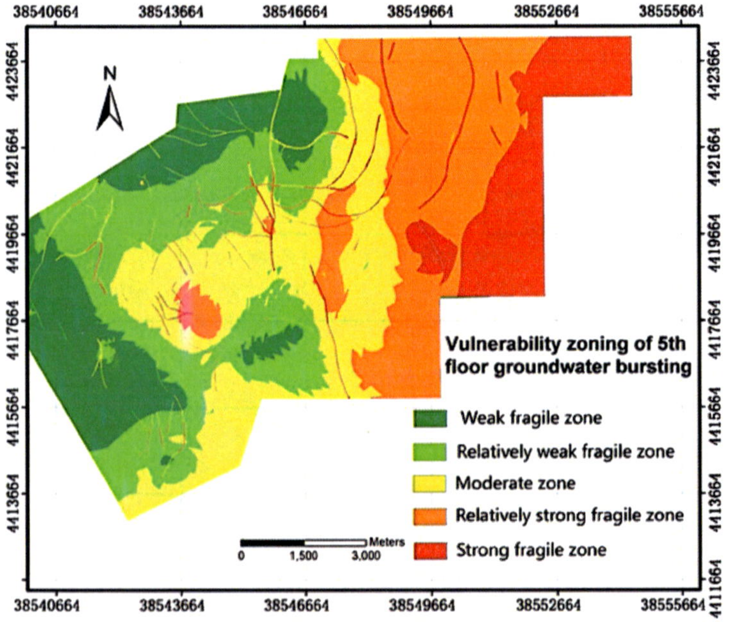

Fig. 2.19 Risk zoning map of water inrush

model of water inrush vulnerability is established to calculate the vulnerability indexes of each evaluation cell, producing a zoning map for the water inrush risk. Figure 2.19 shows an example. The accuracy of forecasting and evaluation results are tested by using locations of known water inrush points or locations of water inrush after mining.

2.4.3 Three-Map and Two-Prediction Method

In this method, the three maps are: zoning map for water abundance of water filling aquifers in coal seam roofs, safety zoning map for roof caving, and comprehensive zoning map for water gushing (inrush) conditions of roofs (Wu et al. 2016, 2000). According to the method, the necessary and sufficient condition for the water filling in coal seam roofs is that water-flowing fractured zones in roofs formed by coal mining connect overlying direct water filling aquifers and the aquifers have large water abundance in the caving zone of the mining face. In zones with low water abundance, serious water disasters are not likely to happen even water-flowing fractured zones connect aquifers. Severe water disasters only take place in zones with large water abundance under the condition that water-flowing fractured zones connect aquifers.

The two predictions consist of groundwater flux rate prediction for the working face prior to mining and groundwater flow rate prediction for the working face during mining. The method can favorably predict water disasters in coal seam roofs, and the concrete steps are as follows:

- Determine water abundance of water filling aquifers in coal seam roofs. On the basis of considering the main controlling factors influencing water abundance of water filling aquifers in roofs, weights of the factors are solved according to the controlling effects of each factor on the water abundance. Then, the zoning map for water abundance of water filling aquifers in roofs is developed through overlay analysis of multi-source geo-information, as shown in Fig. 2.20.
- Develop the zoning map for roof caving. The developing height of caving and fractured zone after mining is calculated by using the empirical formula in Coal Mine Safety Regulations. Afterwards, the safety zoning map for roof caving is plotted according to the distance from the height of the above zone to the water-resisting layer between coal seam roof and aquifer floor. Figure 2.21 shows an example.

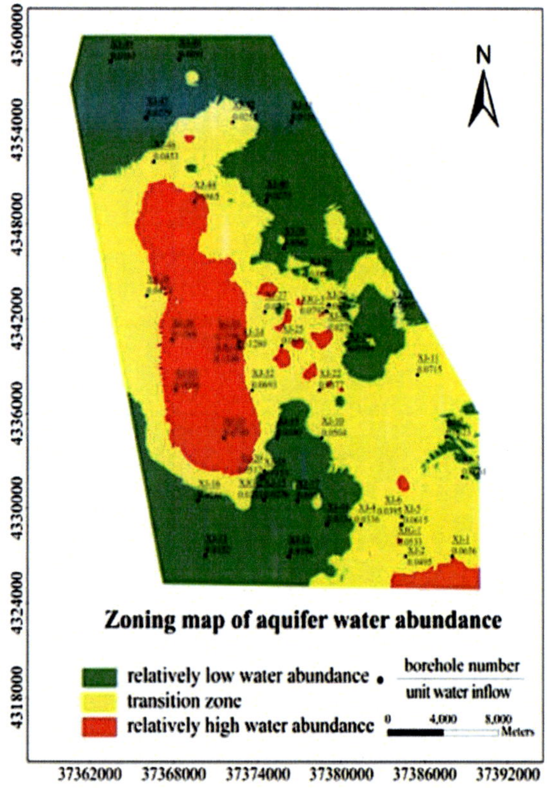

Fig. 2.20 Zoning map of aquifer water abundance

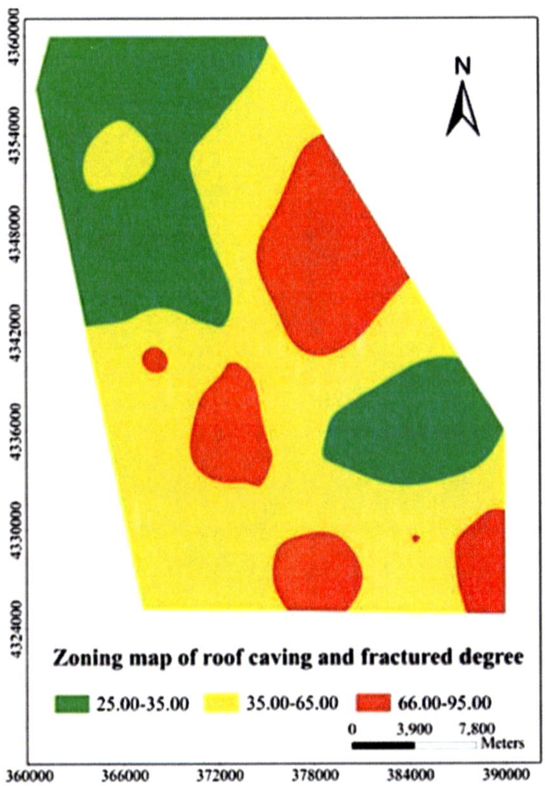

Fig. 2.21 Zoning map of roof caving and fractured degree

- Through the overlay analysis of the generated zoning map for water abundance of water filling aquifers in roofs and safety zoning map for roof caving, the comprehensive zoning map for water gushing (inrush) conditions of roofs is developed. Figure 2.22 shows the risk evaluation results.
- The two predictions refer to the mine water inrush quantity predictions in natural and man-made conditions. The natural conditions mean that the geological and hydrogeological conditions are not disturbed by mining activities while the man-made conditions mean that the geological and hydrogeological conditions are disturbed by mining or transformed by some engineering measures such as groundwater drainage and reinforcement the fractured strata by grouting. The two predictions are based on mine-specific conceptual site model, three-dimensional groundwater flow field, calibration of model parameters, and stress distributions during mining. Figure 2.23 shows the predicted water flow rate in a mining face in natural conditions, whereas Fig. 2.24 shows the predicted water flow rate in the same mining face in man-made conditions.

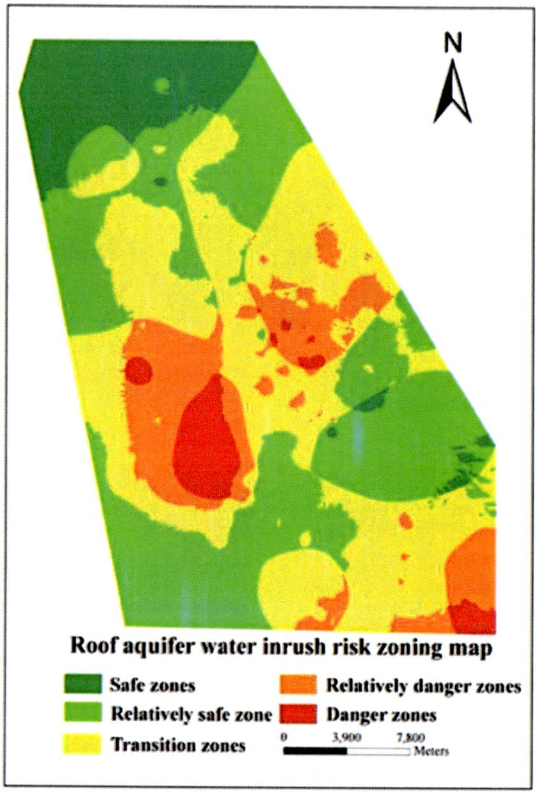

Fig. 2.22 Risk zoning map of roof aquifer water

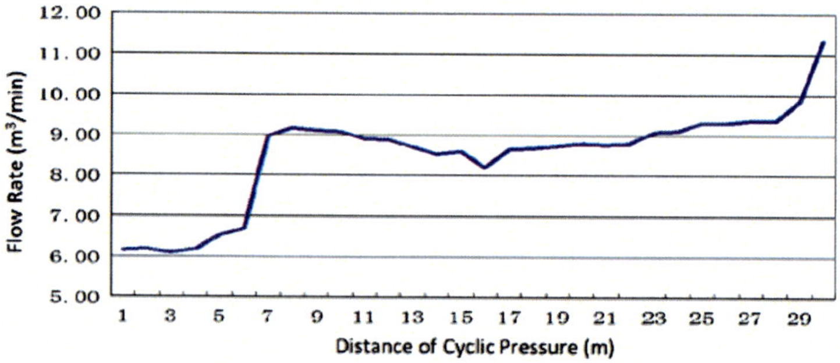

Fig. 2.23 Dynamic inflow in mining face under natural state

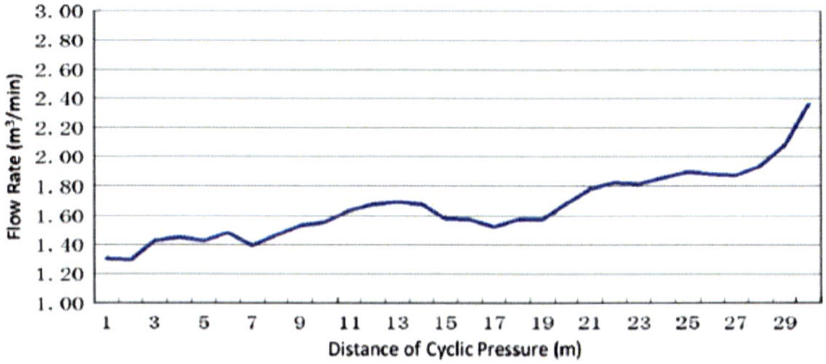

Fig. 2.24 Transient inflow in mining face under mining state

2.4.4 Five-Map and Two-Coefficient Method

The five-map and two-coefficient method is another coal floor water hazard evaluation method. The five maps consist of contour map of coal floor protection layer failure depth, contour map of coal floor protection layer thickness, contour map of water head above coal floor, contour map of effective protection layer thickness, and evaluation map of mining above confined aquifer. The two coefficients are water inrush coefficient and the coefficient of mining above the confined aquifer. Water inrush coefficient is the ratio of effective protection layer thickness to water head value. Coefficient of mining above the confined aquifer is the index that represents hydraulic pressure resistance per meter of stratum. This method can favorably predict coal seam floor water disasters, and the detailed steps are as follows:

- Develop contour map of coal floor protection layer failure depth. In the process of mining, coal floor is fractured to a certain depth, also referred to as floor failure depth, which induces water flow in the rock stratum. The floor failure depth can be obtained with experiments and empirical calculations.
- Develop contour map of coal floor protection layer thickness. The rock stratum between coal floor and aquiclude roof is named as protection layer. It is the barrier that stops confined water from flowing into the mining space. Its thickness and variation pattern can be obtained from exploratory boreholes.
- Develop contour map of water head above coal floor. The contour map is based on groundwater monitoring data. The confined water heads are different for different floor elevations.
- Develop contour map of effective protection layer thickness. The effective protection layer thickness is calculated by subtracting the floor failure depth from the coal floor protection layer thickness. It is a portion of the protection layer that possesses real ability of water head pressure resistance and safety protection function.
- Finally, develop the evaluation map of mining above confined aquifer based on the existence and thickness of effective protection layer in accordance with the

comprehensive analysis on two-coefficient and three-class evaluation. The evaluation map shows the relative water inrush risks of mining above the confined aquifer.

References

Dong SN (2010) Some key scientific problems on water hazards frequently happened in China's coal mines. J China Coal Soc 35(1):66–71

Guan ET (2012) Origin of water bursting coefficient and process of modification. Coal Geol China 24(2):30–32

He MC, Xie HP, Peng SP, Jiang YD (2005) Study on rock mechanics in deep mining engineering. Chin J Rock Mechan Eng 24(16):2803–2813

Kesseru Z (1997) Risk and safety assessment for mining and tunneling in karst environment. J Mine Water Environ 16(2):67–82

Kuznetsov SV, Trofimov VA (2002) Hydrodynamic effect of coal seam compression. J Min Sci 39(3):205–212

Li BY (1999) "Down Three Zones" in the prediction of the water inrush from coalbed floor aquifer-theory, development and application. J Shandong Inst Min Technol (Nat Sci) 18(4):11–18

Qian MG, Miao XX, Xu JL (1996) Theoretical study of key stratum in ground control. J China Coal Soc 21(3):225–230

Qian MG, Miao XX, Xu JL (2000) Key strata theory in ground control. China University of Mining and Technology Press, Beijing

Santos CF, Bieniawski ZT (1967) Floor design in underground coalmines. Rock Mech Rock Eng 22(4):249–271

Wang ZY, Liu HQ (1993) Mining in confined aquifer. China Coal Industry Publishing House, Beijing

Wolkersdorfer C, Bowell R (2005) Contemporary reviews of mine water studies in Europe. Mine Water Environ 24(1):2–37

Wu Q, Huang XL, Dl D, Yin ZR, Li JM, Hong YQ, Zhang HJ (2000) "Three maps-two predictions" method to evaluate water bursting conditions on roof coal. J China Coal Soc 25(1):62–67

Wu Q, Wang M, Wu X (2004) Investigations of groundwater bursting into coal mine seam floors from fault zones. Int J Rock Mech Min Sci 41(4):557–571

Wu Q, Yang L, Zhu B (2009) Application of vulnerability index method in coal floor water bursting evaluation in Zhaogezhuang coalmine. Coal Geol China 21(6):40–44

Wu Q, Li B, Liu SQ (2013a) Vulnerability assessment of coal floor groundwater bursting based on zoning variable weight model: a case study in the typical mining region of Kailuan. J China Coal Soc 38(9):1516–1521

Wu Q, Zhao SQ, Sun WJ, Cui FP, Wu C (2013b) Classification of the hydrogeological type of coal mine and analysis of its characteristics in China. J China Coal Soc 38(6):901–905

Wu Q, Xu K, Zhang W (2016) Further research on "three maps-two predictions" method for prediction on coal seam roof water bursting risk. J China Coal Soc 41(6):1341–1347

Xie GX, Chang JC, Hua XZ (2007) Influence of mining velocity on mechanical characteristics of surrounding rock in fully mechanized top-coal caving face. Chin J Geotech Eng 29(7):963–967

Zhou WF (1997) Paleocollapse structure as a passageway for groundwater flow and contaminant transport. Environ Geol 32(4):251–258

Zhou W, Li GY (2001) Geological barrier—a natural rock stratum for preventing confined karst water from flowing into mines in North China. Environ Geol 40(8):1003–1009

Chapter 3
Modeling of Groundwater Flow in Karst Aquifers for Mine Water Control

Groundwater modeling, based on realistic conceptual site models, is a powerful tool for solving groundwater related problems resulting from both mining operation and mine closure. Modeling applications for mine water control requires that specific features of groundwater flow in the environment disturbed by mining activities be reflected in the analytical or numerical models. The groundwater flow is dynamics because of dewatering activities, and the modeling domain in formations that are disturbed by mining. The groundwater flow regime can be turbulent, and water-flowing media consist of inherent porosities in the rock matrix, structural discontinuities (fractures and faults), and large open mine voids. These unique features of groundwater flow in mines are analogous to those in well-developed karst aquifers. As a prolific source of groundwater and one of the most important recharge sources to water flow in mines, karstic carbonate aquifers have been extensively studied over the last three decades. They have been recognized as one of the most challenging geological media in terms of groundwater and engineering problems. Topics such as yield of water and petroleum to wells, contaminant fate and transport, contaminant monitoring, and remediation involve many uncertainties because of the extreme heterogeneity with a distribution of hydraulic conductivity that spans many orders of magnitude. Karstic aquifers have highly efficient interface with surface water through swallets and springs. They often contain open conduit flow paths with hydraulic characteristics more like surface streams than groundwater.

To develop a more consistent understanding of karst terrains and their variability, various attempts were made employing classification schemes that are based on descriptions of the various flow phenomena in the field. The first comprehensive conceptual model of karst aquifers was that of White (1969) who focused attention on the variety of geologic setting and their controlling influence on groundwater flow patterns. Based only on the type of permeability, Shuster and White (1971) divided groundwater flow in karst aquifers into conduit flow and diffuse flow. After recognizing the inadequate of this binary classification, Smart and Hobbs (1986) added recharge source and aquifer storage as two additional 'coordinates' to construct a

S. Dong et al., *Methods and Techniques for Preventing and Mitigating Water Hazards in Mines*, Professional Practice in Earth Sciences,
https://doi.org/10.1007/978-3-030-67059-7_3

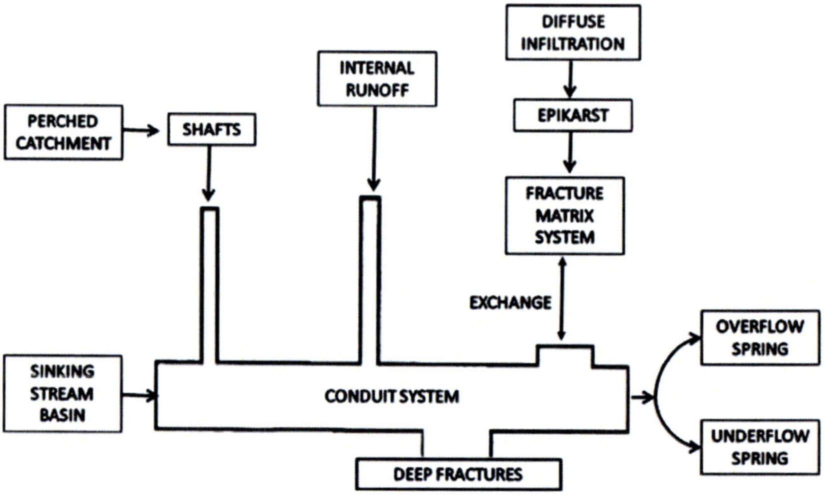

Fig. 3.1 Conceptual model for karst aquifers

three-parameter aquifer classification. Ray et al. (1994) added a transport parameter—dispersion to this classification to arrive at a combined risk factor for plotting groundwater sensitivity maps in karst areas. A more recent conceptual model was constructed by White (1999), which focused on the essential components of the karst aquifer flow system, as shown in Fig. 3.1.

Although these conceptual models are very useful in providing a framework for the description of various observations made in karst areas, there still remains a need for a more quantitative analysis of the field data commonly available. The demand for quantitative assessment of groundwater flow and solute transport has increased as the focus of many karst studies has switched from water resources development aspect towards a water resources protection point of view. In contrast, flow in karst terrains is still considered hardly quantifiable. Predictions, based on classical hydrogeological investigation methods seem to fail quite frequently, similarly most attempts to employ standard mathematical models did not prove satisfactory so far (Teutsch and Sauter 1991).

Instead of trying to pigeonhole aquifers into specific categories along the lines of conduit—diffuse or the three-dimensional scheme of Smart and Hubbs (1986), quantification of aquifers requires definition of important parameters that can characterize karst aquifers. The presence of the parameters and their relative importance distinguish one aquifer from another. Parameters recognized in karst aquifers include area of groundwater basin, aquifer permeability and dispersion coefficients, conduit response time to recharge, conduit/fracture coupling function, and spring discharge (White 1999). Obviously, there are more parameters in karst aquifers than those in granular aquifers and most of the parameters are different from those used in granular aquifers. Therefore, different approaches have to be taken to determine them.

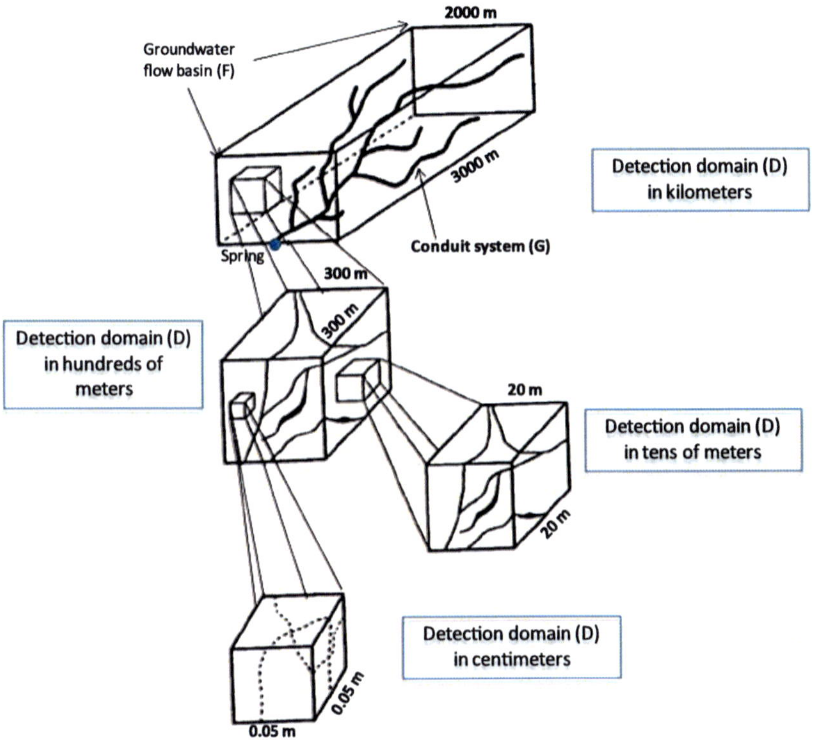

Fig. 3.2 Scale effect in karst investigation

Another challenge in analyzing groundwater flow or solute transport in karst terrains results from the variation of hydraulic characteristics with spatial scale. In general, three scales are involved in most karst investigations—the groundwater basin, the flow-dominating conduit or fracture system, and the volume covered by a detection method. Figure 3.2 shows a schematic example of the heterogeneous system with the illustrations of F, G and D, where F is the size of the catchment, H the conduits or fracture system, and D the size of the cone of depression during a pumping test. The heterogeneity of the system is extremely high at the scale of individual fractures but certainly less visible at the larger scale D of the pumping test. Any movement of the detection window D, i.e. conducting the pumping test at different location, may lead to 'smooth' variation of hydraulic parameters but would not to an abrupt change in values. A log-linear distance-drawdown relationship is expected to show a straight line at the steady state, indicating homogeneous subsurface conditions at the given detection scale. Consequently, due to the size of the detection window one would not expect to locate individual fissures or conduits. On the other hand, any more selective, i.e. a smaller scale, detection method like for example a borehole slug-test or a borehole tracer test would produce a much higher parameter variability.

The spring hydrograph reflects the response of the entire flow system to aerial recharge events. It describes the spatially averaged unsteady-state characteristics of the aquifer at a catchment scale. Because the detection scale is equivalent to the catchment scale, the distribution of the hydraulic parameters within the catchment is not detectable. The system heterogeneity is not visible at this detection scale. When a karst system consists of several distinct subsystems, the information concerning their internal connections may be available but the individual structure within each subsystem is unknown. The average distribution of the hydraulic parameters in the whole system can be detected but not within the individual subsystems.

What is generally described as the characteristics of a karst system depends very much on the type of the investigation method used and on the scale of the flow domain. A detection method with a large averaging volume will produce parameter fields that appear to be almost homogeneous, whereas from small scale measurements the same aquifer may appear highly heterogeneous. An investigation method may be more selective in the analysis of the high or low hydraulic zones. Therefore, application of a singular method to karst investigation is not adequate to characterize a karst system. Only the combination of different methods at core sample, borehole, intermediate scale, and catchment scale may provide a somewhat complete view of the relevant processes.

3.1 Inputs to Karst Hydrogeological Systems

A karst hydrogeological system can be envisaged as an open system with a boundary defined by the catchment limits and with input, throughput and output fluid flows and solute transports (Fig. 3.1). It is generally recognized that recharge conditions in karst terranes are more favorable than non-karst terranes (Li 1988). Burdon and Papakis (1963) distinguished two types of recharge: primary infiltration and secondary infiltration. The primary infiltration is the normal recharge in contrast to the secondary infiltration that occurs when waters from karst springs quickly pass over the permeable rock and re-infiltrate. The same authors also drew attention of distinguishing diffuse infiltration and concentrated infiltration. They believe that the infiltration capacity of open karst is of the order from 5 to 10 mm per hour. It is only in unusual cases that the intensity of precipitation exceeds infiltration capacity. Burdon and Papakis' (1963) notation of diffuse and concentrated recharge has been termed percolation input and allogenic water (or swallet stream) input by Drew (1970) and Smith et al. (1970), or 'infiltration retardee' and 'rapidee' by Mangin (1975). By comparing the chemical analyses in ground waters from different karst springs, Bakalowicz (1977) mentioned that this two end-member recharge mechanism is perhaps too simple. Thus, a three end-member spectrum, i.e. recharge by diffuse flow, fissure flow and conduit flow, was suggested by Atkinson (1985). Smart and Hobbs (1986) elaborate this classification, proposing a model in which the three fundamental attributes of recharge are ranged between these end members.

Considering different recharge sources, Ford and Williams (1989) classified the recharge into autogenic recharge and allogenic recharge. Autogenic recharge is derived solely from precipitation falling directly onto the karst outcrops. It is often quite diffuse, down many fissures across the karst outcrop. Allogenic recharge is derived from neighboring or overlying non-karst rocks, which normally occurs as concentrated point inputs of sinking streams. Both water chemistry and recharge volume per unit area are different in these two styles of recharge, with considerable consequences for the scale and distribution of the development of secondary permeability (Palmer 1991). The capacity of the input passage is the ultimate regulator of the volume of recharge. Thus, if infiltration rate is too great, then ponding occurs, giving rise to overflow via surface channels or to surface flooding in blind valleys and poljes. Lloyd et al. (1981) demonstrated that the recharge conditions are related to the overlying soil. If the soil is less permeable than the rock beneath, then it provides a recharge regulator, limiting recharge to the infiltration capacity of the soil. Permeable rock formations overlying the karst, referred to as epikarst zone, also act as percolation "governors" in much the same way, their vertical saturated hydraulic conductivity being the principal control.

Epikarst is a key factor in the absence of surface streams in most karst areas. As shown in Fig. 3.3, the epikarst consists of the upper most portion of the unsaturated rock where significant fracturing, solutional enlargement, and storage may occur.

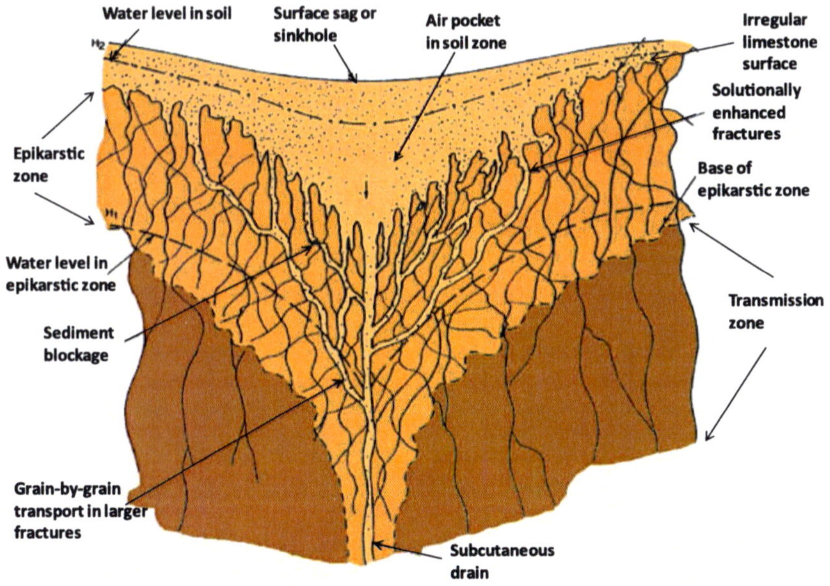

Fig. 3.3 Recharge processes through epikarst

It is often separated from the water table by a relatively less permeable rock that is sporadically breached by subcutaneous drains and vadose shafts (Quinlan 1989). When the volume of water entering the epikarst begins to exceed the drainage capacity of the zone, a hydraulic gradient may develop with a subcutaneous drain acting as the lowest point of hydraulic head, similar to a pumping well. Lateral flow and transport to subcutaneous drains may exceed several hundred meters and allow for rapid flow rates.

The recharge mechanism in karst terranes has also been extensively studied by British hydrogeologists, but from a different prospect. The thermonuclear tests in the atmosphere in the 1950s and early 1960s increased the tritium content of rainfall and provided a tracer with which groundwater recharge mechanism could be studied. As the chalk was considered a karst aquifer, the revelation from the early analyses of the low tritium content in rivers derived from chalk springs came as a surprise (Downing and Headworth 1989). Dry coring of the chalk revealed that most of the tritium was stored in the matrix of the upper part of the profile (Smith et al. 1970), as shown in Fig. 3.4. Since then, the importance of the matrix in groundwater recharge processes has been firmly established.

The initial studies implied that infiltration was moving down through the unsaturated zone at about 1 m/year by an apparent piston displacement. Only about 15% appeared to move rapidly through fissures to the water table (Smith et al. 1970).

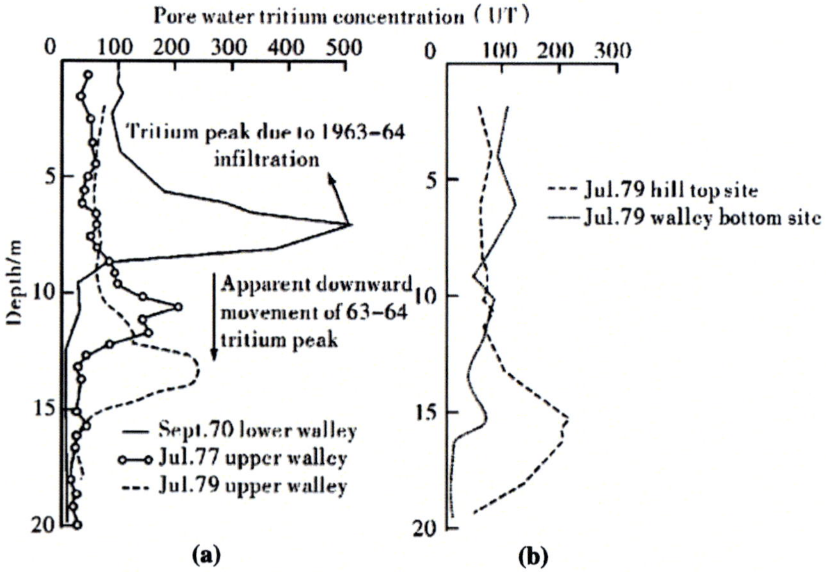

Fig. 3.4 Sequential tritium profiles from the unsaturated zone of the chalk at the dorset site of great Britain

Foster (1975) explained the initially surprising result by suggesting that fissure flow recharge did occur, but that tritium diffused from the water in the fissures into the chalk matrix. This view was supported by the evidence from the investigation in East Yorkshire of England, based upon the rapid response of water table to rainfall and the absence of surface water runoff at times of prolonged or heavy rainfall (Geake and Foster 1989). Bacterial contamination of the chalk aquifer is also the evidence for the existence of fissure flow (Zhou et al. 1992). Microscopic observations show that the inter-granular porosity is moderate to high (0.25–0.45) but the pore sizes are so small (mostly 0.05–1 μm diameter) that the hydraulic conductivity is rarely exceeding 0.005 m/d (Bell et al. 1989). Gravity drainage of the matrix will be almost negligible, and thus, in the unsaturated zone, most of the pores remain virtually saturated. However, the chalk is transversed by frequent fissures and discontinuities, and the secondary porosity provides potential pathways for relatively rapid flow in the unsaturated zone, if the matrix tension falls sufficiently. However, Chilton et al. (1989) believed that water flow is via the closely spaced primary fractures rather than through matrix pores nor through secondary porosity, as the diameters of the matrix pores are insufficient to permit significant movement and the secondary fissures are generally near horizontal and therefore are unlikely to contribute significantly to downward movement.

However, in a series of experiments over six years at Bridgest Farm of England, Wellings (1984) was able to observe the behavior of the water flow in the unsaturated zone of the chalk. He found that the dominant recharge flow mechanism was matrix flow rather than fissure flow because the matrix potentials were generally in the range of −200 to −400 cm during the winter recharge period. Only once during heavy rainfall period were the potentials observed to rise to above −50 cm, the fissure flow occurred. Recharge during the summer normally consists of residual drainage from below the zero-flux-plane at a decreasing rate that allows the water table to fall steadily. The water appears to move downwards at an approximate velocity of 0.8 m/year (Wellings 1984; Wellings and Bell 1980). In an investigation on the effects of the recharge of sewage effluent upon the quality of the chalk groundwater, Montgomery et al. (1984) and Beard and Montgomery (1981) demonstrated the removal processes for Nitrogen, Organic matter and bacteria observed in unconsolidated aquifers occurred also in the chalk. This result also implies the matrix diffuse flow in the unsaturated zone. The fissure flow recharge is rare and when it does occur, the whole fissure system is not active. Daily rainfall total more than a few millimeters per day appears to be stored temporarily in the soil and weathered chalk near to the surface (Gardner and Cooper 1989). This ‘buffer capacity’ allows drainage to be released to the water table at rates that are often low enough to be conducted by the matrix alone.

Thus, recharge of the karst aquifers is a very complicated issue, depending on the precipitation intensity and their morphological, hydrogeological conditions. Recharge into the aquifers can be (1) direct infiltration through sinkholes, ponors, estavelles; (2) diffusion infiltration from surface, temporary natural ponds through

Table 3.1 Recharge mechanisms in karst aquifers

		Recharge sources	
		Autogenic	Allogenic
Recharge passageways	Swallets, sinking streams, drain shafts		Concentrated flow recharge
	Sinkholes, surface depressions	Concentrated flow recharge	
	Subcutaneous and enlarged fissures	Both concentrated flow recharge and fissure flow recharge, with fissure flow recharge dominated	
	Fissures and joints	fissure flow recharge and diffuse flow recharge, with diffuse flow recharge dominated	
	Intergranular matrix and microfissures	Diffuse flow recharge	
	Boreholes	Concentrated flow recharge	Concentrated flow recharge
	Permeable non-karst rock overlying limestones		Mainly diffuse flow recharge

matrix and fissures; (3) direct percolation of rainfall and snowmelt water through the matrix and the networks of fissures; (4) direct inflow from neighboring or overlying non-karst rocks; and (5) artificial inflow through boreholes, tunnels. Both rapid transit of water in sinkholes, fissures and piston displacement in microfissures and macropores may exist. The solutes moving in the fissures diffuse into the matrix and move down by continuous exchange between the matrix and the micro-fissures by diffusion. When the infiltration over a period of time exceeds a threshold value (suggested as 4 mm/d in the British chalk) flow occurs in macrofissures, although with average rainfall the chalk matrix is able to absorb the water in the unsaturated zone (Downing et al. 1978; Downing and Headworth 1989). Table 3.1 expresses our present understanding of the recharge mechanisms in karst terranes. It shows the basic relationships between the recharge sources, recharge passages, and recharge flow types. Further modifications will be inevitable as our knowledge improves.

3.1.1 Discharge-Storage Method

In the karst areas where autogenic recharge dominates, the amount of infiltrated water, groundwater level and groundwater discharge vary mainly with precipitation (Fig. 3.5). For the whole karst catchment, the variation frequency in groundwater level and discharge is similar to the variation frequency in infiltration. A schematic diagram is given in Fig. 3.6 to show the components of a typical spring hydrograph after one rainstorm. The recharge due to the rainstorm is calculated by the sum of the spring discharge increment and the storage increment (Aiguang 1988).

$$V^1{}_r = V^1{}_c + V^1 SUBd = \int_{t_{1-1}}^{\infty} (Q_1 - Q_{1-1}) dt$$

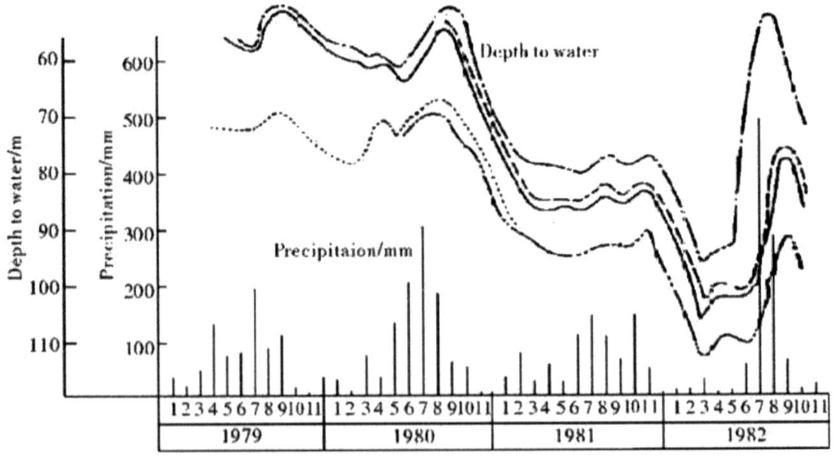

Fig. 3.5 Precipitation versus groundwater hydrograph

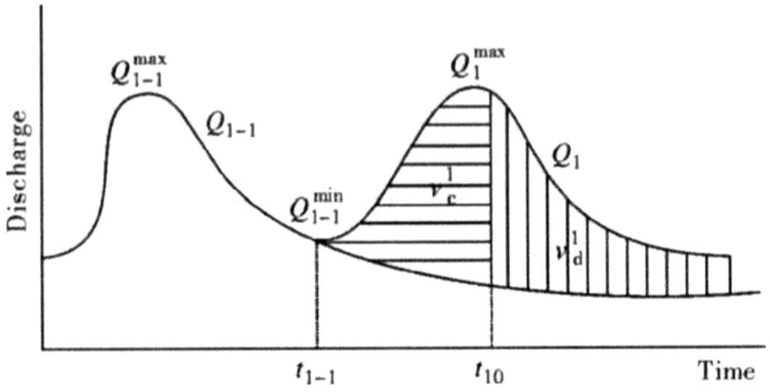

Fig. 3.6 Components of a spring discharge hydrograph

where

V_r^1 Recharge volume in one rain-storm;
V_c^1 Increasing amount of water of the spring discharge corresponding to the rain storm;
V_d^1 Storage increment;
Q_1 Discharge after rain-storm;
Q_{1-1} Discharge prior to rain-storm

The storage increment in the aquifer gives rise to a groundwater level increase. If the groundwater level returns to its previous value after a certain time, the aquifer storage change can be assumed to be zero. Then the total recharge over a rain-storm period can be estimated from the total discharge (natural discharge plus extraction) over the same period with the following procedures (Li and Zhou 1990).

1. Plot groundwater level hydrographs for different observation wells H(t)—t, where H is the groundwater level, t is the time.
2. Select a point $a_i(t_i, H_a)$ in any one curve and draw a line parallel to the time axis through this point. After one cycle, this line intercepts the same curve at point $a_{i+1}(t'_{i+1}, H_a)$, where t_i is the repeated time and t'_{i+1} is the initial repeating time.
3. Repeat step (2) for all the observation wells and cycles. According to the distances of the wells away from the discharge point, draw groundwater cross section for different repeated times. The other groundwater level cross sections at times t'_{i+1} and its neighboring time t''_{i+1} (arbitrarily selected around time t'_{i+1}) can be obtained by interpolation. When the interpolated groundwater level cross section at time t''_{i+1} overlaps the water level section at time t_i, then $t''_{i+1} = t_{i+1}$, where t_{i+1} is the repeating time.
4. Plot the discharge hydrograph Q_t—t, the total recharge (discharge) V_r from t_i to t_{i+1} can be calculated by the following integration:

$$Vr = \int_{ti}^{ti+1} Qtdt$$

In order to demonstrate how to use this hydrograph method to calculate the total recharge, an example is given in an isolated karst catchment. The groundwater level hydrographs in 6 observation wells and the discharge hydrograph are shown in Fig. 3.7. During the rainy season, 3 peaks were observed with groundwater levels of 27.60 m, 33.65 m and 38.95 m respectively. The first cycle lasts 522 h, the second 396 h and the third 360 h.

From the upper part of Fig. 3.7, four pairs of data points were selected: t_1(198.0 h) – t'_2(582.0 h); t_3(369.0 h) – t'_4(888.0 h); t_5(648.0 h) – t'_6(966.0 h); t_7(1038.0 h) – t'_8(1273 h). Figure 3.8 shows the groundwater level cross sections at various repeated and repeating times (0 represents the discharge point). The step by step procedure of calculating the total recharge is listed in Table 3.2 and the results are given in Table 3.3.

In the hydrograph method, the infiltration rate is expressed as a ratio between total volume of water discharged from the drainage area (by hydrograph integration) and the total precipitation in it. It does not consider the discharge of aquifer reserves being accumulated over a long period of time, prior to present hydrological conditions. Some large karst aquifers continue to discharge even after a year or more without precipitation. In many cases instant increase of hydrograph is the consequence of pressure propagation not the actual contribution of new water to the total discharge.

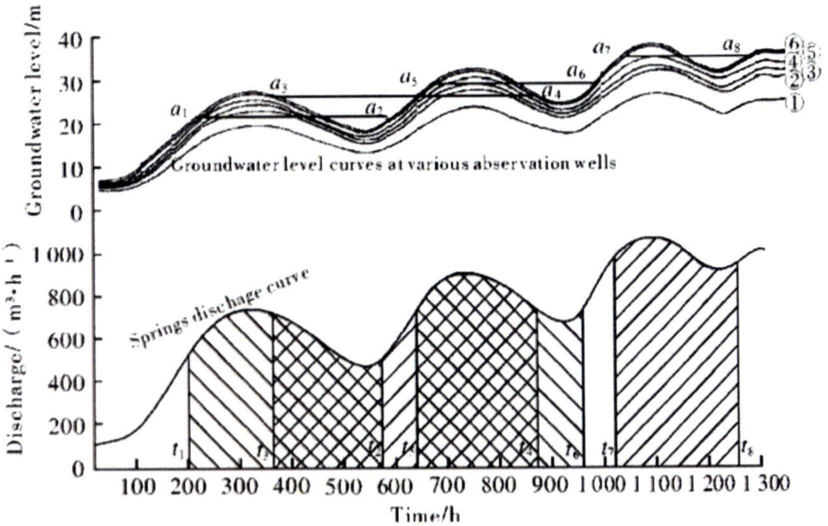

Fig. 3.7 Hydrographs of spring flow and groundwater level in calculated area

Fig. 3.8 Procedures to estimate recharge using recession-curve-displacement method

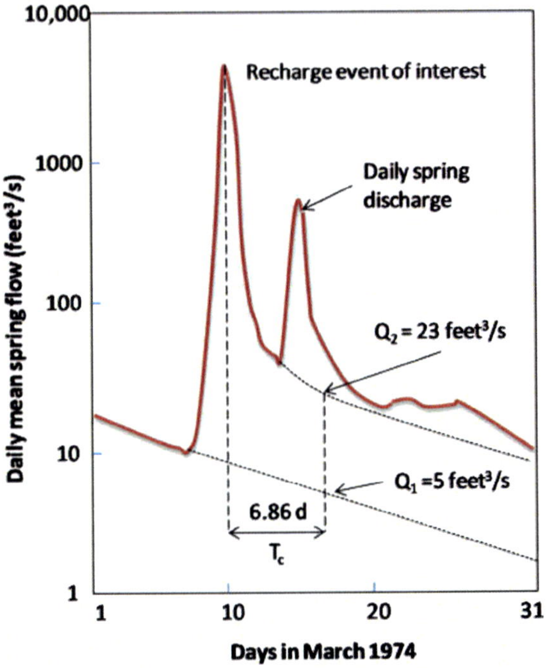

Table 3.2 Interpolated groundwater levels (m) at different repeating times

			Observation well					
			1	2	3	4	5	6
Time (h)	t_1 (repeated time)	198.0	15.00	17.90	18.65	20.00	21.70	21.90
	t'_2 (initial value of t_2)	582.0	15.20	18.10	18.80	20.20	21.70	21.90
	t''_2 (neighboring value of t'_2)	579.0	14.95	17.80	18.50	19.80	21.30	21.50
	t_2 (repeating time)	580.5	15.08	17.95	18.65	19.98	21.50	21.70
	t_3 (repeated time)	369.0	19.28	22.73	23.50	24.70	26.00	26.20
	t'_4 (initial value of t_4)	888.0	19.40	22.85	22.60	24.80	26.05	26.50
	t''_4 (neighboring value of t'_4)	894.0	19.2	22.65	23.40	24.50	25.80	26.00
	t_4 (repeating time)	885.0	19.30	22.75	23.50	24.65	25.93	26.08
	t_5 (repeated time)	648.0	20.50	24.50	25.60	27.30	29.50	29.80
	t'_6 (initial value of t_6)	966.0	20.30	24.25	25.25	26.90	29.10	29.25
	t''_6 (neighboring value of t'_6)	972.0	20.75	24.80	25.80	27.50	29.60	29.95
	t_6 (repeating time)	969.0	20.53	24.53	25.53	27.20	29.35	29.6
	t_7 (repeated time)	1,038.0	25.90	30.90	32.20	34.20	36.60	37.00
	t'_8 (initial value of t_8)	1,269.0	25.65	30.75	32.00	34.10	36.25	36.89
	t''_8 (neighboring value of t'_8)	1,278.0	26.10	31.25	32.45	34.50	36.9	37.49
	t_8 (repeating time)	1,273.5	25.88	31.00	32.23	34.30	36.58	37.19

Table 3.3 Calculated recharges in the karst catchment using the discharge-storage method

	Cycle number			
	1	2	3	4
Starting time—ending time $t_i - t_{i+1}$ (h)	198.0–580.5	369.0–885.0	648.0–969.0	1,038.0–1,273.5
Time length (h)	382.5	516.0	312.0	235.5
Recharge volume V_r ($m^3 \times 10^6$)	2.8725	4.42028	3.19669	2.54508

3.1.2 Recession-Curve-Displacement Method

This method is based on the upward shift in the recession curve of spring discharge that occurs as a result of recharge. It is intended for analysis of flow systems that are driven by diffuse recharge. Recharge will increase the total groundwater discharge at the spring, V, which is the total volume of water that will drain from the spring if allowed to do so for an infinite time without further recharge. Based on the linear relation between the logarithm of spring discharge and time, V can be expressed as (Rutledge and Daniel 1994):

$$V = Q_0\alpha/2.3026$$

where Q_0 is the initial spring discharge and α is recession index, the time required for spring discharge to decline through one log cycle.

Rorabaugh (1964) expressed groundwater discharge as a complex function of time after recharge. The function, however, can be approximated after 'critical time', T_c, by an equation that expresses the logarithm of groundwater discharge as a linear function of time. Rorabaugh and Simons (1966) obtained a formulation that gives critical time as a function of the recession index:

$$T_c = 0.214\alpha$$

Therefore, the total spring discharge at critical time after the peak flow is equal to approximately one-half of the total volume of water recharged the system. The total recharge is calculated by:

$$R = 2(Q_2 - Q_1)\alpha/2.3026$$

where Q_1 is the spring discharge at critical time as extrapolated from the recession curve preceding the peak and Q_2 is the spring discharge at critical time as extrapolated from the recession curve following the peak.

The recession-curve-displacement method can be executed by the following steps, as shown in Fig. 3.8, considering the large peak to represent a recharge event:

1. Compute recession index, α (32 d/log cycle);
2. Compute critical time, T_c (0.2144α, or 6.86 d);
3. Locate time that is 6.86 days after the peak;
4. Extrapolate pre-event recession to Q_1 (5 ft^3/s);
5. Extrapolate post-event recession to Q_2 (23 ft^3/s);
6. Compute total recharge ([2 × 18 × 32/0.3026] × 86,400 = 4.32 × 10^7 ft^3).

3.1.3 Meteorological Model

In impermeable catchments the resources available can be determined by monitoring discharge flows. Knowledge of rainfall and evaporation are an added luxury. In permeable catchments the situation is less easy since discharge flows may not represent the whole resource available and disparities between surface and groundwater catchments means that one is not certain of the area the river drains. Then estimates of areal evaporation and rainfall are essential to calculate recharge to permeable catchments.

The standard method to permit this calculation was developed by Penman in 1949 and extended by Grindley (1967). No better method has been developed since for regional studies. In his early work Grindley (1967) had shown how potential evaporation could be calculated from the four meteorological parameters of temperature, humidity, sunshine and wind speed. However, potential evaporation (or evapotranspiration) is concerned with losses from a crop which is able to draw freely on

Table 3.4 Infiltration estimates at the chalk catchments of England (Geake and Foster 1989)

Site	Gussage dorset	Fleam dyke cambridge	West norfolk
Mean annual rainfall (mm)	880	600	700
Estimated mean annual recharge for root constant of 25–100 (mm/year)	374–450	110–190	200–280
Stratigraphy	Upper chalk	Middle chalk	Middle/Upper chalk
Chalk porosity	0.4–0.47	0.33–0.43	0.3–0.4
Chalk intergranular hydraulic conductivity in saturated zone (m/day)	Not determined	10^{-4}–10^{-3}	1.5×10^{-3}
Predominant type of unsaturated flow	Intergranular	Microfissure	Microfissure
Land use	Heath	Arable	Arable
Profile date	Sep. 1970 Jun. 1977 Jul. 1979	May 1979 Oct. 1980 Feb. 1981 Sep. 1981	Feb. 1976 –Mar., 1977 Nov. –Dec., 1979 Oct. –Nov., 1984

water, whereas in normal circumstances evaporation is curbed by a build-up in a soil moisture deficit. The method propounded by Penman and Grindley allows actual evaporation to be calculated having regard to soil and vegetation type. These are expressed as a 'root constant', commonly 75–200 mm in the British chalk, which represents the water which can be freely drawn by a crop before it starts to wilt, and transpiration is curtailed. Table 3.4 shows some infiltration estimates for the chalk catchments, using the Penman method.

Water balance analyses for chalk catchments in Hampshire suggested that the root constant range (75–200 mm) may lead to underestimation of water infiltration rates and a root constant of 22 mm would be more appropriate (Headworth 1970). Lysimeter experiments at Cambridge chalk catchment also suggested that the root constant should be reduced. However, soil physics measurements of water fluxes at Hampshire site imply that a root constant of 175 mm might be suitable (Wellings 1984). Water balance calculations based on weekly or daily data also suggested that root constants greater than 75 mm should be preferred (Rushton and Ward 1979). A more detailed discussion of the root constant selection was given by Headworth (1970).

It is important to recognize that the fate of the waters which have infiltrated into underground is determined by the unsaturated flow regimes prevailing above the regional water table. Only that water which actually reached the water table, and in this way became part of saturated zone should be considered as recharge. In 1980, Wellings and Bell introduced the 'Zero-Flux-Plane' method to measure the unsaturated water flux in the upper 3 m of the chalk to overcome this shortcoming. But, in addition to the numerical difference between water table recharge and infiltration, the former lags in time behind the latter due to the travel through the unsaturated zone.

3.2 Groundwater Flow in Karst Hydrogeological Systems

No single aspect of karst aquifers has received more discussion than the concept of the 'karst water table'. There was an argument in Europe beginning in the latter years of nineteenth century about whether the concept of a water table had any meaning in karst. One school argued that sinking streams descend swiftly through the vadose zone and terminate in an integrated groundwater body with a well-defined water table. The other school argued, mostly from alpine examples, that underground karst streams were independent with no elevational relationship between each other or with a regional water table. The anti-water table school was fond of pointing out places where air-filled caves passed beneath flowing surface streams.

In the moderate relief terrain of eastern United States, the karst aquifers have well-defined water-table surface as measured from static water level in wells at equilibrium state. Much of the confusion over the karst water table arises from two sources: the highly dynamic response of the conduit system to storm events and the low permeability of the bedrock in the absence of solution openings. During low-flow conditions, the conduit may carry only a small free surface stream marking the bottom of a groundwater trough. It is quite possible for perched surface stream to overlie the air-filled conduit due to the impermeable bedrock rock and channel alluvial deposits. During a storm event, the sinking stream and internal runoff inputs carry water rapidly into the base-level conduit system. As a result, cave passages flood rapidly, often in minutes. There may be a mounding of the water table above the conduit because of the hydrostatic head provided by the upstream reaches and the relatively slow response of wall rock. After the storm, the conduit drains rapidly, returning to pre-storm conditions on time scales from days to weeks.

A karst observation well is extremely unlikely to encounter a primary conduit. Then, the question is to what extent the water level observed at the well characterizes the aquifer. Figure 3.9 shows a model consisting of two fundamental components linked in a simple two-dimensional orthogonal array (Smart 1999). Horizontal water-filled cylindrical pipes convey water between larger radius, variable volume, vertical pipes, or reservoirs. Flow through each pipe segment is governed by the Darcy-Weisbach equation:

$$Q = 2\pi g^{0.5} R^{2.5} (I/f)^{0.5}$$

where

g Gravity acceleration;
I Hydraulic gradient, $I = \Delta h/Ll$;
Ll Length of the conduit element;
R Radius of the pipe; and
f Darcy-Weisbach friction factor. Model results are relatively insensitive to f and l, therefore, they are held constant at 0.05 and 10 m, respectively.

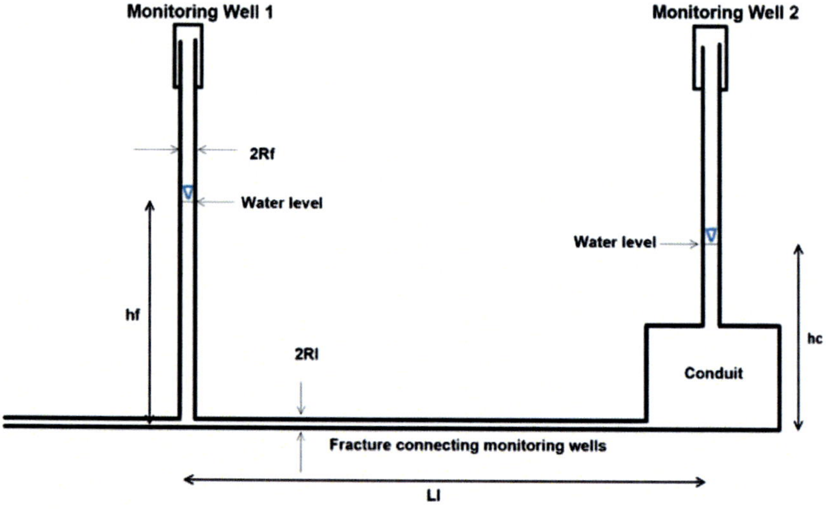

Fig. 3.9 Schematic diagram for observation well and conduit

A time-series approach is taken to investigation the response of karst system to hydrological forcing. Recharge is modeled as an arbitrary characteristic hydrograph form, projected as an independent time sequence of hydraulic head in a major conduit. The head change in the observation well is computed from the volumetric water balance of free-surface storage reservoirs:

$$\Delta h = \left(\sum Q_{in} - \sum Q_{out}\right)/(\pi R_w^2)$$

where Δh is then applied to the current head to obtain the subsequent head.

Figure 3.10 gives the simulation results, together with the necessary parameters for the calculation.

Water flux through the link is limited, and therefore the head in the observation well lags behind and has lower amplitude than that in the conduit. The departure of the observation-well signal from the conduit signal can be simply defined by differences in peak height (amplitude error) and time lag (phase error). Accordingly, these discrepancies are affected by the link and well radii (R_l and R_w), as shown in Fig. 3.11. The current conduit flood amplitude of ~8 m translated through 1 mm radius link has a phase error of ~4 h, peak amplitude 2 m less than that in the conduit, and a maximum head difference of approximately 5 m. However, those differences are negligible when $R_l \geq 5$ mm. This modeling practice indicates that the water level in observation wells will characterize the karst aquifer if the intercepted water-conducting fractures or conduits are no smaller than 1/10 of the well radii. Otherwise, a significant period of time is required for the water head in the observation well to equilibrate with that in the conduit.

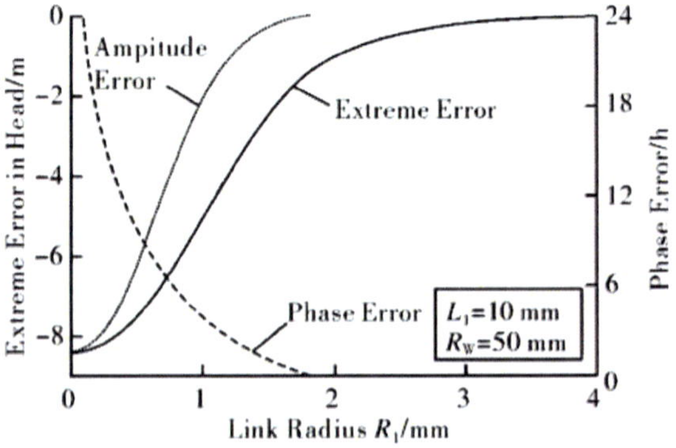

Fig. 3.10 Head time series for a conduit and observation well (The conduit head series is a known condition, and the head series for the observation well is simulated.)

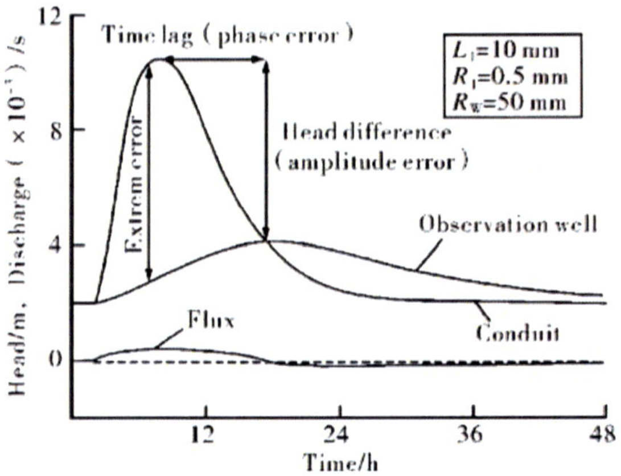

Fig. 3.11 Phase, amplitude, extreme errors between observation well and conduit against the link radius

3.2.1 Groundwater Flow Patterns in Karst Aquifers

Groundwater flow patterns are reflective of groundwater flow characteristics through geological media. They provide obvious and schematic information for analyzing the groundwater flow and the permeability of the media. Rock type and tectonic structure are two main factors controlling karst development and groundwater flow. Table 3.5 lists main flow patterns according to aquifer type. Table 3.6 lists main flow patterns according to tectonic structure.

Flow patterns can also be obtained by observing water level change during drilling processes. The water level change reflects the hydrogeological conditions at different

Table 3.5 Flow patterns in karst according to aquifer type

Aquifer type	Illustration	Main characteristics
Continuous aquifer	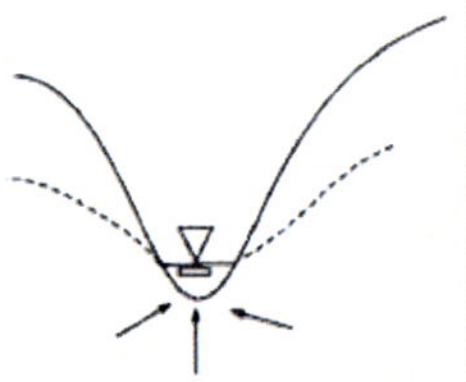	Karst is very well developed. Deep karst may be found. Groundwater has a continuous surface. Conduit flow may exist
Double-aquifer	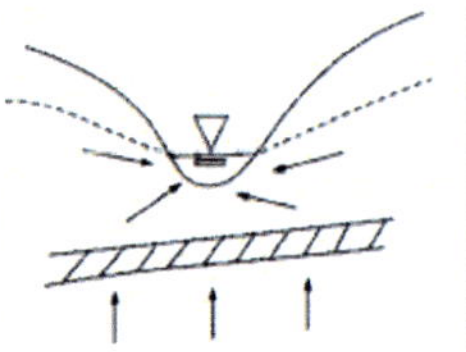	Rock has different solubility. Some part acts as aquitard. The karst at the upper is well developed. The karst development is restricted below the aquitard layer. Two aquifers exist. The lower one is confined
Multi-aquifer	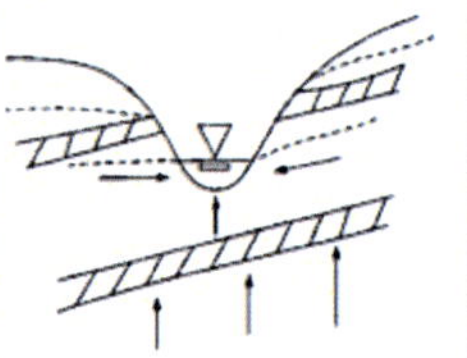	Rock is more heterogeneous, and several aquitard layers exist. There are several aquifers in the same system
Complex-aquifer	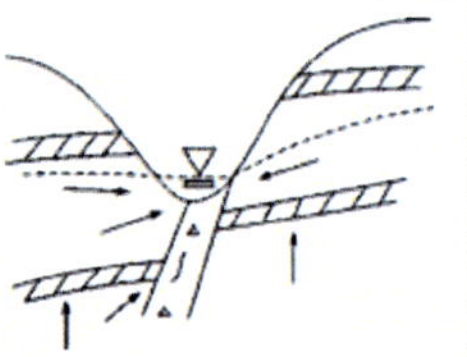	Rock was faulted. Different aquifers have hydraulic connection. Deep karst may be found along faults

depths. Table 3.7 summarizes six commonly observed curves in karst terrains and their corresponding flow patterns.

The flow patterns discussed above represent specific hydrogeological conditions and flow properties. In a complicated karst terrain, many those flow patterns may overlap each other. A conceptual composite flow pattern is given in Table 3.8.

3.2.2 *Influenced Flow Patterns*

Due to recharge or discharge of groundwater, the natural flow fields are changed, for example, a cone of water level depression is formed when groundwater is extracted from a well. If X represents the length of the cone along the axis x and Y represents the length of the cone along axis y, then $a = X/Y$ represents the non-uniformity of the

Table 3.6 Flow patterns in karst according to tectonic structure

Tectonic structure	Illustration	Main characteristics
Inclined karst terrain	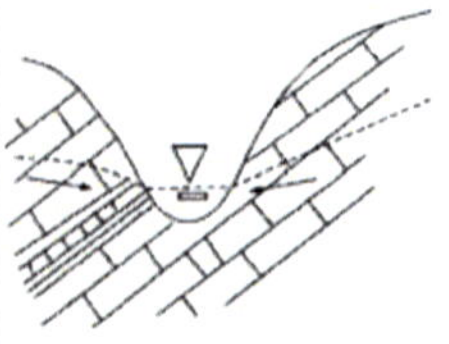	Karst is developed along rock beddings and fissures. Horizontal conduits may be found, especially around river valleys. Water flows towards valleys or discharges as springs
Anticline	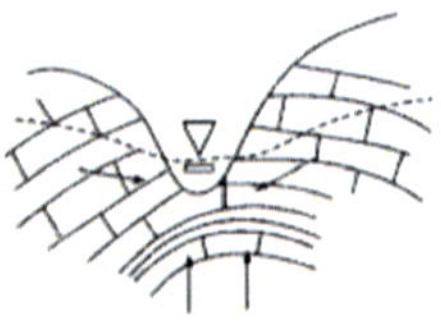	Karst is developed along anticline axis and associated tensile fissures. Water flows towards valleys and may discharge as springs along the valleys
Syncline	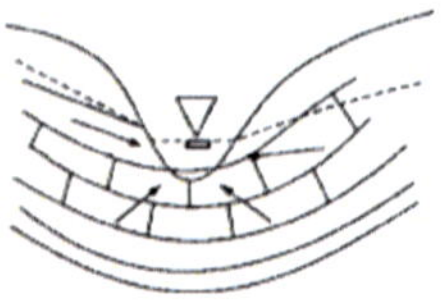	Karst is developed along axis and associated tensile fissures. Water flows normally towards valleys. In some circumstances, water discharges as springs
Faulted terrain	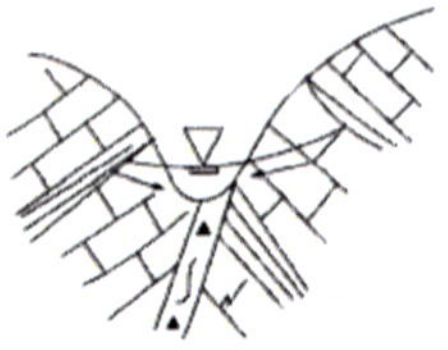	Karst is developed along faults. Water flow condition is complicated. A continuous water level may be found. Flow pattern is site dependent

depression cone. According to the values of a, Table 3.9 summarizes five influenced flow patterns.

Based on the flow pattern analysis, the main characteristics of groundwater flow in karst hydrogeological systems are summarized as follows:

1. Groundwater and solutes move in soluble rocks, such as limestone, dolomite, gypsum and salt.
2. In the soluble rocks, many fissures, caves, and even complicated conduits are developed. The hydraulic permeability is highly heterogeneous on the scale hierarchy.
3. Groundwater flow patterns are characterized by the co-existence of steady and unsteady flows, laminar and turbulent flows, isolated groundwater flow and continuous groundwater flow.
4. Groundwater circulation is deeper and stronger. Both horizontal and vertical circulations exist. Its circulation depth can be several hundreds of meters or even several thousands of meters.

Table 3.7 Water level change patterns during drilling process

Flow characteristics	Water level change pattern	Hydrogeological significance
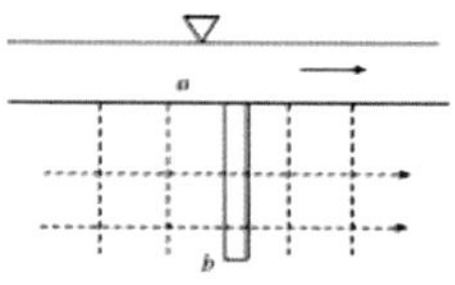		Rock is uniform, but permeability is small. Karst development is weak but uniform. Hydraulic gradient is small. Water level contour is nearly vertical
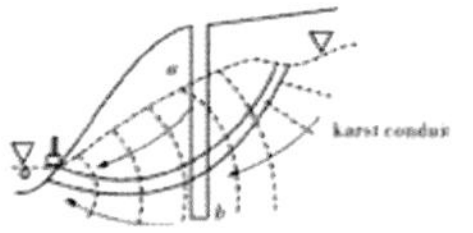	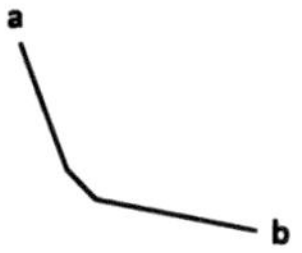	Karst develops at different layers. Hydrogeological conditions are complicated. Abrupt decrease in water level means interception of karst caves or conduits. Water level contour is complicated
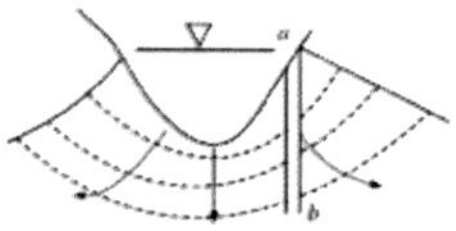		Karst develops well. River water recharges the aquifer. Flow lines radiate from the river
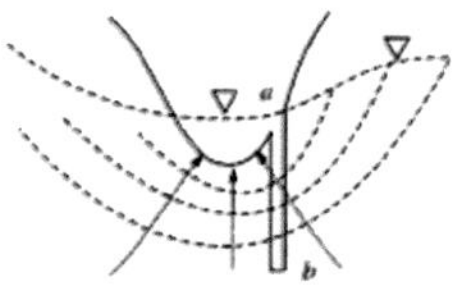		Karst development is poor. Karst water recharge river. Flow lines converge towards river
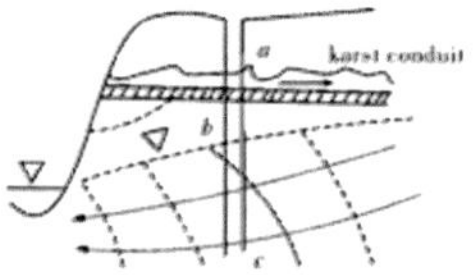		Perched underground streams many be found. Water level is discontinuous
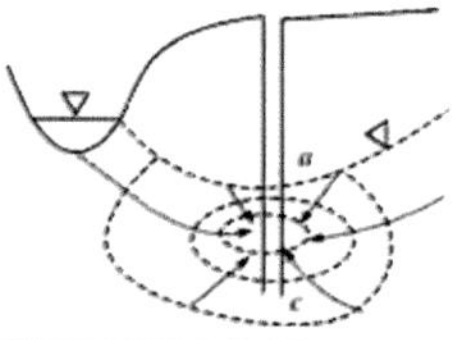		Karst development is deep. Drilling intercepts highly permeable rock or conduits. Water level contour is closed

5. The flow field is more sensitive to water recharge, extraction, and the environment is more vulnerable to contamination.

3.2.3 *Confluent Flow in Karst Aquifers*

For laminar seepage flow in porous media, the basis for quantitative hydrogeological analyses is the Darcy's law, which can be expressed as:

Table 3.8 Composite flow pattern in karst valleys

Hydrodynamic zone	Illustration	Flow properties
I: Vertical infiltration	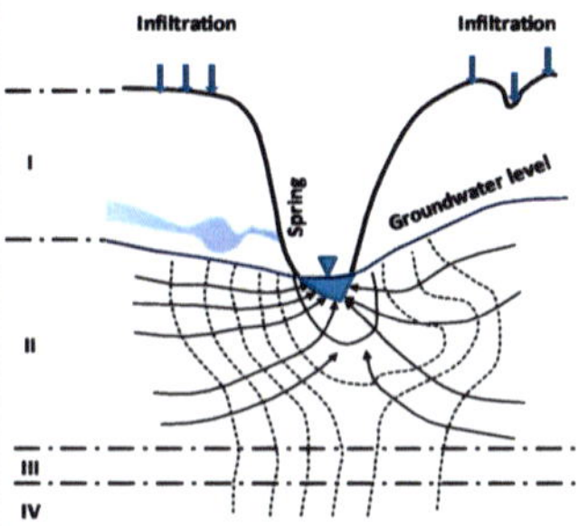	Groundwater flows downwards through fissures, pores, sinkholes. No continuous water level is formed. Some conduits may be found in this zone
II: Horizontally dominant flow		This is called shallow saturated zone. Flow lines are nearly horizontal. Horizontal conduits or groundwater streams, enlarged fissures are well developed
III: Vertically dominant flow		In this area, flow lines converge towards to the river. Siphon conduits may be developed
IV: Deep circulation		This is called deep saturated zone. Groundwater flows towards a far-reaching drainage point. Flow lines are horizontal and deep karst may be developed

$$v_s = K_s \frac{\Delta H}{l} = K_s I$$

where

v_s Seepage flow velocity;
l Groundwater flow distance;
ΔH Water potential difference;
K_s Hydraulic conductivity of Darcy flow and $K_s = k\rho g/\mu$;
ρ Specific gravity of the fluid;
μ Fluid viscosity;
K Permeability;
I Hydraulic gradient;
G Gravitational acceleration.

However, one may challenge the applicability of Darcy's law to karst aquifers because of the wide range of water-flowing passages and flow types, In an early experiment, Bocker (1973) noticed that Darcy's law is not valid when water flows through a 3 mm fissure with hydraulic gradient of 0.01. In another series of experiments, Ewers (1978) showed that the law applies when solutional proto-conduit up to 1 mm diameter is first extending through a fissure. It ceases to apply once

Table 3.9 Influenced flow patterns in karst terrains

Flow field	Illustration	a = X/Y	Flow properties
Uniform flow field	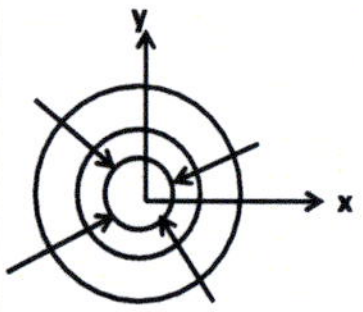	1–1.5	Mainly fissure flow. Karst develops uniformly. Plane view of the depression cone is nearly circle
Relatively uniform flow filed	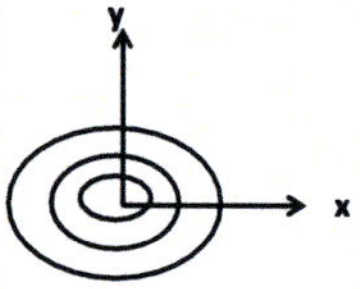	1.5–3.0	The plane view of the depression cone is an ellipse. Karst aquifer is relatively homogeneous but anisotropic
Non-uniform flow field	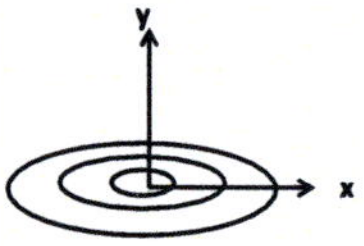	3.0–5.0	The plane view of the depression cone is an elongated ellipse. Karst develops mainly in x direction
Extremely non-uniform flow field	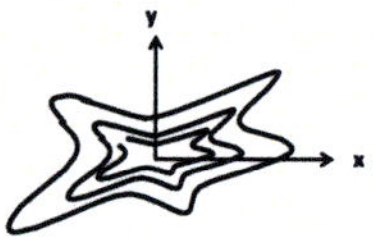	>5.0	Flow patterns are complex. Karst develops in several directions

the extension is completed and the proto-conduit is connected with others. Mangin (1975) also suggested that in karst, the range of conditions under which Darcy's law can be considered valid is restricted, especially in well-developed karst systems. In well-developed karst areas, Yuan (1978) found that the flow regime is turbulent when the flow velocity in a 3 cm wide conduit exceeds 0.001 m/s. Under the condition of turbulent flow, the Chezy non-linear law has been suggested, which is expressed by:

$$v_b = K_b(I)^{\frac{1}{2}}$$

where v_b is the turbulent flow velocity, and K_b is the hydraulic conductivity of turbulent flow.

The problem associated with application of the Darcy;s law and Chezy's law to karst aquifers is that of flow regime. The Darcy's law is valid only when the flow is laminar and the medium is assumed to be homogeneous (Bear 1972; Freeze and Cherry 1979) and Chezy's law is applicable when the flow is turbulent. If the Darcy's law were valid, a plot of specific discharge or velocity against hydraulic gradient would reveal a straight line, and if Chezy's law is valid the same plot would be a straight line in a log-log coordinate system.

Figures 3.12a and b show the basic concept of groundwater flow in karst aquifers from both vertical and horizontal prospects respectively, where flow occurs in

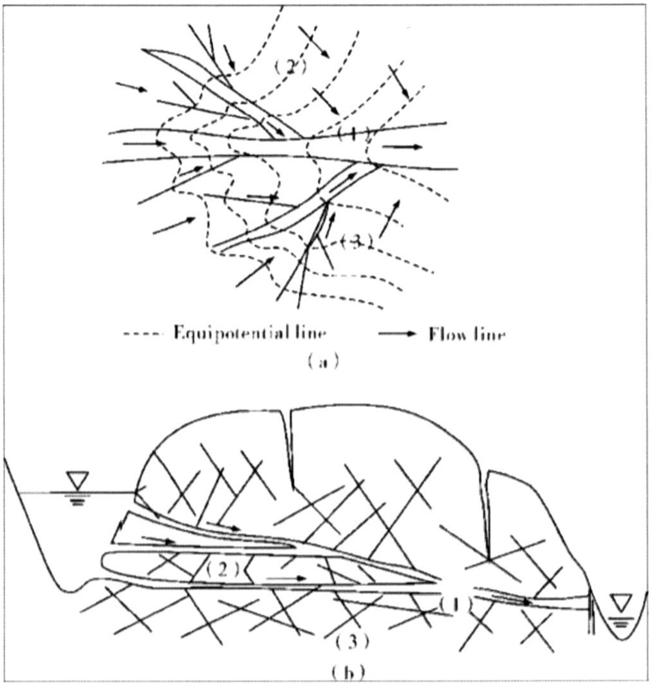

Fig. 3.12 Schematic model showing confluent flow in karst aquifers **a** Plane view; **b** Cross-sectional view. (1) Conduit flow; (2) Conduit-fracture flow; (3) Fracture-matrix flow

fissures, enlarged fissures and conduits. As water flow follows the least-resistance principle, groundwater in small fissures or porous matrix tends to flow towards the enlarged fissures and conduits. Between enlarged fissures and conduits, the conduits have higher conductivity and thus groundwater in enlarged fissures tends to flow towards the conduits. As a result, groundwater is collected by and flows through the main conduits. We call this phenomenon of water accumulation into the largest conduits confluent flow. In confluent flow, karst media may be divided into three groups, i.e. (1) storage media constituted of matrix voids and small fissures, mainly operating a function for water storing; (2) transport media that consist of enlarged fissures providing a passage for water circulation; (3) control media connecting caverns or conduits throughout the entire karst system.

Similar to seepage flow, the unsteady confluent flow in karst aquifers is expressed by:

$$u_i = f(I)$$

where u_i is defined as the confluent flow specific velocity, i.e. water flow velocity per unit change of water potential.

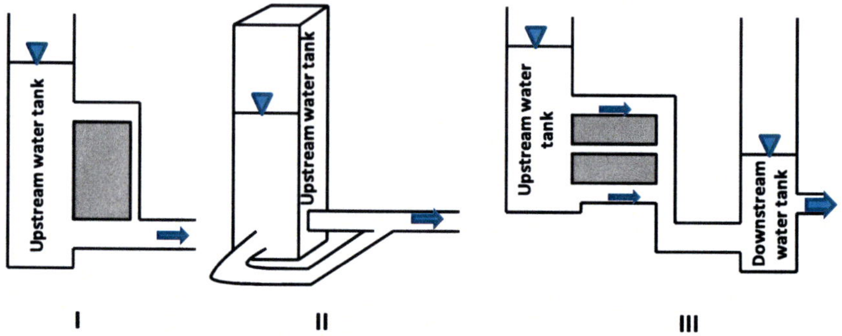

Fig. 3.13 Analog models simulating confluent flow in karst aquifers

Figure 3.13 shows three experimental structures used for simulating the confluent flow (Zou 1994). The upstream water tanks control the recharge. Plastic pipes with different diameters represent the karst conduits. The discharge is either free or controlled by a downstream water tank. The first model is designed to simulate the hydrodynamic properties in a karst system with two conduits at different levels. The second model is to study the effect of two converging horizontal conduits on groundwater flow. The third one is similar to the first but has three conduits and the discharge point is confined by a downstream water tank. The dimensions of the pipes and the initial conditions are listed in Table 3.10.

The representative experimental results are given in Table 3.11 and the relationship between u_i and I is shown in Fig. 3.14, which can be described by:

$$u_i = 0.146e^{-1.05I}$$

The relationship between u_i and I has also been studied by quantitative groundwater tracer tests. Figures 3.15a and b show the typical u_i-I curves found in two of the major karst areas of south China. Three useful functions have been found and they are:

$$u_i = 1/(a + b/I)$$

Table 3.10 Dimensions of selected pipes and the initial conditions for confluent flow simulations

Analog model	I	II	III
Inner diameter of pipes (cm)	1.00	0.6	0.8
Cross section area (cm^2)	0.78	0.28	0.50
Length (cm)	85.00	100.00	160.00
Initial discharge (L/h)	70.40	40.4	156.10
Initial water level (cm)	31.11	22.8	37.78
Number of pipes	2	2	3

Table 3.11 Results of confluent flow simulations

Analog model	I		II		III	
Reading number	Water level (cm)	Discharge (L/h)	Water level (cm)	Discharge (L/h)	Water level (cm)	Discharge (L/h)
1	31.11	70.40	22.90	30.40	37.78	117.60
20	30.42	69.20	22.21	29.60	36.45	114.70
30	26.83	53.60	18.62	27.10	34.72	104.40
35	26.14	53.10	17.93	26.60	34.14	99.00
40	25.36	52.60	17.42	26.10	33.39	89.60
50	25.06	52.10	16.85	25.60	30.37	80.40
75	22.28	49.70	14.07	23.20	26.27	72.60
100	18.82	47.10	10.61	19.80	24.88	69.70
150	11.86	40.80	3.65	11.20	28.33	56.30
200	11.02	39.50	2.81	9.50	27.32	44.90
300	10.20	38.30	1.99	7.90	22.17	39.80
400	10.25	38.40	2.04	8.00	22.00	39.50
500	9.74	37.30	1.53	6.40	20.74	38.00

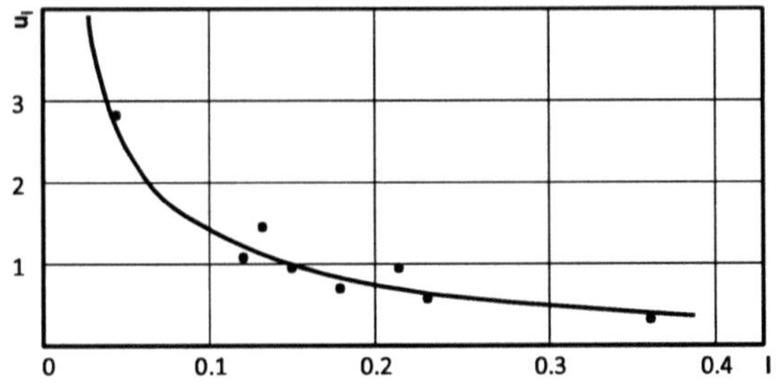

Fig. 3.14 Experimental relationship between u_i and I

$$u_i = ae^{bt}$$

$$u_i = a_0 + a_1 I + a_2 I^2 + a_3 I^3$$

where a, b, a_0, a_1, a_2, a_3 are statistical parameters.

Of the above equations, the first equation is often suitable for strong conduit flow conditions in karst-developed areas. The second equation is more suitable for enlarged fissure flow conditions in moderately karst-developed areas, whereas the third equation may be applicable to any flow conditions. Unlike Darcy's law, the relationship between u_i and I is not linear. Instead, u_i may decrease with the increase of I. This is the main difference between confluent flow and seepage flow.

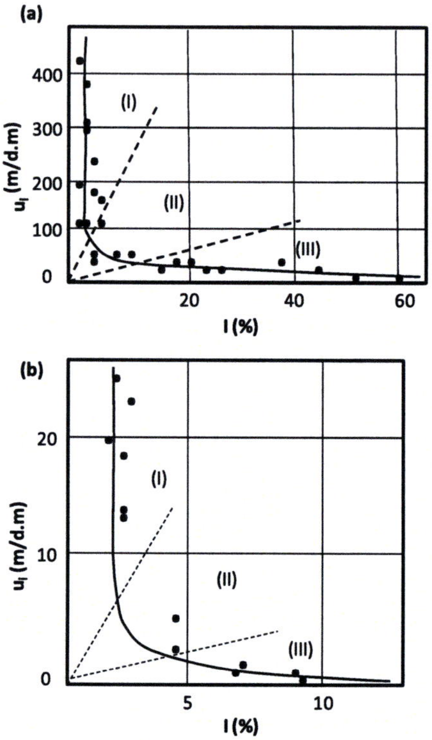

Fig. 3.15 Relationship between u_i and I from tracer tests. **a** Zhaihuangkou project; **b** Yantan project. I—Fast flow; II—Intermediate flow; III—Slow flow

According to u_i values, three sub-regimes can be divided in the u_i-I curves, as shown in Fig. 3.15. Different sub-regimes represent different flow conditions and empirical criterion values are given in Table 3.12.

Table 3.12 Flow properties of karst groundwater based on u_i and I

Flow type	I (%)	u_i (m/d m)	Flow properties	Fissure size (cm)
Fast flow	<4	>200	Fast conduit flow. Turbulent flow dominates	>20
Intermediate flow	4–10	200–60	Intermediate conduit-fissure flow. Darcy's law is applicable under certain conditions	5–20
Slow flow	>10	<60	Slow fissure-matrix flow. Darcy's law is applicable	<5

3.2.4 Siphon Karst Flow

Siphon spring is another unique feature, manifesting the complexity of groundwater flow in karst terrains. People have known siphon springs for a long time, but the boundary conditions for its occurrence and the quantitative analysis of the siphon spring flow is still an active research topic. Figure 3.16 shows a schematic device for siphon spring simulation. Water is injected through the main conduit A and then it is diverted to containers B_1, B_2 and B_3 through inlets a_1, a_2 and a_3, respectively. When water levels in the containers sequentially reach c_3, c_2 and c_1, water starts to flow through the outlets b_3, b_2 and b_1. As soon as the water levels exceeds c_3', c_2' and c_1', the siphon functions and the discharges through the drain outlets increase dramatically. As the water levels decrease the discharge flows stop when the water levels are lower than c_3'', c_2'' and c_1'' in these three containers. If recharge continues, water levels in the containers increase again and another cycle will start. If we assume that the recharge flow rate is Q_r, discharge Q_d, and siphon discharge Q, the necessary condition for the siphon flow to occur is:

$$Q_{d,max} < Q_r < Q_{min}$$

If $Q_r \leq Q_{d,max}$, the water levels in the containers is unable to reach c_1', c_2' and c_3', water flows through the outlets but no siphon flow occurs. If $Q_r \geq Q_{min}$, siphon flow will never stop, and it gives rise to a confined spring.

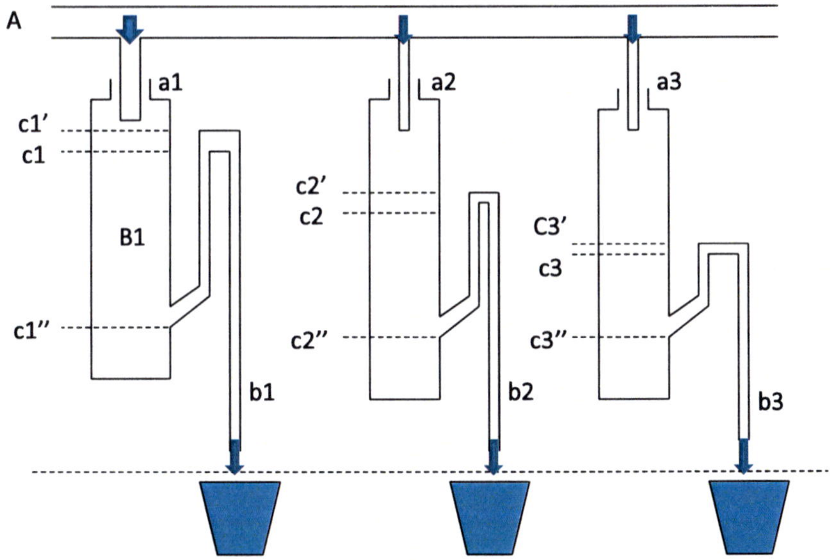

Fig. 3.16 Simulation of siphon springs

Fig. 3.17 Recharge rate versus siphon period

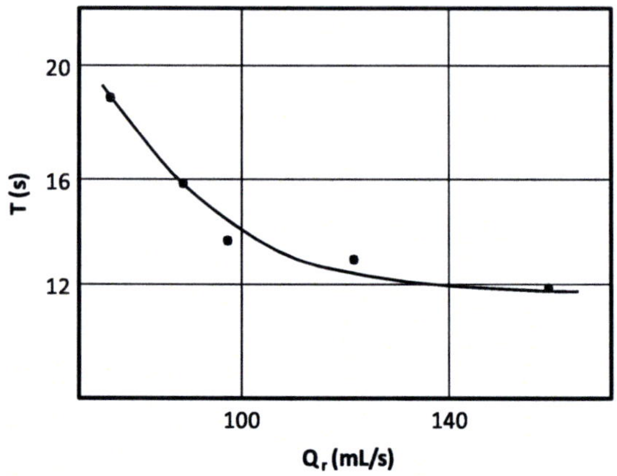

The volume of the container is 1.341 L; the volume of container below c″ is 0.2 L; and the diameter of the drain outlet is 2.5 cm. Simulation results revealed the relationship between the period (T) and the feeding recharge rate (Q_r), as shown in Fig. 3.17. If W represents the accumulated siphon discharge, it increases with time in a S-shaped pattern during one cycle, i.e. W increases with time slowly first, then quickly, and slowly again. The average recharge over one cycle (Q_{ave}) can be calculated by:

$$Q_{ave} = W_T/T$$

where W_T is the total volume of water discharged in the period of T.

At a given recharge rate of 0.15 L/second, Fig. 3.18 shows the experimental discharge hydrograph due to siphon flow. Three functions are given in Fig. 3.18 to match the measured discharges. From 0 to 4 s, the discharge increases exponentially with time ($Q = ae^{bt}$); from 4–8 s the discharge decreases exponentially with time ($Q = ae^{-bt}$); and from 8–10 s, the discharge decreases linearly with time (Q = a-bt).

3.3 Water Budget Analyses

For water resources estimation or planning management it is essential to be able to answer such questions as how much water can be used; where it comes from; and what physical parameters characterize the aquifer. Determination of the structure and main physical properties of anisotropic and heterogeneous karst aquifers poses more challenging to hydrogeologists than any other aquifers do. The complicated relationship between the catchment scale, flow dominating scale and the detection scale in karst aquifers makes it very difficult to obtain representative data. Several flow regimes may coexist, and their characteristics are poorly defined. But at least there is sometimes the opportunity to make direct observations underground, in caves,

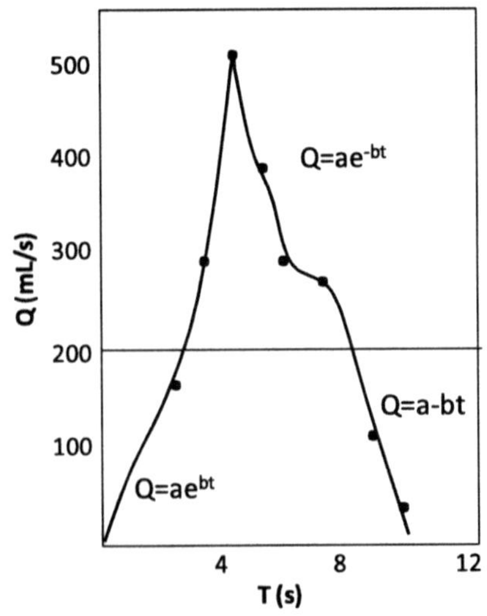

Fig. 3.18 Simulated discharge hydrograph for siphon springs

through accessible passages penetrating a part of any given karst terrain. With direct observations limited to caves, boreholes, inputs and outputs, the rest of the aquifer properties must be deduced.

Defining of the water budget elements is a classical method in calculating available water resources. It has been widely applied in the karst hydrogeology. The focuses are on the definition of the ratio between the water entering the aquifer and the water leaving it. Even though many factors associated with surface water, soil water, fissure water and conduit water, need to be considered in this analysis, it mainly consists of the following two phases:

1. Exact determination and measuring of as many as possible water balance elements such as Evaporation, transpiration, infiltration, surface, and subsurface flow.
2. Establishment of a physically approved relationship between these elements through the parametric analysis.

In a controlled karst area, the water balance equation over a period can be expressed as:

$$
\begin{aligned}
&\text{Precipitation in controlled area over the period of time} = \text{Evapotranspiration} \\
&\quad + \text{Total surface water outflow from area} + \text{Infiltration} \\
&\quad + \text{Total subsurface water outflow from area} \\
&\quad + \text{Total groundwater extraction} \\
&\quad - \text{Total surface water inflow from upstream} \\
&\quad + \text{Total subsurface water inflow from upstream} \\
&\quad \pm \text{Storage change in the aquifer}
\end{aligned}
$$

To determine the budget elements is not an easy work. It requires the experience of the hydrogeologists involved. If the conditions permit, in situ test fields should be established. When infiltration conditions are heterogeneous several water-balance equations may be required. Measuring of surface stream sinking rates is usually not a problem, provided the sinking is concentrated and not impounded by the polje retention. Recharge from other aquifers cannot be measured directly, as well as the subsurface outflow. Evapotranspiration cannot be determined by direct measurements. Instead, it is estimated by empirical formulas (Grindley 1967). It may also be expressed by the estimated relation to the directly measured evaporation from the free water surface. The storage change may be approximately estimated from water level change in observation wells. However, when long-term hydrological data are collected, the storage change is assumed to be zero over the balance period.

Infiltration is an important but difficult parameter. The methods to calculate it are discussed in Sect. 3.1 of this Chapter. However, in some cases, calculation of the infiltration coefficient is more straightforward for water balance analysis. An example of infiltration coefficient calculation for large-scale karst water balance estimation is given in Han et al. (1993). Three subsystems are defined in their studied catchment according to the stream flow distribution. In each subsystem, β_1, β_2, and β_3 were used to represent the infiltration coefficients for the outcrop, semi-buried and buried limestones, respectively. Correspondingly, three hydrological stations were setup to monitor various budget elements in the three subsystems. The water balance equations were:

$$\begin{array}{ll} 45.96b_1 + 399.0b_2 + 868.04b_3 = 233.8 & \text{sub} - \text{system 1} \\ 817.28b_1 + 189.8b_2 + 192.92b_3 = 261.89 & \text{sub} - \text{system 2} \\ 1181.08b_1 + 189.8b_2 + 365.84b_3 = 375.81 & \text{sub} - \text{system 3} \end{array}$$

The infiltration coefficients were obtained by solving the above equations and $\beta_1 = 0.235$; $\beta_2 = 0.201$; $\beta_3 = 0.164$.

Very often, the prevailing features of groundwater circulation, confluent flow, and the complex relationship between ground water and surface water make it difficult to determine various flow terms used in the budget equation. In addition, the catchment area and the geometry of a karst aquifer may change in both space and time, especially in a period of heavy rainfall. Because a karst aquifer often discharges its concentrated flow through a large springs, the basic, and maybe the only reliable measurable (or with the existing recordings) elements are the gross precipitation, the spring or surface stream discharge, and the water level fluctuation. The shape of the outflow hydrograph at a spring is a reflection of the response of the whole aquifer to recharge, whereas the shape of the groundwater level hydrograph provides information of the aquifer storage variation around the observation well. Therefore, the hydrograph analysis offers considerable potential insight into the nature and operation of karst systems. The information obtained will be considerably improved when the hydrographs are analyzed along with chemographs (water quality variation).

3.3.1 Discharge Hydrograph

It is generally accepted that karst aquifers have a rapid response to recharge by precipitation if in direct contact with the surface. However, depending on the predominant type of the aquifer porosity, transmitting passages and the recharge type (autogenic or allogenic), significant variations in discharge hydrographs may be found in different karst areas. When the main accumulation of groundwater takes place in the uniformly distributed fissure network, the influence of direct infiltration of precipitation may be significantly attenuated, with a resultant longer hydrograph peak and retardation time. If, however, large conduits and cavities predominate, the reaction of a drainage area may be shorter than one day, and with a manifold increase of discharge rate. Figure 3.19 shows schematically three typical discharge hydrographs, based on different internal structures of the karst system.

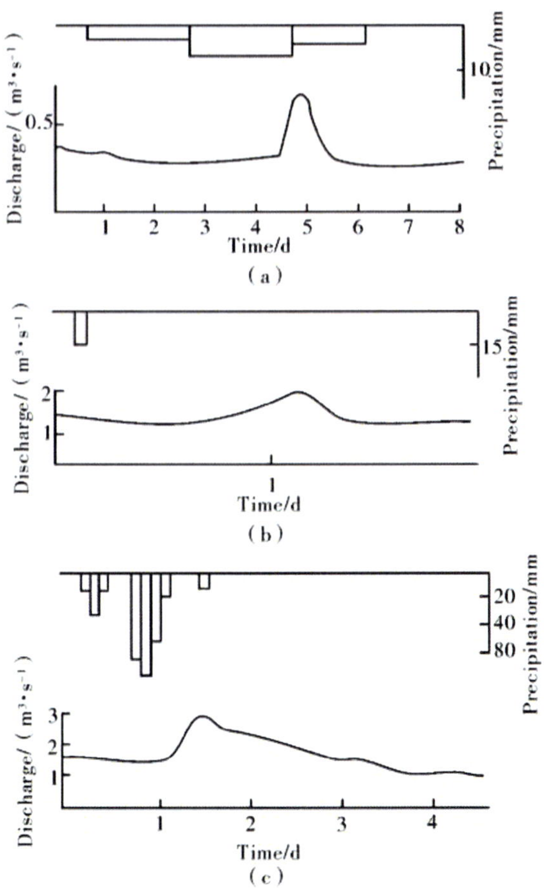

Fig. 3.19 Three typical spring hydrographs. **a**—Conduit dominated; **b**—Fractured dominated; **c**—Combination of conduit and fracture system

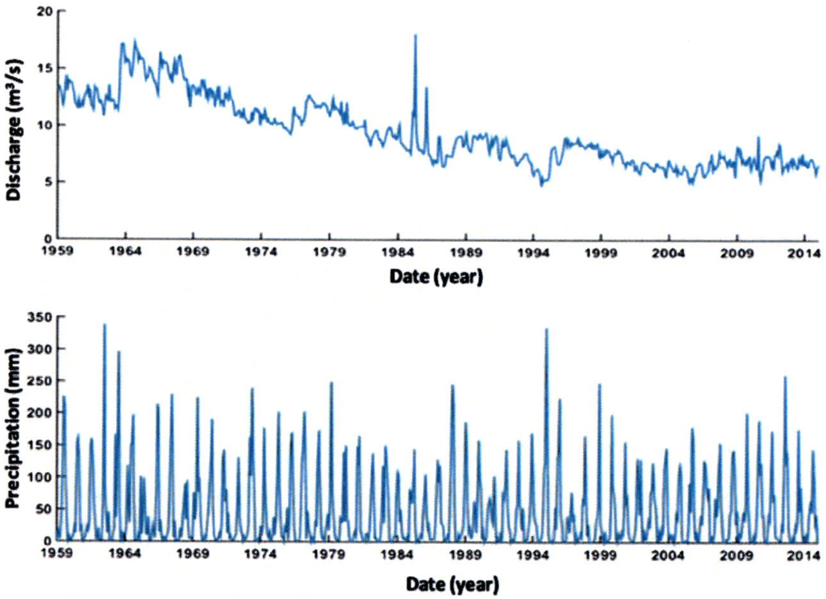

Fig. 3.20 Discharge hydrographs at Niangziguan spring, China

In practice, the hydrograph is far more complicated than those shown in Fig. 3.19. Figure 3.20 shows the discharge hydrograph at Niangziguan spring, which drains a complicated karst system. The spring discharge is closely related to precipitation but with 2–3 months' delay. More than one peak is observed after one rainstorm, implying more than one-recharge sources or multiple water flow passages. The ratio between characteristic discharges (maximum, average, and minimum) indicates a fissure or enlarged fissure-dominated system. The discharge at a spring that receives stable recharge far away from the discharge and the aquifer and has a uniformly distributed porosity may have poor relation with the precipitation. There are other springs that are sensitive to precipitation, but their discharges remain relatively constant because of river recharges.

3.3.2 Discharge Recession Analysis

In most discharge hydrographs, the discharge has a certain time delay in responding to precipitation before it increases to its peak value and then, decreases to its pre-storm outflow as a recession process. When the hydrograph is at its peak, storage in the karst system is at its maximum. The rate of withdrawal of water from that storage is indicated by the slope of the subsequent recession curve.

Quantitative analysis of the receding spring discharge derives particularly from the work of Maillet (1905), who proposed that the discharge of a spring is a function of the volume of water held in the storage and thus, it can be described by an exponential relation:

$$Q_t = Q_0 e^{-\alpha(t-t0)}$$

where

Q_t Discharge at time t;
Q_0 Discharge at time t_0 (normally set $t_0 = 0$);
t Time elapsed between Q_t and Q_0;
e Base of the Napierian logarithms;
α Recession coefficient, which can be calculated by:

$$\alpha = (lgQ_0 - lgQ_t)/[0.4343(t - t_0)]$$

During the recession process, the water volume in the system continuously changes with time. As there is no inflow into the aquifer during the recession period, so we have:

$$Q_t dt = dV$$

where dV is the equivalent water volume change in time dt. Integration of both equations lead to the water volume at time $t(V_t)$:

$$V_t = (Q_0/a)\, e^{-a(t-t0)}$$

Since $Q_0\ e^{-\alpha(t-t0)} = Q_t$, the volume of the karst aquifer at time t is:

$$V_t = Q_0/\alpha$$

It is obvious that the above equations are applicable only to the homogeneous system. For a karst system, the internal structure is not uniform, and the discharge may consist of contributions from several sub-regimes. The shape of the recession curve indicates predominant types of porosity being drained, as shown in Fig. 3.21. Therefore, the recession coefficient α may not be a constant. Many authors including Milanovic (1981) and Li and Zhou (1990) have proposed different recession equations. As groundwater flow in karst aquifers may consist of conduit flow, conduit-fissure flow (enlarged fissure flow), and fissure-matrix flow as discussed in Sect. 3.2, the suggested expression for karst spring discharge is:

$$Q_t = Q_{p1}(1 - a_1 t) + Q_{p2} e^{-a2t} + Q_{02} e^{-a3t}$$

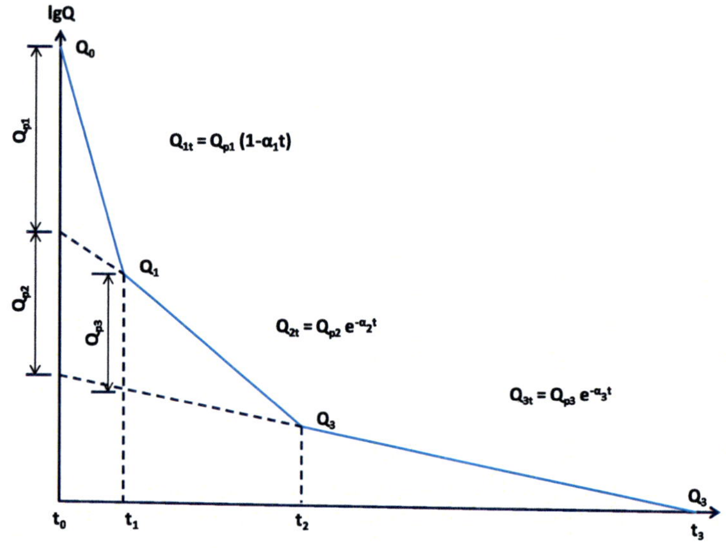

Fig. 3.21 A typical discharge recession curve

$$\text{When } t > t_1 \quad Q_t = Q_{p3}e^{-a2t} + Q_{12}e^{-a3t}$$
$$\text{When } t > t_2 \quad Q_t = Q_2e^{-a3t}$$

where $Q_{p1} = Q_0 - Q_{01}$; $Q_{p2} = Q_{01} - Q_{02}$; $Q_{p3} = Q_1 - Q_{12}$; and a_1, a_2, a_3 are the recession coefficients at different sub-regimes.

Each sub-regime of the recession curve reflects a certain type of hydrodynamic properties in the karst drainage system. In the above equations, a_1 reflects rapid outflow from conduits, caves, the amount of water that filled these conduits emptying very quickly and the water flow may be turbulent. Coefficient a_2 is interpreted as a parameter characterizing the outflow of a system of enlarged integrated fissure system. a_3 is considered a response to the drainage from the fissure network and pores, including that in rocks and soils above the water table, as well as from sand and clay deposits in caves. Figure 3.22 shows the recorded recession curve in an underground stream in South China. Based on the data from October 4, 1978 to November 13, 1978, three sub-regimes i.e. conduit flow sub-regime, enlarged fissure flow sub-regime, and fissure and pore flow sub-regime are identified. The corresponding recession equation is:

$$Q_t = 60.27(1 - 0.245t) + 16.14e^{-0.0868t} + 7.69e^{-0.0239t}$$

Let V_1, V_2, V_3 represent water volumes discharged from conduits, enlarged fissures, and fissures and pores. The total volume of water at the beginning of the recession V_0 is:

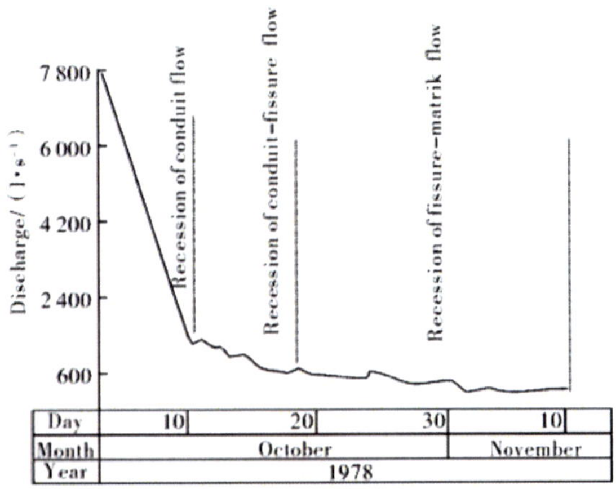

Fig. 3.22 Recession curve analysis at Shuiyuandong underground stream

$$V_0 = V_1 + V_2 + V_3$$

At the end of the recession process, the water volume in the karst aquifer is Q_3/α_3. The total volume of water discharged over the recession period is $V_0 - Q_3/\alpha_3$. The contribution of each component to the total volume of discharged water is expressed by r_i and calculated by:

$$r_i = V_i/V_0 \quad i = 1, 2, 3$$

The calculated results for the recession curve in Fig. 3.22 are listed in Table 3.13. The conduit- dominated flow lasts only 5 days, accounting for only 2% of the total flow. Table 3.14 lists the calculated recession coefficients and their corresponding components for other underground streams in South China.

Table 3.13 Recession analysis in Shuiyuandong underground stream, China (discharge in m^3/s; volume in 10^6 m^3)

	$Q_1 = 60.27(1\text{-}0.245t)$			$Q_2 = 16.14e^{-0.0868t}$			$Q_3 = 7.69e^{-0.0239t}$			Total		
Time (day)	Q_{p1}	V_1	r_1	Q_{p2}	V_2	r_2	Q_{02}	V_3	r_3	Q_t	V_0	%
t = 0	60.3	0.88	2	16.14	16.07	36	7.69	27.8	62	84.1	44.75	100
t = 5	9.3	0.14	0	8.62	8.58	26	6.85	24.65	74	24.74	33.37	100
t = 10				3.95	3.93	15	6.05	21.87	85	10	25.8	100
t = 15				1.11	1.1	5	5.37	19.41	95	6.48	20.52	100
t = 20							4.77	17.24	100	4.77	17.24	100
t = 25							4.23	15.29	100	4.23	15.29	100
t = 30							3.75	13.56	100	3.75	13.56	100
t = 35							3.33	12.04	100	3.33	12.04	100
t = 40							2.96	10.7	100	2.96	10.7	100

Table 3.14 Calculated recession coefficients and their sub-regimes in some of underground streams of south China

Underground streams	Conduit sub-regime			Enlarged Fissure sub-regime			Fissure and pore sub-regime			Total		
	α	V (10^6 m^3)	r_1 (%)	α	V (10^6 m^3)	r_2 (%)	α	V (10^6 m^3)	r_3 (%)	Q_t (m^3/s)	V (10^6 m^3)	r (%)
Disu, Guanxi	0.079	10	10	0.049	49.8	48	0.027	42.4	42	60.11	102.17	100
Fengxiang, Guangxi	0.024	1.35	15	0.011	0.93	10	0.0094	6.8	75	1.66	9.08	100
Shuiliandong, Guanfxi	0.0245	0.88	2	0.0868	16.07	36	0.0239	27.8	62	84.1	44.75	100
Bengbengdong, Hunan	0.151	0.08	8.3	0.039	0.116	12	0.0165	0.755	79	0.59	0.951	100
Houzhai, Guizhou	0.048	1.47	2	0.0066	6.15	10	0.001	54.43	88	2.75	63.59	100

3.3.3 Discharge Chemograph

Discharge variations at a spring are often accompanied by changes in water quality. Characteristics that may vary include ions in solution, electrical conductivity, environmental isotopes, suspended sediment and temperature. Although some of these are physical rather than chemical attributes of water, it is convenient to consider plots of any of these water quality aspects against time as chemograph.

When using the chemograph to separate the water entering the aquifer (new water) from the water already in it (old water), we assume that their compositions are different. From analysis of 14 karst springs in the central Appalachians, Shuster and White (1971) demonstrated the chemical composition difference in the 'new' water and the old water. The reality of old water in storage being pushed out by new recharge water was confirmed by the work of Bakalowicz et al. (1974). Using ^{18}O variations in three karst springs, they showed that a precipitation event can cause 'old' water in different stores in a karst system to be flushed out and mixed in different proportions (Ford and Williams 1989). In a recent investigation, Dreiss (1989) measured the Ca, Mg and HCO_3 concentrations at Maramec Spring in southern Missouri. Large fluctuations in Ca and Mg concentrations occur in the discharge after major storm events. She believed that rapid flow of relatively dilute, storm-derived water through conduits causes the fluctuation in chemistry. The additional conditions for the applications of hydrochemical separation are:

- Concentrations of the chemical constituents in the rainwater, chosen for monitoring, are uniform in area and in time.
- Corresponding concentrations in the pre-storm water (old water) are also uniform in space and time.
- The effects of other processes in the hydrologic cycle during the episode, including other recharge sources, are negligible.
- The concentration and transport of elements are not changed by chemical reactions in the aquifer.

The last condition is the most delicate and assumes minor dissolution of carbonate rocks during the flow of new water through karst conduits and fissures. The first two conditions are limiting factors for the application of natural isotope separation. However, regarding the Ca and Mg ions, they are acceptable as the ion concentrations in precipitation are very low relative to those in carbonate groundwater.

Considering simple mixing of old, pre-storm water, Q_{old}, and new water from recent precipitation or run-off, Q_{new}, the total spring discharge, Q_s is the sum of the two components:

$$Q_s = Q_{old} + Q_{new}$$

The two components are shown schematically in Fig. 3.23. Without chemical reactions, the mass balance of a specific ion in the spring discharge is:

Fig. 3.23 Decomposition of spring flow into old and new waters

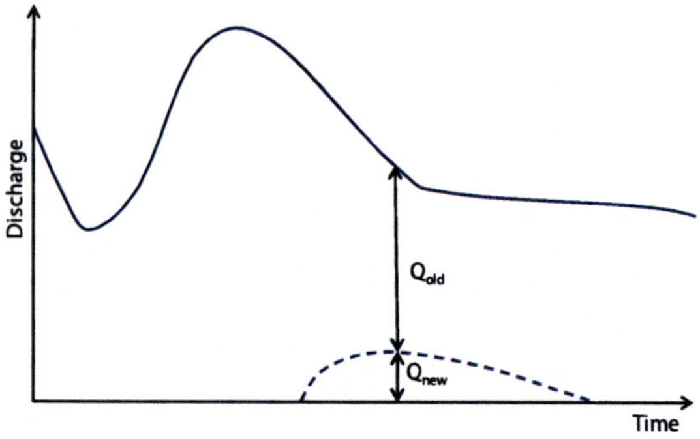

$$Q_sC_s = Q_{old}C_{old} + Q_{new}C_{new}$$

where C_s is the measured mass concentration of the ion in the spring discharge; C_{old} is the ion concentration in pre-storm water; C_{new} is the ion concentration in recent storm-derived water as it enters the conduit system.

If C_{new} is small relative to C_s and C_{old}, the mass flux of ions contributed by recent precipitation is small relative to the mass flux from pre-storm water.

$$Q_{old}C_{old} \gg Q_{new}C_{new}$$

Thus,

$$Q_{old} = (Q_sC_s)/C_{old}$$

Combining the equations above leads to:

$$Q_{new} = Q_s - (Q_sC_s)/C_{old}$$

If C_{old} is known, the component of the spring discharge that enters the aquifer as rapid recharge to the conduit system after a storm can be estimated from field measurements of Q_s and C_s. C_{old} can be determined by monitoring the ion concentration in the spring discharge. After a certain time of no excess precipitation, the ion concentration tends to remain constant, and the constant value can be taken as representative of pore water composition in the secondary porosity of the aquifer (Dreiss 1989).

3.3.4 Groundwater Level Hydrograph

The analysis of piezometric levels with respect to their spatial distribution and temporal variation provides important information about the local response of the aquifer to the external action, for example the storage change in a rainstorm. This is especially true for the heterogeneous karst aquifer. Four types of hydrographs are listed in Table 3.15 according to the response of groundwater level to one rainstorm.

When the groundwater level hydrograph has a similar pattern as the discharge hydrograph, the recession analysis is applicable to it. The general recession equation is:

$$H_t = h_{p1}e^{-a1t} + h_{p2}e^{-a2t} + H_{02}e^{-a3t}$$

where $h_{p1} = H_0 - H_{01}$; $h_{p2} = H_{01} - H_{02}$; H_0, H_1, H_2 are the initial water levels at the beginning of each sub-regime; and a_1, a_2, a_3 are the corresponding recession coefficients.

However, depending on the local hydrogeological conditions, the three components are not necessarily manifested in all the hydrographs. Figure 3.24 shows three

Table 3.15 Different types of response to one single rainstorm in groundwater level

Illustration (h—amplitude, t—delay time)	Relative amplitude	Relative delay time	Hydrogeological significance
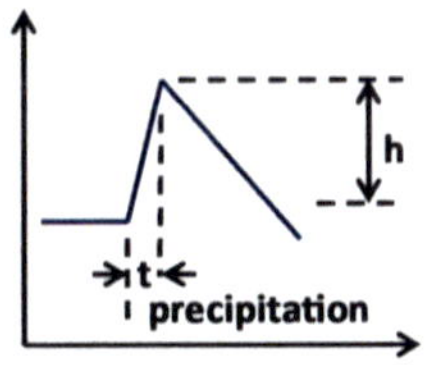	>5	1 (reference value)	Strongly permeable around the well
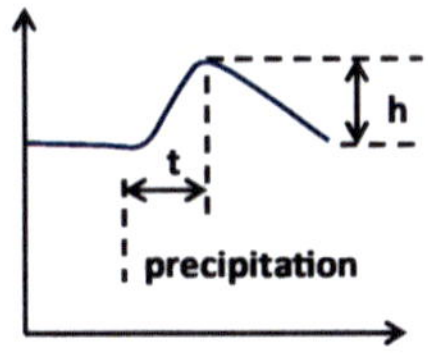	3–5	1–2	moderately permeable around the well
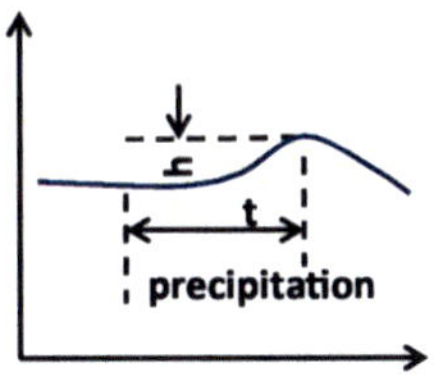	1–3	2–3	Less permeable around the well
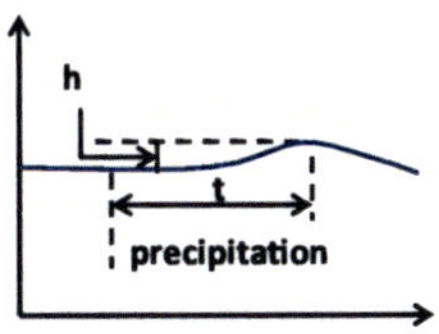	1 (reference value)	>3	Nearly impermeable around the well

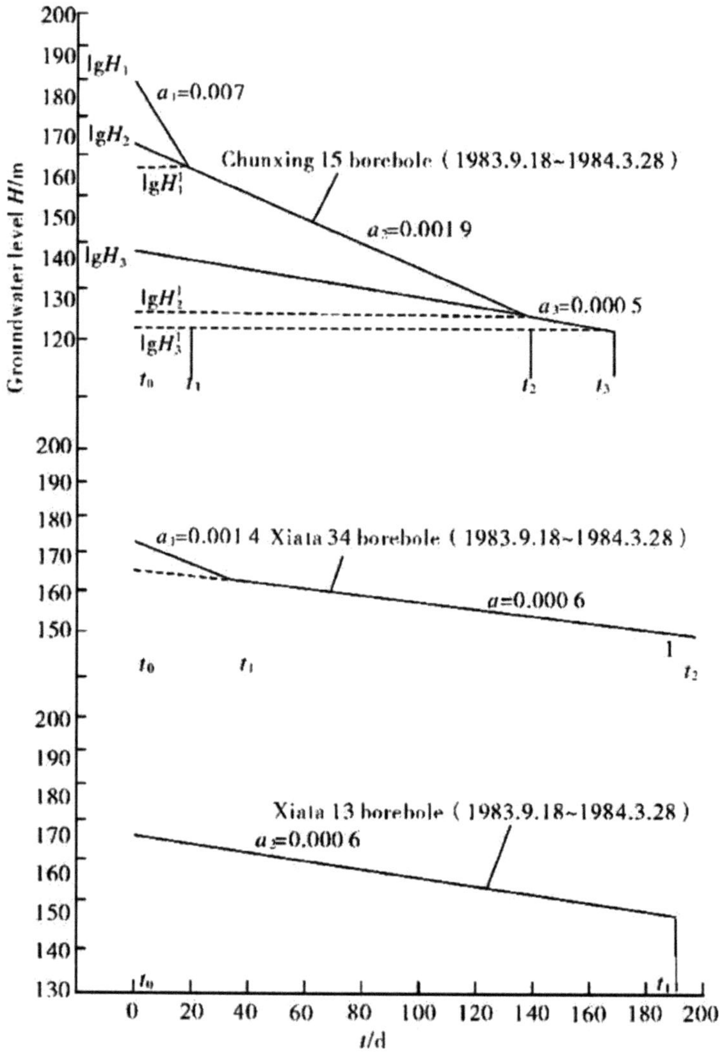

Fig. 3.24 Groundwater level recession curves at Dahua Reservoir, China

groundwater level hydrographs and their analyses. The top hydrograph has three components, indicating that the well is located in or near conduits (underground streams) and it is close to the discharge point. The middle and the bottom hydrographs have respectively two and one component, showing fissure dominated flow and the wells are some distance away from the discharge point.

3.4 Statistical and Stochastic Methods

In hydrological research, usual objects of statistical calculations are different time series analyses in the characteristic discharge, groundwater level fluctuation, and daily or yearly precipitation. The probability distribution that describes a certain set of measured hydrologic data is controlled by numerous factors varying from place to place. The basic requirements for the correct statistics analysis are sufficient data and testing of the obtained empirical probability function by several tests including Chi-square, Student's test (t-test) and F-test.

3.4.1 Regression Analysis

Connecting of measured hydrologic data of different time series that represent behavior of a certain karst hydrogeological system by linear or non-linear regression is common in karst research. Though non-linearity of karst aquifer processes is evident, linear regression is more often applied for its relatively simple calculation. Moreover, when non-linear approach is chosen, the establishing of corresponding equation is usually performed through the linearization of its predictor variable, thus having the same calculation procedure. The general equation of linear regression has the following form:

$$Y = a + bX + e$$

where Y is the dependent variable; X is the independent variable called predictor; and ε is the random variable. Parameters a and b are the parameters which can be estimated by means of the least square method to fit the observed data where

$$b = \frac{\sum X_i Y_i - \frac{1}{N}\sum X_i Yi}{\sum X_i^2 - \frac{1}{N}(\sum X_i^2)} = \frac{\sum(X_i - \overline{X})(Y_i - \overline{Y})}{\sum(X_i - \overline{X})^2}$$

$$a = \overline{Y} - b\overline{X}$$

and X_i and Y_i are observed data at the *ith* set, N is the total number of data set.

In practice, a thorough investigation of groundwater recharge and discharge is required before a regression equation is established. Take Fig. 3.20a as an example. The spring discharge is closely related to the precipitation. Some data were extracted and listed in Table 3.16.

As the spring drains a large catchment, one year's discharge may be contributed by several years' precipitation. Assume that the spring discharge at year t is linearly related to the average of precipitation over m years before year n, we get the following regression equation:

Table 3.16 Annual spring discharge and precipitation (based on Fig. 3.20a)

Year	Annual precipitation (mm)	Spring discharge (m^3/s)	Year	Annual precipitation (mm)	Spring discharge (m^3/s)
1955	514		1967	605	14.8
1956	690		1968	445	14.4
1957	392		1969	615	13.3
1958	570		1970	539	12.8
1959	707		1971	595	12.4
1960	568		1972	255	11.3
1961	593		1973	703	10.8
1962	632		1974	378	10.8
1963	802	13.6	1975	535	10.2
1964	787	15.8	1976	594	10.2
1965	399	15.2	1977	722	11.2
1966	696	14.5	1978	481	11.3

$$Q_t = a + \frac{b}{m} \sum_{i=t-n}^{i=t-m-n+1} P_i$$

where

Q_t Is the average spring discharge at year t (m^3/s);
a, b Are the parameters to be determined;
P Is the annual precipitation (mm);
N Is selected to be 0, 1 and 2 years;
M Is selected to be 2, 3, 4,…11 years

$$R_{QP} = \frac{1}{m-n} \sum_{i=1}^{i=m-n} P(t)Q(i+n)$$

The correlation coefficient R_{QP} is a function of m and n. They can be expressed as:

The maximum correlation coefficient is obtained when n = 0 year and m = 7 years and it equals 0.97. The regression equation for the data in Table 3.16 is:

$$Q_t = -6.36 + \frac{0.0329}{7} \sum_{i=t}^{i=t-6} P_i$$

The parameters a and b are calculated by assuming that all precipitation contributes to the discharge. In fact, when the precipitation is very small, no water reaches the groundwater table due to the effects of evaporation and moisture deficiency (Avdagic 1988). Only when the precipitation is big enough does the precipitation contribute to the discharge. With respect of groundwater recharge, we call this kind of precipitation as 'effective precipitation', in which water reaches the groundwater level and has an effect on the discharge, while the precipitation that does not contribute to spring discharge is called 'non-effective precipitation'. Non-effective precipitation is an important parameter in establishing regression equations between the precipitation and discharge. In this calculation, monthly precipitation that is less than 10, 20, 30, 40, and 50 mm is considered as non-effective, the newly calculated coefficients and regression coefficient are given in Table 3.17 ($n = 0$; $m = 7$).

It seems that considering of the non-effective precipitation significantly changes the value of a, whereas the value of b and the regression coefficient are slightly affected. As smaller value of a means better correlation of discharge and precipitation, the monthly non-effective precipitation is about 30 mm in this example.

If we consider the annual precipitation as an independent variable, we can conduct the multiple regression analysis. Previous analyses indicate precipitation over the seven previous years have a great effect on the present discharge, so the multiple regression equation has eight parameters and can be written as:

$$Q_i = a + b_1 P_i + b_2 P_{i-1} + b3\, P_{i-2} + b_4 P_{i-3} \\ + b_5 P_{i-4} + b_6 P_{i-5} + b_7 P_{i-6}$$

where

Q_i	Is the total annual spring discharge at year i;
a, b_1, b_2,... b_7	Are the parameters to be estimated;
P_i	Is the precipitation at year i

The total annual spring discharges and effective annual precipitations are calculated from Fig. 3.20a and listed in Table 3.18. The calculated equation is:

Table 3.17 Effect of non-effective precipitation on regression equation

Monthly non-effective precipitation (mm)	Parameter a	Parameter b	Regression Coefficient
0	−6.36	0.0329	0.970
10	−3.35	0.0332	0.971
20	−1.91	0.0350	0.969
30	−0.45	0.0362	0.970
40	0.74	0.0376	0.970
50	1.90	0.0390	0.969

Table 3.18 Annual discharge and effective precipitation

Year	Spring discharge (10^8 m^3/year)	Effective precipitation (mm)	Year	Spring discharge (10^8 m^3/year)	Effective precipitation (mm)
1980	3.41	353	1967	4.68	555
1979	3.65	429	1966	4.56	600
1978	3.76	424	1965	4.79	319
1977	3.58	651	1964	4.97	755
1976	3.20	552	1963	4.28	803
1975	3.22	503	1962	3.83	483
1974	3.42	314	1961	4.00	519
1973	3.41	646	1960		568
1972	3.57	190	1959		680
1971	3.94	522	1958		542
1970	4.09	465	1957		311
1969	4.19	602	1956		665
1968	4.52	398	1955		434

$$\begin{aligned} Q_i = & -1.2736 + 0.0009P_i + 0.0018P_{i-1} \\ & + 0.0015P_{i-2} + 0.0014P_{i-3} + 0.0018P_{i-4} \\ & + 0.0015P_{i-5} + 0.0008P_{i-6} \end{aligned}$$

The regression coefficient is 0.99; F-test is 63.03; and the residual standard deviation is 0.20. The current year's and the 7th year's precipitation do not contribute very much to the discharge. The current year's discharge mainly comes from the precipitation infiltration of 2–6 years ago, accounting for about 91%. Because these models describe the delayed response of the system to input, they are often called moving average models (MA) in stochastic analysis.

Since the fluctuations in groundwater level or spring discharge are very often dependent upon antecedent corresponding values, an auto-correlation coefficient λ can be introduced into the multiple regression equation (Houston 1983). If the dependence upon prior conditions is very strong, the auto-regression terms are dominant, and the stochastic model may be built solely of these elements. Such model is called auto-regressive (AR) with the order defined by testing the significance of autocorrelation function. Combination of AR and MA models leads to ARMA models (Auto-regressive moving average), which has been used in many hydrological analyses. The general form of ARMA models is:

$$\begin{aligned} Y_k = & V_k + a_1 Y_{k-1} + a_2 Yk - 2 + \ldots + a_r Y_{k-r} \\ & + b_1 X_k + b_2 X_{k-1} + \ldots b_s X_{k-s+1} + \varepsilon_k \end{aligned}$$

where

- Y_k Is the output (spring discharge or groundwater level) at time k;
- X_k Is the input (rainfall or groundwater extraction) at time k;
- a_i Is the auto-regressive parameters (total of r);
- b_i Is the moving average parameters (total of s + 1);
- V_k Is a constant at year k;
- ε_k Is the error.

AR, MA and ARMA models are applicable to linear and stationary systems. However, most hydrologic processes have certain periodicity or trend, and such seasonal time series have to be deseasonalized before nonseasonal ARMA models are fitted to the data. The most common methods are (1) subtraction of the seasonal mean; and (2) subtraction of the seasonal mean and divided by the seasonal standard deviation (Kresic 1993). Analysis by Fourier series has been applied as well, in which

$$Y_i = \frac{a_0}{2} + \sum_{i=1}^{i=N} (a_j\, cos(2\pi ji/N) + b_j\, sin(2\pi ji/N))$$

where Y_i is the sequence of average spring discharge or water level over a time interval (i = 0, 1, 2,…, N); a_j, b_j are Fourier coefficients; and j is the number of cycles and j = 0, 1, 2,…, L;

$$\begin{cases} L = N/2 & \text{when N is an even number;} \\ L = (N-1)/2 & \text{when N is an odd number;} \end{cases}$$

$$\begin{cases} a_j = \frac{2}{N} \sum_{i=1}^{i=N} Q_{icos}(2\pi ij/N) \\ b_j = \frac{2}{N} \sum_{i=1}^{i=N} Q_{isin}(2\pi ij/N) \end{cases}$$

3.4.2 Kernel Analysis

Kernel analysis, also called linear systems analysis or 'black box' method, is another application of time series analysis to karst hydrogeology. It has been used for many years in hydrology to describe rainfall-runoff relationships and several studies have proposed the use of linear transfer functions to investigate regional scale solute transport in fractured and highly heterogeneous aquifers (Rinaldo and Gambolati 1987; Duffy and Harrison 1987). Linear systems analysis was first applied to karst aquifers by Knisel (1972) and Dreiss (1982) for short-term isolated rainstorms. The precipitation was considered as an input time series and spring discharge as an output series. If the karst system is assumed to be linear and time invariant, the observed fluctuations in spring discharges, groundwater levels or chemical compositions are the output response of a series of input stimuli by means of a linear filter. From the measured input and output time series, a kernel function can be identified to reproduce the observed behavior of the physical system, as shown in Fig. 3.25.

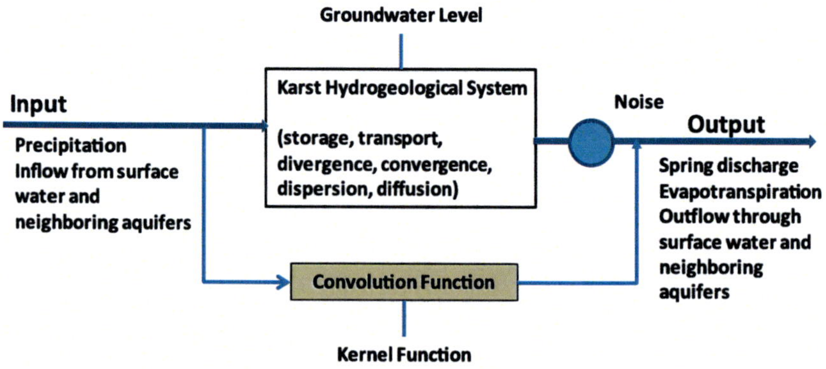

Fig. 3.25 Concept of kernel analysis for karst aquifers

The behavior of a linear, time-invariant system is described by the convolution integral:

$$Y(t) = \int_0^t h(t-\tau)X(\tau)d\tau = \int_0^t h(\tau)X(t-\tau)d\tau$$

where

Y(t)	Is the continuous system output;
X(τ)	Is the input series;
h(t-τ) or h(τ)	Is the kernel function (Dreiss 1982).

For discrete, noisy measurements, the above equation becomes

$$Y_i = \Delta t \sum_{j=0}^{j=i} X_j\, h_{i-j} + e_i \quad i = 0, 1, 2, \ldots, N$$

for N + 1 output data points collected at intervals of equal time length Δt. Y_i are the output values at time iΔt. The X_i and h_i represent mean input and kernel function values over time step interval Δt. The error term e_i is residual due to measurement errors, or if model assumptions are not strictly met, nonlinearities and temporal changes in the system. If any error is involved in the data, the identification problem becomes one of finding an optimum kernel function using all available information.

By minimizing the sum of the square of the errors E, the kernel function values h_{i-j} can be identified:

$$\min E = \sum_{i=0}^{i=N} e_i^2$$

which is subject to the constraints that the discrete kernel is non-negative:

$$h_k \geq 0 \quad k = 0, 1, 2, \ldots, m$$

where m is called the memory length of the system as defined by Dreiss (1982). When the total volumes of the input and output series are equal the sum of the kernel function values is unity.

$$\Delta t \sum_{k=0}^{k=m} h_k = 1$$

The data from Fig. 3.20a are used to test the linear systems analysis. The discharge error between calculated and measured values is evaluated by varying the memory length and the non-effective precipitation. As shown in Table 3.19, the minimum error

Table 3.19 Calculated discharge error for effective precipitation and memory length

Monthly non-effective precipitation (mm)	Minimum discharge error (m^3/s)	Memory length (year)
0	−0.908	4
10	0.073	5
20	−0.263	6
30	−0.246	7
40	−0.002	8
50	0.306	9

is obtained when the memory length is 8 years, and the non-effective precipitation is 40 mm.

For different time intervals, Table 3.20 gives the calculated kernel function values. Large time interval may be suitable when the karst spring has a distant recharge and small interval is able to find out the effect of local recharge on the discharge. As shown in Table 3.20, the discharge is mainly related to the precipitation over the last 8 years when a year interval is selected, which reflects a distant regional recharge and slow groundwater flow. However, when the time interval reduces to 3 months, h_0 has a higher value (0.00411), reflecting the existence of a local recharge source.

In the context of the kernel for a karst aquifer, the h_k describes the response of the conduit system to an instantaneous unit input. The zero^{-order} moment (μ_0) is the integral of the kernel function, which is described in the equation for discrete values. The first order moment of h_k is approximated by:

$$\mu_1 = \Delta t \sum_{k=0}^{k=m} k h_k$$

The first moment is the centroid of the area under the kernel. The quantity μ_1/μ_0 has dimensions of time and is the mean residence time of water travelling to the spring. The amount of spreading or mixing in the system response is described by the variance (σ) of the response, which is the second central moment about the mean residence time, t_m:

$$\sigma^2 = \Delta t \sum_{k=0}^{k=m} (k - tm)^2\, h_k$$

or non-dimensional coefficient of variation C_v, where

$$C_v = \sigma / tm$$

Values of σ^2 and C_v are related to the distribution and interconnectedness of travel pathways in a system. The shape of the kernel function depends on the distribution

Table 3.20 Calculated kernel function values using data from Fig. 3.20a

Time interval	Kernel function value (discharge in m^3/s; precipitation in mm) ($\times 10^{-2}$)					Memory length (year)	Discharge error (m^3/s)
One year	$h_0 = 0.364$	$h_1 = 0.781$	$h_2 = 0.646$	$h_3 = 0.476$	$h_4 = 0.522$	8	−0.002
	$h_5 = 0.524$	$h_6 = 0.423$	$h_7 = 0.238$				
Half year	$h_0 = 0.406$	$h_1 = 0.821$	$h_2 = 0.648$	$h_3 = 0.741$	$h_4 = 0.654$	9	−0.084
	$h_5 = 0.605$	$h_6 = 0.481$	$h_7 = 0.733$	$h_8 = 0.475$	$h_9 = 0.510$		
	$h_{10} = 0.377$	$h_{11} = 0.319$	$h_{12} = 0.410$	$h_{13} = 0.252$	$h_{14} = 0.244$		
	$h_{15} = 0.038$	$h_{16} = 0.229$	$h_{17} = 0.075$				
4 months	$h_0 = 0.230$	$h_1 = 0.722$	$h_2 = 1.029$	$h_3 = 0.466$	$h_4 = 0.796$	7.7	0.138
	$h_5 = 0.988$	$h_6 = 0.408$	$h_7 = 0.692$	$h_8 = 0.526$	$h_9 = 0.401$		
	$h_{10} = 0.637$	$h_{11} = 0.613$	$h_{12} = 0.696$	$h_{13} = 0.424$	$h_{14} = 0.337$		
	$h_{15} = 0.729$	$h_{16} = 0.374$	$h_{17} = 0.198$	$h_{18} = 0.541$	$h_{19} = 0.181$		
	$h_{20} = 0.314$	$h_{21} = 0.277$	$h_{22} = 0.262$				
3 months	$h_0 = 0.411$	$h_1 = 0.892$	$h_2 = 0.982$	$h_3 = 0.842$	$h_4 = 0.544$	8.8	0.013
	$h_5 = 0.686$	$h_6 = 0.856$	$h_7 = 0.805$	$h_8 = 0.531$	$h_9 = 0.717$		
	$h_{10} = 0.437$	$h_{11} = 0.589$	$h_{12} = 0.383$	$h_{13} = 0.783$	$h_{14} = 0.509$		
	$h_{15} = 0.715$	$h_{16} = 0.588$	$h_{17} = 0.431$	$h_{18} = 0.359$	$h_{19} = 0.388$		
	$h_{20} = 0.563$	$h_{21} = 0.434$	$h_{22} = 0.350$	$h_{23} = 0.325$	$h_{24} = 0.492$		
	$h_{25} = 0.117$	$h_{26} = 0.399$	$h_{27} = 0.183$	$h_{28} = 0.188$	$h_{29} = 0.007$		
	$h_{30} = 0.147$	$h_{31} = 0.074$	$h_{32} = 0.115$	$h_{33} = 0.092$	$h_{34} = 0.042$		

of pathways in the aquifer. The variance of h_k will be larger for aquifers where flow pathways are highly interconnected. Similarly, the variance of h_k will be small for conduit-type karst aquifer, where large quantities of water are transmitted rapidly in discrete conduits. Thus, the variance should be larger for springs in less mature karstic terrains and in aquifers with large recharge areas.

The skewness of h_k can be expressed as the third central moment Γ

$$\Gamma = \Delta t \sum_{k=0}^{k=m} (k - tm)^3 h_k$$

or non-dimensional skewness coefficient γ,

$$\gamma = \Gamma / \sigma^3$$

Skewness results from at least three physical phenomena: (1) increasing dispersion with time as water travels to the spring, (2) a non-systematic distribution of travel distances, and (3) delay effects from storage in relatively immobile portions of the aquifer. The first of these becomes less important with increasing fluid flow velocities. The second depends on the distribution of distance between points of rapid recharge and the spring outlet. The third develops as water enters and leaves dead-end solution conduits, fine pores, and fissures in response to changes in fluid pressure in the conduits.

Although the linear kernel analysis has proved to be very useful in karst hydrogeology research, it may cause large errors when the karst system has obvious non-linear behavior. An alternative to the non-linearity is to use the second order Volterra kernel function (Zhan 1994):

$$Y(t) = \int_0^t h(\tau)X(t-\tau)d\tau \int_0^t \int_0^t g(\tau,\delta)X(t-\tau)X(t-\delta)d\tau d\delta$$

where g(τ, δ) is the non-linear second order kernel function.

Identification of the related parameters using the non-linear kernel function is very complicated and has been discussed by Rao and Rao (1987). A relatively simple stochastic model for non-linear karst systems is introduced in the following section.

3.4.3 *Threshold Autoregressive Analysis*

The above discussed regressive models usually require a linear system and a large sample collection or documentation of a long sequence to avoid the potential errors. Some karst systems demonstrate strong non-linear behavior, and the spring flow measurements may not be long enough to guarantee the proper equation establishment. Under such circumstances, one of the alternatives is to use the threshold autoregressive model (TAR). The TAR model is a non-linear model that is linear in segments. The time series can be non-stationary. The discrimination of model conducted with the Akaike Information Criterion (AIC) criterion can avoid large deviation that may happen for small sample collection if other stochastic methods are used (Dai et al. 1988). Assume the spring discharge series is {Y_i} and the precipitation series is {X_i},), we write the TAR model:

$$a_0^{(1)} + \sum_{i=1}^{i=r_1} a\,i^{(1)}\, Y_{k-i} + \sum_{i=1}^{i=s_1} b_i^{(1)}\, X_{k-i+1} + \varepsilon_k^{(1)}\, X_{k-m} \le P'$$

$$\mathrm{Y}_k = a_0^{(2)} + \sum_{i=1}^{i=r_2} a\,i^{(2)}\, Y_{k-i} + \sum_{i=1}^{i=s_2} b_2^{(2)}\, X_{k-i+1} + \varepsilon_k^{(2)}\, X_{k-m} > P'$$

where

r_1, r_2, s_1, s_2	Are positive integers;
$a_i^{(j)}$, $b_i^{(j)}$	Are parameters to be estimated, j = 1, 2;
$\{\varepsilon_k^{(j)}\}$	Is the white noise with the mean of zero and quadratic deviation of $\sigma_t^{(j)2}$;
m	Is called the delay parameter in the analysis;
P'	Is the threshold value.

In parameter calculation, the maximum step L and the maximum delay parameter D of the two segments are given. Let $n_0(D) = \max\{d, L\}$, d = 1, 2,…, D. For each of the fixed r and d, the data $Y_{n0+1}, Y_{n0+2}, \ldots, Y_N$ are grouped into two categories. If $X_{n0\text{-}d+i} \le P'$ (i = 1, 2,…, N-n_0), then Y_{n0+i} belong to the first category; otherwise the second one. Substituting the first category data into the first equation, the parameters can be obtained by the least square estimation. The same method is used for the second category data to obtain the associated parameters. For given P′ and d, the parameters r_j and s_j are determined by the AIC criterion:

$\mathrm{AIC}(r_j, s_j) = \min\{N_j \ln(\sigma_t^{(j)2}) + 2(r_j + s_j + 2)\}, j = 1,2;\ 1 \le r_j \le L;\ 1 \le s_j \le L$
By varying the values of P′ and d, we get a series of AIC values. The best P′ and d are those values corresponding to the minimum AIC (r_j, s_j).

3.5 Mixing-Cell Models

Both the regression analysis and systems analysis belong to the 'black box' model. They are unable to describe and control the physical process of the karst aquifers. As a result, the estimated parameters in the regression analysis and kernel functions in systems analysis will depend on the type of storms. Different storms will result in different impulse responses because they cause different flow regimes in the system. It is not satisfactory for studying or simulating both the structure of a karst water system and hydrodynamic field. Therefore, such a treatment of the karst aquifer is often far removed from physical reality and tells us little about the structure of the system and how it operates. In addition, application of the black box model needs a lot of observation data of the spring or stream discharges, which is difficult to obtain especially in some newly developed areas. The treatment that is more realistic is a 'grey-box' approach, which uses such information as is available on sub-terrain conditions to clarify the structure of the system and to help explain its observed operation. Although each karst aquifer is unique in its individual characteristics, some structure components such as boundary conditions, input and output locations, spring discharges, interior structure of linkages and stores, system response to recharge, are widely recognized.

For several decades, hydrologists have used mixing cells and discrete reservoirs as a basis for modeling hydrologic systems. Chemical engineers also have been using mixing cell models to study the efficiency of mixing in reactor vessels. For the most part these models were solved analytically and assumed constant flow, constant cell volumes, and simple cell networks, all of which placed constraints on flexibility. In recent years, more complicated mixing cell models have been developed, such as those developed by Adar and Neuman (1988) and Gieske and de Vries (1990), in which the flow is non-steady and the network is in three dimensions. Coupling with the three-dimensional finite-difference groundwater flow model (MODFLOW), a three-dimensional mixing cell was also developed to predict chloride concentration changes in the city of El Paso, Texas (Rao and Hathaway 1990).

3.5.1 Discrete-State-Compartment Model

The discrete-state-compartment (DSC) model is a mixing cell model, which was introduced to karst hydrogeology in 1972 by Simpson (1988). The karst system is represented by a set of interconnected cells, through which water and dissolved materials are transported in accordance with physical laws. Transport of water and dissolved materials is governed by a set of recursive equations, which describe the physical system as a series of discrete states. The mass conservation is the law to derive the recursive equations. For any given cell, the basic equation of the DSC model is:

$$S(i) = S(i-1) + [V_r(i) * C_r(i)] - [V_d(i) * C_d(i)] \pm R(i)$$

where

$S(i)$ Cell state at iteration i, the mass or amount of tracer in the cell;
$V_r(i)$ Boundary recharge volume at iteration i, the input volume of water to the cell;
$C_r(i)$ Boundary recharge concentration at iteration i, the input concentration of tracer;

$V_d(i)$ Boundary discharge volume at iteration i, the output volume of water from the cell;
$C_d(i)$ Boundary discharge concentration at iteration i, the output concentration of tracer;
R(i) Tracer source or sink within the cell.

The word tracer indicates any measurable substances dissolved in water whose distribution in nature will aid in the calibration of a DSC model. It can be a numerical tracer, whose magnitude is related to the mass or concentration of the real-world tracer. Tracers in the model and in the real-world are assumed to occupy zero volume.

The above equation is applied sequentially to each cell in the network, so boundary discharge volumes and concentrations from 'upstream' cells become boundary recharge volumes and concentrations to the 'downstream' cells. In the equation, the only unknown on the right-hand side is $C_d(i)$. This quantity can be obtained by specifying one of two mixing cell rules: the simple mixing cell rule or the modified mixing cell rule. The former rule simulates perfect mixing within a cell; the latter simulates some regime between perfect mixing and pure piston- flow. For the simple mixing rule, the expression for $C_d(i)$ is:

$$C_d(i) = [S(i-1) + \{V_r(i) * C_r(i)\}]/\left[V_g + V_r(i)\right]$$

where V_g is the volume of groundwater in the cell.

For the modified mixing cell rule, the expression for $C_d(i)$ is:

$$C_d(i) = S(i-1)/V_g$$

When the inputs to or outputs from the model are from or to the outside environment, they are designated V_{rb}, C_{rb}, V_{db}, and C_{db}. Real-time is related to iteration number by taking the average fluid input to the system, Q_r, over a given interval, Δt. If Q_r is a function of time, we have:

$$V_{rb}(i) = Q_r(t) * \Delta t$$

For cases where the DSC model is applied to free-surface pools in karst, the discharge from a cell will be a function of the volume of cell, then

$$V_d(i) = V_g(i)/\chi$$

where χ is a constant of proportionality for any individual cell. Its value is usually determined by trial and error calibration (Simpson 1988).

The volume of water in a cell can remain constant (i.e. $V_d = V_r$) or vary with time (Simpson and Duckstein 1975). Each cell in the DSC model represents a region of the physical system. The region represented by each cell is determined from the uniformity of the sub-region of the system and the data availability. Cells may be arranged in any manner in space, so that the DSC model may represent one-, two-, or three-dimensional system. Within each cell, the DSC model can simulate either perfect mixing or some regime between perfect mixing and pure piston-flow. It should be noted that even though pure piston-flow may be specified for individual cells, this does not imply pure piston-flow for the system. Mixing occurs among the various cells of the model. In addition to the above-mentioned type of mixing, it is possible to simulate the diffusion of a tracer by using special cells known as 'dead cells' (Simpson and Duckstein 1975). Dead cells may exchange tracer with adjacent cells without requiring a net transfer of water.

The DSC model requires a tracer for calibration. Tracers may represent real substances which are dissolved in the water or they may serve as numerical labels to identify water as it passes through the model. The amount of tracer in a cell is called the cell state. The concentration of the tracer in a cell is uniform within a single cell, while gradients may exist between individual cells. The DSC model allows the flow paths between the cells and the discharge from the system to be specified by the modeler. This requires an initial estimate of the flow system to obtain the initial set of specifications. These specifications are adjusted by the modeler in the calibration process to obtain agreement between the simulated and observed tracer concentrations.

DSC models are not hydraulic models because they do not utilize the rigorous equations of flow and do not describe the physical properties of the system. Nor are they predictive models; rather, they are interpretive models. From properly calibrated DSC models, one can obtain such information as recharge rates, residence times, storage volumes and general flow directions. DSC models are approximations and therefore do not yield unique solutions. The modeler must determine which solution best approaches the real system. Thus, success of the modeling effort depends heavily on the skill and tuition of the modeler.

Figure 3.26 shows a hypothetical three cell model, consisting of steady volume of 2 for each cell, i.e. $V_g = 2$. Flow input is steady for all N ($V_{rb} = 0.01$). A pulse (numerical tracer) is injected at $i = 1$ ($C_{rb} = 10{,}000$ for $i = 1$; $C_{rb} = 0$ for all $i > 1$). A(i) is the fraction of tracer elements in a cell as a function of i. Since fluid input

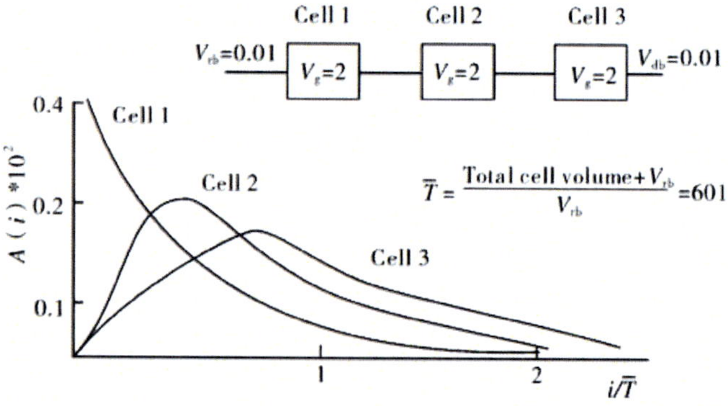

Fig. 3.26 Groundwater age distribution in a DSC model

is constant, after a sufficient number of iterations the curves of Fig. 3.26 represent the age distributions of the fluid elements in each cell. Campana and Mahin (1985), combining spatial and temporal tritium data from precipitation, stream water, and groundwater with river recharge and spring discharge, constructed a 34-cell model for the Edwards limestone aquifer (−260 km × 30 km) of south-central Texas, U.S.A. This model was non-steady state, constant volume, and was calibrated by tracking the progress of the bomb-induced tritium pulse. Kirk and Campana (1990) used a similar model to estimate the groundwater residence times, vertical flows and regional annual recharges in a complex carbonate-alluvial system.

A quantitative dye tracing conducted at eastern Tennessee during a storm event was used to illustrate the application of DSC models. About twenty-four grams of Rhodamine WT (RWT) were introduced into a sinkhole. The sinkhole drains approximately sixty acres of stormwater runoff. The stormwater and the injected RWT joined the existing groundwater in the underlying conduit and passed through two small karst windows, and subsequently discharged at a spring approximately 130 m away from the sinkhole (Stephenson et al. 1997).

Dye recovery was monitored at the spring and water samples were analyzed with a fluorometer. Figure 3.27 shows the RWT breakthrough curve and the spring and runoff hydrograph over the period of dye trace. The breakthrough curve is the result of dye mixing with the groundwater. However, the 40 min lag of the peak concentration of RWT behind the peak discharge of the spring flow indicates imperfect mixing. The immediate response of spring discharge to the stormwater runoff reveals the presence of a significant segment of phreatic (completely water-filled) conduit. The hydrostatic head imposed by the stormwater inflow in the sinkhole was instantaneously propagated through the conduit and affected the spring discharge. The turbidity that obscured the dye color at the spring indicates turbulent flow within the conduit.

Non-steady state, constant volume mixing cell models were constructed to interpret the tracing test. The average base flow of the spring was approximately 235 l/h,

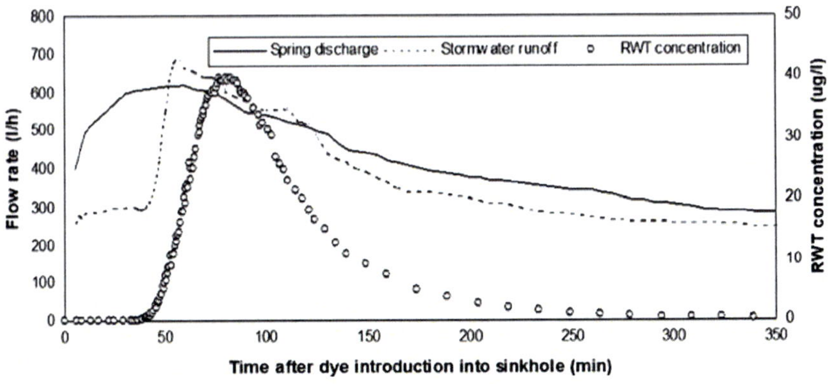

Fig. 3.27 Results of a tracer test at a spring in Tennessee, USA

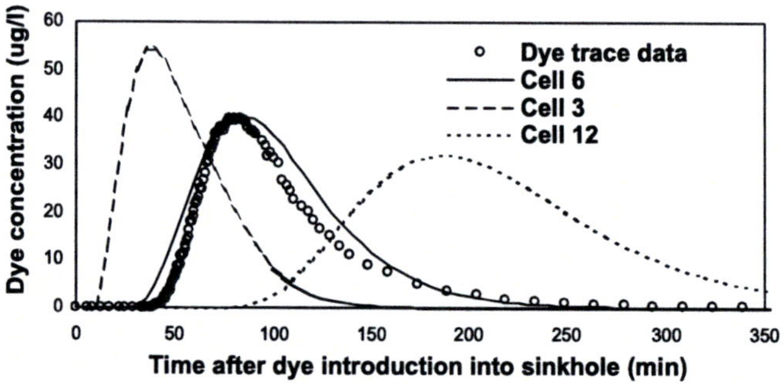

Fig. 3.28 Simulation results from DSC model

which was calculated by graphic interpretation of the spring discharge hydrograph (Viessmann et al. 1989). The tracer concentration and the spreadness of the breakthrough curve were interactively affected by the water volume assigned to each cell and the number of cells in the model. The time step also affected the accuracy of the simulation and the dye breakthrough time. By assuming a complete mixing in each cell, Fig. 3.28 compares three models consisting of 3, 6 and 12 cells, respectively. The water volume of each cell is 75 m^3 and the time step used is 5 min. The model with 6 mixing cells gives the best fit to the observed RWT breakthrough curve. Thus, the total conduit volume is about 450 m^3. For a linear passage, the cross-section of the conduit is about 3.5 m^2, which is not uncommon in the general area.

3.5.2 *Water Tank Models*

A water tank model with a cluster of tubes of various diameters is proposed in this section. Because this method is closely related to our understanding of the hydrogeological conditions of the karst system, we take the Beishan karst system, Guangxi Province of China as an example to show the process of constructing the model and its application.

Beishan karst system is located in the mountainous area of northern Guangxi of China, where a large lead-and-zinc mine has been in operation for many years. The mining area may be divided into two parts along the fault F6 as shown in Fig. 3.29. The northern part is a mountainous area made of strongly karstified reef limestones. Channels have been found near No. 6 cave between the south and north parts. The karst water from the north flows as cascade into the south discharge area during mean water period and discharge partly through No. 4 spill cave during flood period, with the maximum discharge of 18 m^3/s. The karst water in the south, however, has a unified hydrodynamic filed, issuing from No. 1 springs in dry season and flowing out at different levels in rainy season (Fig. 3.30).

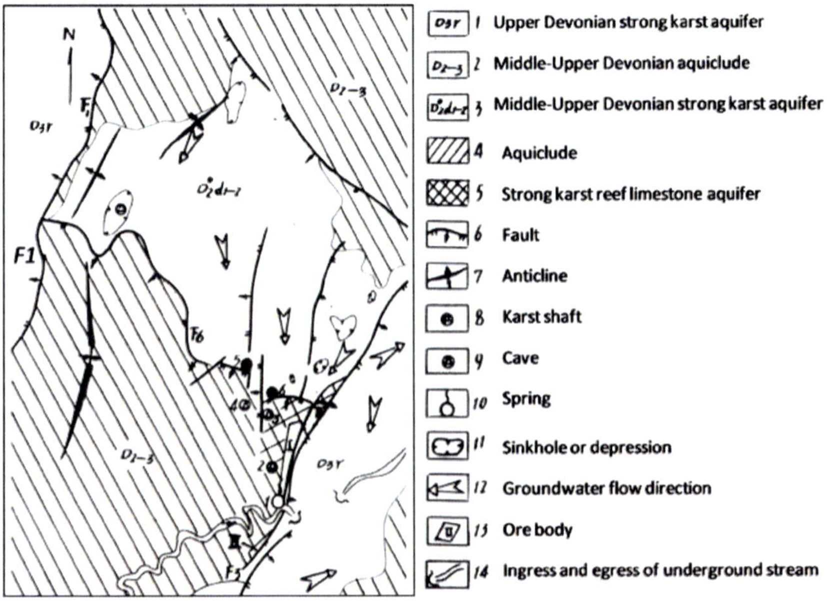

Fig. 3.29 Schematic hydrogeology of Beishan karst system, south China (not in scale)

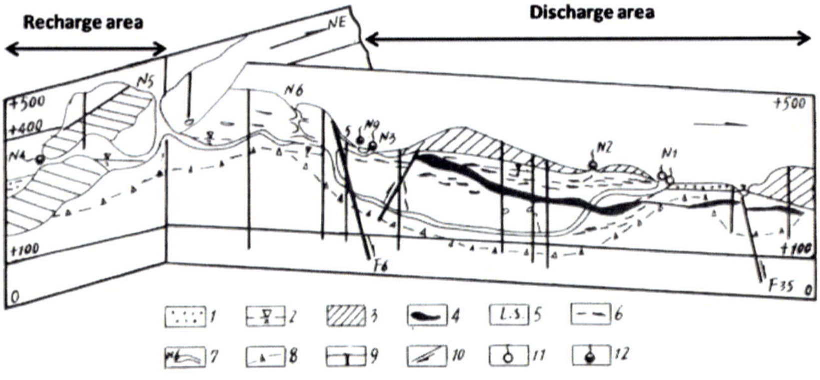

Fig. 3.30 Flow routs in Beishan karst system

Figure 3.31 shows the generalized water tank model for this area. The subsystem in the north is generalized as the first order water tank (T_1) and the subsystem in the south as the second order water tanks (T_2, T_3). The water levels in the representative boreholes at both ends of the channels are used as water levels of tanks, and the parameters of the flow capacity for the channels are obtained through simulating with the following equations:

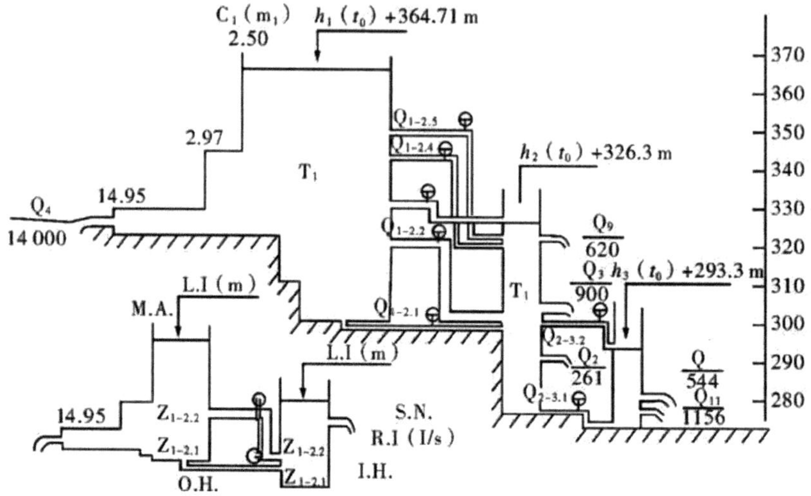

Fig. 3.31 Water tank simulation model

$$T_1: \quad C_1(h_1)\frac{dh_1}{dt} = Q_p(t) + Q_4(t) - \sum_{j=1}^{j=n} Q_{1-2,j}(t)$$

$$T_2: \quad C_2(h_2)\frac{dh_1}{dt} = Q_p(t)\sum_{j=1}^{j=n} Q_{1-2,j}(t) - \sum_{j=1}^{j=p} Q_{2-3,j}(t) - Q_3(t) - Q_2(t) - Q_9(t)$$

$$T_3: \quad C_3(h_3)\frac{dh_1}{dt} = \sum_{j=1}^{j=p} Q_{2-3,j}(t) - Q_1(t)$$

With change of water levels, flow in the channels may be perennial, intermittent, and terminated. The mathematical expression is:

$$Q_{1-2,j}(t) = \begin{cases} \alpha_{1-2,j}[h_1(t) - h_2(t)]^{1/m} & \text{when } h_1(t) > Z_{1-2,j};\ h_2(t) > Z_{1-2,j'} \\ \alpha_{1-2,j}\left[h_1(t) - Z_{1-2,j}\right]^{1/m} & \text{when } h_1(t) > Z_{1-2,j};\ h_2(t) < Z_{1-2,j'} \\ 0 & \text{when } h_1(t) < Z_{1-2,j} \end{cases}$$

where:

h_1, h_2, and h_3	Representative water level functions in three tanks respectively (m);
$C_1(h_1)$, $C_2(h_2)$, and $C_3(h_3)$	Section area functions in the three tanks with change of water level (m^2);
$Q_p(t)$	Effective rainfall of recharge mountainous area (m^3/day);
Q_1, Q_2, Q_3, Q_4 and Q_9	Spill cave discharges or spring flow of each simulated tank (m^3/day);

$\Sigma Q_{1\text{-}2,j}(t)$ and $\Sigma Q_{2\text{-}3,j}(t)$	Sum of discharge of the karst channels between T_1 and T_2, T_2 and T_3 (m^3/day);
n and p	Conditions of channels;
j	Order number of channel;
$\alpha_{1\text{-}2,j}$	Tank model parameter and the j-th channel's discharge coefficient, its dimension depends on the state of flow (m), $1 \le m \le 2$;
$Z_{1\text{-}2,j}$ and $Z_{1-2,j'}$	Elevation of the j-th channel's outlet and inlet between T_1 and T_2

The simulation results are shown in Fig. 3.32. Five channels at different levels were recognized between T_1 and T_2. The relative errors of fitting water discharges are 4–6%. The example indicates that water tank model is based on good understanding of the structure of karst water systems, their reasonable generalization and their correct division into subsystems. The accuracy is dependent of the correct conceptualization of the subsystems and the dimensions of their connection tubes.

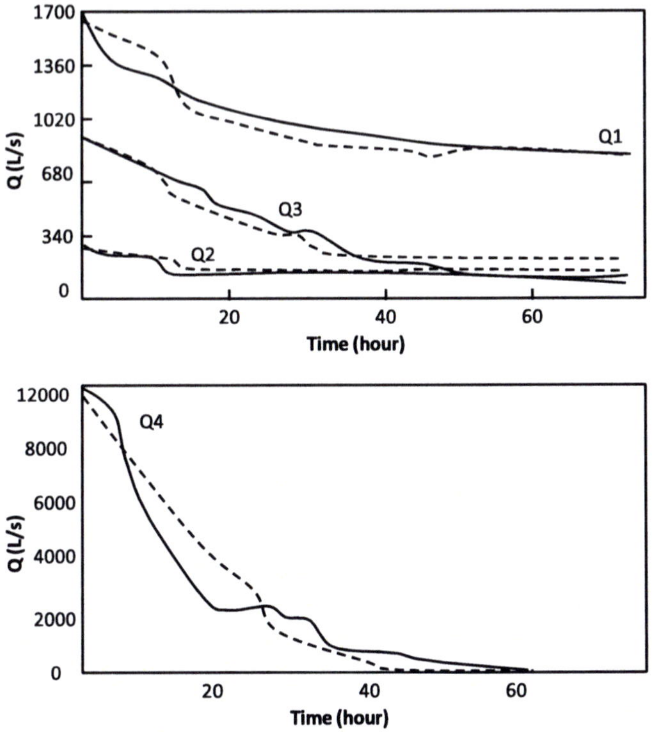

Fig. 3.32 Simulated spring discharge from water tank model (solid line—measured spring discharge; dotted line—simulated spring discharge)

3.6 Physics-Based Models

The majority of above-mentioned methods are primarily established and evaluated without physical defining of karst aquifer properties such as its saturated depth and hydraulic conductivity, jointly known as transmissivity, and effective porosity or storage coefficient. A mathematical transfer function or a regression equation describes the karst system. Due to the non-physical basis, these transfer functions have only very limited predictive properties. On the other hand, due to the high risk of groundwater contamination in karst areas there is a growing demand for reliable protection measures. This has led to a series of studies in developing groundwater flow models in karst terrains in many countries including Germany, China, USA, and Yugoslavia. All methods based on physical laws of groundwater flow are a part of the physically based approach. On the other hand, physical models may imply considerable parameter identifications when applied to practical field conditions in karst terrains, where hydraulic measurements are usually scarce and highly variable. The difficulty of identifying parameters in karst areas results from the fact that data is not available at the scale of the flow dominating heterogeneities, i.e. at the scale of the main conduits. Until now there is no standard investigation method available that permits the reliable detection of the major conduits within a given catchment area. The chance to encounter a conduit in a borehole is extremely small if no surface indication like a sinkhole or a dry valley is available. In a natural karst system, the hydraulic conductivity between a hardly karstified zone and a fully developed conduit system could differ as high as 5–7 orders of magnitude (Zhou 1990). Attempt to quantify groundwater flow in a karst catchment using a physically based model appears almost impossible at first.

Most karst aquifers cannot be treated as porous media for direct application of standard equations describing groundwater flow in hydrogeology. In absence of a better tool, however, the equations derived on strict assumptions such as homogeneity, full aquifer penetration, simplified boundary conditions, intergranular porous media and so on, are still used with various modifications. In general, these modifications include replacing intergranular media by fissured or karstified fissure media with corresponding hydraulic conductivity (conductivity tensor).

The equations of groundwater flow are solved in two approaches—analytic and numeric. In the first case, usual bases of calculation are radial flow towards one or several wells, and planar flow towards linear drains that have been discussed by many researchers including Bedinger et al. (1988), Walton (1987) and LaMoreaux (1989). In the second case, entire flow domain is under the calculation by discretizing it into several relatively uniform areas and the groundwater flow equations are solved by finite-element or finite-difference or boundary element method. The major limitation for the application of both groups of solutions is the question of how representative the field data are to identify hydrogeological parameters of a karst aquifer. In other words, the scale effect in karst hydrogeological system research incapacitates generalization of the results obtained in a portion of aquifer. A good example is the monitoring of

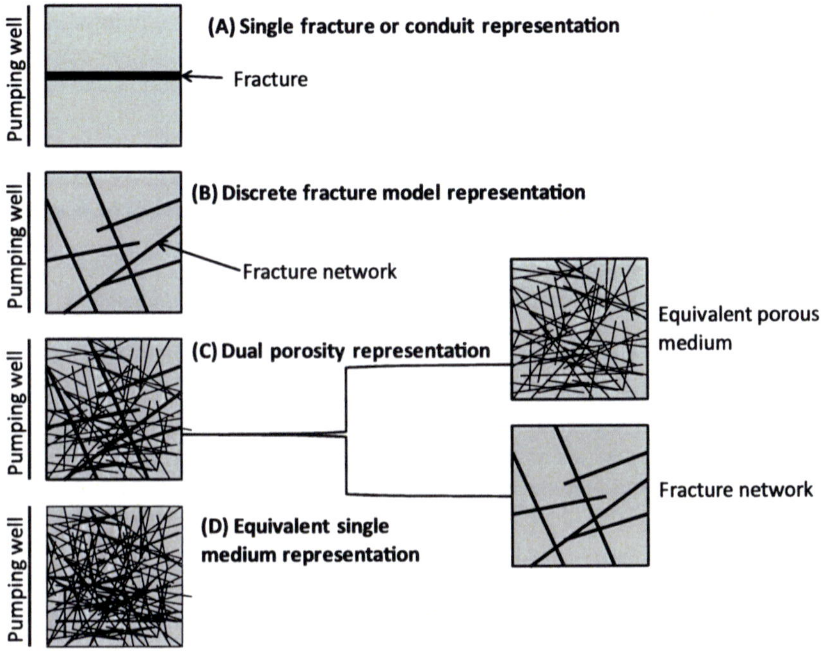

Fig. 3.33 Model representations of karst aquifers at different scales

head drop in several observation-wells during a pumping test. One observation-well may be located entirely in homogeneous rock matrix, the other one in highly permeable karst conduits, and the third one in dead-end cavity. The data have to be carefully analyzed before they are used to calibrate models. Generally, both porous medium and fissure medium approaches have been proposed for the simulation of the hydraulic behavior of karst aquifers. Figure 3.33 illustrates the possible model representations and their corresponding scales. Three main types are equivalent-porous-media models, dual-porosity models and discrete-fracture models, while the single fissure model is the basis of the discrete-fracture models.

3.6.1 *Equivalent-Porous-Medium Models*

In principle, the mathematical model that describes groundwater flow through a porous medium can be stated for every point within the considered phase at the microscopic level (Bear 1972). The problem is that this model cannot be solved at this level, since the detailed geometry of the surface that bounds the fluid phase is not known and/or is too complex to be described. Furthermore, we cannot measure values of variables at points within a phase in order to validate a flow or transport model and to determine model parameters. Consequently, the complete description and solution of a fluid flow problem at the microscopic level is impossible. To circumvent these difficulties, the fluid flow problem has to be transformed from the microscopic level to a macroscopic one, at which the fluid flow is reformulated in terms of averages of the microscopic values. These average values are measurable quantities. We usually refer to this approach as the continuum approach. In the continuum approach, the real porous medium domain is replaced by a model in which each phase is assumed to be present at every point within the entire domain. Each phase, thus, behaves as a continuum that is present in the entire domain (Bear 1993). For every point, average values of phase and component variables are taken over an elementary volume, centered at the point, regardless of whether this point falls within the considered phase or not. The averaged values are referred to as macroscopic values of the considered variables. At the same time, if a sample centered at a point is to represent what happens at that point and at its close neighborhood, it is obvious that the size of the sample should not be too large. The volume of a sample that satisfies these conditions was defined as a representative elementary volume (abbreviated REV) of the considered porous medium domain at the given point (Bear 1972). By traversing the porous medium domain with a moving REV, we obtain a field of the averaged values for every variable. The values are continuous and differentiable functions of the space coordinates and of time. With the introduction of the concept of the REV, the mathematical model has become an important potential tool in the investigation of the fluid flow problem that is not directly observable.

To what extent the karst aquifer can be represented by porous medium models is an issue open to discussion. In principle, the karst system can be considered a continuum only when a REV is defined. Long et al. (1982) used the theory of flow through fissured rock and homogeneous anisotropic porous media to determine when a fissured rock behaves as a continuum. In their opinion, a fissured rock can be said to behave like an equivalent porous medium when (1) there is an insignificant change in the value of the equivalent permeability with a small addition or subtraction to the test volume and (2) an equivalent permeability tensor exists which predicts the correct flux when the direction of a constant gradient is changed. Long and Witherspoon (1985) further studied the effects of the fissure density, scale measurements, aperture distribution and fissure orientation on the behavior of fissured rock. They proposed that the porous medium model is acceptable provided a plot of the measured directional permeability has the form of an ellipsoid. Pankow et al. (1986) compared two contaminated sites in fissured rock and concluded that the combination of characteristics made only one suitable for modeling using the equivalent porous medium model. Khaleel (1989) considered fissured basaltic rocks where the fractures are defined by the columns of matrix material. For uniform fissures he concludes that the porous medium model is applicable at a scale of about six times the column diameter, but for a lognormal aperture distribution this could be 20–30 times the diameter. When modeling the groundwater flow in the British chalk aquifer, Keating (1982) found that it was difficult to define the REV due to the existence of a highly permeable zone at the upper part of the aquifer. Application of the equivalent-porous-medium model to simulate the response of the water table in the Candover catchment and the flows of the Candover streams to both normal winter recharge and the long-term pumping test was unsuccessful. The simulation of water table fluctuations required a relatively high storativity (of the order of 5%), while the simulation of the stream flows necessitated storativity of the order of 1% (Keating 1982). Thus, the simulations of water table fluctuation and stream flows are not compatible with each other by using the single continuum model.

Fractured rock tends to behave more like an equivalent porous medium when (1) fissure density is increased; (2) fissure apertures are constant rather than distributed; (3) fissure orientations are distributed rather than constant; and (4) larger domain sizes are simulated. Considering different sizes of fissures and caves present in karst systems, definition of a REV is very unlikely, especially in mature karst terrains. If the REV can be defined, it is likely that it varies with the scale and for large conduit karst systems, and the REV tends to be very large (Dreiss 1989). For the purpose of regional groundwater resources assessment and management, the continuum assumption is often acceptable, particularly when hydraulic gradient is low and storage very large (LaMoreaux et al. 1989). In this case, the fissures and conduits are considered as part of the primary porosity, ignoring the difference in porosity scale between the fissures and the rock matrix. This approximation seems to be reasonable as at the scale of grains, even the porous medium can be considered as fractured (Barker 1991). In the porous medium approximation, the mathematical models are based on Darcy's law and the mass continuity (Gelhar and Axness 1983; Hsieh et al. 1985). The equations describing one-dimensional groundwater flow and solute transport (non-reactive) are:

$$\frac{\partial}{\partial x}\left(K_x M \frac{\partial h}{\partial x}\right) = S\frac{\partial h}{\partial t}$$

and

$$\frac{\partial}{\partial x}\left(D_x \frac{\partial C}{\partial x}\right) = v_x \frac{\partial C}{\partial x} + \frac{\partial C}{\partial t}$$

where

K_x Hydraulic conductivity;
D_x Hydrodynamic dispersion coefficient;
S Storage coefficient;
h Water head potential;
M Aquifer average thickness;
v_x Apparent groundwater velocity;
C Solute concentration.

Many commercial computer software has been developed and their applications have been well documented (Hawnt and Joseph 1980; Baca et al. 1984; Zhou 1990).

3.6.2 Discrete-Fracture Models

Though karst aquifer by no means should be considered as a composition of discrete fissure, certain interest for the background theories may arise during recent and future combined approaches to modeling groundwater flow in karst. The major objection to fissure approach is that assumption on linear (non-turbulent) flow in discrete fissures or conduits can be easily rejected in the case of karst aquifers. However, in certain cases where karst is not well developed and the involved fissures are limited or the scale of interest is small, an equivalent study based on discrete fracture models may lead to satisfactory results. Based on the distributions of fissure aperture, fissure length, fissure spacing, and fissure orientation, a statistical model for the whole fissure network is generated. For non-layer fractured rock, there are three basic configurations of fracture geometry. They are (1) random fracture location (the Baecher model, randomly located fractures of a single size distribution); (2) structured fracture location (the Levy-Lee fractal process to produce cluster of small fractures around widely scattered, large fractures); and (3) clustered or semi-deterministic fracture location (the nearest-neighbor model, a semi-probabilistic pattern-based model that simulates the tendency of fractures to be clustered around major joints and faults by preferentially producing new fractures in the vicinity of earlier fractures, or the War Zone model that simulates the geometry of shear zones by preferentially producing fractures in the region between sub-parallel neighboring fractures). Figure 3.34 shows several examples of fracture network.

In karst aquifers, bedding planes should be taken into consideration as well. In order to make the statistical models approach reality as much as possible, they may be conditioned by fissure data measured either on the bedrock outcrop (Dershowitz et al. 1991) or in boreholes. Unlike the continuum approach that emphasizes the effect of the equivalent parameters on fluid flow, the discrete fracture approach emphasizes the effect of the medium structure, in particular the geometry of fissures on fracture flow (Dershowitz et al. 1992). Discrete representation of fissure aquifers has been used to simulate single-phase flow (Long and Witherspoon 1985; Endo et al. 1984), solute transport (Schwartz et al. 1983; Rasmuson 1985; Casas et al. 1990a, b; Dverstorp et al. 1992), and multiple-phase flow (Rasmussen 1991; Kwicklis and Healy 1993; Zhou 1995). In most of the developed discrete-fracture models, the investigators start with the common assumptions that individual fissures are the basic unit for the fissured medium and that each fissure is presented by a parallel-plate flow model assuming a constant aperture. The average flow rate through a single fissure is (linear-laminar flow) (de Marsily 1986):

$$v_s = \frac{\rho g}{12\mu} b^2 I$$

where v_s–Seepage flow velocity in the fissure and b is the fissure aperture size.

Comparison of the above equation with Darcy's law reveals the following isotropic relationship between fissure permeability, k^f, and aperture size:

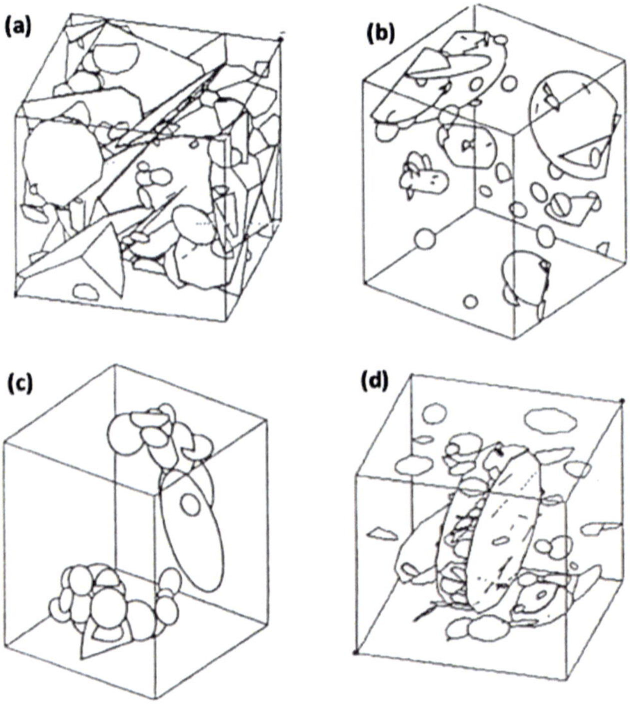

Fig. 3.34 Discrete-fracture models **a** Enhanced Baecher model based on stationary Poisson point process; **b** Nearest neighbor model based on non-stationary Poisson point process; **c** Levy-Lee fractal model; **d** War zone model

$$k^f = \frac{b^2}{12}$$

Witherspoon et al. (1980) studied the validity of the above equation and they found that the equation seems to be valid even for a closed fissure under stress. Until the 1980s, researchers in the field of fissure flow considered a single fissure as a pair of parallel plates separated by a distance b, in describing fluid flow and transport in fissures (Tsang 1992).

However, more recent experimental and theoretical work focusing on the aperture variability within fissures and the impact of this on flow and transport, has indicated that the parallel-plate model seems to be inadequate in describing either the fluid flow or solute transport processes due to the roughness of the natural fissures and variability of the aperture sizes (Brown and Scholz 1985; Pyrak-Nolte et al. 1988). Fluid flow does not appear to be evenly distributed within the fissure plane. Large areas within a fissure plane did not appear to carry water and flow was isolated to only 5–20% of the fissure plane, which is often called the channeling flow (Heath 1985; Bourke 1987). Apparently, using the parallel-plate fissure representation fails to recognize the spatial heterogeneity in fissure planes. The constant aperture assumption in the discrete fracture models limits the ability of these models to represent real-world behavior of fissure flow regimes. Considering the strong roughness in fissures and conduits and the variability of their sizes in karst aquifer (Thraikill 1986), we believe that some improvements are required before the discrete fracture models are applied to karst systems.

3.6.3 Double-Continuum Models

A more complex conceptualization of fissured media involves distinguishing between the rock-matrix porosity and the fissure porosity. Although fluid mainly flows through fissures in fissured rock, a detailed study of the interaction of the fissure and the matrix block is essential. A significant proportion of the total storage capacity of the system is provided by the rock matrix, while the fissures provide the dominant path for regional transport (Barker 1991). In double-continuum models, each continuum region is often treated as an independent porous medium flow system but is related to the other through pressure and mass continuity expressions (Murphy and Thomson 1993).

For groundwater flow:

$$\frac{\partial}{\partial x}(K_x^m M \frac{\partial h^m}{\partial x}) = S^m \frac{\partial h^m}{\partial t} + (-1)^{m+1} \alpha(h^m - h^{m+1}) \quad m = 1, 2$$

For solute transport:

$$\frac{\partial}{\partial x}(D_x^m \frac{\partial C^m}{\partial x}) = v_x^m \frac{\partial C^m}{\partial x} + \frac{\partial C^m}{\partial t} + (-1)^{m+1} \alpha^*(C^m - C^{m+1}) \quad m = 1, 2$$

where

$$\alpha^* = \alpha \frac{(h^1 - h^2)}{n_e M}$$

Symbol m is the medium index representing either fissure continuum or matrix continuum; n_e is the effective porosity; and α is the exchange coefficient.

The double-continuum models discussed here consist of both dual-porosity models and dual permeability models. In dual-porosity models, the hydraulic conductivity or permeability in fracture network is often considered much larger than that in the matrix blocks (Bibby 1982). The dual-permeability concept reported by Hill and Thomas (1985) allows both matrix-to-matrix and fracture-to-fracture flow between blocks. Dual permeability models require much greater computing time than the dual porosity models. Gilman and Kazemi (1988) used the dual permeability idealization to develop an efficient algorithm to account accurately for the gravity effects in both fractures and matrix. They also accounted for viscous displacement in matrix blocks caused by pressure gradient in the fracture network.

During the last few years, double-continuum models have become a recognized tool for the prediction of flow and transport in karstified limestone aquifers (Baoren and Xuming 1988; Teutsch and Sauter 1991). This modeling approach manages to improve the representation of physical reality. The karst aquifer heterogeneity is approximated by two overlapping continua. The most karstified areas (conduit- or fissure-flow), yielding high groundwater flow velocities and the rapid response to rainfall events represent one continuum. The second continuum is represented by the moderately karstified aquifer zones (fissure- or matrix-flow) with lower hydraulic conductivity and higher storativity. The scale hierarchy for the double-continuum-porous-equivalent would be $G \ll D < F$ for the fissure or matrix flow continuum and $G \leq D \leq F$ for the conduit or fissure flow continuum (Fig. 3.2). This approach is favorable for many practical karst aquifer studies where data is usually scarce but detailed enough to show a single continuum model cannot be applied (Sauter 1991).

However, two assumptions have to be made in order to apply the double-continuum approach to karstified aquifers. The first assumption is the laminar flow in both continua. This is probably not true for the fast conduit-flow system, especially after strong recharge event. The second assumption is a potential continuum in the highly permeable conduit-flow system, at an intermediate scale smaller than the modeled domain. This is probably not true for those aquifers comprising extremely large active conduits where $G = F$. Under such circumstances, the double-continuum model may seem to be too simple. Where large-scale information is available on the location and the extent of the major conduits, a combination of the double-continuum approach and the discrete fracture approach or three-medium model might be most adequate. The fissure and matrix flow system could be represented by a double-continuum equivalent and the conduits system through discrete fracture elements. Under such a circumstance, the parameter identification is mostly reduced to the detection of the geometry of the large conduit system (Yusun 1988).

Figure 3.35 shows a simplified rectangular karst aquifer that is 2600 m long, 1100 m wide and 30 m high. The rock mass was discretized into blocks with various

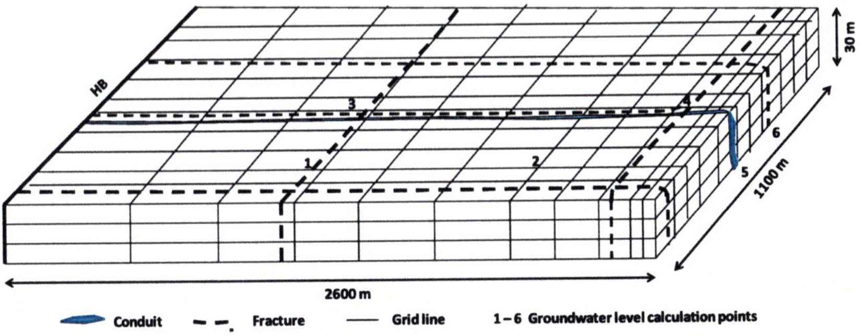

Fig. 3.35 Schematic model of a karst aquifer consisting of conduits, fissure, and matrix blocks

dimensions by two sets of vertical fissures, which intersects each other perpendicularly. One set consists of three transverse fissures with different distances; the other consists of three longitudinal fissures with equal distance. An underground stream was developed along the middle longitudinal fissure. Hydraulic conductivities of 800 m/d and 4.0 m/d are given to the fissure and matrix media, respectively. The storage coefficient of the matrix medium is 0.05. For simplicity, both the top and the bottom of the aquifer are impervious so that the aquifer is confined. Three lateral sides are impervious; only one side is associated with surface water.

Two numerical models are considered in the calculation, i.e. the double-continuum model and the triple-medium model. In the first model, fissure and matrix systems are treated as two overlapping continua without considering the effect of the underground stream. In the second one, apart from the two overlapping continua, the stream is treated as a separate medium. Finite element method is used to solve equations for groundwater flow. Each matrix block is divided into several elements to express storage function of the matrix void. The double-continuum model has 363 matrix elements, 198 fissure elements and 576 nodes. To make the triple-medium model be similar to the double-continuum model, its 11 fissure nodes along the middle longitudinal fissure are exchanged by 11 stream nodes. The water levels are controlled by the boundary water level as shown by the curve HB in Fig. 3.36. The slope of the stream flow is 2%. The simulation time step is 10, 10, 10, 20, 20, 20 days and then 30 days after.

Figures 3.37 and 3.38 show the calculated equipotential lines of karst water flow from the double-continuum model and triple-medium model. Figure 3.37 reveals that all fissure water comes from the matrix storage and the water level distribution is significantly different from that when the underground stream is considered. Curves 1, 2, 3, and 4 show the variations of water level with time at different monitoring points from the triple-medium simulations. The water drainage in the matrix blocks lags that in fissures.

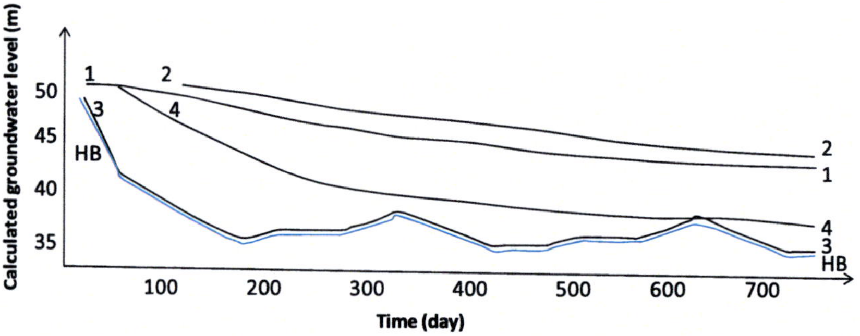

Fig. 3.36 Variation of calculated water levels with time

Fig. 3.37 Calculated water level (m) after 450 days from double-medium model

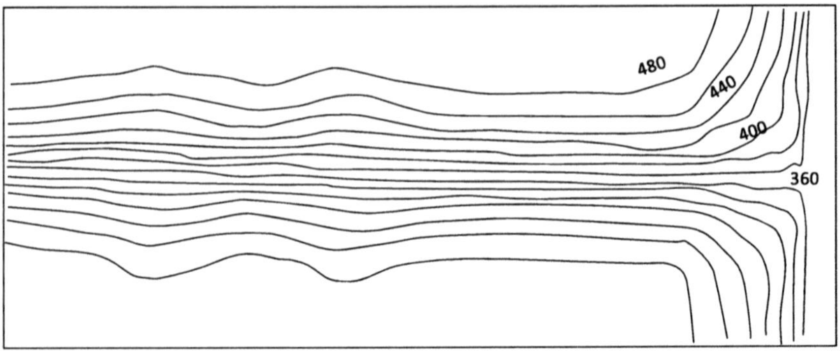

Fig. 3.38 Calculated water level (m) after 150 days from triple-medium model

Double-continuum or triple-medium approaches are based on the physics of groundwater flow and capable of simulating spatial variations in flow and transport phenomena. These advantages, however, require on the other hand the knowledge of flow and transport parameters of the conduit flow, conduit-fissure flow, and fissure-matrix flow systems at appropriate spatial resolution scales. The hydraulic parameters dedicated to the respective discretized unit (element/cell), is a result of the sum of the parameters of the identified sub-system, weighted by their respective proportions within the particular model unit. When a numerical model accommodates several flow sub-systems employing multiple-medium approach, the representative parameters at the respective scales must be available.

3.6.4 Determination of Hydraulic Parameters at Respective Scales

The hydraulic parameters (hydraulic conductivity and storage coefficient) of the matrix (micro-fissures) continuum can be measured in the laboratory on cores with dimensions of several centimeters. Due to the heterogeneity of karst aquifers, the opportunity of obtaining representative samples is limited. Hydraulic conductivities may vary by several orders of magnitudes, depending on the sampling location. For the intact limestone rocks, the most often cited range from 10^{-9} to 10^{-8} m/s.

3.6.4.1 Well Tests

The fissure continuum of the karst system could be visualized as joints, spaced at decimeter interval (10–30 cm) and it is frequently measured in quarries. The hydraulic conductivity is evaluated when the higher permeable set of fractures and the solution-widened fissures are spaced apart between one and several meters. The multiple-medium system is investigated at this level by double packer tests, where the packers enclose a test interval of less than one meter. Slug and injection tests with a test interval of several meters and radius of investigation of several tens and pumping tests with the radius of investigation of several hundreds of meters can identify the relevant parameters at their respective scales. The well tests are performed in single boreholes such as slug tests, packer tests or between two or more boreholes such as pumping tests. Input signals are quantity of flow, pressure (water level) or a combination of both. Well test designs and data analyses can be found in Bedinger et al. (1988).

Because slug or packer tests involve displacement of only a limited amount of water compared with pumping tests, they "sample" only a relatively small zone of the aquifer. Figure 3.39 contrasts the results from slug tests and pumping tests for Grand Bahama. Over 60% of the pump test results have higher hydraulic conductivities than the maximum recorded in the slug tests, but the two data sets give similar minimum values. This suggests that in pumping tests the cone of depression expand to intersect dissolution conduits, spacing of which is so wide that they are unlikely penetrated directly by random boreholes. At high hydraulic conductivities, turbulent flow may develop, and the values determined using the Darcy's equation are erroneous.

It is also important to bear in mind the difference in flow geometry between porous medium and karst aquifers when applying the flow equations derived in porous medium aquifers to karst aquifers. Figure 3.40 illustrates different flow fields generated by pumping water in both aquifers. Analysis of the flow geometry in well tests results in values of flow dimension and hydraulic conductivity or transmissivity. In a uniform system, the flow dimension is proportional to the flow-through area (A_c) divided by the distance from the pumping source (r). In other words, a classic Theis-type aquifer has a flow dimension of 2 since the through-flow area grows proportional to the distance from the pumping well. The symmetry of the flow dimension shows

Fig. 3.39 Comparison of hydraulic conductivity data from pumping and slug tests

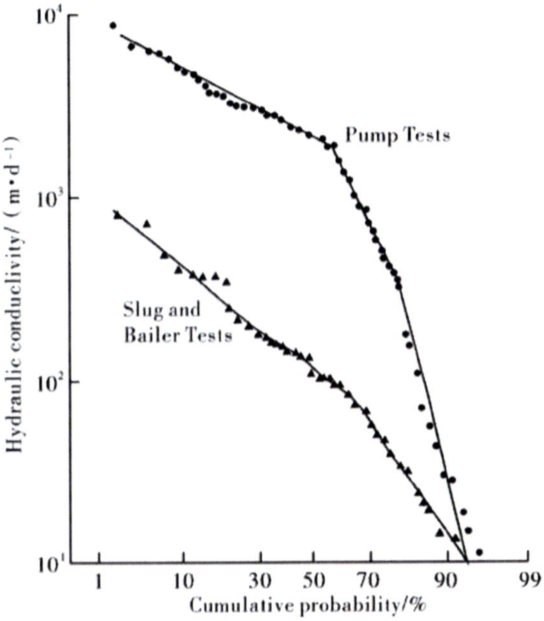

that through-flow area is directly proportional to the distance from the source raised to the power of flow dimension minus 1. Integer dimension is not a prerequisite. A branching network of fissures or conduits may have a dimension greater than 1 but not as large as two. Flow dimensions are closely related to the number of fractures intercepted in the boreholes (Fig. 3.41). In packer tests, for example, if the two-packer interval is intercepted by more than 30 fissures, flow dimension is probably in the neighborhood of 3. If the intervals intercept less than two fissures, the flow dimension is close to 1.

Therefore, well test design to evaluate karst aquifer characteristics is not so straightforward. Furthermore, how fissure or conduit structures affect flow fields is still s topic under investigations. Multi-packer connectivity test is one of the recently developed techniques to characterize fractures aquifers (Fig. 3.42). This method assumes that the transmissivity of a test zone is equal to the sum of the transmissivities of the conductive fissures that intersect that test zone:

$$T_i = \sum_{j=1}^{j=n_i} T_{ij}$$

where

T_i Is the apparent transmissivity of the ith packer interval;
n_i Is the number of conductive fissures in the ith interval;
T_{ij} Is the transmissivity of the jth conductive fissure within the ith packer interval

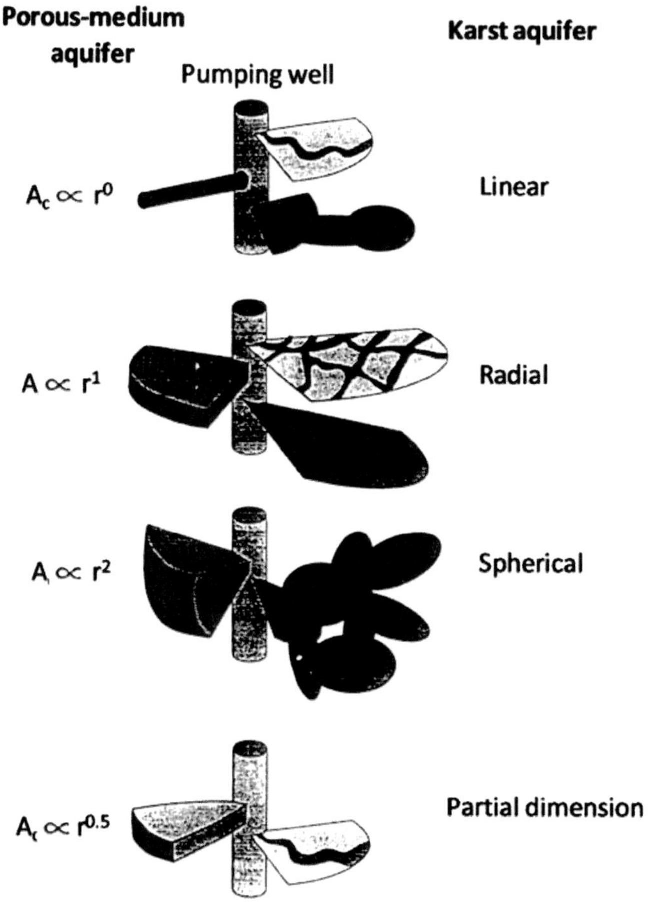

Fig. 3.40 Flow geometries in porous-medium and karst aquifers

Within any given interval, the number of conductive fissures n_i is assumed to be a random number defined by a Poisson distribution:

$$f_n(n) = \frac{\overline{n}^n e^{-n}}{n!}$$

where n is the Poisson process rate, which is equal to the expected value of n. The conductive fissure frequency is given by:

$$f_c = \frac{\overline{n}}{L_i}$$

where L_i is the length of the test zone.

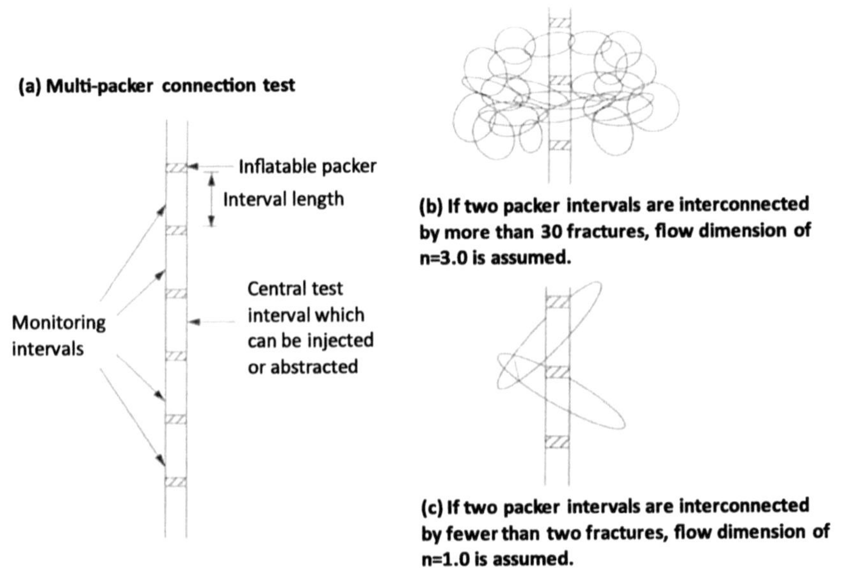

Fig. 3.41 Schematic illustration of packer testing

The distribution of fissure transmissivity is assumed to be independent within each packer interval with a given distribution form. The distribution of T_i is the sum of a random number of random events and is therefore a compound Poisson process. In this approach, the mean number of fissures in a given interval is defined by the Poisson distribution rate parameter n, and the distribution of the fissure transmissivity T_{ij} is described by a log-normal distribution with a mean and standard deviation μ_{logT} and σ_{logT}. For any given set of parameters describing the distribution of fissure transmissivity $f(T_{ij})$ and conductive fissure frequency f_c, the distribution of packer interval transmissivity $f(T_i)$ can be found by Monte Carlo simulation. The intensity and transmissivity distribution for the conductive fissures are estimated by finding the best match between the observed distribution of packer interval transmissivity $f(T_i)$ and the distribution of test zone transmissivity found by simulation for a given fissure frequency and single fissure transmissivity distribution.

3.6.4.2 Catchment Analysis

Slug tests, packer tests and pumping tests, have a limited testing radius, which is usually in the order of several tens to hundreds of meters, depending on the type of aquifer material. The only signal, exciting larger regions of aquifer system is natural pulses, such as rainfall events, which can be evaluated by applying recession analysis to the output signal, i.e. the spring discharge. This method has the advantage that information is gained over a large area. Because the registration of the output signal

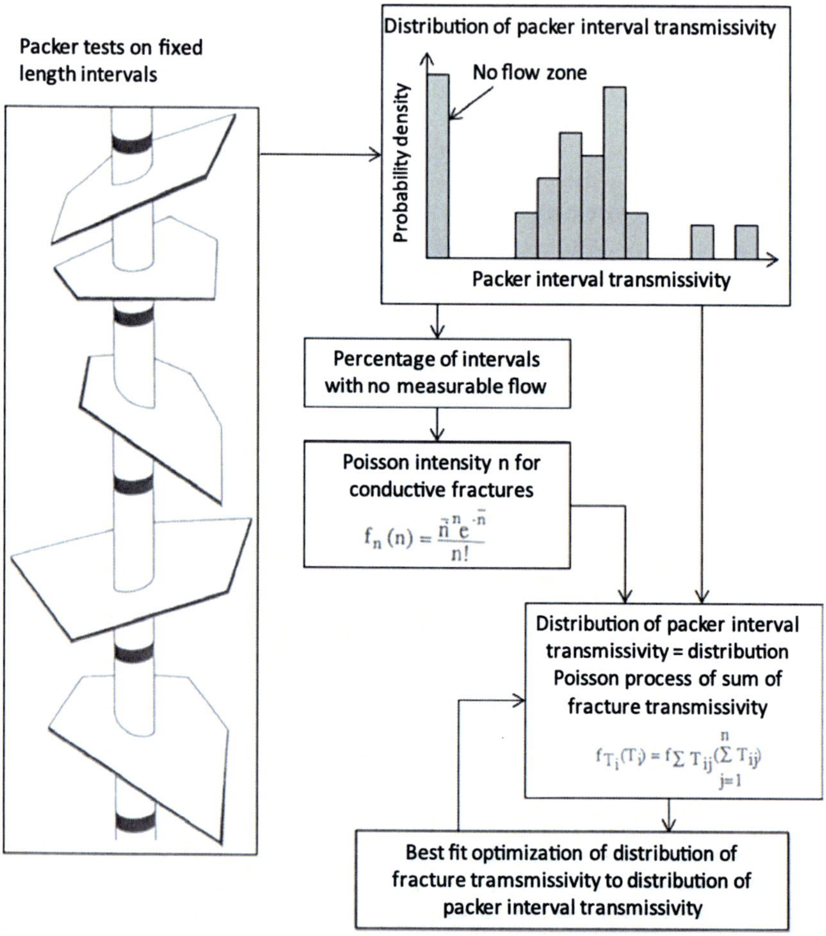

Fig. 3.42 Multi-packer connectivity tests

can only be measured at a spring, a differentiation into regionally varying parameters is generally not possible. However, if the regional hydraulic conductivity varies laterally and with depth, the value obtained depends on the intensity of the signal (recharge depth) and only reflects the hydraulic characteristics of the flow-dominant portion of the aquifer.

The approach taken for the evaluation of the regional hydraulic conductivity assumes that there is no more recharge to the aquifer with knowledge of the respective discharge and the saturated thickness. Figure 3.43 shows a schematic catchment with a convergent spring flow. Under the condition that seepage flow dominates in the karst system, Application of Darcy's law lead to:

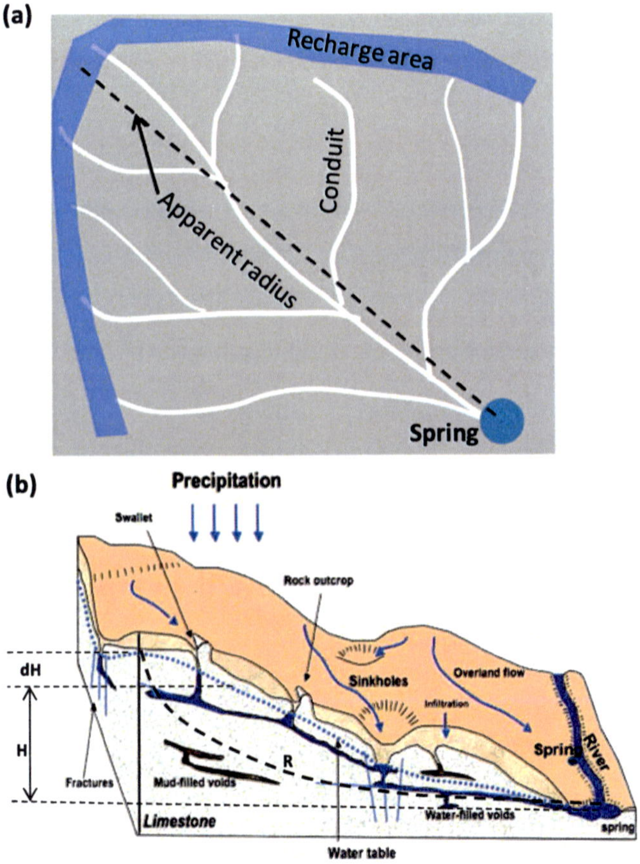

Fig. 3.43 Radial-flow model for spring discharge

$$Q_t = \frac{2\pi \frac{\beta^0}{360^0} K_s M H t}{\ln \frac{R}{r_s}}$$

where

- Q_t Volumetric flow rate at time t;
- H_t Groundwater level difference between the water level at R and the elevation of the spring;
- K_s Hydraulic conductivity;
- R Length of seepage;
- M Thickness of the aquifer;
- r_s Fictitious radius of the spring;
- β^o Convergent angle of radius flow

The spring discharge can also be expressed by:

$$Q_t = -2\pi \frac{\beta^o}{360^o} RMS \frac{dHt}{dt}$$

where S is the specific yield.

By the principle of mass conservation, we have

$$Q_t = Q_0\, e^{-\alpha_1 t}$$

where Q_0 is the initial spring discharge.

The radial flow recession coefficient of laminar flow can be expressed as:

$$\alpha_1 = K_s/[SRln(R/r_s)]$$

Similarly, we can derive the radial spring discharge equation for turbulent flow from Chezy Equation.

$$Q_t = Q_0(1 - \alpha_2\, t)$$

Then, the radial flow recession coefficient of turbulent flow can be expressed by:

$$\alpha_2 = K_b \Big/ \left[2SR(1/r_s - 1/R)^{1/2} H_0^{1/2}\right]$$

where H_0 is the initial water level; and K_b is the hydraulic conductivity of turbulent flow.

Expressions of the recession coefficient indicate their relation to the hydraulic conductivity and specific yield of the aquifer. Based on the calculated values of the recession coefficient from the methods discussed in Sect. 3.3, the hydraulic parameters in either conduit or fissure system can be estimated from the above equations.

For lateral groundwater flow discharging into streams, Rorabaugh (1964) showed that the slope of base flow recession curves of water released from bank storage after flood event plotted on a log-linear diagram (discharge—log) is proportional to aquifer hydraulic conductivity. With a knowledge of the groundwater basin geometry (L—average distance to the groundwater divide) and the storage coefficient (S), hydraulic conductivity can be computed according to the following equation.

$$K_s = (4aSL^2)/(Mp^2)$$

Atkinson (1977) and Sauter (1991) successfully applied this method to evaluate regional average hydraulic conductivity in karst aquifers. Hydraulic conductivity determined from catchment analyses often represents the upper limit of an aquifer system. However, horizontal variations in aquifer characteristics cannot be detected

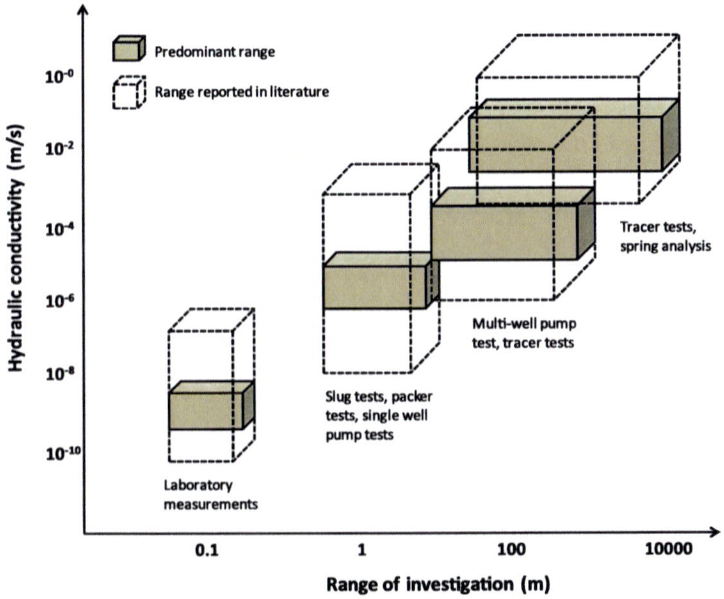

Fig. 3.44 Effect of investigation scale on hydraulic conductivity in karst aquifers

with this method because the dependent variable—discharge can only be measured at one single point. Different behavior of the recession curve in responding to flow conditions may reveal the temporal variation of the aquifer characteristics.

Figure 3.44 shows the common measured ranges of the hydraulic conductivity from different investigation methods. The hydraulic conductivity increases with an increasing scale of investigation. Lowest values are determined in the laboratory and the highest on a catchment scale. Hydraulic conductivities obtained in boreholes cannot be directly used as input parameters for a regional-scale model. The selection of the appropriate hydraulic parameters requires a careful analysis of the scale. The values are applicable at the location of the parameters to the respective part of the system, which consists of a fast transit (conduit), an intermediate (conduit-fissure) and a slow (fissure-matrix) flow sub-system.

3.7 Quantitative Analysis of Tracer Tests

Tracer test is an indispensable tool in hydrogeological investigation in karst terranes. The results provide information on the aquifer characteristics between the tracer injection point and the tracer recovery point. Depending on the type of test and the scale of investigation, the test data reflect the properties of either conduit system, or fissure system, or both. If, for example, the input location is in a sinkhole, the transport characteristics of conduits are most likely examined (neglecting the effect of the transport through the unsaturated zone). With a forced gradient tracer test between two boreholes, test data represent more likely the fissure system (considering the probability of encountering a directly connecting fissure by a borehole is limited). Tracer tests, with the input in boreholes and the recovery measured at a spring, may be influenced by fissure and conduit systems. The relative proportion of each system is determined by the distance that the tracer has to travel from the borehole to reach the draining conduit. Parameters determined from a tracer test characterize the aquifer on that particular scale. Moreover, vigorous analysis of the breakthrough curve will not be able to overcome problems related to poor design of tracer tests and inadequate implementation of procedures during the tests (refer to Mull et al. 1988 for procedures and techniques).

3.7.1 Tracer-Breakthrough Curves

Quantitative interpretation of tracer-breakthrough curves provides parameters controlling the tracer transport and flow conditions in the aquifer. The intended quantitative analysis must be taken into account in planning the tracer test. An adequate quantitative interpretation of tracing tests depends on:

1. Conservative behavior of the tracer;
2. Thorough inventory of sampling locations;
3) Adequate quantity of tracer injected;
4. Appropriate injection technique;
5. Sufficient monitoring frequency at sampling locations;
6. Precise discharge measurements at sampling locations; and
7. Sufficient monitoring length.

A typical tracer-breakthrough curve is shown in Fig. 3.45. Transport parameters pertinent to the breakthrough curve analysis at a sampling point are:

T_L—elapsed time to the arrival of the leading edge of the tracer-breakthrough curve;
T_p—elapsed time to the peak concentration C_p of the tracer-breakthrough curve;
T_c—elaplsed time to the centroid of the tracer-breakthrough curve;
T_t—elapsed time to the tailing edge of the curve.

The mean travel time is the difference in elapsed time of the centroids of the tracer-breakthrough curves defined upstream and downstream on the same streamline:

$$t_c = T_{c(n+1)} - T_{cn}$$

where n is the sampling location. Similarly, the travel times of the leading edge, peak concentration, and tailing edge are, respectively:

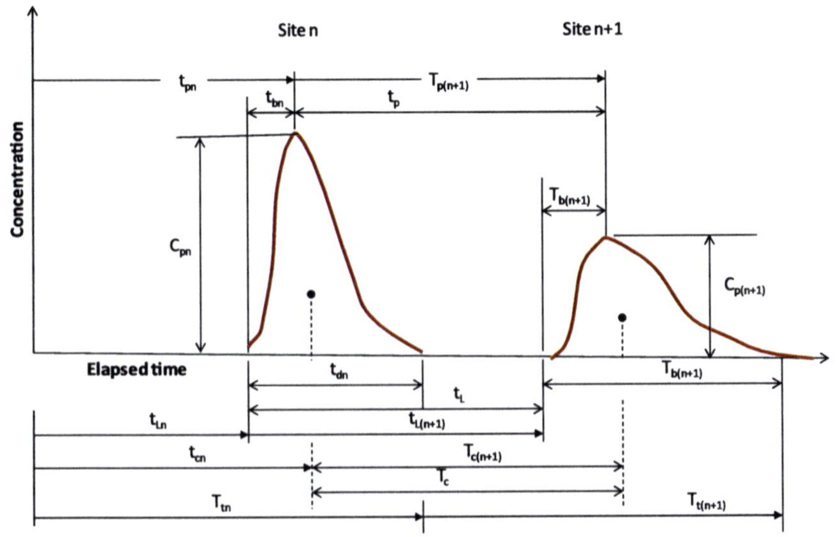

Fig. 3.45 Tracer-breakthrough curves along a tracer streamline from an instantaneous tracer injection

$$t_L = T_{L(n+1)} - T_{Ln}$$
$$t_p = T_{p(n+1)} - T_{pn}$$
$$t_t = T_{t(n+1)} - T_{tn}$$

3.7.2 *Estimation of Hydraulic Parameters of Karst Conduits*

Hydraulic parameters for karst conduits and fractures are estimated by the method of moments. Moment analysis is actually to determine the area under the tracer-breakthrough curve, weighed by the discharge hydrograph. Assume that the spring discharge (Q_i) and tracer concentrations (C_i) are measured at time t_i and the time interval is Δt_i, the mass of tracer recovered over the period of Δt_i ($Q_i\ C_i\ \Delta t_i$) is plotted against time t_i. The zeroth order moment of the recovered mass, μ_0, is computed by:

$$\mu_0 = \sum_{i=1}^{i=n} Q_i (C_i - C_0) \Delta t_i$$

where n is the number of sampling interval which is equal to the total number of samples minus 1 and C_0 is the background tracer concentration, measured at the time of injection.

The zeroth order moment is equal to the total recovered mass of tracer. The first moment (μ_1) is the centroid of the area under the recovered mass-time curve, which is expressed by:

$$\mu_1 = \sum_{i=1}^{i=n} t_i\, Q_i(Ci - C_0)\Delta\, t_i$$

Then, the mean residence time of the tracer travelling to the sampling point, $\bar{t}$ is the ratio of μ_1 and μ_0, i.e.

$$\bar{t} = \mu_1/\mu_0.$$

The second moment (μ_2) represents the spreading or mixing in the recovered mass, which is:

$$\mu_2 = \sum_{i=1}^{i=n} (t_i - \bar{t})^2\, Q_i(C_i - C_0)\Delta\, t_i$$

The standard deviation (σ) is calculated by $(\mu_{2/}\mu_{0)}^{1/2}$.

The mean travel time is the length of time required for the centroid of the tracer mass to traverse the entire length of the aquifer system. It can be used to calculate the mean velocity of groundwater flow (u) by $x/\bar{t}$, where x is the measured straight-line distance between the input and sampling point. A straight-line assumption for karst conduit is unrealistic because of its sinuosity. Worthington (1991) recommends a correction factor of 1.5. Parameter 'x' should be placed by 'x_s' where $x_s = 1.5x$.

The mean residence time allows for a rough estimation of the volume of conduit or fracture (*Vol*) traversed by the tracer cloud (Atkinson et al. 1973):

$$Vol = \sum_{i=1}^{m} Qi\Delta ti$$

where m is the number of monitoring intervals for spring discharge until the mean residence time.

With the knowledge of the volume and length of the conduit system, the cross-sectional area (Area) can be estimated by:

$$Area = Vol/xs$$

By assuming a cylindrical karst conduit, its diameter is estimated by:

$$Diameter = 2\sqrt{Area/\pi}$$

3.7.3 Evaluation of Dynamic Dispersion in Karst Aquifers

Similar to hydraulic conductivity, the dispersion coefficient is another scale-dependent parameter. A dispersion means spread, which is the integrated result of mechanical dispersion and molecular diffusion. The dispersion coefficient measures the rate at which concentrated solutes spread out along their flow path and can be calculated from the standard deviation of the tracer-breakthrough curve. With the assumption of uniform flow and constant velocity for the entire duration of the tracer test, Fisher (1968) presented an equation to calculate the dispersion coefficient (D_{tracer}) in open channels:

$$D_{tracer} = \frac{\bar{u}^2 \sigma^2}{2\bar{t}}$$

Chatwin (1971) developed an alternative method that applies to open-channel flow and closed-conduit flow as well. The dispersion coefficient is given by:

$$\left[t \, ln\left(\frac{Ap}{Ct^{0.5}}\right)\right]^{0.5} = \left[\frac{xs}{2(D_{tracer}^{)0.5}} - \frac{\bar{u}t}{2(Dtracer)^{0.5}}\right]$$

The constant of proportionality A_p can be estimated from (Day 1975):

$$Ap = Cp\sqrt{tp}$$

The first term on the right-hand side of the above equation is the y intercept, while the second term on the right-hand side is the gradient of the line. Either term on the right-hand side allows for solution for dispersion coefficient when a plot of the left-hand side against early-time data reasonably falls as a straight line (Day 1975).

More complicated Multi-Dispersion-Model (MDM) has also been used to evaluate tracer-breakthrough curves for large-scale tracer tests where convection and dispersion processes dominate (Maloszewski 1992; Kass 1998). MDM is an extension of one-dimensional convection-dispersion model. As shown in Fig. 3.46, the resulting breakthrough curve is seen as a composite outcome of different flow paths. The concentration contributed by an individual flow path is calculated by:

$$Ci(t) = \frac{Mi}{Qi} \frac{1}{t_{0i}\sqrt{4\pi PDi\left(\frac{t}{t_{0i}}\right)^3}} \exp\left[-\frac{\left(1 - \frac{t}{t_{0i}}\right)^2}{4PDi\left(\frac{t}{t_{0i}}\right)}\right]$$

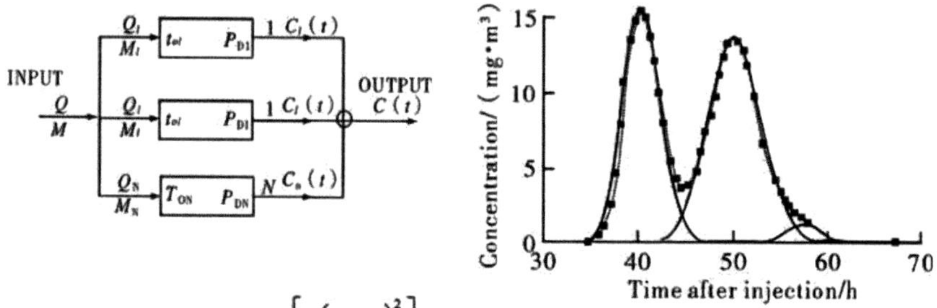

$$C_i(t)=\frac{M_i}{Q_i}\cdot\frac{1}{t_{0_i}\sqrt{4\pi P_{D_i}\left(\frac{t}{t_{0_i}}\right)^3}}\cdot\exp\left[-\frac{\left(1-\frac{t}{t_{0_i}}\right)^2}{4P_{D_i}\left(\frac{t}{t_{0_i}}\right)}\right]$$

$$C(t)=\sum_{i=1}^{N}C_i(t)$$

	Peak I	Peak II	Peak III
t_0 [h]	40	50	57
PD [-]	0.0012	0.0015	0.0004
α_L [m]	14	18	5

Fig. 3.46 Components of MDM

where

- C_i Is the tracer concentration;
- M Is the tracer mass;
- Q Is the discharge;
- T_0 Is the mean residence time;
- P_D Is the dispersion parameter:

$$PD=\frac{D}{\bar{u}x}=\frac{\alpha}{x}$$

where D is the dispersion; $\bar{u}$ is the mean velocity; α is the dispersivity; and i is the index of the flow path.

The measured concentration is the superposition of all the individual paths, i.e.:

$$C(t)=\sum_{i=1}^{N}Ci(t)$$

3.8 Application of Dual-Porosity Model to Groundwater Simulation in the Ordovician Limestone in Jiaozuo Coalfield, China

3.8.1 Introduction to Jiaozuo Coalfield

The dual-porosity model is applied to a complex karst aquifer in Jiaozuo Coalfield of North China. Jiaozuo Coalfield lies at the toe of an inclined alluvial plain before Taihang Mountain in Henan Province. It consists of over 10 coal mines and all the coals are mined underground. To the north of the coalfield is the recharge area, where the Cambrian-Ordovician carbonate rocks outcrop and encompass an area of approximately 1,168 km^2. The total estimated natural recharge rate is 480 m^3/m. The study area is well-known for its richness in water resources and for its complicated hydrogeological conditions (Wu et al. 2009).

Commercially valuable coal seams are within the carbonate rock sequences. Carbonate rocks often form the roof or bottom of the tunnels and stopes. The most hydrogeologically significant carbonate rock formation—Ordovician limestone, underlies the major coal seams. The average thickness of this formation is 400 m and karst features such as conduits, caves, and springs are well developed. The karst aquifer is confined in the coalfield and its potentiometric pressure is well above the elevation of the major coal seams. Water inrushes from the underlying karst aquifer can suddenly often occurred through faults, improperly abandoned boreholes, and paleo-collapse columns. More than 700 water inrushes have been recorded in the mining history of this coalfield. More than 60 water inrush incidents have water flow more than 10 m^3/min, and the maximum is 320 m^3/min. The water inrushes have flooded the coal mines on 14 occasions and severely undermine the mining safety and productivity. Because of the intimate relationship between karst water and the coal exploitation, mining operations in this area have been associated with karst water investigations for several decades. Understanding the groundwater flow in the karst aquifer is essential to properly manage the water resources and help minimize its threat to the mining operations.

3.8.2 Karst Conduit Distribution

The groundwater flow in Jiaozuo coalfield is controlled by an active and integrated conduit network. All the conduits are developed from dissolution of major faults. Figure 3.47 shows the locations of the major conduits and their connections. The conduit network divides the study area into 15 fissure systems. Each fissure system is hydrogeologically unique and they all include highly interconnected joints, fractures, and bedding planes that are less than 1 cm in their openings. The majority fractures have apertures from 0.5 to 2.5 mm, as measured on the outcrops. Fracture orientation

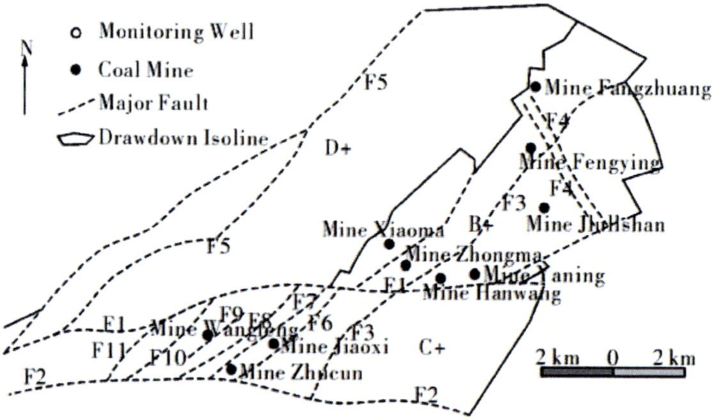

F1–Fenghuangling Fault;F2–Zhucun Fault;F3–Jiulishan Fault;F4–Fangzhuang Fault;
F5–Zhaozhuang Fault;F6–39# Fault;F7–3# Fault;F8–Tiangguanqu Fault;
F9–Zhongzhan Fault;F10–Wangfeng Fault;F11–Fengfeng Fault

Fig. 3.47 Distribution of karst conduits in Jiaozuo Coalfield

varies from N 20° E to N 160° E. The density is approximately 4—10 fractures/m^2. Visual observations in the mine tunnels and exploratory boreholes indicate that the carbonate rocks are highly fractured in the subsurface. The matrix blocks contain primary pores but the permeability tests in cores indicate they are not permeable and water release in them is constrained by the capillary effects. The fissure system in this investigation also includes transmissible faults with lengths ranging from several meters to 100 m and their offsets are less than 10 m.

Hydrogeological investigations and water inrush data in the coal mines indicate that the conduits are those faults that are more than 1000 m long and have displacements of over 30 m. Table 3.21 lists the major faults that are included in the conduit network and their hydraulic properties. The conduit network in the study area has the following characteristics:

- The conduits are the main flow paths of the groundwater and the flow direction is dominantly along the conduits.
- Constant water exchange occurs between the conduits and their surrounding fissure systems. However, the two fissure systems separated by one conduit can have different water heads in them and the difference can be 70–200 m.
- The conduits can transmit groundwater vertically, which makes the groundwater flow three-dimensional.
- The groundwater flow in the conduit network is highly heterogeneous and anisotropic.

Table 3.21 Conduits and their properties in Jiaozuo Coalfield

Fault	Conduit name	Orientation	Length (m)	Displacement (m)	Hydraulic properties
F1	Fenghungling	EW	70,000	200–300	This conduit is approximately 10–20 m wide. It is highly conductive. Exploratory boreholes revealed caverns up to 1 m in diameter. Pumping tests indicate that the permeability ranges from 17 to 42 m/d. Numerous water inrush incidents have occurred along the conduit. The maximum flow rate for the water inrushes is 4 m^3/s
F2	Zhucun	EW	20,000	700–1,000	The conduit is highly conductive along its strike. Karst features are well developed. Its permeability ranges from 13 to 18 m/d
F3	Jiulishan	NE	70,000	300–1,000	Many karst springs emerge along this conduit. The total discharge is up to 12 m^3/s. The cone of depression prolongs along this conduit after water inrushes
F4	Fangzhuang	NW	10,000	200	Many water inrush incidents took place along this conduit. The maximum flow rate for the water inrushes is 1.2 m^3/s
F5	Zhaozhuang	NE	35,000	300	Water levels are different on two sides of the conduit. Groundwater flows along the conduit
F6	No.39 well	NE	9,500	170–300	Karst features are well developed. Exploratory boreholes in the conduit often encountered cavities
F7	No.3 well	NE	4,600	279	This conduit consists of three adjacent but parallel faults. It is highly conductive
F8	Tianguanqu	NE	10,000	120–300	Highly conductive
F9	Zhongzhan	NE	5,000	60–110	It is conductive in the vertical direction. Water inrushes occurred along this conduit via collapse columns
F10	Wangfeng	EW	8,000	60–120	Many water inrush incidents occurred along this conduit. Water sources included the underlying thick limestone and the overlying thin-bedded limestone of the coal seams
F11	Fengfeng	NE	9,000	80–120	Highly conductive

3.8.3 Calibration of the Dual-Porosity Model

A large scale multi-well pumping test was used to calibrate the developed model and to determine the initial hydrogeological parameters in the two systems. The pumping test lasted 28 days with a constant pumping rate of 1 m^3/s. The maximum drawdown at the end of the pumping test is 13.3 m. It influenced groundwater flow up to 20 km along Fenghuangling conduit (F1) and Jiulishan conduit (F3). The cone of depression prolonged mainly along these two faults. The measured water level drawdowns are contoured in Fig. 3.48, and they are compared to the drawdowns calculated from the calibrated parameters (Fig. 3.49). Figure 3.50 shows the distribution of parametric regions to represent the heterogeneous system. Table 3.22 gives the calibrated parameters for the conduit network and fissure system.

The calibrated parameters were further used to reproduce the drawdowns measured on March 9, 1979 when a major water inrush occurred in one of the coal mines (Fig. 3.51). In this water inrush incident, the average flow rate was 4 m^3/s. The inrush caused the water level in the Ordovician limestone to drop 0.4-1.7 m. The groundwater level along the Fenghuangling fault dropped more than 1.0 m. As shown in Fig. 3.52, the calculated drawdowns from the dual-porosity model and the calibrated parameters match the measured drawdowns (Fig. 3.51) reasonably well.

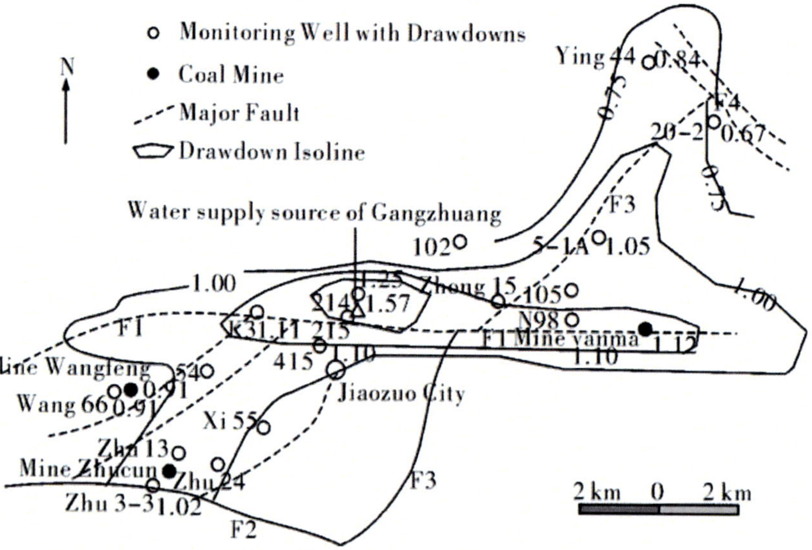

Fig. 3.48 Water level drawdown measured in response to a multiple-well pumping test in the water supply source of Gangzhuang

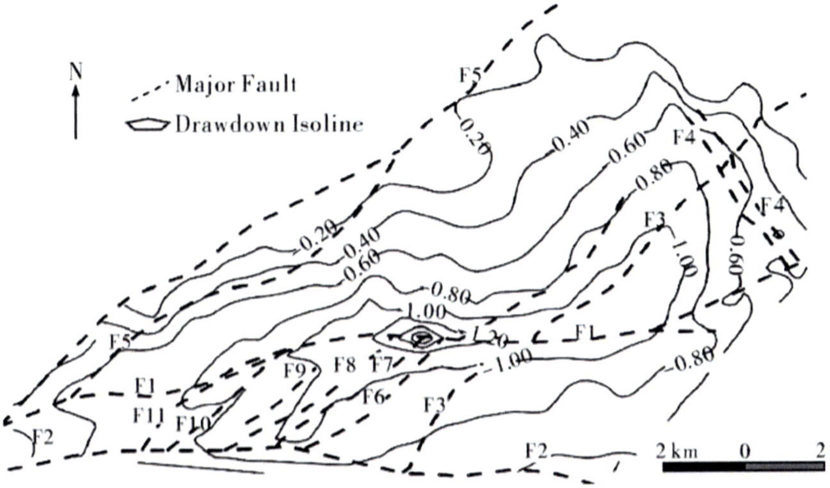

Fig. 3.49 Calculated drawdowns by the dual-porosity model in response to the multiple-well pumping test, as shown in Fig. 3.48

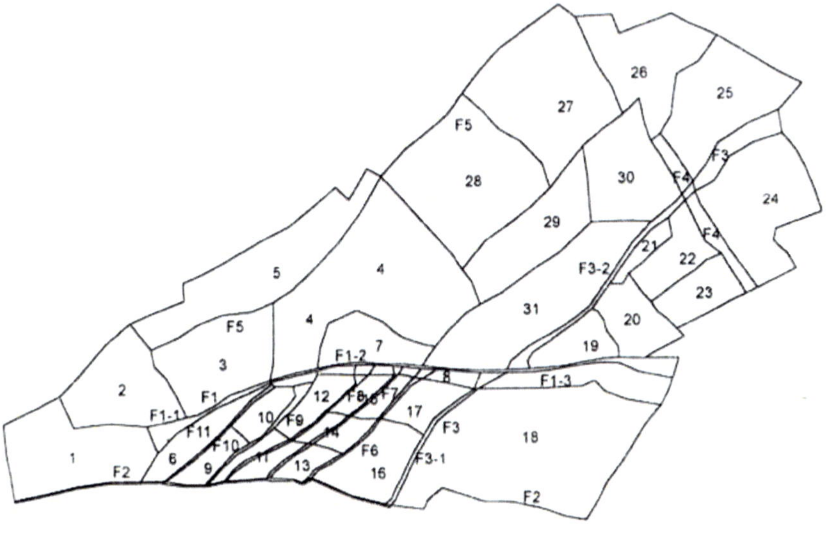

Fig. 3.50 Calibrated parametric conduits for conduit system and parametric regions for fissure system

Table 3.22 Calibrated hydraulic parameters for the modeling domain

Conduit system

Conduit	K_x (m/d)	K_y (m/d)	K_z (m/d)	S ($\times 10^{-4}$)	Conduit	K_x (m/d)	K_y (m/d)	K_z (m/d)	S ($\times 10^{-4}$)
F1-1	165	37	86	37.5	F1-2	186	68	86	45.8
F1-3	147	36	141	24.5	F2	114	76	97	45.2
F3-1	132	117	87	35.7	F3-2	147	105	75	35.2
F4	45	135	57	24.5	F6	57	45	54	14.8
F7	54	46	45	15.2	F8	34	34	44	12.3
F9	54	54	56	12.1	F10	45	38	32	21.8
F11	38	34	54	11.2					

Fissure system

Region	K_x (m/d)	K_y (m/d)	K_z (m/d)	S ($\times 10^{-4}$)	Region	K_x (m/d)	K_y (m/d)	K_z (m/d)	S ($\times 10^{-4}$)
1	47	32	46	1,900	17	76	76	35	11.5
2	32	32	32	1,100	18	10	10	10	1.3
3	46	46	46	1,300	19	124	124	124	42.5
4	25	48	36	1,100	20	112	112	112	31.4
5	11	8	8	40	21	67	67	102	270
6	6	6	6	80	22	46	86	74	10.2
7	48	65	54	9.5	23	9.5	9.5	8.9	2.36
8	67	24	42	14.2	24	14	14	14	0.14
9	14	14	21	6.4	25	18	18	18	0.87
10	72	67	54	10.6	26	36	36	36	1,700
11	81	81	70	6.4	27	26	26	26	2,400

(continued)

Table 3.22 (continued)

Fissure system									
Region	K_x (m/d)	K_y (m/d)	K_z (m/d)	S ($\times 10^{-4}$)	Region	K_x (m/d)	K_y (m/d)	K_z (m/d)	S ($\times 10^{-4}$)
12	98	87	58	7.96	28	18	18	18	1,100
13	21	21	36	5.5	29	37	37	37	2,400
14	24	24	30	9.5	30	57	80	42	16.7
15	83	64	65	7.65	31	36	36	47	13.8
16	26	26	12	1.45					

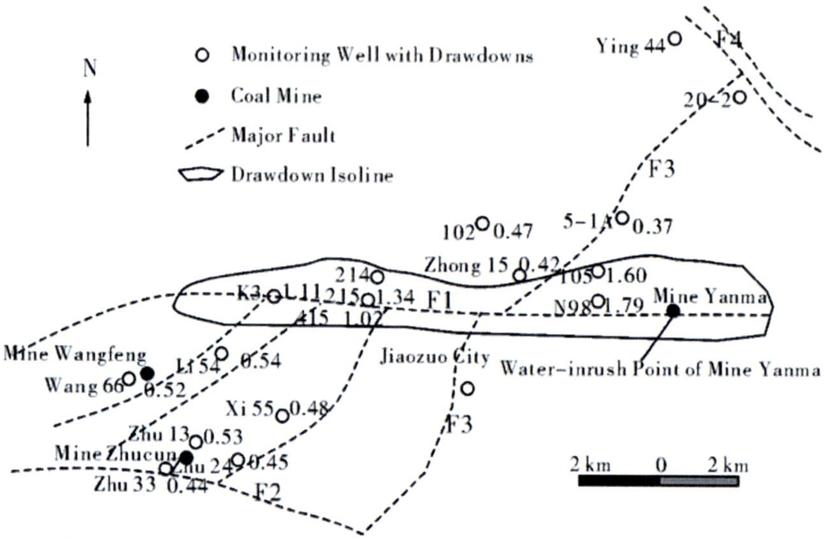

Fig. 3.51 Groundwater level drawdowns measured in response to a water inrush incident in Yanma Mine

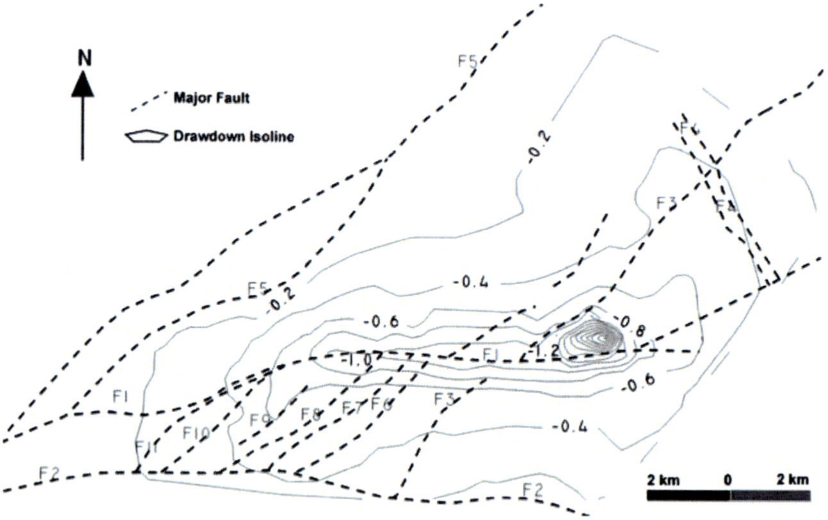

Fig. 3.52 Calculated drawdowns by the dual-porosity model in response to the water inrush incident shown in Fig. 3.51

References

Adar EM, Neuman SP (1988) Estimation of spatial recharge distributions using environmental isotopes and hydrochemical data. J Hydrol 97:279

Aiguang C (1988) Stochastic modelling of the karst spring of Xin'an in Shanxi Province. In: Proceedings of 21st IAH Congress, Guilin, China, 621

Atkinson TC (1977) Diffuse flow and conduit flow in limestone terrain in the Mendip Hills, Somerset (Britain). J Hydrol 35:93

Atkinson TC (1985) Present and future directions in karst hydrogeology. Ann Soc Geol Belgique 108:293–296

Avdagic I (1988) Hydrological definition of karst areas. In: Proceedings of 21st IAH congress. Guilin, China, p 612

Baca RG, Arnett RC, Langford DW (1984) Modeling fluid flow in fractured-porous rock masses in finite-element techniques. Int J Num Meth Fluids 4:337–348

Bakalowicz M (1977) Etude du degre d'organisation des ecoulements souterrains dans les aquiferes carbonates par une methode hydrogeochimique nouvelle. C.R Acad Sc Paris 284D:2463–2466

Bakalowicz M, Blavoux B, Mangin A (1974) Apports du traacage isotopique naturel a la connaissance du fonctionnement d'un system karstique-teneurs en oxygene-18 de trois systemes des Pyrenees, France. J Hydrol 23:141–158

Baoren C, Xueming Z (1988) The mathematical simulation of hydrological regime of Chaohua spring in Mixian county. In: Proceedings of the 21st IAH congress. Guilin, China

Barker JA (1991) Transport in fractured rock. In: Downing and Wilkinson (eds) Applied groundwater hydrology. Oxford Science Publication, pp 199–216

Bear J (1972) Dynamics of fluids in porous media. Dover, Mineola, NY

Bear J (1993) Modelling flow and contaminant transport in fractured rocks. In: Bear J, Tsang CF, de Marsily G (eds) Flow and contaminant transport in fractured rock. Academic Press, Inc

Beard MJ, Montgomery AC (1981) Survival of Bacteria in a sewage effluent discharged to the chalk. Water Pollut Control

Bedinger MS, Reed JE, D'Lugosz J (1988) Practical guide to aquifer-test analysis, environmental monitoring systems laboratory office of research and development. U.S. Environmental Protection Agency, Las Vegas, Nevada, USA

Bell FG, Cripps JC et al. (1989) Chalk fabric and its relation to certain geotechnical properties. In: International chalk symposium. Thomas Telford, London

Bibby R (1982) Mass transport studies of solutes in dual-porosity media. Water Resour Res 17:1075–1081

Bocker T (1973) Dynamics of subterranean karstic water flow. Karszt-es Barlangkutatas 8:107–145

Bourke PJ (1987) Channeling of flow through fractures in rock. In: Proceedings of GEOVAL-87 international symposium. Swedish Nuclear Power Inspectorate, Stockholm

Brown SR, Scholz CH (1985) Broad bandwidth study of the topography of natural rock surfaces. J Geophys Res 90(B14):12575–12582

Burdon DJ, Papakis N (1963) Handbook of karst hydrogeology, UN Special Fund. Institute for Geology and Subsurface Research, Athens, Greece

Casas MC, Ledoux E, de Marsily G, Tillie B, Barbreau A, Durand E, Fuega B, Peaudecerf P (1990a) Modelling fracture flow with a stochastic discrete fracture network: calibration and validation, 1, The flow model. Water Resour Res 26:479–489

Casas MC, Ledoux E, de Marsily G, Barbreau A, Calmels P, Durand B, Fuega B, Margritta R (1990b) Modelling fracture flow with a stochastic discrete fracture network: calibration and validation 2, The transport model. Water Resour Res 26:491–500

Campana ME, Mahin DE (1985) Model-derived estimates of groundwater mean ages, recharge rates, effective porosities and storage in a limestone aquifer. J Hydrol 76:247–264

Chatwin PC (1971) On the interpretation of some longitudinal dispersion experiments. J Fluid Mech 48(4):689–702

Chilton PJ, Lawrence AR, Barker JA (1989) Chlorinated solvents in chalk aquifers: Some preliminary observations on behavior and transport. In: Chalk, 605-610, Proceedings international chalk symposium. Thomas Telford, London
Dai ED et al. (1988) Threshold autoregressive model applied to prediction of karst spring flow. In: Proceedings of the 21st IAH congress. Guilin, China
Day TJ (1975) Longitudinal dispersion in natural channels. Water Resour Res 11(6):909–918
de Marsily G (1986) Quantitative hydrogeology. Academic, San Diego, Calif
Dershowitz W (1992) The role of the Stripa phase 3 project in the development of practical discrete fracture modelling technology. In: 4th international symposium on the OECD/NEA Stripa Project. Stockholm
Dershowitz W, Roberds W, Black J (1991) Application of discrete fracture analysis to site characterization. Geotechnical Engineering Congress, Boulder, Colorado
Downing RA, Headworth HG (1989) Hydrogeology of the chalk in the UK: the evolution of our understanding. In: International Chalk Symposium. Thomas Telford, London
Downing RA, Pearson FJ, Smith DB (1978) The flow mechanism in the chalk based on radioisotope analysis of groundwater in the London Basin. J Hydrol 40
Dreiss SJ (1982) Linear kernels for karst aquifers. Water Resour Res 18(4):865–876
Dreiss SJ (1989) Regional scale transport in a karst aquifer. Water Resour Res 25(1):117–134
Drew DP (1970) The significance of percolation water in limestone catchment. Ground Water 8(5):8–11
Duffy CJ, Harrison J (1987) The statistical structure and filter characteristics of tritium fluctuations in fractured basalts. Water Resour Res 23(5):894–902
Dverstorp B, Andersson J (1989) Application of the discrete fracture network concept with field data: Possibilities of model calibration and validation. Water Resour Res 25:540–550
Dverstorp B, Andersson J, Nordqvist W (1992) Discrete fracture network interpretation of field tracer migration in sparsely fractured rock. Water Resour Res 28:2327–2343
Endo HK, Long JCS, Wilson CR, Witherspoon PA (1984) A model for investigating mechanical transport in fracture networks. Water Resour Res 20:1390–1400
Ewers RO (1978) A model for the development of broad scale networks of groundwater flow in steeply dipping carbonate aquifers. Trans Brit Cave Res Assoc 5:121–125
Fisher BH (1968) Dispersion predictions in natural streams. J Sanit Eng Div Am Soc Civ Eng 94:927–943
Ford D, Williams P (1989) Karst geomorphology and hydrology. Unwin Hyman, London
Foster SSD (1975) The chalk groundwater tritium anomaly-a possible explanation. J Hydrol 25:159–165
Freeze RA, Cherry JA (1979) Groundwater. Prentice-Hall Inc, Englewood Cliffs, NJ
Gardner CMK, Cooper JD (1989) Hydrology of the unsaturated zone of the chalk of South East England. In: International chalk symposium held at Brighton polytechnic
Geake AK, Foster SSD (1989) Sequential isotope and solute profiling in the unsaturated zone of British Chalk. Hydrol Sci 34(1)
Gelhar LW, Axness CL (1983) Three dimensional stochastic analysis of macrodispersion in aquifers. Water Resour Res 19:161–180
Gieske A, de Vries J (1990) Conceptual and computational aspects of the mixing cell method to determine ground water recharge components. J Hydrol 121:277
Gilman JR, Kazemi H (1988) Improvements in simulation of naturally fractured reservoirs. Soc Pet Eng J 695–707
Grindley J (1967) The estimation of soil moisture deficit. Meteorol Mag 76:97
Han XR, Lu RG, Li QS (1993) Karst water system—a study on big karst springs in Shaanxi. Geology Publishing House, Beijing, China (Chinese)
Hawnt RJE, Joseph JB (1980) Experience with borehole recharge in the Lee Valley, Unpublished Paper for a meeting of South-Eastern Section of British Hydrological Society
Headworth HG (1970) The selection of root constants for the calculation of actual evaporation and infiltration for chalk catchments. J Inst Water Eng 24:431

Heath MJ (1985) Solute migration experiments in fractured granite, South West England. In: Proceedings of a joint CEC-NEA workshop. Rotterdam, Netherlands, pp 191–200
Hill AC, Thomas GW (1985) A new approach for simulating complex fractured reservoirs. Paper SPE 13537 presented at the SPE Reservoir Simulation Symposium, Dallas
Houston JFT (1983) Groundwater systems simulation by time-series techniques: U.S., Groundwater 21(3):301–310
Hsieh PA, Neuman SP, Stiles GK, Simpson ES (1985) Field determination of three dimensional hydraulic conductivity tensor of anisotropic media, 2. Methodology and application to fractured rock. Water Resour Res 21:1667–1676
Kass W (1998) Tracing technique in geohydrology. A.A. Balkema, Brookfield, 581 pp
Keating T (1982) A lumped parameter model of a chalk aquifer-stream system in Hampshire, UK. Groundwater 20(4):430
Khaleel R (1989) Scale dependence of continuum models for fractured basalt. Water Resour Res 25:1847–1855
Kirk ST, Campana ME (1990) A deuterium-calibrated groundwater flow model of a regional carbonate-alluvial system. J Hydrol 119:357
Knisel WG (1972) Response of karst aquifers to recharge, Hydrol Pap 60, Colo State Univ, Ft. Collins
Kresic NA (1993) Review and selected bibliography on quantitative definition of karst hydrogeological systems, Annotated Bibliography of Karst Terranes. In: LaMoreaux PE (ed) International contributions to hydrogeology
Kwicklis EM, Healy RW (1993) Numerical investigation of steady liquid water flow in a variably saturated fractured network. Water Resour Res 29:4091–4102
LaMoreaux PE (1989) Water development for phosphate mining in a karst setting in Florida—a complex environmental problem. Environ Geol Water Sci 14(2):117–153
LaMoreaux PE, Hughes TH, Memon BA, Lineback N (1989) Hydrogeologic assessment—Figeh Spring, Damascus, Syria. Environ Geol Water Sci 13(2):73–127
Li JK (1988) Mine water control and management in karst terrains. Coal Industry Publishing House, Beijing, China (Chinese)
Li JK, Zhou WF (1990) The studies on the control and management of groundwater from ordovician limestone in collieries, North China. Shaanxi People's Publishing House, Shaanxi, China (Chinese)
Lloyd JW, Harker D, Baxendale RA (1981) Recharge mechanisms and groundwater flow in the chalk and drift deposits of southern East Anglia. Q J Eng Geol 14:87
Long JCS, Witherspoon PA (1985) The relationship of the degree of interconnection to permeability of fractured networks. J Geophys Res 90(B4):3087–3098
Long JCS, Remer JS, Wilson CR, Witherspoon PA (1982) Porous media equivalents for networks of discontinuous fractures. Water Resour Res 18:645–658
Maillet E (1905) Essais d'hydraulique souterraine et fluviale. Hermann, Paris
Maloszewski P (1992) Mathematical modelling of tracer transport in different aquifers: Results from ATH test fields. In: Hotzl H, Werner A (eds) Tracer hydrology. Proceedings of 6th internarial symposium water tracing. 21–26 Sept., Karlsruhe, Rotterdam, Balkema, pp 25–30
Mangin A (1975) Contribution a letude hydrodynamic des aquiferes karstiques, DES Thesis, University Dijon, France
Milanovic PT (1981) Karst hydrogeology. Water Resour Publ
Montgomery AC, Beard MJ et al (1984) Effects of the recharge of sewage effluent upon the quality of chalk groundwater. Water Pollut, Control
Mull DS, Liebermann TD, Smoot JL, Woosley Jr. LH (1988) Application of dye-tracing techniques for determining solute-transport characteristics of ground water in karst terranes, United States Environmental Protection Agency
Murphy JR, Thomson NR (1993) Two-phase flow in a variable aperture fracture. Water Resour Res 29:3453–3478
Palmer AN (1991) Origin and morphology of limestone caves. Geol Soc Am Bull 103:1–25

Pankow JF, Johnson JP, Hewetson JP, Cherry JA (1986) An evaluation of contamination migration patterns at two waste disposal sites on fractured porous media in terms of the equivalent porous medium (EPM) model. J Cont Hydrol 1:65–76
Pyrak-Nolte LJ, Cook NGW, Nolte DD (1988) Fluid percolation through single fractures. Geophys Res Lett 15:1247–1250
Quinlan, JF (1989) Ground-water monitoring in karst terranes: recommended protocols and implicit assumptions: Las Vegas, Nev., U.S. Environmental Protection Agency, Environmental Monitoring Systems Laboratory, EPA/600/X-89/050, 100 p.
Rao BK, Hathaway DL (1990) A three-dimensional mixing cell solute transport model and its application. Ground Water, 28:509–516
Rao SG, Rao AR (1987) Recursive estimation of kernels of nonlinear rainfall-runoff models. J Hydrol 95:341–364
Rasmuson A (1985) Analysis of hydrodynamic dispersion in discrete fracture networks using method of moments. Water Resour Res 21:1677–1683
Rasmussen TC (1991) Steady fluid flow and travel times in partially saturated fractures using a discrete air-water interface. Water Resour Res 27:67–76
Rasmussen TC, Huang CH, Evans DD (1985) Numerical experiments on artificially-generated, three-dimensional fracture networks: an examination of scale and aggregation effects. Mem Int Assoc Hydrogeol 17th 676–682
Ray JA, Webb JS, O'dell PW (1994) Groundwater sensitivity regions of Kentucky: map sheet, kentucky department of environmental protection
Rinaldo A, Gambolati G (1987) Basin-scale transport of dissolved species in groundwater. Paper at advanced research workshop. NATO, Lisbon, Portugal
Rorabaugh MI (1964) Estimating changes in bank storage and groundwater contribution to streamflow. In: International association of scientific hydrology. Publication 63, pp 432–441
Rorabaugh MI, Simons WD (1966) Exploration of methods relating groundwater to surface water. Columbia river Basin—second phase. U.S. Geological Survey Open File Report, 62 p
Rushton KR, Ward C (1979) The Estimation of Groundwater Recharge, Lecture Notes, Birmingham University, UK
Rutledge AT, Daniel CC III (1994) Testing an automated method to estimate groundwater recharge from stream records. Ground Water 32(2):180–189
Sauter M (1991) Assessment of hydraulic conductivity in a karst aquifer at local and regional scale. In: Proceedings of the third conference on hydrogeology, ecology, monitoring, and management of groundwater in Karst Terranes, Maxwell House, Nashville, TN, USA
Schwartz FW, Smith L, Crow AS (1983) A stochastic analysis of macroscopic dispersion in fractured media. Water Resour Res 19:1253–1265
Shuster ET, White WB (1971) Seasonal fluctuations in the chemistry of limestone springs: a possible means of characterizing carbonate aquifers. J Hydrol 14:93–128
Simpson ES (1988) The discrete state compartment model and its applications to flow through karstic aquifers. In: Proceedings of the 21st IAH congress. Guilin, China
Simpson ES, Duckstein L (1975) Finite state mixing-cell models. Karst Hydrology and Water Resources, Dubrovnik
Smart CC (1999) Subsidiary conduit system: A hiatus in aquifer monitoring and modeling. In: Palmer AN, Palmer MV, Sasowsky I (eds) Proceedings of karst modeling, charlottesville. VA, pp 146–157
Smart PL, Hobbs SL (1986) Characterization of carbonate aquifers: a conceptual base. In: Proceedings of the environmental problems in karst terranes and their solution conference, national water well association. Bowling Green, Kentucky, USA
Smith DB, Wearn PL et al. (1970) Water movement in the unsaturated zone of high and low permeability strata using natural tritium. In: Isotope hydrology. IAEA, Vienna
Stephenson BJ, Zhou WF, Beck BF, Green TS (1997) Highway stormwater runoff in karst areas—Preliminary results of baseline monitoring and design of a treatment system for a sinkhole in

Knoxville, Tennessee. In: Proceedings, the sixth multidisciplinary conference on sinkholes and the engineering and environmental impacts of Karst: Springfield, Missouri, pp 173–181

Teutsch G, Sauter M (1991) Groundwater modelling in karst terranes: scale effects, data acquisition and field validation. In: Proceedings of the third conference on hydrogeology, ecology, monitoring, and management of groundwater in Karst Terranes, Maxwell House, Nashville, TN, USA

Thraikill J (1986) Models and methods for shallow conduit-flow carbonate aquifers. In: Environmental Problems in Karst Terranes and their solutions conference. Bowling Green, Kentucky, pp 17–31

Tsang YW (1992) Use of "equivalent apertures" for rock fractures as derived from hydraulic and tracer tests. Water Resour Res 28:1451–1455

Viessmann, Jr W, Lewis GL, Knapp JW (1989) Introduction to hydrology (3rd edition). Harper & Row Publishers, New York, 780 p

Walton WC (1987) Groundwater pumping tests: Design and analysis. Lewis Publisher Inc

Wellings SR (1984) Recharge of the upper chalk aquifer at a site in Hampshire, England. J Hydrol 69:259

Wellings SR, Bell JR (1980) Movement of water and nitrate in the unsaturated zone of the upper chalk, near Winchester, Hants, England. J Hydrol 48:119

White WB (1969) Conceptual models for limestone aquifers. Groundwater 7:15–21

White WB (1999) Conceptual models for karstic aquifers. In: Palmer AN, Palmer MV, Sasowsky ID (eds) Proceedings of the symposium of karst modeling, charlottesville, virginia. Karst Water Institute Special Publication 5, 10–16

Williams PW (1985) Subcutaneous hydrology and the development of doline and cockpit karst. Z Geomorph 29(4):463–482

Witherspoon PA, Wang JSY, Iwai K, Gale JE (1980) Validity of cubic law for fracture flow in a deformable rock fracture. Water Resour Res 16:1016–1024

Wu Q, Zhou W, Pan GY, Ye SY (2009) Application of a discrete-continuum model to karst aquifers in North China. Ground Water 47(3):453–461

Yuan DX (1978) Heterogeneity of groundwater flow in karst terrains. Geology Publishing House, Beijing, China (Chinese)

Yusun C (1988) The media and movement of karst water. In: Proceedings of 21st IAH congress, Guilin, China, p 555

Zhan BS (1994) The Volterra kernel functional simulation of karst water of underground river system in the southern part of Puding County, Guizhou Province, Human Activity and Karst Environment. Science and Technology Publishing House, Beijing, China

Zhou WF (1990) Control of groundwater from Ordovician limestone in North China's coal fields (in Chinese). J China coal Soc 15:55–66

Zhou WF et al (1992) Effects of protective layers in karst water inrush in coal mines–an example from the Handan Wangfeng Mine (in Chinese). Carsologica Sinica 11:2–14

Zhou WF (1995) Behavior of Dense Non-aqueous Phase Liquids in Fractured Rock, PhD Thesis, Imperial College, London, UK

Zou CJ (1994) Hydro-electric engineering geology in karst terrains. Hydro-electric Publishing House, Beijing, China (Chinese)

Chapter 4
Prevention and Control of Mine Water Hazards from Underlying Aquifers

4.1 Water Prevention and Control Technology in Mining Lower Coal Seams Under Potentiometric Pressure in Xingtai Dongpang Mine

4.1.1 Mine Background

Xingtai Dongpang Mine of Jizhong Energy Co., Ltd. is located approximately 10 km southwest of Neiqiu County, Hebei Province. The #2 coal seam is the primary mining target, while the #9 coal seam is the secondary target. The mine has been in operation for more than 30 years since it was completed in December 1983. The designed production capacity is 1.8 million t/year, and the current approved production capacity is 3.35 million t/year. Dongpang Mine has administrative responsibility for its Main Shaft, Northern Shaft, and Xipang Shaft. Among them, the Main Shaft mines the coal resources to the southern boundary. The Northern Shaft mines the lower coal seams under the northern wing of the mine field. The current mining elevation is −220 m below mean sea level (bmsl), and the approved production capacity is 570,000 t/year. The Xipang Shaft mines the lower group of coal, with an approved production capacity of 400,000 t/year.

The 9103 working face is for trial mining of the #9 coal seam, which belongs to the lower coal group in the mine. The elevation of the upper roadway floor ranges from −110 to −147 m bmsl, and the elevation of the lower roadway floor is between −124 and −153 m bmsl. The distance between the #9 coal seam floor and top of the Ordovician limestone varies from 30 to 42 m. The potentiometric pressure elevation of the Ordovician limestone aquifer is between 47 and 75 m above mean sea level (amsl) in the mine. The measured potentiometric pressure elevation of the Ordovician limestone ranges from 59 to 63 m amsl in the working face. The water inrush coefficient of the working face is greater than the critical water inrush coefficient of 0.06 MPa/m in the section where mining is also affected by geologic

S. Dong et al., *Methods and Techniques for Preventing and Mitigating Water Hazards in Mines*, Professional Practice in Earth Sciences,
https://doi.org/10.1007/978-3-030-67059-7_4

structures. There is a possibility that Ordovician limestone water inrushes occur during the mining process.

To the western boundary of Dongpang Mine, there were in the past seven small coal mines that extracted #9 coal seam in the vicinity of and above the 9103 working face. Seven water inrushes occurred from the Ordovician limestone with a maximum water flow rate of 2000 m^3/h. The working face is in the relatively water-rich area of the Ordovician limestone aquifer in the southern wing of the mine field. The electrical resistivity imaging results indicate existence of relatively water-rich sections below the #9 coal floor barrier. The 9100 mining area, where the working face 9103 is located, is adjacent to the Xipang Shaft. The stratigraphic dip of the working face changes more drastically than that in the Xipang Shaft. Small folds are present near the corners or turning points of the working face. Areas where the strata are folded, or the dip angle changes drastically are prone to development of fissures that connect the Ordovician limestone aquifer to the working area, inducing water inrushes.

The main factor affecting the mining safety of 9103 face are the Ordovician limestone aquifer and water pooling in the goaf of the seven small coal mines. The potential pathways in which the Ordovician limestone water gushes into the coal mining face include water-conducting faults, water-conducting fissure zones, and water-conducting collapse columns, although no collapse columns have been identified in the mine.

4.1.2 *Application of Water Prevention and Control Technology to Mining Under Potentiometric Pressure*

4.1.2.1 Technical Approach of Mining Under Potentiometric Pressure

The water-rich Ordovician limestone aquifer that underlies the #9 coal seam of Dongpang Mine has different structural characteristics from the Feicheng Coalfield and Huaibei-Yongxia Coalfield structures and is representative of the Hanxing mining area. Therefore, mining of #9 coal seam requires to choose a method and approach suitable for water prevention and control in this mine. Comprehensive mine geological and hydrogeological conditions in Dongpang Mine, analysis of mining of 9103 working face under potentiometric pressure conditions and the degree of threat of water hazard, the following water prevention and control measured are adopted to ensure the safe recovery of the #9 coal seam:

- Pre-drain the thin-bedded Daqing limestone aquifer that overlies the Ordovician limestone; set up measures to prevent water inrushes from the pooling water in the small coal mines; and focus the water hazard control investigation on characterization of the Ordovician limestone.
- Carry out mining under potentiometric pressure by taking full use of the water-resistive property of the aquiclude between the #9 coal floor and the top of the Ordovician limestone.

- Improve the effective thickness and integrity of the water-resistive aquiclude and reinforce its water-blocking properties.
- Set up sluice gates to achieve isolated mining; pre-excavate drainage lanes and temporary water storage reservoirs; increase drainage space; install submersible pumps with sufficient capacity; establish safety measures and escape routes in case of an emergency.
- Apply comprehensive exploration methods including geophysics and drilling to explore the structurally weak section of the aquiclude, water-rich section of the aquifer, and potential water-conducting channels (faults, fissures, or collapse columns); and conduct grouting transformation as needed.
- Use drilling engineering to measure three parameters: water quantity, water pressure, and water temperature; calculate the water inrush coefficient; and evaluate quantitatively the water-resistive performance of the aquiclude.
- Use in-situ stress testing technology to measure stress parameters before and during mining; combine with laboratory-tested rock mechanics parameters to calculate and analyze the maximum depth of floor failure caused by mining and the cracking intensity of the water-resisting layer under different stress conditions; and predict the possibility of water inrush.
- Use the water inrush monitoring system to carry out real-time monitoring and forecast on the sections prone to water inrush. Measure and evaluate the destruction depth and strength of the floor caused by mining effect.
- Implement the water control measures sequentially according to the staged characteristics of the excavation project.

4.1.2.2 Comprehensive Exploration Methods

The comprehensive exploration consists of three-dimensional aboveground and underground methods. The primary exploration is conducted underground with supplemental surface exploration. The geophysical exploration is the primary tool, while drilling is auxiliary. The choice of geophysical exploration methods considers the sensitivity to targets of interest and positioning. The exploration drilling considers as much as possible using one hole for multiple purposes.

In terms of the geophysical prospecting methods, the aboveground three-dimensional seismic method is used to explore faults and collapse columns within the #9 coal seam and the hydraulic barrier. The underground geophysical survey uses the electrical resistivity imaging to investigate the water-rich structure in front of excavation. The pit penetration determines the coal seam structure, while the electrical resistivity imaging method and audio-electric perspective are combined to investigate the water-rich areas within the hydraulic barrier of the floor that underlies the working face. The mining practices show that the selection method is reasonable and obviously effective, achieving the purpose of "control" and "guidance".

4.1.2.3 Evaluation of Water Resistance Capacity of the Floor Aquiclude

In-situ tests of the water resistance capacity of the coal seam floor indicate that the confined groundwater level is encountered before the test boreholes expose the Ordovician limestone aquifer, which suggests that the thickness of the protective layer of the floor becomes thinner. Pressurized groundwater level indicates that a water barrier layer is present on the top of the aquifer and has a certain water resistance capacity. Such data is important for the safety evaluation of mining under potentiometric pressure. To further explain the evaluation standard of mining with the pressure, the pressure index is obtained through the relevant test data. The pressure index is used to evaluate the safety of mining under potentiometric pressure.

The water resistance performance of the protective layer of the floor aquiclude can be reflected by the pressure index. The calculation equations of the pressure index are as follows.

Pressure index for each test section:

$$D_{wi} = \frac{P_{w(i+1)} - P_{wi}}{h_{(i+1)} - h_i}$$

Average pressure index of all test sections:

$$\overline{D_w} = \frac{P_{wn} - P_{w1}}{h_n - h_1}$$

Average pressure index for water-bearing zone:

$$\overline{D_{w0}} = \frac{H_0}{hdy} \div 100$$

where

D_w	Segment pressure index (MPa/m)
$\overline{D_w}$	Average pressure index of the entire test sections (MPa/m)
$\overline{D_{w0}}$	Average pressure index of water-bearing zone (MPa/m)
P_{wi}	Potentiometric pressure (MPa) measured for the first time
$P_{w(i+1)}$	The $i + 1$ measured groundwater pressure value (MPa)
h_i	The borehole depth (m) measured for the *ith* time in perpendicular to coal seam
$h_{(i+1)}$	The borehole depth (m) measured for the $i + 1$ in perpendicular to coal seam
i	The number of tests, i = 1, 2, 3, …, n
H_0	Potentiometric pressure measured in elevation at the top of the water-transmissive zone of the floor aquiclude (m)

hdy The thickness of the water resistance section under potentiometric pressure (m).

The discriminant models of different mining modes under pressure are developed according to the spatial distribution of #9 coal in the apparent water zone, water conduction zone and water resistance zone (between the water transmissive zone and the floor destruction zone). Corresponding water control measures are based on the discriminant models.

4.1.2.4 In-Situ Stress Measurement

The theory of rock-water stress relationship summarizes the complex water inrush problem of coal seam floor into the relationship among rock (water resistance section of the floor aquiclude), water (groundwater pressure water against the floor aquiclude), and stress (mining stress and tectonic stress). The process of coal seam floor water inrush is described as follows.

Floor water inrush results from the combined action of mining pressure and bottom water pressure. The mining pressure causes the redistribution of rock stress field and floor seepage field. The result of the interaction between the two makes the minimum principal stress of the floor rock mass less than the pressure of the potentiometric pressure. As a result, hydrofracturing and expansion occur, and water inrush takes place. The water inrush criterion is:

$$I = \frac{P_w}{\sigma_2}$$

where

I Critical index of water inrush. I is dimensionless. When $I < 1$, no water inrush occurs; when $I > 1$, water burst occurs
P_w Potentiometric pressure exerted on the bottom floor rock mass
σ_2 Minimum principal stress of the floor rock mass.

For a mining panel, the potentiometric pressure exerted on the bottom floor rock mass is generally known. The key problem is to determine the magnitude of the minimum principal stress in the floor rock mass and the mining-induced changes.

A total of four boreholes in the 9103 working face are tested for in-situ stress measurements and 30 test points were arranged. Five holes are tested for stress survey. From the in-situ stress test results, it can be seen that during the advancement of the working face, the minimum principal stress is above 4.0 MPa, which is greater than the maximum confined potentiometric pressure of 2.0 MPa against the coal floor. When $I < 1$, water inrush will not occur under relatively complete formation conditions. As the working surface advances, the slope of the pressure-displacement curve gradually decreases. The upper part of the once water resistance aquiclude is affected by mining. The action of advanced support pressure makes its rigidity decreases and gradually soften. As the working panel advances, the minimum principal stress σ_2 gradually decreases. Although some data points are atypical due to the test accuracy, the overall trend is still a decrease. The minimum principal stress σ_2 remains above 4.0 MPa during the advancement of the working face, which is greater than the maximum potentiometric pressure of 2.58 MPa on the bottom floor rock mass. The distance of the first pressure occurrence is 25–30 m.

4.1.2.5 Grouting Reinforcement and Transformation of Coal Floor

The integrity of the aquiclude and sufficient water blocking performance are favorable conditions for the safety of mining under potentiometric pressure. However, due to the influence of sedimentary conditions and tectonic activities, there are generally weak zones or potential water channels even in the aquiclude. The safety of mining poses a threat under such scenarios.

The purpose of grouting reinforcement and reconstruction of coal seam floor is to achieve the safety of mining under potentiometric pressure by making the following modifications:

- Reduce or eliminate the water-rich layers within the coal seam floor,
- Grout the potential water channels,
- Strengthen the weak section of the aquiclude,
- Improve the integrity and water blocking capacity of the water barrier, and
- Increase locally the thickness of the water resistance layer.

Based on the underground tunnel engineering, grouting holes are mainly placed near the opening cut, around the inflection point, and near the end of mining area. The design of various types of boreholes should serve multiple purposes if possible, while their final function is grouting reinforcement. These boreholes are concentrated in drilling sockets and are arranged in a fan shape. Angled boreholes are preferred to expose as many fractures or fissures as possible in the aquiclude. The fissures are blocked through high-pressure grouting to drive the Ordovician limestone water out of the formation and thin or eliminate the water-transmissive zone. The efforts strengthen the floor aquiclude to ensure safe mining under potentiometric pressure, as shown in Fig. 4.1.

The practice of mining under potentiometric pressure at 9103 working face leads to the following experience in the grouting reinforcement and reconstruction of the weak zone of the water barrier:

- Define the scope of grouting reinforcement and reconstruction based on the initial water transmissive zone theory of the aquiclude in combination with the results of comprehensive geophysical prospecting and coal floor water resistance testing.
- Drill and use inclined boreholes underground taking advantage of the roadway engineering of the working face to reduce the invalid footage of drilling from the ground surface and to overcome the shortcomings of vertical drilling in increasing chances to intercept water transmissive geologic features and improve the grouting effect.

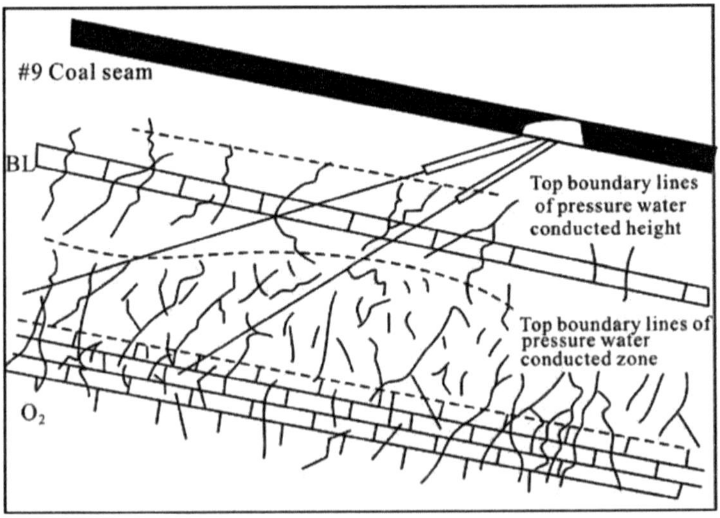

Fig. 4.1 Schematic diagram of reinforcement and reconstruction of the floor aquiclude by grouting (O_2-Ordovician limestone; BL-Benxi bauxite formation)

- Tailor the grouting technology and parameters according to the type of weak zones.

Grouting of the coal floor aquiclude is necessary when groundwater seeps or flows may occur through the aquiclude. High-pressure grouting of low-density slurry is necessary to achieve the purpose of reinforcement where the position of the water transmissive zone is high, and the fracture sizes are small. In implementing grouting transformation in aquicludes that are characterized with the water transmissive zone being close to the mining induced destruction zone and the underlying Ordovician limestone being a strong aquifer, the choice of materials, either low-density slurry, high-density slurry or dual-liquid slurry, relies on the water absorption rate of the boreholes. Partially reforming the conditions of the aquiclude and improving its integrity and water resistance capacity is a robust and an effective measure to ensure the safety of #9 coal mining.

4.1.2.6 Real-Time Monitoring of Water Inrush Risk Zone in Coal Floor

In addition to understanding of the geological and hydrogeological conditions of the working face and determining the water inrush risk zones in the coal floor, the water inrush real time monitoring system is used as another measure to ensure the safe mining of 9103 working face and avoid the accident of water inrush disaster. Sensors for stress, strain, water pressure, and water temperature are installed in the coal seam floor to monitor the height change of the "initial water transmissive zone" that is hydraulically connected to the underlying aquifer and thickness change in the destruction zone of the coal seam floor during the #9 coal mining process. Analysis of changes in these parameters, together with any other indicators, helps forecast the location and place of water inrush occurrences in advance.

The comprehensive analysis of the real-time monitoring data of water pressure, water temperature, stress and strain sensors during the mining of 9103 working face shows no evidence of coal floor water inrush under the potentiometric pressure. The mining takes place safely without an occurrence of water inrush.

Also obtained are the deformation and failure mechanisms of the coal floor rock mass under the current mining conditions of the 9103 working face:

- An overall deformation with vertical tensile stress and tensile strain occurs at 11.7 m (C3t) of the #9 coal floor and below. The rock mass is destroyed.
- Vertical compressive stress and compressive strain and plastic deformation occur at depths between 11.7 and 13.4 m (limestone on top of C2b).
- Changes of vertical stress and strain are minimal at 13.8–15 m (C2b sandstone), suggesting not affected by mining.
- The monitoring data shows that the initial pressure occurs at 27.92 m, and the distance of the following four pressure occurrences are 26.65 m, 20.95 m, 16.69 m and 20.59 m, respectively. The anomalies and appearance of the floor pressure are confirmed by mining of 9103 working face.

Safe recovery of the 9103 trial working face of Dongpang Mine is achieved through the implementation of comprehensive targeted water prevention and control technology and water prevention and control projects. The experience accumulated and lessons learnt during mining of 9103 mining face provide the guidance to coal mining in this mine and other mines with similar geological and hydrogeological conditions.

4.2 Grouting Technology in Thin-Bedded Limestone to Prevent Water Inrushes from Underlying Aquifers in Zhuzhuang Coal Mine of Huaibei Coalfield

4.2.1 Background

The Zhuzhuang Coal Mine of Huaibei Coalfield is in the northwest of Anhui Province. Mining of the first and second levels of coal seams was completed, and the current production level is the third level or lower group at an elevation of approximately −420 m bmsl. Four coal seams are mining targets and they are #3, 4, and 5 coal seams in lower Shihezi Formation and #6 coal seam of Shanxi Formation. The dual-wing layout is used for working face arrangement. The mining adopts the long-wall mining method. Mining by blasting was used before the 1970s and started mechanized comprehensive mining in 1978. Approximately 80% of mining is currently by the mechanized method. The annual output of the mine is approximately 2.1 million tons.

The geological and hydrogeological conditions in Zhuzhuang Mine are complex. Coal mining in the lower group is mainly threatened by the underlying thin-bedded limestone group (#1 limestone to #4 limestone) in the upper Taiyuan Formation and the deep Ordovician limestone aquifer. The thin-bedded limestone of the Taiyuan Formation is characterized with fractures and constitutes fracture karst aquifers. The water yield is very heterogeneous. In large part of the mine, the water yield becomes weak with the increase of mining depth, whereas the potentiometric pressure increases with the increase of the mining depth. In response to the threat of karst water underlying the mining coal seams, Huaibei Mining Group, adopted the following approaches to prevent and control water hazards:

- Mining with potentiometric pressure
- Mining with dewatering and depressurizing
- Targeted or complete underground grouting reinforcement and reconstruction.

Selection of the water hazard control technology is based on analysis and evaluation of the hydrogeological conditions in the specific mining area. Utilization of these methods have successfully completed safe mining of more than 140 working faces, producing more than 100 million tons of coal that is otherwise threatened by the underlying groundwater. These successful practices not only advanced technologies and equipment but also accumulated experience with safe mining of coal reserves that are threatened with pressured underlying aquifers.

The potentiometric pressure of the karst aquifer increases with depth. Hydraulic connection occurs between Taiyuan thin-bedded limestone and the Ordovician limestone where there are vertical water-conductive pathways such as karst collapse column or concentrated fracture zone. Although geophysical techniques including the 3-D seismic method have advanced in recent years, it is still difficult to accurately identify and locate these collapse columns and concentrated fracture zones. There is a potential of water inrush disasters that involve the Ordovician limestone aquifer. Large water inrush incidents cause mine disasters and major losses. For example, on October 24, 1988, a large water inrush occurred when the II617 fully mechanized mining face of Yangzhuang Coal Mine in Zhahe mining area intercepted a vertical water-conductive fault zone. The instantaneous water inflow was 3,153 m^3/h, flooding the second mining level. On February 3, 2013, a water inrush with a maximum instantaneous water flow of approximately 29,000 m^3/h occurred where the 1035 working face of Huaibei Taoyuan Coal Mine encountered a large vertical water-conductive fracture zone. The water inrush incident flooded the entire mine and stopped the mine production. These water hazards show that the traditional technique of transforming the underlying limestone by underground grouting is no longer effective in the prevention and control of floor water inrush in mining of the level III coal seam in Zhuzhuang Coal Mine. Because of complex geologic structures, it is imperative to develop and implement new water control and mitigation technologies.

4.2.2 *Large-Scale Advance Grouting Technology in Transforming Limestone into Water Barrier*

4.2.2.1 Introduction to Technology

In recent years, drilling technology has undergone dramatic development. Near-horizontal directional drilling is getting popular in reinforcing the coal mine floor by the grouting. With this technology, the stabilization section of the borehole can be drilled in the limestone along the bedding planes. The length of the borehole is large and can continuously pass through the limestone. The drilling efficiency and contact area with limestone are high. The directional drilling can expose karst fractures to enhance the effect of grouting transformation to the greatest extent. It is currently the most advanced technology in the transformation of the working face floor.

There are many mine water hazard control methods and each method has its pros and cons. Based on the characteristics of Zhuzhuang Mine, the proposed water hazard control method is to conduct a large-scale advance grouting in the thin-bedded water-conductive limestone, which is underlain by the regional prolific Ordovician limestone aquifer. This is a new concept on water control in mines and involves the following key technologies:

- Directional drilling along bedding planes into #3 limestone (L3) of the Taiyuan Formation
- Scientific placement of boreholes
- Precise control of the drilling trajectory along bedding
- Gradual pressure regulation
- Grout material composed of fly ash and cement at the optimal ratio.

These technologies were applied ahead of mining to the regional water hazard control in the deep level of III63 mining area of Zhuzhuang Mine. The underground drilling and mining practices verified these technologies. The safety effect of water hazard prevention is greatly improved, and the economic benefits are remarkable. This project is the first case study in China of using directional drilling on the ground surface to drill and grout along bedding plane for regional limestone treatment prior to mining. The successful application of this technology is of great significance to the water prevention and control work of similar mines with similar water hazard threats. It helps advance management of coal mine water hazards and enhance the development and innovation of the technical means. It also contributes to theoretical studies on coal seam floor reinforcement and reconstruction.

1. Basic concept

Directional drilling on the ground surface along bedding planes requires precise control technology within target layers. If the drilling direction is under control, the following tasks can be performed more effectively:

- Investigate any suspected water flow points
- Conduct exploration on the fracture network distribution in the thin-bedded limestone that underlies the coal seams
- Perform high-pressure grouting in groundwater pathways to create an effective water blocking plug in the target layer to eliminate the water inrush risk in the target layer and the underlying Ordovician limestone aquifer.

Once the working face is formed, conventional underground drilling is then used to verify the abnormal areas. Supplemental grouting is applied to fractured zones if needed. The combination of surface mitigation and underground verification is effective in ensuring the mining safety.

2. Objective

The objective of these new technologies is to "drill into the targets precisely and grout completely and quickly." Four metrics are used to evaluate the performance: operational process control and mitigation effectiveness. The technical specifications are as follows:

- The drilling length along the bedding plans of the targeted #3 thin-bedded limestone of the Taiyuan Formation is no less than 80%.
- The final pressure at the grouting orifice is no less than 12 MPa. The final pressure may deviate from this specification when water-conductive faults are encountered.
- The water inflow in any single exploratory boreholes drilled into the #3 limestone is no more than 10 m^3/h.
- The total steady water inflow into the working face is no more than 60 m^3/h during the mining process.

The water hazard control measures are successful when all the four above criteria are met.

4.2.2.2 Application

Layout of Boreholes and Construction Structure

The III634 and III636 working faces are used to demonstrate application of the new water hazard control technology. The near-horizontal directional drilling was directed to enter the #3 limestone that underlies the coal seam. As the directional boreholes passed through the bedding planes, fractures or fracture zones in the limestone were investigated. Where fractures were encountered high-pressure grouting was conducted to seal them through the top-down mode. The combination of exploration and grouting construction effectively sealed all intercepted fractures that could act as water inrush channels in the #3 limestone. The #3 limestone was transformed from a fracture aquifer into a water barrier, which effectively blocked water inflow or water inrush from the #3 limestone and the underlying Ordovician limestone aquifer, eliminating water threat to the working faces.

Four drilling groups are arranged for the working faces of III634 and III636. Each drilling group consists of one main borehole and multiple primary and secondary branch boreholes. The borehole arrangement is fan-shaped divergence. The borehole spacing is controlled at 50–80 m. The boreholes are arranged in two ways: one in the strike direction of the stratum and the other in the dip direction of the stratum. The first-level branch hole starts branching out at the appropriate position from the second casing of the main borehole. Similarly, the second-level branch borehole starts branching out at the appropriate position from the first-level branch borehole. These branch boreholes are designed to complete the water control tasks.

The borehole group adopts telescope casing technique with outside casing first, technical casing second and bare borehole on the third stage. The high-pressure grouting is carried out through open bare section of the borehole. Figure 4.2 shows the schematic structure of the directional borehole group.

- The starting borehole diameter is 311 mm with installed casing of 244.5 mm × 8.94 mm. The casing is extended 20 m below the bedrock to isolate the overburden layer.

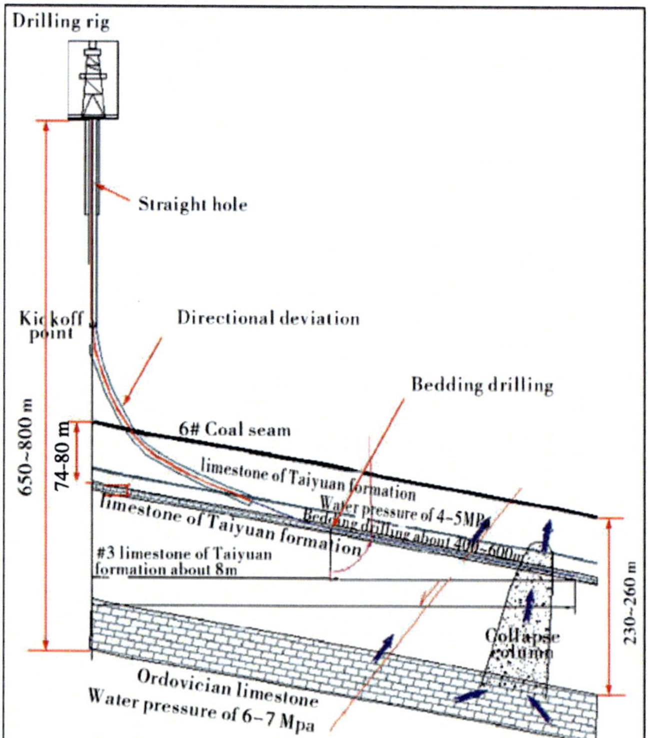

Fig. 4.2 Schematic diagram of drilling structure

- After the first casing, the borehole is drilled at a diameter of 216 mm with installed casing of 177.8 mm × 8.05 mm. The second level casing is extended approximately 30 m below #6 coal seam floor into the top of a mudstone layer.
- Following the second level casing, the borehole is drilled at a diameter of 152 mm. This section is a bare borehole and drilled along the #3 limestone layer.

Drilling and Grouting Operations

To address the water hazards at the working faces of III634 and III636, four drilling groups or fields were constructed. Two drilling groups were arranged along the strike of the strata and they D1 drilling group consisting of one main borehole and five primary branch boreholes and D3 drilling group consisting of one main borehole, five first-level branch boreholes, and one second-level branch borehole. The other two drilling groups are arranged along the dip of the formation and they are named D2 and D4. The D2 drilling group consists of one main borehole and seven first-level branch boreholes. D4 drilling group consists of one main borehole, five first-level branch boreholes, and one second-level branch borehole. These four drilling groups completed a total of 28 boreholes, i.e., four main boreholes, 22 first-level branch boreholes, and two second-level boreholes. They were drilled to reinforce and impermeabilize the #3 limestone for safe mining of the #6 coal seam. The total linear drilling footage is 16,572.15 m, of which 12,206.15 m is along the #3 limestone layer. The total cumulative grouting amount is 42,757 t, consisting of 31,136 t cement and 11,621 t fly ash. Figure 4.3 shows the layout of the boreholes that are drilled into the thin bedded #3 limestone.

- The D1 drilling group was implemented from March 13, 2014 to September 6, 2014, a duration of 178 days. Six boreholes were arranged along the coal seam floor, and a total of 6 boreholes were completed, including one main borehole and five branch boreholes. They are drilled to grout the inner section of III634 working face. The cumulative drilling footage is 3,158.45 m. The cumulative drilling length within the limestone layer is 2,246.45 m. The grouting area is 84,473.8 m^2. The total grouting amount is 11,045 t consisting of 5,152 t cement and 5,893 t fly ash.
- The D2 drilling group started from June 12, 2014 to January 28, 2015. Eight boreholes were designed along the coal seam floor. Eight boreholes were implemented and grouted, including one main borehole and seven first-level branch borehole. The D2 group boreholes are drilled to grout the middle section of the working face III634 and the working face III636. The cumulative drilling footage is 4,906 m. The drilling length in the thin bedded limestone layer is 3694 m. The grouting area is 256,917.6 m^2. The total grout injection volume is 19,406 t, consisting of 14,964 t cement and 4,442 t fly ash.
- The D3 drilling group started from August 14, 2014 to December 14, 2014, a duration of 123 days. These boreholes are for reinforcement of the coal seam floor in the middle section of the III632 working face and the III634 working face. Seven boreholes are design along the coal bed floor. A total of seven grouting boreholes

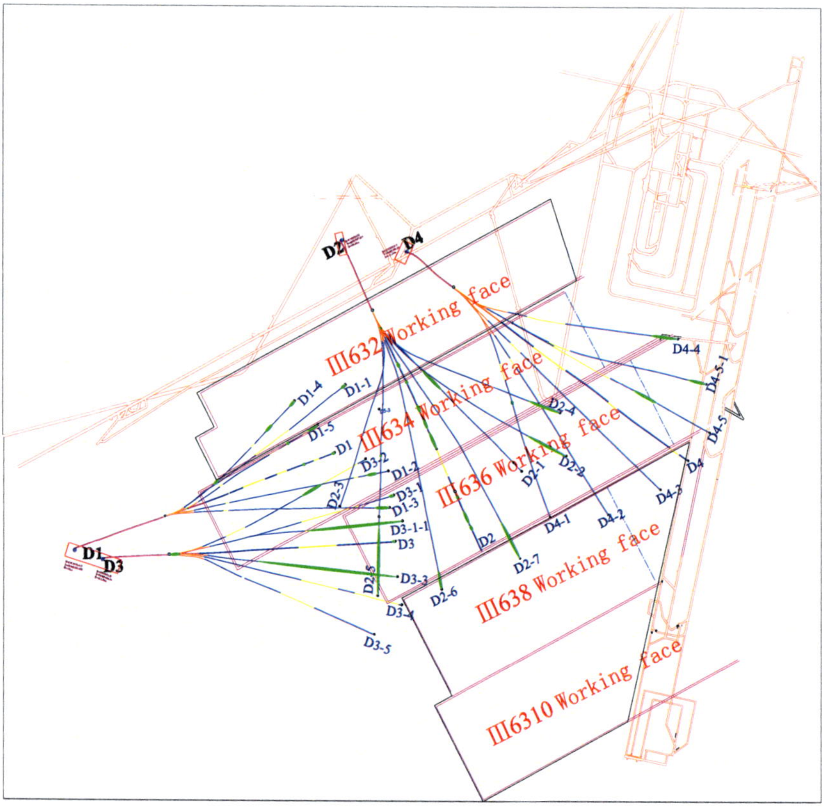

Fig. 4.3 Layout of borehole drilled and grouted into thin bedded limestone

were drilled, including one main borehole, five primary branch boreholes and one secondary branch borehole. The grouting area is 120,539.8 m^2. The cumulative drilling footage is 3,986 m. The drilling length in the limestone layer is 2,957 m. The total grout injection volume is 6,920 t, consisting of 6,413 t cement and 507 t fly ash.

- The D4 drilling group was implemented from October 5, 2014 to April 2, 2015, a duration of 180 days. Seven boreholes were designed along the coal seam floor. A total of seven boreholes were drilled and grouted. The D4 group consists of one main borehole, five primary branch boreholes, and one secondary branch borehole. The D4 drilling unit is drilled to grout the outer section of III634 and III636 working faces. The grouting area is 223,560.6 m^2. The cumulative drilling footage is 4,494.7 m. The drilling in the limestone layer is 3,396.7 m. The total grout injection volume is 5,386 t, consisting of 4,607 t cement and 779 t fly ash.

The drilling workload and grouting workload of each drilling group are shown in Table 4.1.

Table 4.1 Workload summary at III634 and III636 working faces

Borehole number	Depth (m)	Drilling length (m)	Distance to #3 thin-bedded limestone (m)	Section length (m)	Cement (t)	Fly ash (t)	Total grout (t)
D1	996	996	691–996	305	1746	2103	3849
D1-1	988	387	679–988	309	528	1846	2374
D1-2	1,100	499	664–1,100	436	500	50	550
D1-3	1094.45	493.45	665–1,094.45	429.45	1407	538	1945
D1-4	978	377	644–978	334	665	763	1428
D1-5	1,085	433	652–1,085	433	306	593	899
D1 summary	**6,241.45**	**3,185.45**		**2,246.45**	**5,152**	**5,893**	**11,045**
D2	1,123	1,123	620–1,123	503	8,481	3,815	12,296
D2-1	1,036	479	617–1,036	419	1,314	170	1,484
D2-2	1,098	541	631–1,098	467	833	0	833
D2-3	993	436	631–993	362	909	0	909
D2-4	1,044	488	628–1,044	416	306	0	306
D2-5	1,144	587	626–1,144	518	2,782	244	3,026
D2-6	1,180	623	626–1,180	554	130	17	147
D2-7	1,183	626	622–1,183	561	209	216	425
D2 summary	**8,805**	**4,906**		**3,694**	**14,964**	**4,442**	**19,406**
D3	1,062	1124	629–1,062	433	728	507	1,235
D3-1	1,069	511	651–1,069	418	432	0	432
D3-1-1	1,079	341	738–1,079	341	300	0	300
D3-2	1,040	482	637–1,040	403	830	0	830
D3-3	1,067	509	647–1,067	420	230	0	230
D3-4	1,087	529	641–1,087	446	3,643	0	3,643
D3-5	1,048	490	640–1,048	408	250	0	250
D3 summary	**7,452**	**3,986**		**2,869**	**6,413**	**507**	**6,920**
D4	1,182	1,182	670–1,182	512	466	503	969
D4-1	1,082	544	670–1,082	412	1,452	0	1,452
D4-2	1,135	597	675–1,135	460	433	105	538
D4-3	1,171.71	633.71	640–1,171.71	531.71	92	141	233
D4-4	1,059.99	499.99	675–1,059.99	384.99	946	0	946

(continued)

Table 4.1 (continued)

Borehole number	Depth (m)	Drilling length (m)	Distance to #3 thin-bedded limestone (m)	Section length (m)	Cement (t)	Fly ash (t)	Total grout (t)
D4-5	1,190	640	625–1,190	565	180	8	188
D4-5-1	1,148	398	617–1,148	531	1,038	22	1,060
D4 summary	7,968.7	4,494.7		**3396.7**	**4,607**	**779**	**5,386**
Total	30,467.15	16,572.15		**12,206.15**	**31,136**	**11,621**	**42,757**

Performance Evaluation

1. Drilling performance

The average drilling rate in the thin bedded limestone is 83.13%, which meets the technical requirements of the design. The D1 drilling group has a total drilling length of 2,246.45 m and intercepted the #3 limestone of 1,992.45 m. The limestone interception rate is 88.69%. The D2 drilling group is 3,800 m long, and the #3 limestone section is 3,704 m long. The limestone interception rate is 97.47%. The limestone interception rate in the D3 drilling group is relatively low at 61.31% because of the low degree of exploration in this area. The D4 drilling group is 3,262.71 m long, and the intercepted limestone length is 2,667.71 m. The limestone interception rate is 81.76%.

2. Grouting performance

(1) Grouting volume

Based the grouting volume and grouting length, the average unit grouting volume is calculated to 3.51 t/m. The unit grouting volume of individual boreholes reaches 24.45 t/m at D2 borehole, indicating that the fracture density in the #3 limestone section is relatively high. Through high-pressure grouting, the fractures were effectively filled, and the aquifer was impermeabilized to the maximum extent.

Under normal circumstances, the unit grouting volume in the main borehole is the largest. This is because the grouting target of the main borehole is the un-grouted original formation. The limestone formation consists of karst fissures. Therefore, the slurry has a great dispersion range, and the slurry intake is large. The combination of large dispersion range and intake results in large unit grouting volume in the main borehole. The unit grouting volume of the D1 main borehole is 12.62 t/m. The unit grouting volume of the D2 main borehole is 24.45 t/m. The D3 main borehole also conforms to this rule. The unit grouting amount in D3-4 primary branch borehole is larger than that in the D3 main borehole because D3-4 passes through a fault zone, and the grout intake is affected by the fault. The unit grouting volume in the main borehole D4 is not the largest. It may be related to the lower limestone interception rate.

(2) Influence range of grouting

The dispersion range of the slurry is subject to many factors and mainly related to grouting pressure and direction of the dominant fractures in the formation. Karst fissures in the limestone aquifer are typically developed and characterized with extreme heterogeneity and anisotropy. Take the D2 borehole as an example. After the grouting of the D2 borehole is completed, the D2-1 borehole is immediately constructed. The cement fragments are found in the cuttings in D2-1 from 780 m and beyond. The distance from the main borehole is approximately 60 m. There are still cement fragments at 1000 m in D2-1 where the distance to the main borehole exceeds 120 m. In addition, there are also cement fragments in the cuttings of D2-6 drilling at distances from 607 to 880 m where the distance to the main borehole is approximately 50 m. Existence of the cement fragments confirms that the influence range of grouting in the D2 borehole reaches 60–120 m. Based on observations in all the boreholes, the average grouting influence range of at III3634 and III3636 working faces is on the order of 60–80 m.

(3) Groundwater level in boreholes

The groundwater level plays a guidance role in the grouting operations of the boreholes. The development degree of the fissure can be preliminarily estimated from analysis of the groundwater level and results from the pressure water tests. These data can indicate whether there is a connection between adjacent aquifers, whether there is any abnormal water pressure area, and whether there is a fault or geological structures such as collapse columns. The following summarizes the analysis of the groundwater levels observed in each borehole in the III3634 and III3636 working faces.

The depth to groundwater in the Taiyuan limestone ranges from 48 to 55 m in III634 and III636 working faces. The groundwater elevation ranges from −15 to −25 m bmsl. The groundwater level in majority of the boreholes is consistent with that in the Taiyuan limestone, indicating that the target of the boreholes is accurate and reliable. Drilling operations meet the design requirements. However, some of the groundwater levels are abnormal and are often affected by many factors:

- If circulation fluid loss does not occur during drilling of a borehole, the borehole is not considered to be connected to the aquifer. The groundwater level represents stagnant water. This type of boreholes does not have practical values, and the groundwater level does not represent the active groundwater system.
- When the borehole is affected by the geological structures, the groundwater level can be atypical. For example, boreholes D3-3, D3-4, and D3-5 intercepted faults, the measured groundwater level is lower than the typical groundwater level of the Taiyuan limestone.
- When the borehole is affected by water discharge occurrences at the working face, the groundwater level can be lower than the typical groundwater level of the Taiyuan limestone. For example, a water inflow incident occurred during the advancing process on the III6213 working face on October 19, 2014. The initial water flow rate was 15 m^3/h. The water flow rate increased to 120 m^3/h

on October 21, 2014. The groundwater level in the grouting section of the D4 borehole is affected between 528 and 632 m. The depth to the groundwater is approximately 87 m.

- When the borehole is affected by mining, the groundwater level can be lower than the typical groundwater level of the Taiyuan limestone. An example is the groundwater level of D4-4 borehole that is affected by the mining of III520 working face. The depth to groundwater is approximately 158.90 m. In addition, the groundwater level should be considered when designing initial grouting composition.

Verification of Grouting Effect

1. Evaluation by geophysical surveys

In March 2015, Xi'an University of Science and Technology was contracted to conduct an audio electrical imaging to investigate the water richness of the floor rock layer at III634 working face. The apparent conductivity value of the rock layer in the section between 40 and 80 m varies from 1.0 to 8.6 Siemens (S)/m. The average value is 4.6 S/m, and the standard deviation is 1.3 S/m. Two anomalies with apparent conductivity values greater than 6 S/m were identified in this interval, and they are numbered anomaly B-1 and anomaly B-2.

The anomalies B-1 and B-2 are controlled by boreholes III634-4, III634-6, III634-2 and Fang 7, respectively. Of the four boreholes, both III634-2 and Fang 7 intercepted #2 and #3 limestone layers, and no groundwater was encountered. There was no water at boreholes III634-4 and III634-6 in #2 limestone, and the flow rate in #3 limestone is 4 m^3/h. These observations indicate that the area has obvious impact by the ground grouting.

2. Verification and evaluation by underground drilling

The design requires that a total of 23 verification boreholes (13 in the machine lane and 10 in the wind lane) be drilled to evaluate the grouting effect and drain any residual water if present. Of them, five boreholes in the machine lane were drilled into #2 limestone, and the water inflow was 3–20 m^3/h. These boreholes are used to dewater #1 and #2 limestone. The other 19 boreholes did not encountered water in #2 limestone, and they were continued to advance into #3 limestone. Five of the 19 boreholes encountered groundwater in #3 limestone with the maximum water flow rate of 6 m^3/h. The distance between the end points of the drilled boreholes varies from 50 to 100 m. The borehole spacing meets the verification design requirements.

(1) Verification of blind spots by directional drilling on the ground

The borehole influence radius is assumed to approximately 35 m. The III632 working face was divided into four areas #1, #2, #3, and #4. Area #1 is verified with the ground boreholes, where areas #2, #3 and #4 are verified with three boreholes III634-4, III634, and III634-1. Two of the three boreholes did not encounter groundwater in #3 limestone, while the water flow in the other borehole is merely 3 m^3/h.

(2) Verification of areas not controlled by drilling and grouting

There are areas where the ground drilling failed to follow the bedding interval after entering #3 limestone. These areas are not apparently controlled by drilling and grouting. Three areas fall into this category. Five exploratory and dewatering boreholes, Fang 9, Fang 10, Fang 11, Fang 12 and Fang 13 were drilled for verification. All the five boreholes are dry.

(3) Verification of water-rich areas

The section between D2-6 and the D2-7 branch boreholes is a high water-yield area in #3 limestone. Drilling of D2 main borehole identified high frequency of fractures. One exploratory and dewatering borehole Fang 4 was drilled for verification, and no groundwater was encountered.

(4) Verification of fault zones

Construction of the machine and air circulation lanes revealed that faults are mainly concentrated in the middle of the working face. Extensive investigations were conducted in this area, and many faults with large displacements were encountered. This area is verified with seven boreholes III634-2, III634-3, III634-8, Fang 4, Fang 5, Fang 6, and Fang 7. All these boreholes were drilled into #3 limestone and none encountered water, which verifies that the fractures were filled by the ground grouting.

3. Evaluation by underground artesian groundwater testing

The underground artesian water releasing test was conducted at borehole Fang 7. Two borehole Fang 1 and Fang 2 were used as observation wells. After 20 days of artesian discharge, the potentiometric pressure at Fang 1 was reduced from 2.2 to 1.9 MPa, and the water at Fang 2 was reduced from 2 to 1.7 MPa. The groundwater level in #2 limestone had a drawdown of 30 m, indicating that the #3 limestone was fully impermeabilized by ground grouting. The hydraulic connection between #2 limestone and #3 limestone was cut off, and there was no hydraulic connection between them. In addition, the static potentiometric pressure at boreholes drilled in the #3 limestone is approximately 2.5 MPa, which is lower than the pressure of 3.98–5.29 MPa that acts on the coal floor. These data indicate that there is no hydraulic communication between the #3 limestone, and the ground grouting is effective.

4. Comparison of groundwater flow between different working faces within the same hydrogeological unit

The III631 working face and the III634 working face are in the same hydrogeological unit. The III631 working face is not treated by surface directional drilling and grouting. A total of 35 boreholes are drilled underground in III631 working face. All the 35 boreholes encountered groundwater in #3 limestone with the maximum water flow at 400 m^3/h. Two boreholes intercepted groundwater in #2 limestone with the maximum flow rate at 80 m^3/h. This comparison clearly indicates that the ground grouting transformed the limestone from an aquifer to an aquitard.

4.3 Utilization of the Top of the Ordovician Limestone in the Sangshuping Mine of Hancheng and the Underground Grouting Transformation Technology

4.3.1 Mine Background

Sangshuping Coal Mine is a large-scale mining enterprise affiliated to Shaanxi Hancheng Mining Co., Ltd. It is in the northern part of the Hancheng mining area in the Weibei Coalfield, which is on the west bank of the Yellow River. The main coal seam is #3 coal seam. Due to the high methane content and high risk of release in this coal seam, it is decided to extract the coal seam in advance. The original design of Sangshuping Coal Mine was to mine #2 coal seam, which is overlain by #3 coal seam. Mining of #2 coal seam is challenged with its relatively thin thickness and vulnerability to methane outburst. It is decided to first mine #11 coal seam, which is underlain by #3 coal seam. Mining of #3 coal seam follows mining of #11 coal seam. The distance between the bottom of the #11 coal seam and the underlying Ordovician limestone is relatively small. The water inrush coefficient does not meet the safe mining requirements as stipulated in the Coal Mine Water Control Regulations (2009). There is a risk of water inrush from the Ordovician limestone during mining.

The 3105 working face is used to illustrate the grouting and utilization technology as applied to the top of the Ordovician limestone. The size of the 3105 working face is 180 × 713 m. The coal mining height is 2.60 m. The coal seam floor elevation varies from 256 to 328 m amsl. The mining method is uphill mining. The geological reserve of the working face is approximately 476,200 tons.

The distance from #11 coal to the Ordovician limestone is between 16.5 and 21.5 m with an average of approximately 17.0 m, as shown in Fig. 4.4. Groundwater level of the Ordovician limestone is 375 m amsl. The hydrostatic pressure exerted on the aquiclude in the floor of the working face ranges from 0.64 to 1.36 MPa. The water inrush coefficient of the Ordovician limestone in the 3105 working face varies from 0.042 to 0.72 MPa/m. The contour of the water inrush coefficient is shown in Fig. 4.5. The lithology of #11 coal seam floor includes conglomerate and mudstone (aluminum mudstone and argillaceous mudstone).

The structure of the 3105 working face is relatively simple. The contours of the coal floor elevation indicate that the working face is characterized with a monoclinic structure. The coal seam inclines approximately 5.3°. No faults and collapse columns with large displacement are reported.

Although the structure of the 3105 working face is relatively simple, the risk of water inrush is still present because of thin aquiclude below the coal seam, high potentiometric pressure of the Ordovician limestone, and mining induced damage to the aquiclude.

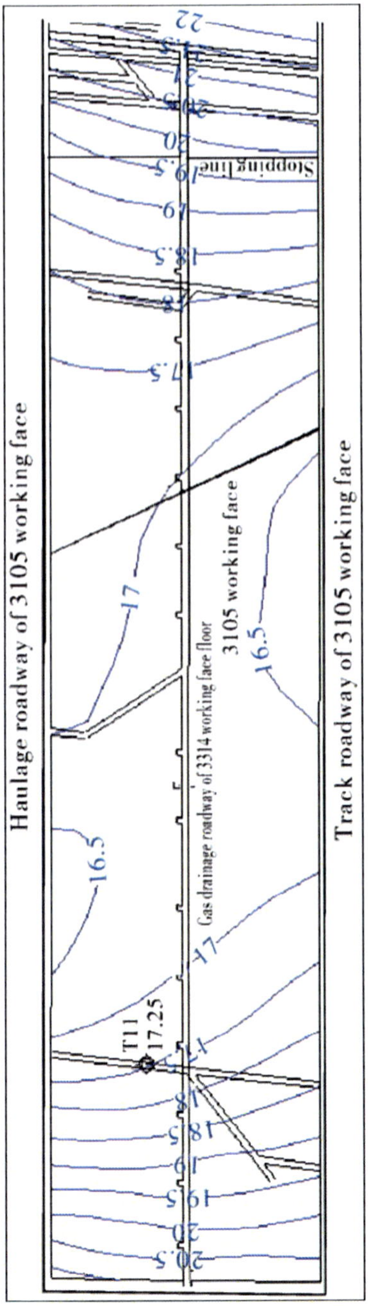

Fig. 4.4 Distance from #11 coal seam to the Ordovician limestone in 3105 working face

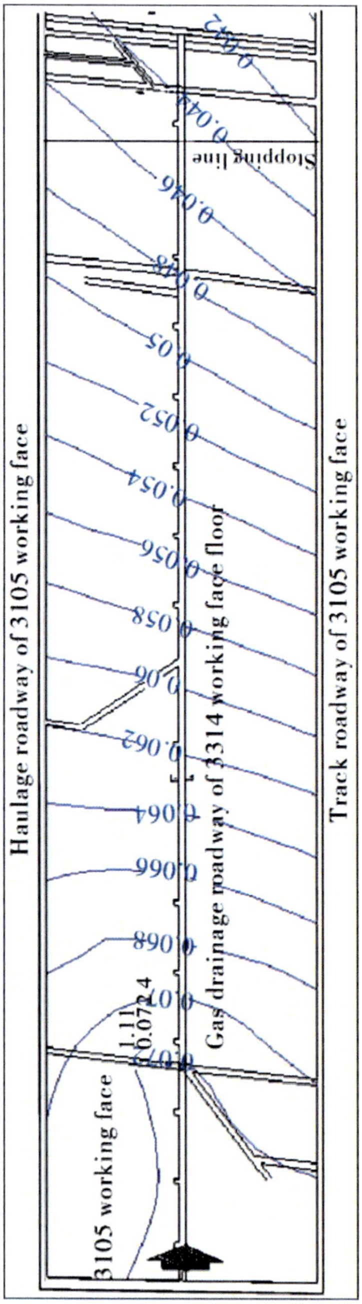

Fig. 4.5 Distribution of water inrush coefficient in 3105 working face

4.3.2 *Utilization of Top of the Ordovician Limestone and Underground Grouting Transformation Technology*

4.3.2.1 Technical Approach to Control Water Hazards from the Ordovician Limestone

Sangshuping Coal Mine 3105 working face is characterized with short distance between the #11 coal seam and the Ordovician limestone, thick aquiclude as the hydraulic barrier, and the critical water inrush coefficient being greater than 0.06 MPa/m. The overall technical approach for prevention and control of the Ordovician groundwater is to utilize the top of the Ordovician limestone. If the top of the Ordovician limestone is absent of geologic structures and karst fractures, this interval of the Ordovician limestone can be directly used as a water resistance layer. If the geologic structure in the top of the Ordovician limestone is complex and karst fractures are well developed with strong water richness, this interval of the Ordovician limestone can be impermeabilized by grouting, resulting in an additional layer of hydraulic barrier. Mining of the working face can be completed by making use of a top section of the Ordovician limestone.

Mining of the working face can only proceed after the water inrush threat from the Ordovician limestone is mitigated. The specific implementation plan is as follows:

- Use directional drilling to explore the geologic structures and water richness in the top of the limestone layer under the working face.
- Use the directional boreholes as grouting holes where the geologic structures are encountered, and the water yield is high.
- Once the roadway system is in place for the working face, conduct electrical resistivity imaging survey along the roadways, at the open cuts and ends of the working face to investigate the groundwater conditions of the rock layer underlying the roadway.
- Conduct audio electrical imaging to investigate the groundwater conditions in the formations underlying the working face.
- Conduct radio frequency survey to investigate the geologic structure within the coal seam.

4.3.2.2 Underground Drilling Engineering

Selection of Key Parameters

1. Borehole spacing

The grout hole spacing in directional drilling for impermeabilization of the underlying aquiclude is determined by the dispersion range of each grout hole. Based on empirical experience with grouting project in China, the single-hole grouting dispersion range is between 25 and 30 m. Therefore, the drilling interval is determined to be approximately 40 m when the directional drilling is conducted.

2. Target formation of drilling

The choice of the target formation for drilling is based on the difficulty of drilling and whether the thickness of the underlying water resistance layer water-insulating layer after completion of the grouting meets the requirements of the safe water inrush coefficient as stimulated in the Coal Mine Water Control Regulations (2009). The water resistance layer below #11 coal seam is thin in Sangshuping Coal Mine. The required thickness of the water resistance layer is back-calculated according to the provisions of the safe water inrush coefficient. The calculated water inrush coefficient cannot meet the requirement even if the entire floor rock formations are assumed to be effective water resistance. Grouting reconstruction layer is determined to be at the top of the Ordovician limestone, and the grouting reconstruction layer is 10–15 ft into the Ordovician limestone.

3. Borehole angle

Four main factors are considered in the design of borehole inclination: borehole cutting discharge, construction of directional slanting section, borehole drilling difficulty, and borehole maintenance. The hole inclination is designed to be between -15 and $-20°$ within 3105 working face.

4. Borehole structure

The directional drilling for grouting the formations that underly the coal seam purpose is composed of casing section, rotary drilling section, directional oblique section, and directional stable oblique section. The number of casing levels is determined by the site-specific geological and hydrogeological conditions. The general borehole structure is shown in Fig. 4.6. The drilling structure should be tailored to the actual geological and hydrogeological conditions when applying to other mines. In this construction, the hole formation is relatively stable, and the water richness of the formation through which the drilling starts is weak. Most of the boreholes adopt the two-level casing structure. The tertiary casing is used where underlying rock formation is severely fractured. The casing consolidation length should not be less than 50 m. And, the distal end of the casing should not be less than 20 m from the bottom of the coal seam.

Analysis of Investigation Results

Six drilling platforms consisting of a total of 15 directional boreholes are designed for 3105 working face. In addition, one exploration borehole is completed at the open cut. The total drilling footage is 7424 m. The drilling started in March 2013, and all construction was completed in March 2014. The plan view of the actual drilling trajectory of directional drilling is shown in Fig. 4.7. Hydrogeological observations

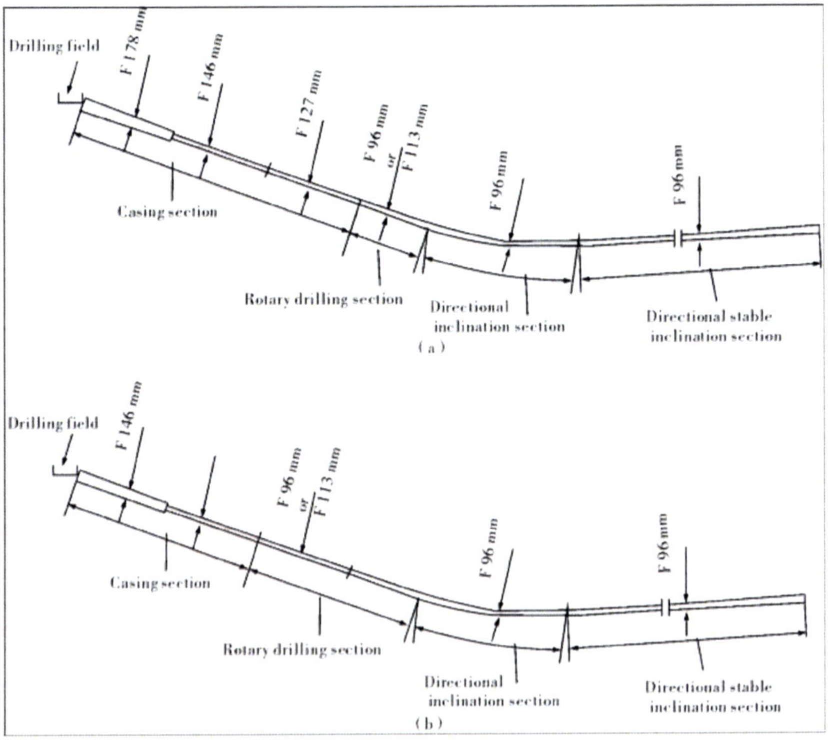

Fig. 4.6 Schematic diagram of borehole structure

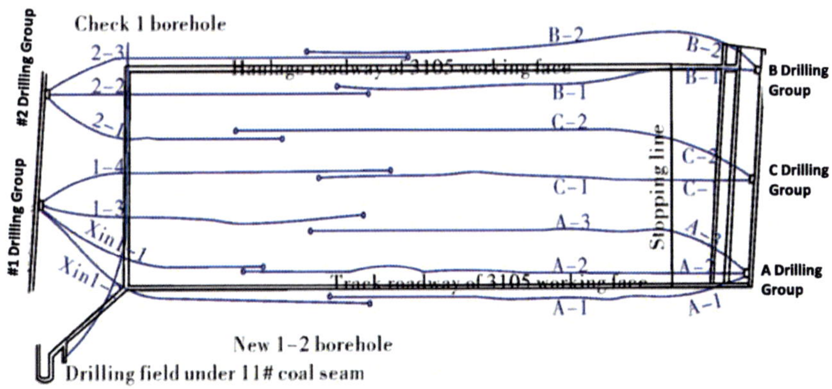

Fig. 4.7 Layout of directional boreholes

were made during the drilling process, and the water injection test was conducted at termination of the boreholes. The results are summarized in Table 4.2.

In general, no significant groundwater was encountered during drilling. No large circulation liquid loss was observed. The amount of water produced by the drilling was small after entering the Ordovician limestone in both the inclined section of the coal floor and the stable inclined section. The water intake rate of each borehole in the water injection test is very small. The unit water intake rate of the rock layer in the boreholes was less than 0.001 L/min m m. These data show that the fractures in the top of the Ordovician limestone are not developed, which is beneficial to mining under potentiometric pressure.

Table 4.2 Hydrogeological observation and water injection test results

Borehole	Hydrological observation	Water intake rate (L/min m m)
1-1	Approximately flow rate of 2 m^3/h prior to encountering the Ordovician limestone	
New 1-1	Water flow rate of 5 m^3/h at depth of 117 m, potentiometric pressure at 0.12 MPa; typical potentiometric pressure in the Ordovician limestone at 0.89 MPa	0.00079
1-2	No water	
New 1-2	Flow rate of 2.1 m^3/h at depth of 110 m, potentiometric pressure at 0.12 MPa; common water pressure in Ordovician limestone at 0.89 MPa	0.00082
1-3	Water flow at 4.3 m^3/h at depth of 113 m, water pressure at 0.12 MPa; typical potentiometric pressure in the Ordovician limestone at 0.89 MPa	0.00082
1-4	Water flow rate of 2.1 m^3/h at depth of 99 m, potentiometric pressure at 0.12 MPa; typical potentiometric pressure in the Ordovician limestone at 0.89 MPa	0.00082
2-1	No water	0.00082
2-2	No water	0.00082
2-3	No water	0.00084
A-1	No water	0.00079
A-2	No water	0.00077
A-3	No water	0.00078
B-1	No water	0.00074
New B-2	No water	0.00078
C-1	No water	0.00075
C-2	No water	0.00074

4.3.2.3 Grouting Engineering

The target layer for grouting is between 10 and 15 m in the top of the Ordovician limestone. The top down method is used for grouting with multiple steps. Grouting stops at the grouting orifice. A combination of continuous and intermittent grouting is adopted. When the water flow in the borehole is less than 30 m^3/h, the grouting started every 100 m of advancement in the Ordovician limestone. If the water flow in the borehole is greater than 30 m^3/h, the drilling stopped at any time to start grouting. If the drilling process does not encounter groundwater and the circulation fluid loss is normal, there is not specific requirement on length of the grouting section.

The #32.5 ordinary Portland cement is used for grouting. Its quality meets the national GB175-92 standards. Moisture and agglomerated cement or expired cement is not allowed. The water used for cement mixing must meet the national quality standards for concrete mixing. In particular, the SO_4^{-2} concentration should be <1%, and pH should be >4.

The grouting effectiveness of each grout hole is evaluated with two metrics: grouting pressure and grouting volume. In actual operations, the grouting can be stopped when both metrics are met at the same time. According to the hydrogeological conditions, requirements of these two metrics are: grouting pressure at the orifice reaches 3 MPa, and the unit water intake rate of the borehole is less than 0.005 L/min m m.

The directional drilling of the 3105 working face did not encounter significant groundwater inflows. Grouting was performed at completion of each borehole. The grouting follows strictly the requirements. A total of approximately 60 tons of cement was injected into the 3105 working face.

4.3.2.4 Underground Geophysical Prospecting

Geophysical Prospecting Layout

At completion of the roadways excavated for the working face, electrical resistivity imaging was used in the transportation lanes, track lanes, and cutouts, whereas audio electric imaging and radio wave imaging surveys were carried out in the working face area.

Analysis of Exploration Results

1. Electrical resistivity imaging

Figure 4.8 shows the results of the electrical resistivity imaging in the transport lane, track lane, and cutouts. No significant geophysical anomalies were identified. Integration of the electrical resistivity imaging results with those from the near-horizontal directional drilling and grouting the anomalies rules out the possibility

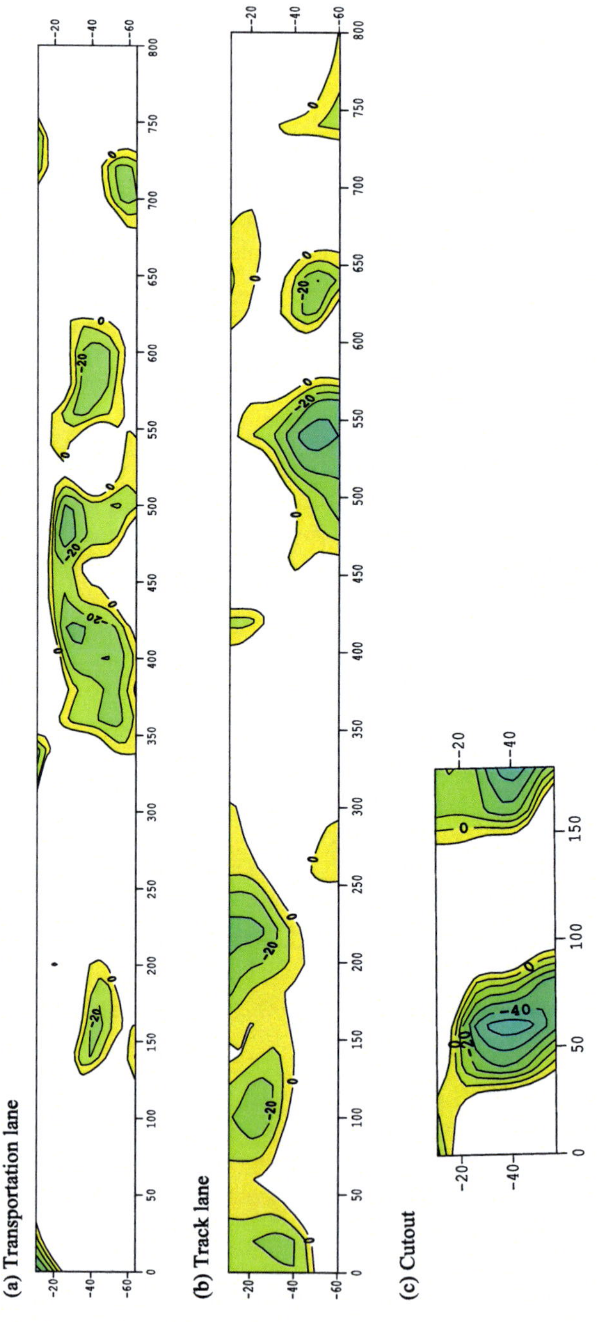

Fig. 4.8 Cross-sectional view of apparent resistivity measured by electrical resistivity imaging

of the groundwater flow from the Ordovician limestone. Geophysical anomalies identified on the surveys at the open cuts were investigated by exploratory boreholes. However, the depth and extent of the intrusive investigation were limited. There is still a possibility of atypical groundwater conditions in these anomalous areas.

2. Audio electric imaging

The results of audio electric imaging are shown in Fig. 4.9. The resistivity lows are shown in the plan view in the range of 0–30 and 30–60 m. The results indicate that a large and high amplitude anomaly near the cutout. No hydrogeological anomalies were identified from verification boreholes. However, the depth and extent of the borehole investigation were limited. There is still a possibility of atypical groundwater conditions in these anomalous areas. The geophysical anomalies identified within the working face are small. All these anomalies were investigated by directional drilling. No atypical hydrogeological conditions were observed.

3. Radio wave imaging

The radio wave imaging survey results in interpretation of three anomalies, as shown in Fig. 4.10. Based on analysis of the exposed tunnel cuts in the transportation lanes and track lanes, the anomalies are likely all caused by large changes in the coal thickness, high gangue content, and twisted geological structures. Within the working face, no collapse columns and large faults were reported.

4.3.2.5 Verification Project

The geophysical exploration results suggest a possibility of anomalous hydrogeological conditions near the cutout. Based on to the underground conditions, one underground borehole, Inspection-1, was drilled into the anomalous area for verification.

In course of drilling borehole Inspection-1, water flow occurred at advancement of 52 m and 129 m with the flow rate at approximately 3.2 m^3/h and 4.5 m^3/h, respectively. The potentiometric pressure measured at completion of the borehole was 0.19 MPa. Water injection test was conducted, and the unit water intake rate was 0.00178 L/min m m. Because the amount of water inflow from the borehole is relatively small, and the potentiometric pressure of the other borehole is low, and the potentiometric pressure deviates from that in the Ordovician limestone by approximately 1 MPa, it was concluded that this area is hydrogeologically isolated with stagnant groundwater. The possibility of water hazards from the Ordovician limestone can reasonably be rules out.

4.3.2.6 Effectiveness Analysis

The comprehensive analysis of the post-grouting investigation results indicates that the top 10–15 m of the Ordovician limestone can be used the added aquiclude for

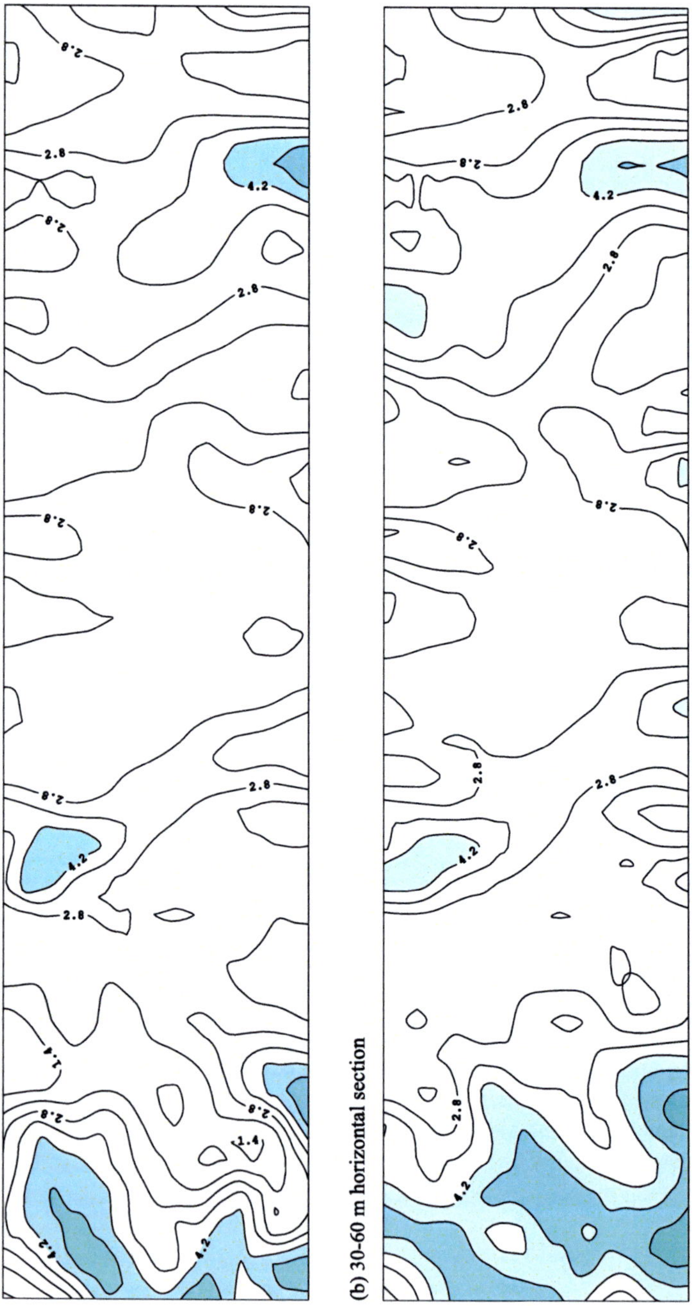

(a) 0-30 m horizontal section

(b) 30-60 m horizontal section

Fig. 4.9 Plan view showing audio electric imaging anomalies

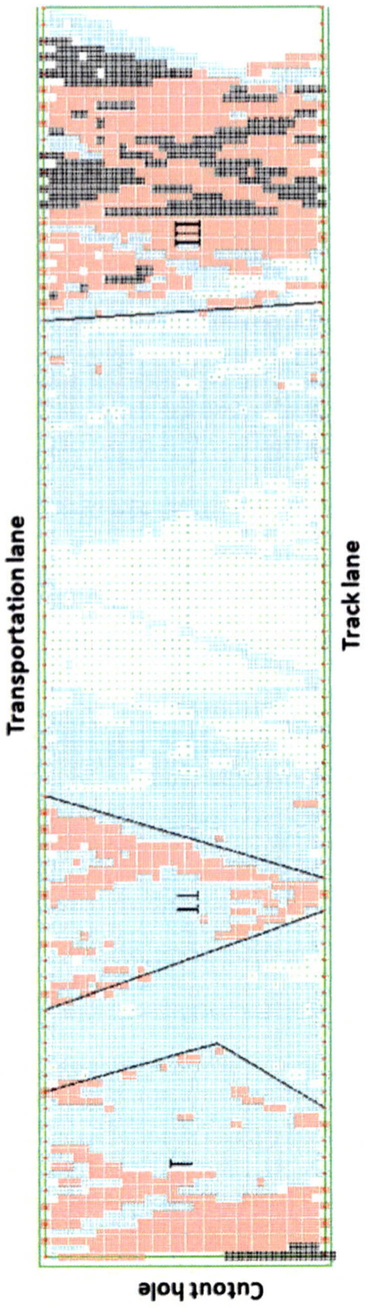

Fig. 4.10 Radio wave imaging anomalies

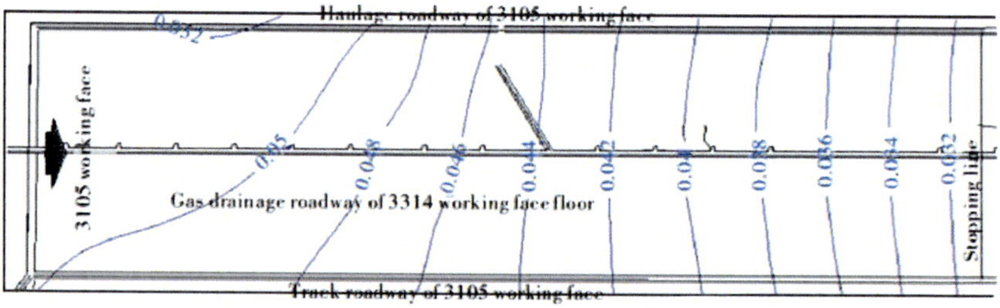

Fig. 4.11 Revised water inrush coefficients at 3105 working face

3105 working face. The karst fractures are not developed, and the water richness is weak. Adding the extra layer of approximately 10 m to the thickness of the effective aquiclude leads to revised water inrush coefficient distribution on the working face, as shown in Fig. 4.11. The revised water inrush coefficient is between 0.03 and 0.052 MPa/m. The 3105 working face was safely mined without water inrush incidents from the underlying aquifer.

The underground directional drilling-dominated grouting technology is a breakthrough in the mine water hazard control, which makes it possible to utilize the top of the Ordovician limestone as additional aquiclude to prevent water inrush from the underlying Ordovician limestone. This technology has the potential to be applied to other working faces in the Hancheng mining area and other coal mines that are threatened by the Ordovician limestone.

4.4 Emergency Mitigation Technology of Water Inrush Induced Mine Flooding in Luotuoshan Coal Mine in Wuhai Energy Co., Ltd.

4.4.1 Overview

4.4.1.1 Overview of Water Inrush Incident

At 7:29 on March 1, 2010, a huge water inrush disaster occurred through a collapse column in the 16th coal seam air return lane of Luotuoshan Coal Mine in Wuhai Energy Co., Ltd. Based on the flooded volume, the maximum water inflow was estimated to be 65,000 m^3/h. 31 people were trapped underground. The water inrush source is the underlying Ordovician limestone aquifer with a potentiometric pressure of approximately 2.3 MPa.

In response to the water inrush, Luotuoshan Coal Mine implemented an emergency plan at 7:31 on March 1, 2010 to evacuate the underground workers. The principals of Shenhua Group and Wuhai Energy Corporation quickly mobilized equipment and personnel and started the emergency rescue.

At 19:00 on March 1, 2010, then Vice Premier Zhang Dejiang arrived the scene of the accident, listened to the rescue plan, and made eight important instructions on the rescue work. Director of the State Administration of Safety Supervision (Luo Lin), Deputy Director of the State Administration of Safety Supervision and Director of the State Administration of Coal Mine Safety (Zhao Tiehao), Deputy Director and Chief Engineer of the State Administration of Coal Mine Safety (Wang Shuhe) and other working group experts arrived at the scene of the accident in the afternoon of March 1. The experts immediately evaluated the accident situation. Working with the Inner Mongolia Autonomous Region Government and Shenhua Group, they studied the rescue procedures and operations as well as restoration process.

The rescue headquarters determined three engineering measures that were implemented currently to rescue the trapped miners underground. The three measures are: construction of communication and supply paths by drilling, dewatering to reduce the water level in the flooded mine, and plugging the water inrush point. The drilling task was the top priority. The emergency rescue work was carried out by organizing and mobilizing the personnel and equipment of multiple organizations.

The drilling was designed and conducted in three areas: 0901 working face conveyor belt along the open cut (Area A), air return lane (Area B) and #16 coal seam air return lane (Area C). In Areas A and B, drilling was targeted the underground tunnels to check on lives and provide oxygen and nutrients for trapped people in the underground. In Area C, the drilling was to grout and seal off the #16 coal seam air return lane to cut off the connection with the water source. This helped emergency dewatering to create conditions to implement underground rescue. After no life indication was found in Areas A and B tunnels, the focus of the drilling was shifted to the water blocking and dewatering. The dewatering effectiveness and underground rescue execution rely on the successful blockage of the #16 coal seam air return lane in Area C.

The emergency water blocking project started on March 5 in Area C and ended on April 28, 2010. It lasted 54 days. Eight boreholes and one Ordovician limestone observation well were drilled. The total drilling footage is 3947.94 m. The eight boreholes were used as grouting holes. The grouting materials included 75.18 m^3 stones and sand, 322.42 m^3 double-slurry consisting of cement and water glass, 8444.97 m^3 slurry, and 6502 tons #42.5 cement.

The pumping test that was conducted on April 5, 2019 confirmed that the connection with the water inrush source is blocked, and conditions are created for dewatering. The dewatering started on April 8. On April 14, the water level in the mine is lowered adequately so that rescue team could enter the mine safely. On May 13, the bodies of 31 victims were found, which signaled the end of the rescue effort. As of 12:00 on May 10th, the emergency response pumped a total of 1.44 million m^3 of water.

The emergency water plug that was constructed in Area C was tested for effectiveness. Field measurements indicate that the residual flow passing through the plug was 48 m^3/h, which suggests that the sealing effect was sufficient. The structural stable of the water plug and poor permeable nature ensured that the dewatering lowered the water level so quickly and the rescue was executed. However, from engineering technical perspective, the water inrush pathway, i.e., the collapse column, must be sealed to eliminate future water inrush accidents during mining and to ensure security of water supply.

The technical approach of sealing the collapse column involved in engineering measures on the ground surface. Exploratory and grouting boreholes were drilled on the ground surface to investigate the hydrogeological conditions and pump grout to seal the collapse column and the surrounding fractures. The boreholes were advanced to the top of the Ordovician limestone aquifer around the #16 coal seam air return lane. The slurry filled in fractures and disconnect the hydraulic connections between the Ordovician limestone and the mining area. The project started on July 20, 2010 and was completed on November 12, 2010 with a duration of 112 days. Four main grouting boreholes, T1, Z1, Z2, and Z3, 12 branch boreholes, and one grouting inspection borehole (borehole JC) were completed. The drilling footage is 3179.6 m, and the grouted materials include 360.4 m^3 of aggregates and 33,376.27 m^3 of cement slurry.

At the completion of the drilling and grouting project, no groundwater flow flew through the original water inrush point. The results in the inspection borehole JC and the grouting effect analysis provide other lines of evidence that the water-conductive collapse column that caused the water inrush disaster was completely sealed off. The source of causing water inrush was eliminated.

The background and mine geological and hydrogeological conditions are described in the following sections with the emphasis on the technical approaches of investigating the water inrush structures, the key construction techniques and processes applied to the project, as well as the project quality evaluation and water blocking effect inspection technology. The methods and technologies can provide technical reference for mines with similar hydrogeological conditions and types of water inrush disasters.

4.4.1.2 Mine Background

Luotuoshan Coal Mine is a newly built mine, which belongs to Shenhua Group Wuhai Energy Co., Ltd. The designed production capacity of the mine is 1.5 million tons per year. The mine field is 10 km long from north to south, 4–5 km wide from east to west, and encompasses an area of approximately 38.7 km^2. The coal resource reserves in the mine field are 418 million tons, and the recoverable reserves are 247 million tons.

The design of the mine is developed in two mining levels with both inclined shaft and vertical shaft. The first level elevation is 870 m amsl, whereas the second level elevation is 920 m amsl. The mine mainly extracts #9 and 16# coal seams. The current production level is 860 m amsl and divided into 6 mining areas, namely 101, 102, 103, 104, 105, 106. The first mining area is the mining area 101, which is in the northern part of the mine field. The first mining face is arranged in the #9-2 coal seam.

4.4.1.3 Mine Geology and Hydrogeology

Luotuoshan Coal Mine belongs to Zhuozishan coal field. Stratigraphically, it is part of Haibowan region of Zhuozishan–Helan Mountain division of North China stratigraphic area. The main coal-bearing strata are the Upper Carboniferous Taiyuan Formation (C2t) and the Lower Permian Shanxi Formation (P1s).

Topography

Luotuoshan Coal Mine is in the southeast of Zhuozishan coal field. The topography of the mine area is generally high on the east and west sides and relatively low in the middle. The elevation is generally between 1260 and 1225 m amsl. The terrain is relatively flat with low hills. The vegetation is scarce. It is characterized with erosive plateau desert-semi-desert hilly landforms.

Lithology

This area is overlain by Cenozoic strata. The lithology from old to new are:

- Ordovician Lower System:
  - Sandaokan Formation (O11)
  - Luotuoshan Formation (O12).
- Carboniferous Upper System: The thickness of the carboniferous strata varies from 40 to 100 m.
  - Benxi Formation (C2b)
  - Taiyuan Formation (C2t).
- Lower Permian System: The thickness of the Permian strata ranges from 236 to 897 m.
  - Shanxi Formation (P1s)
  - Lower Shihezi Formation (P1x)
  - Upper Permian Upper Shihezi Formation (P2s)
  - Shiqianfeng Formation (P2sh).
- Tertiary Formation (R)
- Quaternary Formation (Q), which has a thickness of up to 22 m.

Geologic Structures

The Zhuozishan coal field is located on the edge of the Ordos group in western Inner Mongolia. It is at the northern end of the composite geologic structure ridge that is formed by the Qi, Lv, and He mountains. The coal seams are in the Carboniferous and Permian formations. Structurally, it consists of an anticline and a large compression-shear fault group oriented in the north–south direction. From east to west there are Qianlishan reverse fault, Kewusu-Moli reverse fault, Qipanjing reverse fault, Aerbasi reverse fault, Gangdeer-Xilaifeng reverse fault, Zhuozishan anticline, and Gangdeer anticline.

Luotuoshan Coal Mine is in the middle section of the west wing of the Zhuozishan anticline in western Inner Mongolia. The structure of the whole area is dominated by monoclinic strata dipping westward. The stratigraphic strike of the rock formations is mainly NW20–15°. The dip angle ranges from 5 to 15°. Only in the southwestern part of the mining area is the S1 syncline that is affected by the Gangdeer reverse fault. The F1 reverse fault is tangent to the north of the #17 exploration line from the Dilibangwusu area. There are also small folds associated with normal faults between F62 and F36. In addition, they are all monoclinic layers dipping to the west. There are 24 faults of various sizes in the mine field. The faults can be divided into two groups. One group consists of near-south-trending reverse faults, which form the western natural boundary of the mining area, i.e., the F1 reverse fault and the Gangdeer reverse fault. The other group consists of near-east-trending normal faults. Some faults are long and have large displacements.

Hydrogeology

The Zhuozishan coal field is bordered by Qianli Mountain in the north, Queergou valley in the south, Gangdeer Mountain in the west and Zhuozishan Mountain in the east. The area is approximately 100 km long from north to south and 5–25 km wide from east to west, encompassing approximately 1000 km^2. The Zhuozishan Mountain and Gangdeer Mountain on the east and west sides of the coal field extend from north to south, forming a mid-high mountain region. The cliffs and the s-shaped valleys are formed by the geologic structures and erosion of flowing water. The terrain in the coal field is dominated by broad valley depressions. The gullies and terraces are developed between valleys. The water flows into the Yellow River from east to west in consistence with the topography. All valleys have seasonal runoff. They are dry during most part of the year. Torrent flows occur in the rainy season, which may cause flooding. However, the torrent flow typically lasts for a short period of time, and the valleys become dry within a few hours after the rain.

Luotuoshan Coal Mine is in the arid and semi-desert region, which is characterized with limited annual precipitation, large evaporation, and dry wind all year round. The perennial surface water flow only occurs in the rainy season and is mostly concentrated in the three months of July, August, and September. Water flows westwards into the Yellow River in the valleys on the western side of the mine field. The topography of the mine field is high in the east and low in the west, and most of it is covered by the Quaternary strata. There are outcrops of old formations outside the western and eastern boundaries.

1. Aquifer

Based on the detailed investigations of Luotuoshan Coal Mine, the water-bearing formations in the mine field can be divided into three types: porous medium aquifer composed of the Quaternary unconsolidated deposits, porous medium—fracture aquifer composed of the hard rock formations, and karst aquifer consisting of the Ordovician limestone. The details are described as follows:

(1) Porous medium aquifer

The lithology of the porous medium aquifer consists mainly of the Quaternary aeolian sand, residual sand on slopes, alluvial sand and gravel layer. Its thickness varies from 0.50 to 25.00 m with an average 6.00 m. The alluvium is distributed in modern dry valleys and terraces. The alluvial deposits are composed of poorly sorted sand and gravel and contains a small amount of pore water. Its burial depth varies from 1.15 to 19.18 m. The pumping tests in water supply wells indicate that the water flow is less than 10 m^3/d in individual wells. The water yield is poor and affected by atmospheric precipitation. The phreatic aquifer does not have hydraulic connection with the underlying confined aquifer but responds directly to the temporary torrent flows in surface rivers. This aquifer is an indirect recharge aquifer to the underground coal mine.

(2) Porous medium—fracture aquifer

The porous medium—fracture aquifer is complex and consists of five water-bearing zones. Two relatively impermeable layers are in between these water-bearing zones.

- Water-bearing zone I: This zone is in Shiqianfeng Formation (P2sh) and Shangshihezi Formation (P2s). It is distributed in the west of the mine field. Its thickness ranges from 6.85 to 328.69 m with an average of 156.40 m. The lithology is interbedded sandy mudstone and fine-, medium-, and coarse-grained sandstone. Groundwater is in the sandstone. During drilling of boreholes, the loss of circulation fluid is generally between 0.05 and 0.06 m^3/h with a maximum of 1.63 m^3/h. The pumping tests are conducted to calculate the specific water yield (q) and hydraulic conductivity (K), which are between 0.109 to 0.115 L/s m and 0.119 m/d, respectively. This water-bearing zone is moderately rich in groundwater and is an indirect recharge source to the underground coal mine.
- Water-bearing zone II: This zone is composed of formations of P1x3 and P1s4. The thickness of the zone is approximately 150 m. The main lithology is gray-white coarse-grained sandstone with gravel, gray-green, gray-purple sandy mudstone, and argillaceous rock. The loss of circulation fluid during drilling is generally less than 0.1 m^3/h. Occasionally, the fluid loss is more than 2 m^3/h, and the groundwater level is mostly reduced. The fissures in this zone are developed locally. Some sections have large water yields. Based on pumping tests, the specific yield q =

0.00129 L/s m, and the hydraulic conductivity K = 0.00118 m/d. The groundwater level elevation is at 1262.25 m amsl. The water quality type is Cl·SO_4-Na·Ca. The zone is poor in water richness and has little hydraulic connection with the underlying formations. This water-bearing zone is an indirect recharge source to the underground coal mine.

- Water-bearing zone III: This zone is composed of P1s2 formation. Its thickness varies from 12 to 16 m. The main lithology consists of dark gray, gray-black sandy mudstone, mudstone, and gray-white sandstone, #5 and #7 coal seams. The loss of drilling fluid is generally less than 0.1 m^3/h with a maximum of 3.84 m^3/h. Based on pumping tests, the specific water yield q varies from 0.00162 to 0.0000267 L/s m, and the hydraulic conductivity K varies from 0.00669 to 0.0000852 m/d. The groundwater level elevation ranges from 1294.93 to 1272.12 m amsl. The water quality type is Cl·SO_4-Na·Ca. This water-bearing zone is characterized with poor water yield and little hydraulic connection with the underlying formations. It is an indirect recharge source to the underground coal mine.
- Water-bearing zone VI: This zone consists of P1s1 formation and coal seam. Its thickness varies from 30 to 50 m. The lithology is mainly dark gray, gray-black sandy mudstone, mudstone, and off-white sandstone, including coal seams #12, #13, #14 and #15. The loss of circulation fluid during drilling is less than 0.05 m^3/h. Based on the pumping test data, the specific water yield q varies from 0.00102 to 0.00182 L/s m, and the hydraulic conductivity K varies from 0.00360 to 0.000229 m/d. The groundwater level elevation is between 1210.67 and 1274.02 m amsl. The water quality type is HCO_3·SO_4-Na or Cl·SO_4-Na. This zone is poor in water yield and has little hydraulic connection with the underlying formations. It is a direct and main recharge source to the underground coal mine.
- Water-bearing zone V: This zone is composed of #16, #17, #18, and #19 coal seams and top of the Ordovician limestone (O_2). Its thickness is 15 m. The specific yield q is 0.00137 L/s m, and the hydraulic conductivity K is 0.0082 m/ d. The groundwater level elevation is 1,276.50 m. The water-bearing zone is poor in water yield and has little hydraulic connection with the underlying aquifer.

(3) Karst aquifer

Lithology of the karst aquifer is dominated by blue-grey limestone. Karst fissures are not developed. The thickness of the limestone formation is greater than 200 m. The aquifer thickness varies from 20.04 to 26.62 m with an average of 23.33 m. The groundwater level elevation ranges from 1,259.21 to 1,269.49 m amsl. The depth to water varies from 18.58 to 23.48 m. The pumping tests are conducted in boreholes L01 and L02. The groundwater level drawdowns are between 60.68 and 60.70 m at pumping rates between 0.0291 and 0.0311 L/s. The specific water yield varies from 0.000480 to 0.000512 L/s m, and the hydraulic conductivity K varies from 0.00162 to 0.00205 m/d. The groundwater quality is characterized with the following: temperature = 10 °C, pH = 7.5, total dissolved solids = 1,960 mg/L, and F concentration = 1.16 mg/L. The groundwater chemical type is SO_4·Cl-Ca·Mg.

The data suggests that the water yield in the limestone is extremely uneven, the aquifer has poor water yield, and karst features are not developed in most of the area. However, the water yield may be strong at geologic structures such as faults.

2. Aquiclude

The aquicludes are present between aquifers. There are mainly two aquicludes as discussed below:

- First aquiclude layer: It is in the Lower Permian Shanxi Formation (P1s3). The thickness is approximately 10 m. The lithology consists of gray-green, gray sandy mudstone and mudstone, which contains #2 and #3 coal seams.
- The second aquiclude layer: This layer is composed of #8, #9 and #10 coal seams and argillaceous rock formations. It has a thickness ranging from 5 to 15 m.

Mine Dewatering Capacity

The main dewatering system of the mine includes the main and auxiliary water silo tunnels with a total length of 427 m. Both tunnels are constructed in the Ordovician limestone. The total volume is 3,015 m^3. Three MDA280-65*7 submersible pumps are installed. The dewatering capacity is 840 m^3/h.

The Article 278 through Article 280 of the Coal Mine Safety Regulations stimulate the requirements on mine drainage capacity.

- At normal water inflow: $Qz = 1.2 \times Q_{normal} = 1.2 \times 212.2 = 254.7$ m^3/h.
- At the maximum water inflow: $Q_{big} = 1.2 \times Q_{max} = 1.2 \times 242.5 = 291$ m^3/h.
- At normal water inflow: $n_{normal} = 254.7/280 = 0.9$ pump, which is set to 1 pump.
- At maximum water inflow: $n_{max} = 291/280 = 1.04$ pumps, which is conservatively set to 2 pumps.
- Spare pump: $n_{spare} = 0.7 \times n_{normal} = 0.7 \times 1 = 0.7$ pump, which is set to 1 pump.
- Inspection and repair pump: $n_{inspection} = 0.25 \times n_{normal} = 0.25 \times 1 = 0.25$ pump, which is set to 1 pump.
- Effective capacity of main water silo: $Q = 8 \times Q_{normal} = 1697.6$ m^3/h.

The calculations show that the effective capacity of the main and auxiliary water solos can accommodate 8 h of normal water inflow. Of the three pumps, one is actively used, one is reserved, and one is under repair and maintenance.

4.4.1.4 Water Inrush Process

At 5:50 on March 1, 2010, a stream of water suddenly gushed out of a blast hole at the elevation of 860 m amsl at the tunneling face of the #16 coal seam air return lane. The artesian water sprayed approximately 4 m and lasted 5 s. At 6:10, the left side of the working face was sheared off, and water gushes out through several blast holes. A water stream in diameter of 20 cm flew out of the left wall of the working face, while water gushed out through 6 blast holes on the right wall. A section the floor with a dimension of approximately 10 m long and 1.2 m wide heaved, and the raised height was between 0.1 and 0.3 m. A crack of approximately 10 m long, 4 cm wide and 5 cm deep was identified on the left side of the heaved bottom.

At 6:30, the bottom formation was lifted by 0.3 m by the underlying potentiometric pressure. At 7:20 when a worker who was repairing a leakage pipe of the dewatering system heard a loud noise of formation breaking. At 7:35, the worker retreated to the staging area at bottom of the shaft. He heard flowing water through the tunnels toward the shaft. At 7:31, the mine administrator sent urgent notification to the underground miners for immediate evacuation. At the same time, the emergency response plan is implemented to prepare emergency rescue.

As of 14:00 on March 1, the groundwater level elevation reached 1079.05 m amsl. The total volume of water flowing into the mine was approximately 100,000 m^3. Except for the 0901 cut-out that is the highest underground work site at the time, all the other roadways and chambers were submerged. At 0:00 o'clock on March 4, the groundwater level in the mine reached the highest elevation at 1093 m amsl. The maximum average water flow rate is estimated to be 65,000 m^3/h.

4.4.1.5 Mitigation Plan

The Luotuoshan Coal Mine water inrush drew attentions from the State Council, the State Administration of Work Safety, the State Administration of Coal Mine Safety, Inner Mongolia and Shenhua Group. The then Premier Wen Jiabao of the State Council issued instructions requesting every possible means to rescue the trapped people and take care of aftermath issues. Vice Premier Zhang Dejiang demanded that the spirit of Premier Wen Jiabao's important instructions be fully implemented and took the following specific measures:

- Take effective measures to rescue the trapped people.
- Make every effort to reduce casualties.
- Treat injured people immediately.
- Cooperate with each other.
- Compensate properly the family members of the dead and injured workers.
- Strengthen leadership and management.
- Scientifically dispatch and carefully organize various aspects of the rescue work.
- Establish an accident investigation team, which is led by the State Administration of Safety Supervision, to conduct the root cause analysis of the accident.
- Adhere to the fact and timely share information with the public.
- Make sure that the rescue work is performed safety and avoid secondary casualties.

The emergency rescue commander required that the tunnels through which the groundwater flew into the mine be sealed off as quickly as possible. The emergency grouting of the water-flowing tunnels must cut off the flow sufficient for the trial dewatering on April 5 so that the rescue dewatering and underground rescue mission can be completed. After completion of the rescue effort, phase II of the water inrush disaster mitigation started. The objective of the phase II is to disconnect the water inrush recharge source by grouting the water inrush pathway at the water inrush point. Xi'an Research Institute of China Coal Industry and Engineering Group accepted the challenged. Together with national experts, they carefully studied the geological, hydrogeological, and engineering geological conditions associated with the water inrush. A phased approach was implemented to mitigate the water hazards:

- Phase I: Seal off the tunnel that is hydraulically connected to the water inrush point so that the rescue effort can be carried quickly to save lives of the workers trapped underground.
- Phase II: Seal off the water inrush pathway at the water inrush point to eliminate water hazards and restore mine production.

The project consists of three parts: emergency rescue and grouting tunnel directly connected to the recharge source, focused investigation of the water inrush point and its structure, and remediation of the water inrush pathway at the water inrush point.

The emergency grouting of the tunnel took advantage of the known location of the target tunnel and the favorable condition that the runnel is a one-end roadway. Directional drilling was conducted on surface to drill precisely into the tunnel, followed by grouting. The aggregates and double slurry consisting of cement and water glass were used upstream to intercept the groundwater flow, whereas cement and sand mixture was used downstream to further impermeabilize the tunnel. Additional grouting and consolidating were conducted in the silt layer at the bottom of the water flowing tunnel. The directional drilling and grouting constructed a water plug with sufficient length and strength so that the water recharge source was disconnected from the rest of the mine completely.

The comprehensive investigation of the water inrush point and its structures started with the ground transient electromagnetic method and high-resolution electrical resistivity imaging to detect the hydrogeological conditions in the water inrush area. In addition, the 3D seismic data collected in this area in 2007 was re-evaluated and re-interpreted. Geophysical anomalies were identified, and verifications were conducted by drilling exploratory boreholes. The technical approach of remediating the water inrush hazard is based on the water inrush structure, location, boundary configuration, and permeability.

The technical approach includes drilling and grouting on the ground surface. The directional drilling directs boreholes to intercept fracture zones around the #16 coal seam air return lane and on the top of the Ordovician limestone aquifer. These fractures are water inrush pathways and are sealed off by grouting. The target grouting led by directional drilling cut off the hydraulic connections between the underlying Ordovician limestone aquifer and coal mining areas.

4.4.2 Emergency Water-Plugging Technology in #16 Coal Seam Air Return Lane

4.4.2.1 Formulation of Plan

Analysis of Hydrogeological Conditions of Water Inrush

1. Water inrush source

The water inrush recharge source is determined to be the Ordovician limestone karst aquifer underlying the #16 coal seam floor. This determination is based on multiple lines of evidence as discussed below.

- Groundwater quality characteristics: The comparison of the chemical composition in the mine drainage and chemical composition in the Ordovician limestone aquifer indicates that their water chemistry type is the same. Both are $HCO_3 \cdot SO_4$-Na·Ca. The various water chemistry indexes are also similar.
- Groundwater temperature characteristics: The water temperature of the mine water is approximately 20.5 °C, which is consistent with the water temperature of the groundwater in the Ordovician limestone. The groundwater temperature in the Ordovician limestone is measured in nearby water supply wells in Pinggou Mine and Hainan District of Wuhai City. Table 4.3 presents the temperature data.
- Dynamic characteristics of groundwater level and groundwater volume: The records of mine water level and dewatering activity after the mine inrush show some unique characteristics.
  - The underground water level recovers quickly once dewatering in the mine is suspended, indicating that the water supply source has strong recharge capacity and high potentiometric pressure.
  - When the water level in the mine is at 1090 m, the pumping rate is maintained at a constant level. However, the groundwater level continues rising slowly for a long time. Such a response indicates that the groundwater continues to be replenished by the recharge source. The persistent water quantity and high potentiometric pressure in the recharge source is not consistent with recharge from an abandoned mine pool, rather it is a typical characteristic of recharge from the Ordovician limestone.
  - At the beginning of dewatering, the groundwater level drops quickly with increase of the pumping rate. As the groundwater level drops in the mine, in potentiometric pressure difference increases between the recharge source and mine. The recharge intensity increases. As a result, the groundwater level drawdown gradually slows down. Again, this observation illustrates the recharge water source is characterized with high potentiometric pressure and strong water recharge capacity.
- Groundwater flow analysis: The maximum water inrush flow is 65,000 m^3/h. As of 12 o'clock on May 10, the emergency rescue dewatering discharged

Table 4.3 Comparison of water temperature measurements

Water source	Mine water after water inrush	Ordovician limestone aquifer	Ordovician limestone aquifer	Ordovician limestone aquifer
Location of temperature measurement	Emergency pump water outlet in Luotuoshan Coal Mine central transportation alley	Luotuoshan Coal Mine 12.17 water inrush point	Pinggou Coal Mine 1070 water inrush point	Wuhai Hainan Water Supply Company water supply well
Water temperature (°C)	20.5	25	14 (19)	20
Temperature measurement time	March 4 and 6, 2010	January 15, 2008	March 6, 2010	March 7, 2010
Water inrush location	Coal seam floor at #16 coal seam air return lane	Coal seam floor in Ordovician limestone	Coal seam floor in Ordovician limestone	Ordovician limestone
Depth of water inrush (m)	400	400	185 (385)	385
Organization	Luotuoshan Mine	Luotuoshan Mine	Pinggou Mine	Water Supply Company

1.44 million m^3. The large water inrush flow and water volume indicate that the water inrush source has a large groundwater reservoir and strong replenishment capacity.

- Water inrush symptom: Per the on-duty worker, a section the floor with a dimension of approximately 10 m long and 1.2 m wide heaved. The raised height was between 0.1 and 0.3 m. A crack of approximately 10 m long, 4 cm wide and 5 cm deep was identified on the left side of the heaved bottom. These symptoms are typical early warning signals that coal seam floor is subject to a strong force from the underlying formation.

2. Water inrush geologic structure

The water inrush occurred strongly. The instantaneous water flow is large. The flow rate increases quickly. The flow is typical of pipe flow in which the flow cross-section is large and free of obstruction. The flow paths are likely larger than regular fractures in bedrocks.

The #9 coal seam air return lane is connected to the air circulation shaft. No fault structures are reported. The previous exploration did not identify faults in this area either. The possibility of a fault as the water inrush pathway is basically ruled out. In addition, empirical experience suggests that the water inrush through vertical features is generally not so large. Based on reported water inrush accidents, the water inrush disaster that occurred in September 1988 in the floor of II617 working face in the

Huaibei Yangzhuang Mine was the largest one in which the flow path was through fractures. The flow rate was 2300 m^3/h, which is significantly smaller than the water inrush flow in this mine. The possibility of water inrush through a karst collapse column is this the highest. This was confirmed in the subsequent investigation of the water inrush channel and the second-phase remediation project.

Mitigation Measures

Three mitigation alternatives are evaluated to perform rapid rescue and recovery of mine production. Alternative 1 is to grout and seal the tunnel that allows the groundwater to flow into the mine. Alternative 2 is to grout and seal the groundwater pathway that connects the recharge source to the mine. Alternative 3 is to grout and seal both the tunnel and flow pathway. Comparison of these three alternatives leads to selection of Alternative 3. Table 4.4 summarizes the comparative analysis.

The first step of grouting and sealing the water flow tunnel and water pathway is to grout and seal the #16 coal seam air return lane. A water plug of a certain length and strength is constructed in the air return lane by injecting aggregate and cement

Table 4.4 Comparative analysis of mitigation alternatives

Alternative	Grout and seal water flow tunnel	Grout and seal groundwater pathway	Grout and seal both water flow tunnel and groundwater pathway
Objective	Block the flow tunnel for emergency rescue	Block water recharge source to eliminate water hazards	Eliminate water hazards and restore the flooded mine
Engineering location	#16 coal seam air return lane	#16 coal seam floor water inrush channel and water inrush zone within 100 m of the Ordovician limestone aquifer	First implement water flow tunnel blockage, and then carry out groundwater pathway grouting
Advantages	• Accurate location information • Short time and less investment • Emergency rescue can be implemented as soon as possible	Eliminate water hazards in the mine	A project to eradicate floods can be implemented as soon as possible
Disadvantages	The Ordovician limestone water threatens the safety of the mine. The threat is not eliminated	• The water inrush channel is unclear, and the exploration project needs to be arranged first • Significant engineering effort	Two constructions, large investment

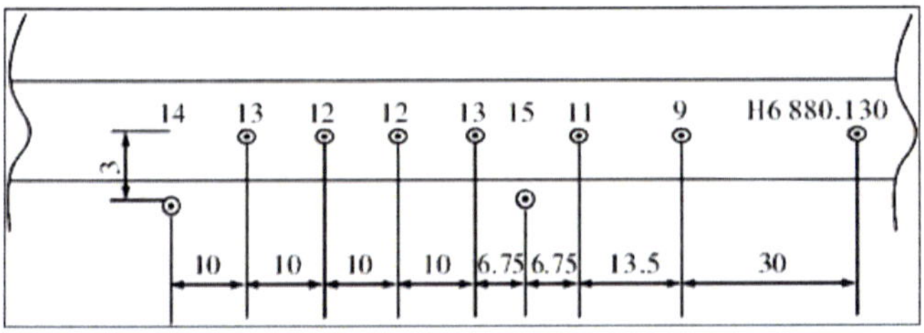

Fig. 4.12 Borehole placements for grouting water inrush pathway

slurry. The connection between the water inrush point and the mine is then cut off to create conditions for the implementation of underground emergency rescue. At the same time, the ground transient electromagnetic method is used to detect the water inrush pathways. After the directional drilling on the ground surface verified the pathways, the grouting is carried out to fill the water inrush channel with cement slurry. The groundwater hazards are mitigated only after the groundwater pathways are completed filled with grout. Figure 4.12 shows the schematic diagram of drilling placements.

The emergency water flow tunnel grouting and sealing project can be divided into the following four stages:

- Stage 1: Use two boreholes #9 and 12#, located upstream and downstream of the tunnel, respectively, to inject double-liquid slurry made of cement and water glass and sand cement mixture. The purpose of this stage activities is to create two water barriers as quickly as possible, one upstream and one downstream, to resist groundwater flow into the coal mine. These two water barrier cross-sections are the bases for construction of the water plug consisting of sand-cement mortar and concrete.
- Stage 2: Use grout holes J1, J2, #13 and J5 with high-concentration cement. The purpose of this stage is to fill and seal any voids within the concrete body that created at Stage 1. The high-volume injection method is used to ensure that the water plug is an integrated, continuous water-blocking wall.
- Stage 3: The bottom of the water plug is a silt layer, and the top of the water plug is in contact with the concrete roof of the tunnel. These contacts are vulnerable to seepage and weak areas to resist pressurized water. The purpose of this stage is to grout the silt layer to increase its strength and water resistance. The grouting is conducted under the dewatering condition. The silt layer at the bottom of the tunnel is replaced with high-concentration cement through the anchor holes.
- Stage 4: The purpose of this stage is to grout the cracks and joints between the water plug and the top of the tunnel. The grouting reinforces the adhesion between the water plug and the surrounding rock as well as increases water resistance capacity to eliminate potential seepage and erosion at contacts. Figure 4.11 shows schematically the drilling and grouting design.

4.4.2.2 Rapid Construction Technology of Water Plug

Construction of Water Plug with Double-Liquid Slurry Under Hydrostatic Condition

The advantages of using the cement-water glass double-liquid slurry in grout hole #9 is that the double-liquid slurry solidifies quickly so that a cross-sectional barrier can be created quickly to block groundwater flow from the upstream. This barrier can also prevent subsequent grouting slurry from spreading toward the water inrush point. The specifications of the double-liquid slurry are water glass concentration at 40Be, water-cement ratio at 0.75:1, and cement-water glass ratio at 1:1.

Two sets of grouting equipment are used with two separate sets of delivery pipelines. The double-liquid slurry is mixed at the bottom of the grout hole with the slurry mixer that is installed at the bottom of the grout hole. This approach effectively prevents the slurry from solidifying in the grout hole and plugging it. In addition, the drilling depth of the grout hole is more than 400 m. The length of the grouting pipes and the weight of the pipes increase the difficulty of lowering the two sets of Φ25 * 2.5 mm grouting pipes. Three pipes are connected to form one long pipe on the ground surface, and the two sets of pipelines are inserted into the grout hole together to overcome the difficulties and save time. To ensure the integrity of the two pipelines, the two pipelines start from the bottom of the hole where the mixer is installed and are welded together at regular intervals. At the ends of each pipeline, two sections of 25 mm inner diameter casings are welded together with a self-made fork to prevent accidental decoupling during lifting of the two-pass pipeline and lowering them into the grout hole. These measures successfully installed the pipes for delivery of the double-liquid slurry effectively and efficiently. The innovative approaches saved precious time in the emergency rescue effort.

Grouting Under Hydrostatic Condition

After the two double-liquid slurry barriers are established, a large volume of single-liquid high-concentration cement slurry is pumped to quickly fill in the tunnel between the barriers. The grouting pump has a capacity of between 600 and 800 L/min. The density of the single-liquid slurry varies from 1.5 to 1.7 t/m^3.

1. Grouting parameters and operational process

The grouting method uses a drill rod that is placed 0.5–1 m above the tunnel floor to directly deliver the cement slurry. This method reduces the interference on the cement slurry during deposition and solidification processes, increases the consolidation strength of the cement slurry, and saves time in construction of the water plug.

- Pre-wetting before grout injection: Pressurized water should be injected in each grout hole before filling and grouting. The purpose is to check the functionality and sealing performance of the grouting equipment and pipeline and establish the unit water intake rate according to the amount of water and pressure. The pre-wetting also helps determine the initial grouting parameters such as slurry displacement, specific gravity, setting time.
- Making slurry: The slurry making process begins after completion of the pressurized water test. The specific gravity of the slurry is controlled between 1.5 and 1.7 g/cm^3, and the slurry making capacity ranges from 36 to 60 m^3/h.
- Grouting operation: The filling and grouting starts with the displacement capacity of 400 L/min and the specific gravity of 1.6 g/cm^3. Appropriate adjustments are made during the grouting process according to the actual situation. The initial stage of filling and grouting is mainly dominated with single-hole large-volume grouting method and combined multi-well large-volume grouting method. During the grouting, the water level changes in other grouting holes and monitoring wells of the Ordovician limestone are monitored and analyzed in time to evaluate the grouting effectiveness. If grouting is effective, grouting continues; otherwise, multi-well intermittent grouting method is adopted to prevent loss of a large amount of slurry. In the later stage of the filling and grouting phase when the water flow cross section is significantly reduced, the single-hole intermittent grouting method is used.
- Grouting termination: The grouting stops when intermittent grouting is required to continue the grouting operation, or the grouting meets the termination standards.

2. Grouting to improve effectiveness

This stage is a key stage where because the coal walls are susceptible to scouring, destruction and breakthrough of some water resistance sections. The self-flow grouting is used without orifice pressure in combination with high-concentration intermittent grouting and multi-well joint interference grouting methods to control the slurry dispersion distance and prevent excessive loss of slurry.

At this stage, grout hole J1 and #13 are mainly used, and grout holes #9, J5, J2 and #12 are supplemental. Under the condition of the hydrostatic condition, the internal space of the water plug is filled with a large flow of single-liquid slurry. The voids between sand-cement mixture are cemented to form a continuous solid wall. If necessary, repeat grouting is carried out to ensure that all voids are filled completed.

A total of 3389.87 m^3 of cement slurry was injected at this stage. The core samples and borehole scanning records confirm that the cement slurry inside the water plug was filled. The cement slurry was deposited at the bottom of each infill hole, indicating that the water plug walls are formed. In grout holes J1 and #13, the slurry is bound solidly with the roof formations.

Grouting Effectiveness Evaluation with Three Step-Drawdown Tests

1. Step-drawdown test design

(1) Test purpose

The purpose of the step-drawdown tests is to evaluate the water resistance performance of the contracted water plug and collect essential data for dewatering design.

(2) Pre-test drawdown

The total drawdown is designed to be 85 m. The pre-test is implemented in three steps. The drawdowns are 35, 25, and 25 m, respectively, which correspond three groundwater level elevation intervals, i.e., from current water level elevation to 1060, from 1060 to 1035, and from 1035 to 1010 m amsl. The recovery groundwater levels are monitoring for 4 h after each drawdown.

(3) Pre-test pumping rate

Two submersible pumps are used with each a capacity of 1000 m^3/h. The expected pumping rate ranges from 3000 to 3500 m^3/h.

(4) Expected dewatering cycle

The expected dewatering frequency 18 h.

(5) Hydrogeological monitoring

a. In addition to the two existing telemetry groundwater level monitoring systems, manual measurements are performed to ensure collection of the essential data.
b. The remotely controlled instrument is calibrated 2 h before the test and confirmed to be accurate. During the test, two personnel, one in the pump house and the other in the duty room, are responsible for its functionality.
c. Dedicated manual water level measurement system is used at both the ventilation shaft and the inclined shaft. Each monitoring group consists of three technicians.
d. Three monitoring wells, i.e., #9, 13#, and J4, are installed in the water plug and Xi'an Coal Research Institute is responsible for water level measurements.
e. The Ordovician limestone aquifer is automatically observed by the water level telemetry system.

(6) Data transmission

The data at the ventilation shaft and inclined shaft are reported by phone calls. Data at other locations are either manually transmitted or automatically transmitted.

The data transmission is as often as needed during the pumping period. The data is transmitted every 30 min during the recovery period.

(7) Data analysis

The data are analyzed in real time by a group of hydrogeologists and engineers to guide the step-drawdown tests.

2. Step-drawdown test

The step-drawdown tests started at 10:00 on April 5, 2010 and ended at 1:00 on April 6 with a duration of 15 h. Three drawdowns and three recoveries were monitoring. The groundwater levels were monitored at six locations including the auxiliary inclined shaft, the ventilation shafts, one monitoring well for the Ordovician limestone (#1 O_2 observation well), three observation wells J4, #13, and #9. The water level at the auxiliary inclined shaft decreased from 1099.69 to 1024.93 m amsl with a total drawdown of 74.76 m.

Because the 1000 m^3/h submersible pump in the ventilation shaft cannot get started, the pumping time is extended. The recovery observation time is shortened from 4 to 2 h to save time. At the same time, two submersible pumps in the auxiliary inclined shaft are started (Table 4.5). Because the mine water inflow data is basically obtained in the first two drawdown tests and the time to lower the groundwater to the design level of 1010 m amsl would be long, it was decided to change the target drawdown to 1024.93 m.

3. Performance analysis

(1) Flow through water plug

Based the recovery groundwater level data in response to the three drawdowns, the volumetric method was used to calculate the residual groundwater inflow into the mine. The water flow rate is obtained by the following equation:

$$Q = (Q_0 + Q_{inrush} + Q_{block})/t$$

Table 4.5 Summary of step-drawdown tests

Step-drawdown test	Ventilation shaft	Auxiliary inclined shaft	
1st	• #1 pump at 1,000 m^3/h • April 5, 10:12 start; 14:00 stop	• #1 pump at 700 m^3/h • April 5 11:06 start; 13:00 stop	• #2 pump at 700 m^3/h • April 5 11:10 start; 14:00 stop
2nd	• #1 pump at 1,000 m^3/h • April 5, 16:00 start; 19:00 stop	• Pump above water level • Lower pump and reinstall	• #2 pump at 700 m^3/h • April 5 16:00 start; 19:00 stop
3rd	• 1# pump at 1,000 m^3/h • April 5 21:00 start; 23:00 stop		• #2 pump at 700 m^3/h • April 5 21:00 start; 23:00 stop

where

Q The average water inflow during water level recovery at a certain elevation

Q_0 The original water inflow above the submerged height of the mine, 104 m^3/h

Q_{inrush} The amount of water inrush at a certain level of the Ordovician limestone inrush point, which changes with the submerged water level, as calculated in the following equation:

$$Q_{inrush} = 81 \times \sqrt{\frac{\Delta S}{190}}$$

where

ΔS The difference between the submerged water level and the Ordovician limestone potentiometric level.

Q_{block} Residual flow through water plug.

The measured mine water inflow and calculated residual flow rate are summarized in Table 4.6.

As shown in Table 4.6, the water plug still allows small amount of water to flow through and the residual flow rate increases with the decrease of water level in the mine.

- The step-drawdown tests show that the structure of the constructed water plug is structurally stable, the residual amount of water passing through is not significant. The overall water resistance capacity is acceptable.
- Some of the grout holes are not completely sealed. The water resistance capacity of the water plug can be improved by further grouting reinforcement so that the residual amount of water is further reduced. Figure 4.13 shows the measured water levels in auxiliary inclined shaft, ventilation shaft, and observation well J4.

Reinforcement Technology of Silt Layer Under Hydrodynamic Condition

After the filling and grouting are completed in the tunnel under the hydrostatic condition, the overall structure of the water plug is constructed. The pumping tests

Table 4.6 Relationship between Ordovician limestone water level and water inflow

Groundwater level elevation (m)	Water level difference between mine and the Ordovician limestone (m)	Flow rate (m^3/h)			
		Q_0	Q_{inrush}	Q_{block}	Q (total)
1058	28	104	34	6	144
1035	51	104	44	44	192
1025	61	104	48	52	204

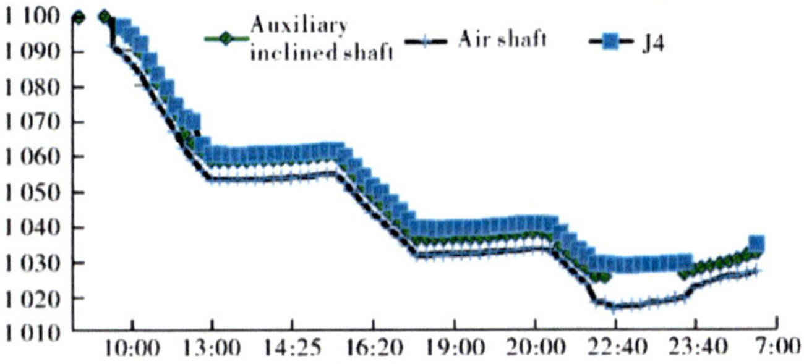

Fig. 4.13 Groundwater level elevation changes in auxiliary inclined shaft, ventilation shaft, and observation well J4

indicate that the water plug is structurally stable but still permeable to groundwater flow. The residual flow appears to come from a layer of silt at the bottom of the tunnel and at top of the plug where fractures are present in the roof formation. The silt layer is approximately 1.4 m thick and is composed of the fills in the collapse column and the residual coal. Its unconsolidated nature suggests that it is permeable to water and its stability is poor. Large water conductive channels can form by erosion under hydrodynamic conditions and even undermine the structural stability of the water plug. Therefore, properly grouting and sealing the silt layer is the key to the sustainable function of the water plug. Restoration of the flooded mine requires lowering the groundwater level to the level of 870 m amsl at the #16 coal seam air return lane. At this level, the water plug would withstand the potentiometric pressure of approximately 2.3 MPa.

The fine particles of the silt layer and poor injectability make the hydrostatic pressure grouting challenge to inject the grout into the silt layer and cement it. Instead, the high-flow grouting method is used with a bottom up grouting process under dynamic water conditions. The principle is that while the flowing water continuously scours and carries the fine particles of the silt layer away, the drilling pipe is inserted into the silt layer. The grouting pressure is applied directly to the silt layer, which replace the silt with cement and consolidates the grout quickly. This way the water plug is integrated with the underlying formation, and any residual water pathways are blocked. Once the silt layer is replaced with cement, the flow path at the bottom of the water plug is sealed off.

1. Specifications and procedures

   - Use the drilling tool to investigate the characteristics and thickness of the silt layer at the bottom of the water plug and collect soil samples for analysis of its properties and strength.
   - Prior to grouting, introduce a small amount of pressurized water to calculate the unit water intake rate. The relationship between the water intake volume and pressure helps determine the grouting parameters such as the slurry injection rate, specific gravity, and setting time.

- The slurry is prepared with 42.5R strength cement. The density of the cement slurry is controlled at 1.6 g/cm^3.
- The cement slurry is injected into the silt layer under pressure by a grouting pump. The multi-well intermittent grouting method is used for 12 h injection and 8 h solidification.
- Accurately control the amount of pressurized water to avoid excessive water entering the silt layer.
- During the grouting, the drilling tool must rotate at a low speed with frequent lifting up and lowering down at regular intervals to prevent the drilling fluid from solidifying and coating on the drilling tool. If grout surfacing such as bubbling or sky lighting at orifice occurs, stop grouting in this hole. Use the process to continue grouting in another grout hole.
- Lower the drilling tool to scan the grout hole 8 h after the injection of each grout hole. The thickness of the cement stones and loss of circulation fluid are monitored to prepare for the next round of grouting until the water plug is constructed into the underlying formations.
- After all boreholes have completed the grouting in the silt layer under hydrodynamic conditions, the grout holes are drilled 5 m into the underlying formation and conduct the grouting to fill any fractures. The operation will ensure that the water plug is rooted in the underlying formation so that any residual flow below the plug is eliminated.

2. Construction summary

This stage started on April 5 and ended on April 28, 2010. The total grout volume is 5,053.10 m^3.

4.4.3 *Comprehensive Investigation Technology of Water Inrush Point*

4.4.3.1 Technical Approach of Water Inrush Point Investigation

The investigation of the water inrush point began in the process of constructing the water plug in the tunnel. Geophysical prospecting including the three-dimensional seismic method, TEM method, and high-resolution electrical resistivity imaging were used to identify the anomalous area of the aquifer. The geophysical interpretations help determine the cause of water inrush and provide the nature and shape of the water inrush structure. Exploratory drilling was then used to verify the anomalies and determine the exact location and boundary of the water inrush point. Form and water-rich aquifer. The data collected through these investigations lay the foundation for the water inrush mitigation plan.

Further analysis of the shape and boundary of the water inrush point is carried out even during the process of water inrush point treatment. Observation of grouting characteristics and the loss of the circulation fluid during drilling provide additional information to better understand fracture distribution around the water inrush point and the heterogeneity of water richness. This information is the basis for the second or third round of grouting effort.

4.4.3.2 Three-Dimensional Seismic Interpretation

In 2007, the Anhui Provincial Coal Geology Bureau conducted the three-dimensional seismic survey in 101 mining area of Luotuoshan Coal Mine. After this water inrush, a more detailed analysis of the data was made at the request of the mine to determine the reason for the water inrush and provide geological data for the mine. The three-dimensional visualization technology was used to re-analyze and re-interpret the seismic data within approximately 200 m around the water inrush point. The following conclusions were obtained:

- No abnormal display is identified on the T9 and T16 reflection waves of the coal seam. The waveform is continuous.
- Reflected waves on the top of the Ordovician limestone have an abnormal amplitude band display.
- Anomalous zone of weak amplitude disturbance is identified within the Ordovician limestone.

The seismic anomaly is interpreted as a small-scale karst collapse column concealed under the #16 coal seam. It has a diameter of approximately 10 m at the interface of the Ordovician limestone. It may extend to the lower part of the coal measure strata. However, the reflection wave of #16 coal seam is continuous and does not show the anomaly. Figures 4.14 and 4.15 show the seismic profiles along two transects.

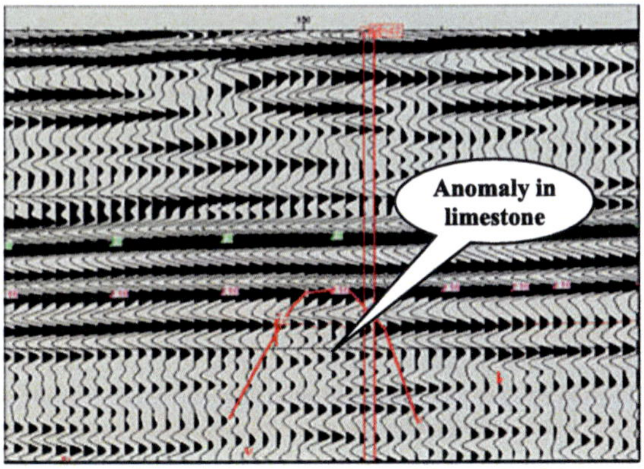

Fig. 4.14 I336 seismic reflection profile

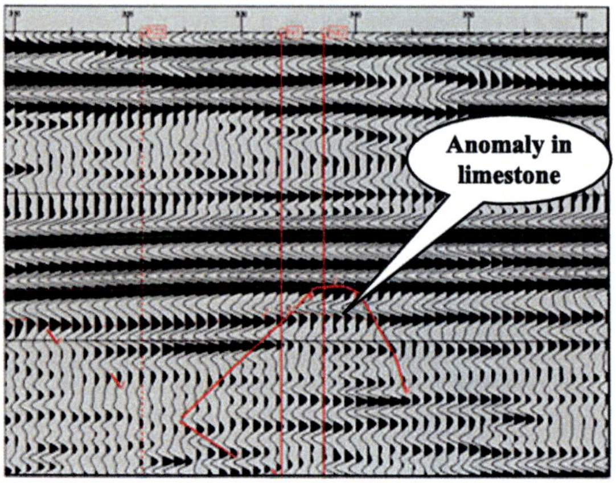

Fig. 4.15 C154 seismic reflection profile

4.4.3.3 TEM Survey

- A total of 21 TEM survey lines and 1701 physical points were completed in this TEM survey.

Figures 4.16 and 4.17 present the survey results. The following conclusions are obtained:

- A near-north-east-southwest low-impedance anomaly is identified within the survey area and is interpreted as a water-rich area in the Ordovician limestone preferential flow zone. Vertically, the extent of the water-rich area increases with the depth in the Ordovician limestone.
- Near the water inrush point, a low-resistance anomaly is identified. The anomaly extends from the Ordovician limestone to the coal-measure strata, partially passing #16 coal seam to reach #9 coal seam.

4.4.3.4 Exploration Drilling Investigation

Purpose

Based on the results of the TEM and three-dimensional seismic interpretations and the characteristics of water inrush process, the exploratory drilling technology is used to determine the nature and location of the water inrush structure for the water inrush mitigation design.

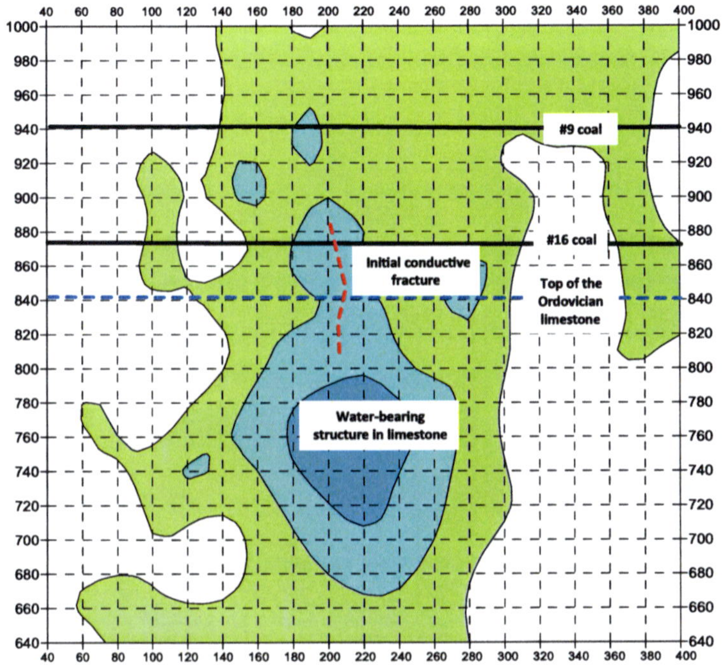

Fig. 4.16 Cross-sectional view of apparent resistivity after 13-line inversion

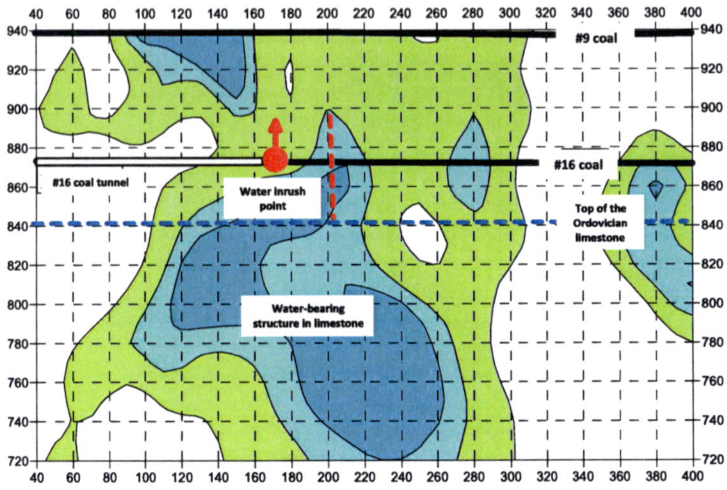

Fig. 4.17 Cross-sectional view of apparent resistivity after 15-line inversion

Exploration Borehole Construction

Directional drilling technology is used to investigate the area in front of the #16 coal seam air return lane. The directional drilling technology has the capability of making branch boreholes from the main borehole. The main borehole T1 was drilled during the period from April 9, 2010. Four branch boreholes were completed on May 12, 2010. The total drilling length is 833 m. Figure 4.18 shows the stratigraphy encountered at T1.

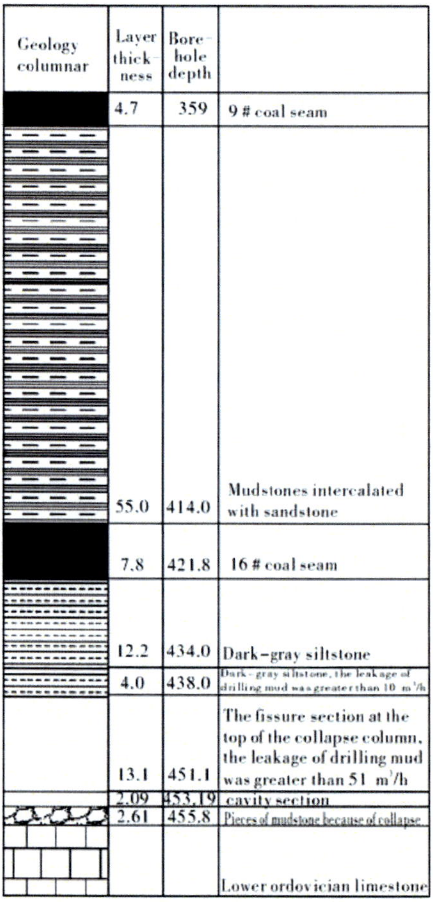

Fig. 4.18 Stratigraphy at main borehole T1

Exploration Results

The drilling revealed that the #16 coal seams are continuous near the water inrush point. The interval between the coal seam and the underlying Ordovician limestone aquifer is 34 m (Fig. 4.18).

During the drilling of the main borehole T1, the loss of circulation fluid was observed to start at 434 m below ground surface or 12.2 m below the #16 coal seam floor. The loss of fluid was 10 m^3/h. The total loss of circulation fluid occurred at 438 m below ground surface or 16.8 m below the #16 coal seam floor. The total loss was greater than the pumping rate of 51 m^3/h. The drilling bit dropped by 2.09 m between 451.10 and 453.19 m below ground surface (Table 4.7). After the drill bit dropped, the core samples were taken to take. The rock core samples are mudstone fragments, which belong to the coal-measure stratum (Fig. 4.19). The surface of the core samples is coated with rust.

Observations of this borehole indicate that karst development in the limestone caused the overlying coal-bearing strata to collapse and form the basis for a collapse column. The collapse column extends upward 4.7 m into the coal-bearing stratum. The water-conductive fractures in the overlying formation is 17.1 m high.

After the drilling bit drop in main borehole T1 the groundwater level elevation that was measured at 18:00 on April 19, 2010 was 1,091.42 m, which was 0.88 m lower than the groundwater level elevation of 1,092.3 m in #1 observation well of the O_2 aquifer. This difference suggests presence of the cone of depression in the Ordovician limestone aquifer after water inrush occurred.

Table 4.7 Summary of exploration borehole results

Borehole		Depth (m)	Data				
			#16 coal seam (m)	Elevation of Ordovician limestone (m)	Loss of circulation liquid, depth (m)	Total loss of circulation liquid, depth (m)	Drill bit drop, depth (m)
					Rate (m^3/h)	Rate (m^3/h)	Height (m)
T1		455	414–421.8		434–438	438	451.1–453.19
			7.8		10	>51	2.09
Branch borehole	T1-1	368–476		455.8		421.8	
		108				>51	
	T1-2	380–465		455.8		416	
		85				>51	
	T1-3	368–465		455.8		415	
		97				>51	
	T1-4	377–465		455.8		414.3	
		88				>51	

Fig. 4.19 Photograph of the core sample collected in the collapse column

After the completion of the main borehole T1, four branch boreholes, T1-1, T1-2, T1-3, and T1-4, were constructed in front of the #16 coal seam air return lane to detect fracture distribution surrounding the collapse column. Loss of circulation fluid occurred at all four branch boreholes at distances of 421.8, 416, 415, and 414.3 m, respectively. The boreholes were drilled 10 m into the Ordovician limestone, and no drill bit drop occurred in any of them. These data indicate that the development height of the water-conductive fissure zone of the collapse column becomes higher toward the tunnel. The vertical depth at T1-4 is 413.2 m after conversion, which suggests that the fractures are in #16 coal seam roof. These fractures are the groundwater flow paths when the water inrush occurred. This is consistent with the water spraying phenomenon in the blast hole just before the water inrush.

4.4.3.5 Exploration Summary

The comprehensive three-dimensional seismic data interpretation, TEM survey, and exploratory drilling determined that the water inrush structure is a small-scale karst collapse column in the Ordovician limestone. Further investigation of the collapse column during mitigation of the water inrush point defined nature and characteristics of the collapse column. The collapse column consists of two high points. The first high point is concealed below the #16 coal seam in front of the tunnel. The boreholes reveal that the collapse column protrudes 4.7 m into the coal measure stratum. The axial length is no more than 10 m. However, water-conductive fractures are well developed around the collapse column. The second high point is revealed by branch borehole Z2-1. This high point has grown into #16 coal seam. Its top is at depth of 409.5 m and located 4.8 m away from the #16 coal seam air return lane on the left side of the tunnel. Figure 4.20 show schematically the morphological development of the collapse column.

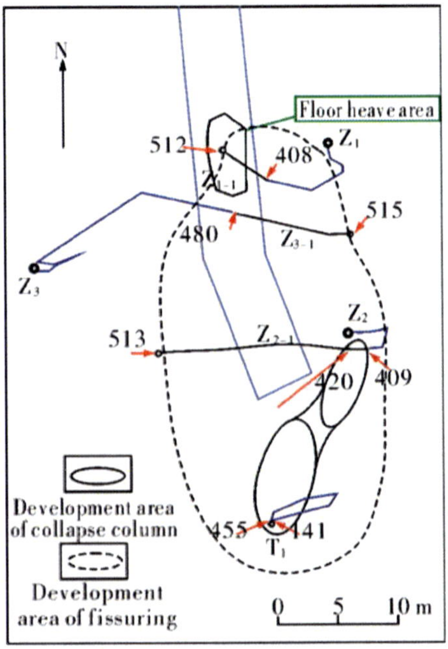

Fig. 4.20 Schematic diagram of morphological development of collapse column

4.5 Characterization and Remediation of Karst Collapse Columns in Renlou Coal Mine, China

4.5.1 Mine Background

Several water inrushes occurred in Renlou Coal Mine. The first water inrush occurred in working panel $7_2 22$ at 380 m deep in 1996. Figure 4.21 shows the lithology in the mine and the water-inrush point. The panel was flooded within 10 h. The final water level was stabilized at 15.59 m amsl, which is approximately the same elevation of the water level in the Ordovician limestone. The maximum water inflow was 19 m^3/s. The water inrush resulted in a water level drop of 7.04 m in an Ordovician limestone monitoring well 16.2 km away. Investigations in response to the water inrush and subsequent grouting confirmed that the water pathway was a karst collapse column. The column was nearly vertical and oval-shaped with long axis of 30 m and short axis of 25 m. Although none of the boreholes reached the bottom of the collapse column, the collapse column was at least 300 m high, having its root in the Ordovician limestone and roof in the Quaternary and Tertiary formations. The top section of the karst collapse column consisted of an open void, which suggests that the collapse column was still actively developing upward. The mine was restored six months later after a successful grouting program.

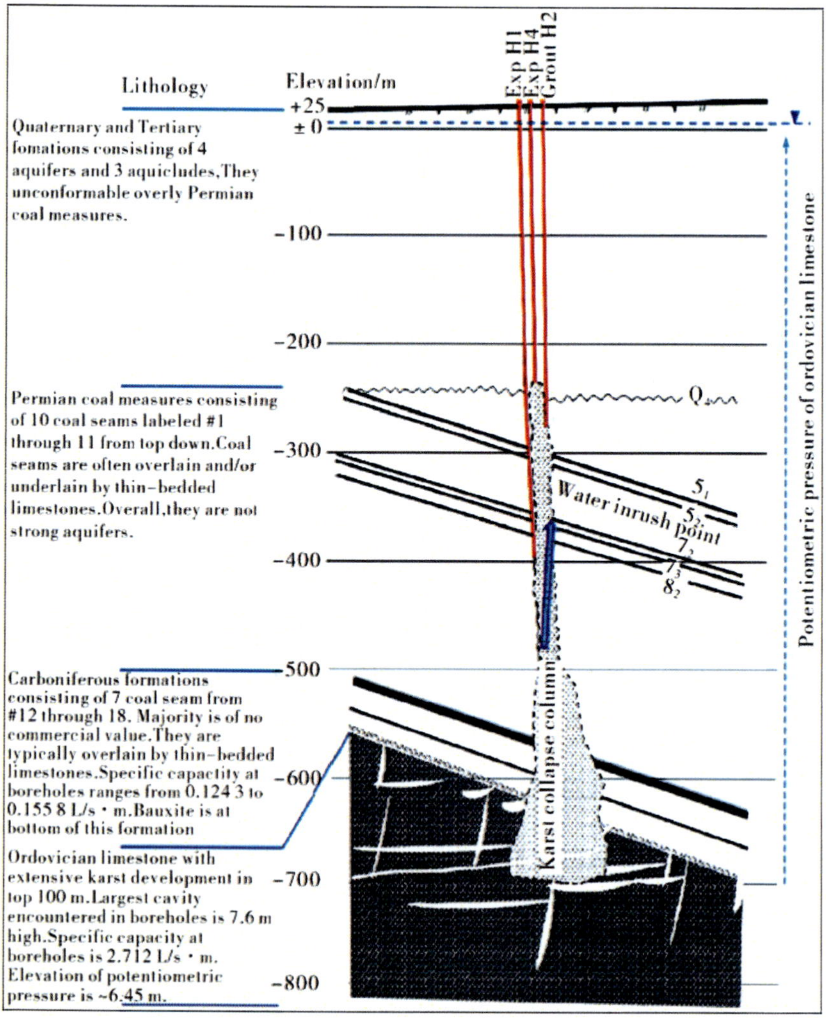

Fig. 4.21 Lithology and 1996 water inrush at Renlou Coal Mine, China

A second relatively smaller water inrush occurred at working panel $7_2 18$ in 1999. Timely recognition of water inrush indicators including pressure increase in front of the tunnel, deformation of tunnel support structures, water flow increasing from 5 to 9 m^3/h allowed quick responses so that the risk for a catastrophic water inrush was mitigated by systematic investigation and grouting. No serious damages occurred at the second encounter with production rate slightly affected. The investigation confirmed that this water inrush was also through a karst collapse column that was oval shaped with its long axis of 40 m and short axis of 30 m. It has developed to a position approximately 20 m below #8 coal seam. This collapse was also actively developing upward.

In June 2010, an anchor hole advanced in II 5_1 Tunnel encountered groundwater. The tunnel was at elevations from −718.6 to −720.1 m bmsl and was excavated for mining #5_1 coal seam. The tunnel was 440 m below the Quaternary and Tertiary formations and at least 300 m above the Ordovician limestone. The stratigraphy encountered by the tunnel consisted of fine sandstone and mudstone with petrified plant parts. Under normal conditions, water in these formations was limited and did not pose safety threat to underground workings. The tunnel also intercepted a normal fault, DF_8, with 5 m displacement. No water seepage was observed through the fault. The 2010 water inrush was discussed in this chapter.

4.5.2 *Water Source Discrimination by Temperature and Hardness Measurements*

As shown in Fig. 4.22, two exploration boreholes, 4-3 and 4-3′, were advanced at different angles to investigate the source and pathway of the groundwater encountered at the anchor hole. At an angle of 47°, borehole 4-3 intercepted fault DF_8, whereas at an angle of 40°, borehole 4-3′ intercepted a high-angled (75°) fracture. The maximum groundwater flow rate at borehole 4-3 was 2 m^3/h, and the maximum flow rate at borehole 4-3′ was 16 m^3/h.

Figure 4.23 shows changes of water temperature and hardness at borehole 4-3 from June 2010 to November 2011. Data at borehole 4-3′ had similar trends. Over a period of 17 months, the water temperature gradually increased from 33 to 41 °C. The normal earth temperature at this elevation in the mine is approximately 35 °C. The higher than normal temperature in the inflow water indicated that the water source was from deeper formations.

Two types of hardness are presented in Fig. 4.23, the total hardness was measured prior to boiling and permanent hardness was measured after boiling. Their unit is degree of General Hardness or German degree (dGH). Both types of hardness show a general trend of increasing. The total hardness increased from 8.34 to 60.58 dGH, whereas the permanent hardness increased from 0 to 48.81 dGH. The persistent increases in the hardness also suggest that the water sources were from the deeper formations. The measured temperature and hardness at end of the monitoring period were similar to those measured in the Ordovician limestone in the mine.

4.5.3 *Geophysical Investigations*

Geophysical surveys were conducted to investigate any geologically and hydrogeologically anomalous areas. Time domain electromagnetic methods (TDEM) and

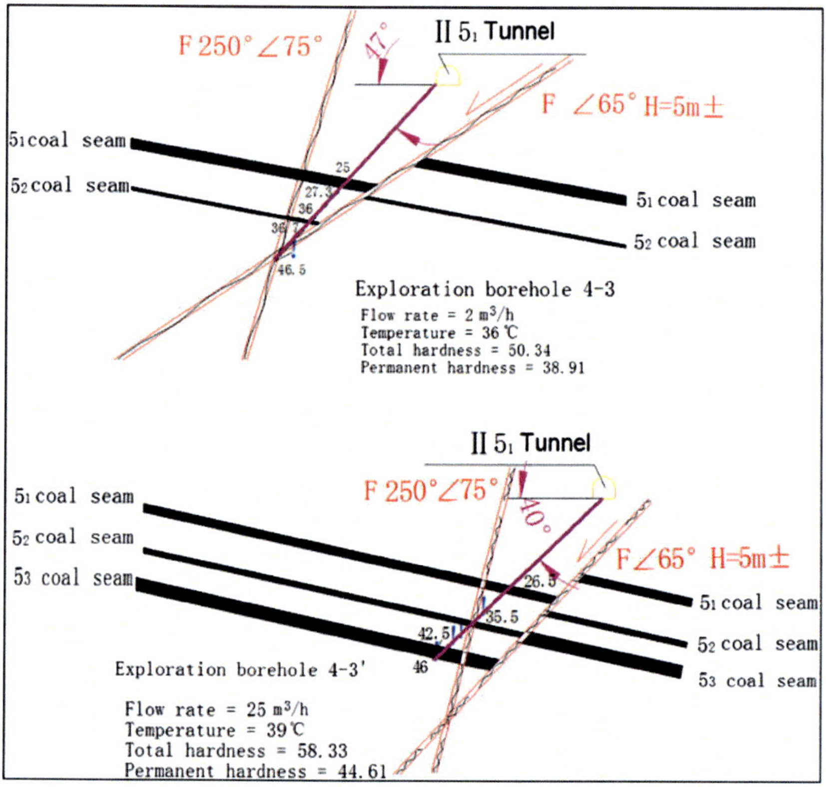

Fig. 4.22 Layout of boreholes 4-3 and 4-3′ for exploration of water sources

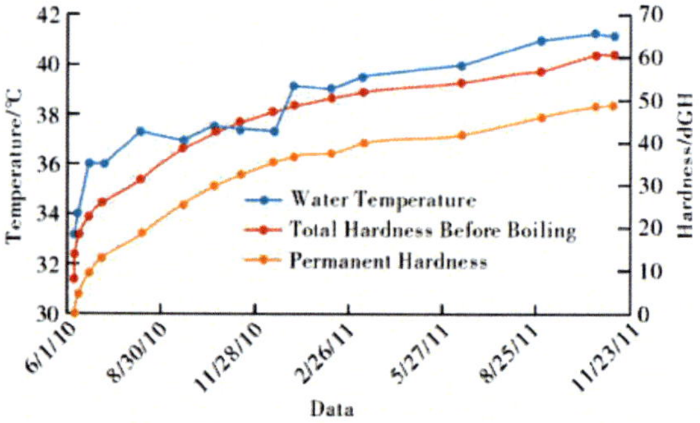

Fig. 4.23 Water temperature and hardness measurements at boreholes 4-3

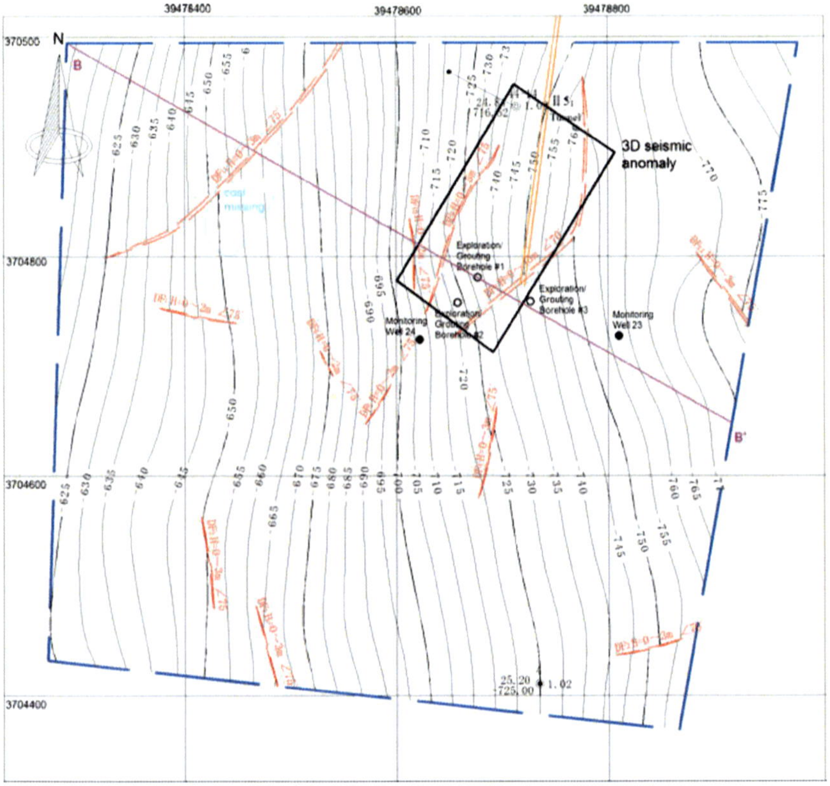

Fig. 4.24 Contours of 5_1 coal seam elevation as interpreted from 3D seismic survey

3D seismic were used on the ground surface, while TDEM, 3D seismic, earth resistivity imaging, and ground penetrating radar were used underground in II 5_1 Tunnel. Anomalies identified and verified from multiple geophysical techniques were considered the targets for further investigations. Figure 4.24 shows an example of 3D seismic interpretation of the #5_1 coal seam elevation. The rectangular is a geophysical anomaly in which the coal seam elevations were distorted. A vertical profile of the 3D seismic is shown in Fig. 4.25. The geophysical anomaly was interpreted to be associated with a karst collapse column.

4.5.4 Borehole Exploration and Grouting

Figure 4.24 shows the locations of three exploratory and grouting boreholes #1, #2, and #3, and two monitoring wells #23 and #24. The monitoring wells were installed to

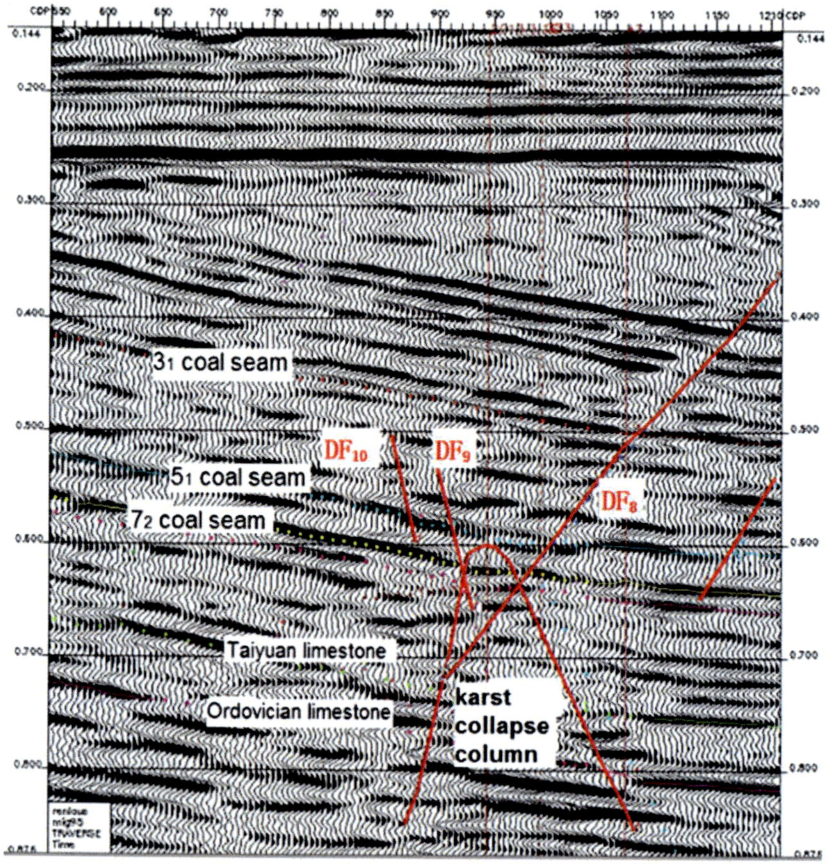

Fig. 4.25 A cross-section of 3D seismic survey (location of the cross-section is shown in Fig. 4.24)

monitor potentiometric pressures of the Ordovician limestone. The exploratory boreholes were drilled in the order of #1, #2, and #3. They were drilled in the geophysically interpreted anomaly to investigate the subsurface conditions, in particular, to confirm presence of the karst collapse column. If the karst collapse column was confirmed, these boreholes were then used as grouting holes to construct a water plug within the collapse feature. Borehole #1 was vertical, while directional drilling was used in both boreholes #2 and #3 to intercept the karst collapse column encountered in borehole #1. Figures 4.26 and 4.27 shows the profiles from borehole #1 through borehole #2 to monitoring well 24 and from borehole #1 through borehole #3 to monitoring well 23, respectively. Table 4.8 summarizes the pertinent observations in these three boreholes.

Both boreholes #1 and #2 encountered drill bit drops and total loss of circulation. The bit drops at boreholes #1 and #2 were 1.5 m and 2 m, respectively. The loss of circulation was greater than 72 m^3/h where the drill bit drops occurred. The drops

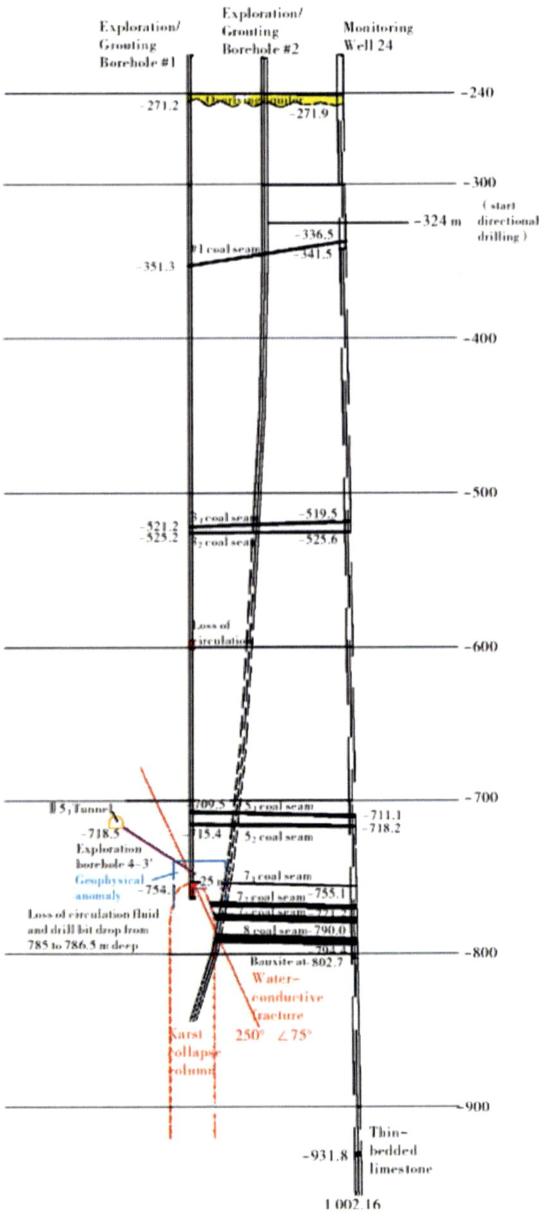

Fig. 4.26 Profile of exploration/grouting boreholes #1 and #2

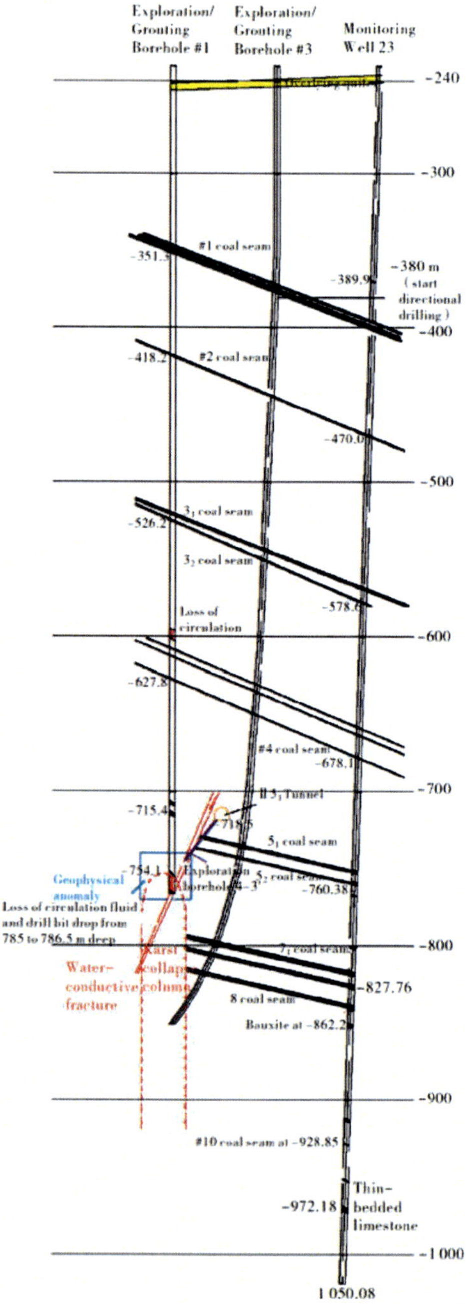

Fig. 4.27 Profile of exploration/grouting boreholes #1 and #3

Table 4.8 Summary of drilling parameters at three exploration/grouting boreholes

Parameters	Exploration/grouting boreholes		
	#1	#2	#3
Total depth (m)	920.48	920.57	986
Directional drilling		At 350 m toward borehole #1	At 380 m toward borehole #1
Depth to bedrock (m)	271.2	273.7	273.1
Depth to #1 coal seam (m)	376.8	360.0–369.6	395.05
Depth to #5 coal seam (m)	740.8	764.9	758.7
Loss of circulation intervals (m)	• 627–694 • 785–787	• 609–619 • 773–775 • 863–865 • 885	739 983
Drilling bit drop (m)	1.5 m from 785 to 786.5; likely entering collapse body	2 m from 773 to 775; likely entering collapse body	
Grouting (t)	1366 t (935 tons at 791 m; 81 t at 820 m; 159 t at 842 m; 191 t at 900 m)	1920 t (445 t at 803 m; 245 t at 835 m; 180 t at 855 m; 110 t at 865 m; 270 t at 865.5 m; 140 t at 885 m; 70 t at 905 m; 420 t at 920.57)	295 t at 911 m
Post-grouting water intake capacity (L/min m m)	0.00099	0.00091	0.00214

occurred at depths between 773 and 787 m. Such characteristics were atypical of the formations at these intervals unless a karst collapse column or large fracture was present. The exploratory and geophysical results suggested that this feature was likely a karst collapse column. The karst collapse column had an oval shape, and its dimensions were estimated to be 55 m in the long axis and 40 m in the short axis. The bottom of the karst collapse column was unknown, and its roof was approximately 20 m below #5 coal seam.

Grouting at boreholes #1 and #2 further confirmed that both boreholes intercepted the top of the karst collapse column. Grouting materials injected at borehole #1 were observed at borehole #2 as well as in II 5_1 Tunnel. Since borehole #3 was advanced after grouting at boreholes #1 and #2 was completed, it encountered cement-filled void and the amount of grout was significantly reduced in borehole #3. Water injection tests were conducted at all three boreholes before completing the grouting. The results (Table 4.8) demonstrated that the water intake capacity was less than the designed value of 0.01 L/min m m. The water intake capacity was calculated by water injection rate (L/min) divided by applied pressure (m of water) and test interval (m). Another indicator of the grouting success was that the post-grouting groundwater flow into II

5_1 Tunnel was minimal and the remaining water seeped from the overlying formations as verified with water quality parameters. The water levels in monitoring wells 23 and 24 were measured to ensure the long-term success of the investigation and remediation effort.

4.5.5 *Summary*

Karst collapse columns are karst features that were often encountered in mines of north China. Because these karst features could be several hundred meters high, they can connect multiple aquifers and lead water to underground workings under the following conditions:

- The karst collapse column is active and permeable to water.
- The karst collapse column is connected to water-bearing media (aquifers or water pools) that can supply a significant water source.
- The water pressure in the water-bearing media is higher than the elevation of the underground working area.

Because of potential damages that can be caused by these structures in mines, proactive detection of any concealed karst collapse columns and determination of their hydrogeological characteristics are essential components of mine water control and prevention programs in China. Multiple techniques including geochemistry, geophysics, directional drilling, grouting, and water injection testing were proved to be effective in detecting and characterizing a concealed karst collapse column in Renlou Coal Mine. The risks posted by the karst collapse column was successfully mitigated by directional grouting.

4.6 Design and Construction of Watertight Plugs in Permeable Karst Collapse Columns in Restoration of Flooded Dongpang Mine, China

4.6.1 *Mine Background*

Dongpang Mine is in Xingtai, China. As a large-scale coal mine, it has an annual production of 3 billion kg. The mine extracts primarily the #2 coal seam in the Shanxi Formation of the Permian Period. Mining takes places on two levels, −300 and −480 m below mean sea level (bmsl), respectively. A water inrush at the maximum flow of 70,000 m^3/h occurred in 2003 at working panel #2903 on the second level (−480 m bmsl) when an undetected karst collapse column was intercepted (Fig. 4.28). The incident flooded the entire mine. Water level and geochemical data indicated that

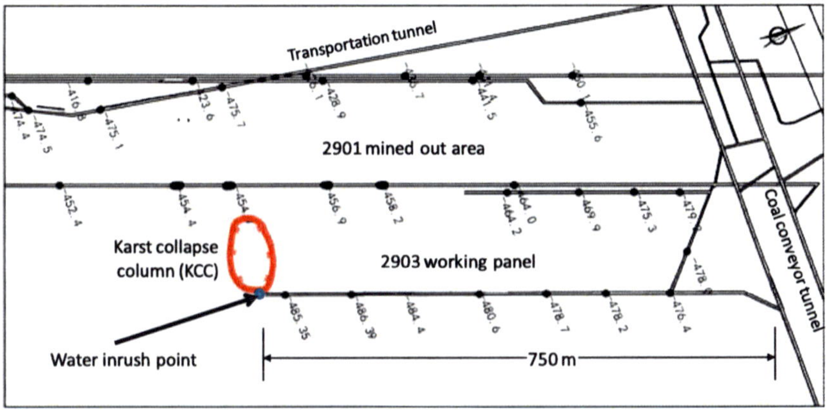

Fig. 4.28 Position of working panel #2903

the Ordovician limestone was the water source. The Ordovician limestone underlies the coal measures and is a highly karstified aquifer. Not only is the limestone several hundred meters thick, but the water in it is confined and pressurized. The karst collapse column acted as a passageway for the pressurized groundwater in the Ordovician limestone to flow upward into the underground working area, resulting in the water inrush (Fig. 4.29).

Characteristics and three-dimensional geometry of the karst collapse column (Fig. 4.29) were investigated with 30 exploratory boreholes. Figure 4.30 shows projected surface locations of these boreholes in relation to the identified karst collapse column. Table 4.9 summarizes the characteristics of the karst collapse column as observed in the borehole logs. As presented in Table 4.9, branch-out and directional drilling was used for target exploration. Seven boreholes, Z1, Z2, Z2′, Z3, Z4, Z5, and Z5′ were master boreholes. Multiple directional boreholes were advanced from these master boreholes at designed kickoff points. For example, Z1 is a master borehole, which consists of three branched-out directional boreholes, Z1-1, Z1-2, and Z1-3. The shape of the karst collapse column was delineated by advancing approximately 8000 m of exploratory drilling. The karst collapse column is shaped like a bowling ball when projected on surface, a larger area in southeast than northwest. The karst collapse column was more than 460 m high with the top at −240.61 m bmsl and the bottom in the Ordovician limestone below −700 m bmsl. Because the surface elevation is approximately 96 m amsl, the karst collapse column is buried more than 300 m below ground surface (bgs). A fracture zone was also identified around perimeter of the karst collapse column.

To restore the flooded mine and prevent future water inrushes, construction of a grout-plug within the karst collapse column was chosen as the preferred alternative to permanently cut off the hydraulic connection. Concrete mixture was pumped into the karst collapse column and the surrounding fractures through the grout holes, which were drilled on the surface. Aquifer tests at directionally drilled boreholes

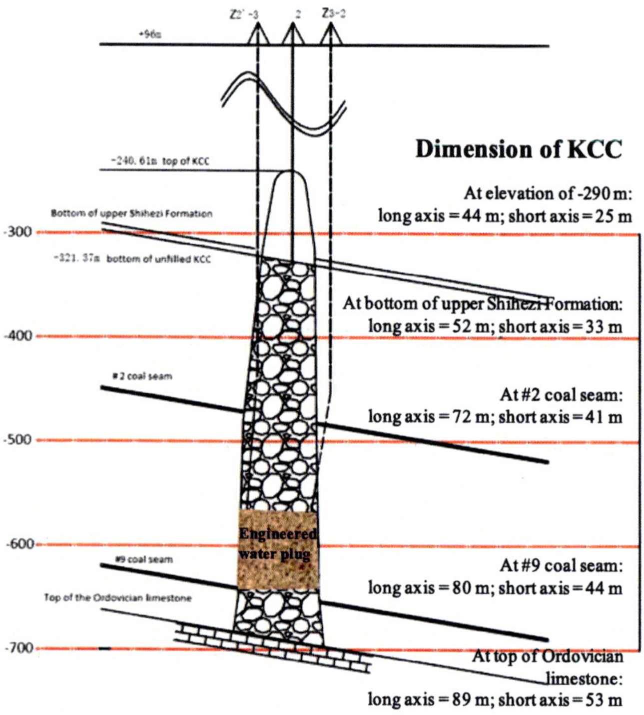

Fig. 4.29 Cross-section of karst collapse column intercepted at working panel #2903

were conducted underground to evaluate quantitatively the effectiveness of the plug. The top of the plug was at −568 m bmsl, while the bottom was at −641 m bmsl. The constructed grout-plug was 73 m high (Fig. 4.29).

4.6.2 *Construction of the Watertight Plug*

4.6.2.1 Grouting Method and Materials

The watertight plug was constructed in stages. The grouting was carried out at three levels: the upper level at −544 m bmsl, the middle level at −568 m bmsl, and the lower level at −641 m bmsl. The boreholes were drilled and grouted sequentially from the upper to the lower levels. Secondary and tertiary grouting holes were drilled after the primary grouting holes until the karst collapse column and its surrounding fractures were completely sealed. Because the grouting was in a collapse feature, there were uncertainties. Therefore, the actual operations were kept flexible. Under certain circumstances when continuous drilling was impossible, for example, large voids were encountered; the drilling tools were deflected; or the drilling fluid was

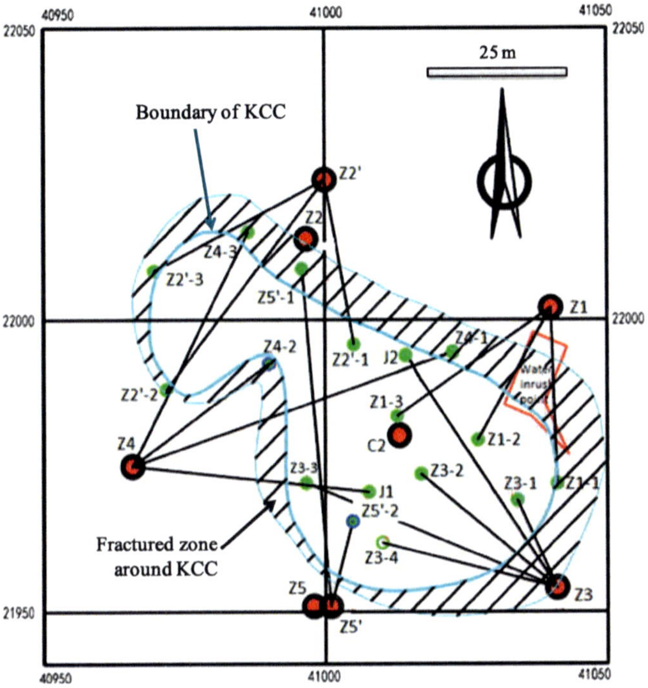

Fig. 4.30 Locations of exploratory boreholes

lost; the grouting target was adjusted accordingly. All the grout holes were thoroughly cleaned within 4 h after injection of single-cement slurry, immediately after injection of double-cement slurry, or immediately when pressure at borehole inlet was reset to zero.

The grouting materials used in this project included standard Portland cement 42.5, water from a local well, and soluble silicate as accelerator with module 3.0 and concentration 45 degree Beaume. Early strength agent of 0.03–0.05% triehanolamine (purity $\geq$ 95%) and 0.3–0.5% cooking salt were added in single-cement slurry. Formulary test was carried out on site to determine the compositions of the cement mixture, specific gravity, gelatification times (initial and final gelatification time), and other performance indices.

4.6.2.2 Control of Grouting Boreholes

Injection of grout was performed through directionally drilled boreholes. On the surface, the injection boreholes were arranged to cover the approximate extent of the karst collapse column as uniformly as possible. Because the plug was constructed at three sections and the boreholes were drilled in angles (Fig. 4.30), each borehole had three projected locations on the plane maps (Fig. 4.31). The projected area of

Table 4.9 Summary of exploratory boreholes

Borehole ID	Borehole parameters (m)		Depth of circulation loss (m)	Characteristics of karst collapse column (m)			Height of detected cavity (m)	Offset at borehole termination (m)	Note
	Depth	Advancement		Depth of encountering karst collapse column	Offset at total circulation loss	Offset when intercepting karst collapse column			
C2	336.61	336.61	325	336.61		0.51	80.76	0.51	Grouting hole for aggregates
Z1	550	550		368					Z1-1, Z1-2, Z1-3 are branched-out boreholes from master hole Z1
Z1-2	737.2	187.2	608.21	630	8	16.5		26.98	Fractures noted at 608.21 m
Z1-3	740.1	215.1						32.34	More circulation loss noted at 708.5 m
Z1-1	737.93	237.93						29.35	
Z2	366.2	366.2	365				1.2		Z2-1, Z2-2 are branched-out boreholes from master hole Z2
Z2-1	342	142						20.58	
Z2-2	391	91	371.5	371.5	20.94		15		Grouting hole for aggregates

(continued)

Table 4.9 (continued)

Borehole ID	Borehole parameters (m)		Depth of circulation loss (m)	Characteristics of karst collapse column (m)			Height of detected cavity (m)	Offset at borehole termination (m)	Note
	Depth	Advancement		Depth of encountering karst collapse column	Offset at total circulation loss	Offset when intercepting karst collapse column			
Z2′	550	550	365						Z2′-1, Z2′-2, Z2′-3 are branched-out boreholes from master hole Z2′
Z2′-2	742.6	192.6	613	640	10.6	17.7		46.19	Fractures noted at 613 m
Z2′-3	738.55	213.55							
Z2′-1	739.3	239.3	640		15.97			28.77	
Z3	550	550						0.86	Z3-1, Z3-2, Z3-3, Z3-4 are branched-out boreholes from master hole Z3
Z3-2	740.18	190.18	572.6	580	0	4.6		32.03	Borehole at boundary of the fracture zone
Z3-3	742.68	217.68						48.04	
Z3-1	738	238						20.06	

(continued)

Table 4.9 (continued)

Borehole ID	Borehole parameters (m)		Depth of circulation loss (m)	Characteristics of karst collapse column (m)			Height of detected cavity (m)	Offset at borehole termination (m)	Note
	Depth	Advancement		Depth of encountering karst collapse column	Offset at total circulation loss	Offset when intercepting karst collapse column			
Z3-4	740.6	230.6						32.63	
Z4	550	550							Z4-1, Z4-2, Z4-3 are branched-out boreholes from master hole Z4
Z4-2	757	207						33.41	
Z4-1	745.5	220.5						59.95	Started circulation loss at 667 m
Z4-3	740	230						40.02	
Z5	716.16	716.16						0.608	Z5-1, Z5-2 are branched-out boreholes from master hole Z5
Z5-1	582	147		528.13		8.91		15.15	Increase in circulation loss at 516.45 m
Z5-2	708.81	208.81	615	631	6.1	8		16.82	

(continued)

Table 4.9 (continued)

Borehole ID	Borehole parameters (m)		Depth of circulation loss (m)	Characteristics of karst collapse column (m)			Height of detected cavity (m)	Offset at borehole termination (m)	Note
	Depth	Advancement		Depth of encountering karst collapse column	Offset at total circulation loss	Offset when intercepting karst collapse column			
Z5′	550	550						0.73	Z5′-1, Z5′-2 are branched-out boreholes from master hole Z5′
Z5′-2	738.06	188.06	699.88					15.04	
Z5′-1	748.2	188.2						58.49	J2 is a branched-out hole from master hole Z3
Z3-J2	742	247						48.27	J1 is a branched-out hole from master hole Z4
Z4-J1	741	246						41.75	
Total		8,788.23							

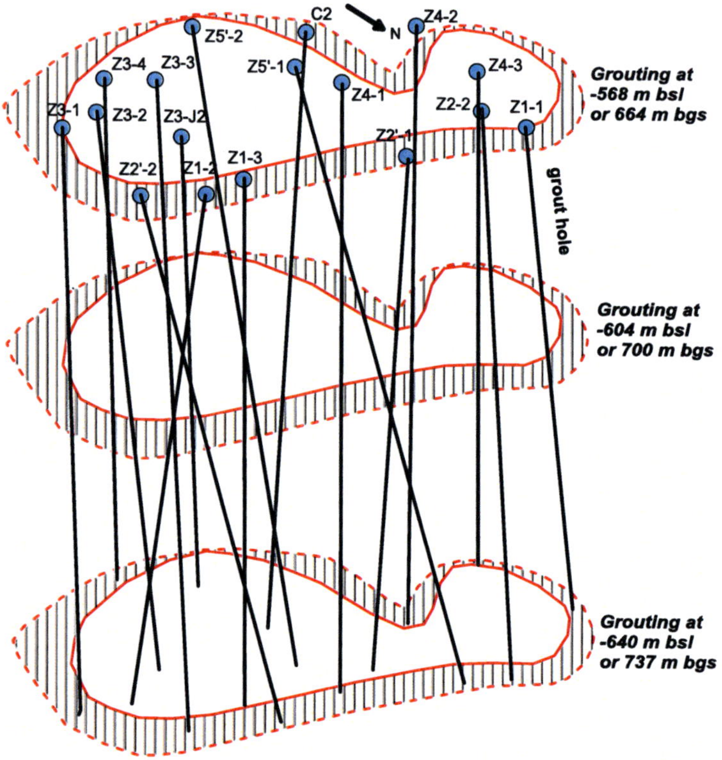

Fig. 4.31 Three-dimensional grout holes on three elevation levels

the karst collapse column on surface was 3546 m^2. A total of 17 grout holes were drilled, which resulted in coverage of 222 m^2 per borehole. Of the 16 grout holes, six were primary ones with borehole spacing from 9.5 to 30 m. The primary holes had a control radius of approximately 14 m. There were five secondary and five tertiary grout holes, respectively, and they were drilled with smaller borehole spacing because of their relatively smaller control radii. Table 4.10 summarizes the parameters of the grout holes.

4.6.2.3 Grouting Stages and Procedures

The grout engineering was divided into the following four stages:

Stage I—Filling Grout: This was the initial stage of each primary grout hole. The main purpose at this stage was to inject a great amount of cement slurry to fill voids or caves. Cement slurry was mainly single slurry of high concentration. This was the initial stage of each primary grout hole. The main purpose at this stage was to inject a great amount of cement slurry to fill voids or caves. Cement slurry was mainly single

Table 4.10 Summary of grout boreholes

Parameter	Primary holes	Secondary holes	Tertiary holes
Number of boreholes (borehole IDs)	7 (Zl-2, Z2′-2, Z3-2, Z4-2, Z5-2, Z5-2′, Z5′-2)	5 (Zl-3, Z2′-3, Z3-3, Z4-1, Z5′-l)	5 (Zl-1 Z2′-1, Z3-1, Z4-3, Z3-4)
Control radius (m)	14.3	10.6	8.8
Spacing between boreholes (plane projection, m)	9.5–30	8–24	5.2–17

slurry of high concentration. Injection was carried out quantitatively and repeatedly under static water pressure without pressure accumulation at the borehole inlet. The main frame of the grout-plug was created by sealing major water passageways.

Stage II—Increasing Pressure: The purpose of this stage was to seal relatively large fractures and solution-enlarged fissures and to consolidate the grouting results of the early stage. Single cement slurry was primarily used. Injection was carried out under static water pressure. During the injection process, the pressure at the borehole inlet was gradually increased. However, the pressure at the borehole inlet was controlled to be no more than 2 MPa.

Stage III—Hydrodynamic Grouting: Each injection borehole was under static water pressure. After completion of stage II, the pressure at the borehole inlet was relatively high as a result of the reinforcement grouting. The directionally drilled aquifer test holes in the −480 m level drift were examined for drainage. The residual inflow rates at these boreholes provided the data to evaluate the water-sealing effect of the early stages. On the other hand, if there was a significant water flows at the holes, the artesian flows produced an artificial flow field around the holes. Under such a hydrodynamic condition, carefully designed grouting procedures were used to seal the water-conducting passages.

Stage IV—Reinforcement Grouting: This was the last stage of all injection boreholes. During the grouting, pressure began to increase at the borehole inlets. High pressure injection sealed small fractures between injection boreholes and increased the strength of grout-plug. Single cement slurry was primarily used at this stage. Grouting under static and hydrodynamic pressures was carried out alternatively.

The Stage I grouting was carried out at boreholes Zl-2, Z2-2, and Z5-2. They were the first boreholes drilled, in which cement slurry was mainly vertically diffused. No cement slurry crossing was observed between boreholes. A total of 8908 m^3 of cement slurry was injected. Appearance of pressure at the borehole inlets from the water-pressing tests was the indicator of the end of Stage I and the beginning of the Stage II grouting. At Stage II grouting, there was no initial pressure at the borehole inlets. Lateral diffusion was also noted at this stage as the pressure at the later drilled Z3-2 was measured at 1.0 MPa.

Many fractures were filled at the late period of the pressure grouting. As a result, the number of open fractures was reduced, thus the cement intake decreased. The height of the slurry column in the boreholes rose continuously. When the boreholes were filled with the cement slurry, the pressure began to rise at the mouth of boreholes. This symbolized the starting of the reinforcement grouting stage. At this stage, the pressure before injection at the borehole inlets was generally more than 2 MPa. The water-pressing pressure at the mouth of boreholes continued to increase with the increase of the injection pressure. The grouting stopped when the water pressure at the mouth of boreholes was more than 3 MPa while the injection pressure was at 2 MPa. Then, the water-pressing tests were carried out to check whether the grout-plug was effective in stopping water flows.

Boreholes Zl-2, Z2-2, and Z5-2 were drilled into the karst collapse column. They were characterized by soft strata, fast footage, and unstable walls. Collapses or cave-ins occurred frequently during drilling. The washing fluid (drilling mud) consumption was all greater than 50 m^3/h. The rocks in the karst collapse column were highly permeable. In order to reduce the washing fluid consumption, the drilling was carried out after grouting in the upper section of the grout-plug. Grouting was performed immediately regardless of the designed target if the borehole was in the grouting section, but the loss of circulation liquid was very serious, and the drilling was too difficult to continue. Grouting of the entire section was often accomplished through multiple intermittent grouting.

For the late drilled primary boreholes Z3-2, Z4-2 ad Z5-2, their drilling conditions were obviously improved, except at borehole Z3-2. A relatively large fluid loss occurred at Z3-2. There was basically no or very little consumption of washing liquid at Z4-2 and Z5-2. Although the drilling was fast and the efficiency was high, the borehole walls were relatively stable. For the last drilled secondary and tertiary holes, the washing fluid consumption decreased significantly. None of them had a continuous drilling liquid loss.

Boreholes Z3-J2 and C2 were used for filling skeletal materials in the upper part of the karst collapse column and also used for water level observation. At the time of grouting, the skeletal material injection was carried out in these two boreholes. With gradual formation of the grout-plug, the water inrush at boreholes Z3-J2 and C2 became more frequent as the water body in the upper karst collapse column was progressively confined and isolated. Because the water level in the Ordovician limestone near the karst collapse column was at elevation of 36 m amsl, the water in the upper karst collapse column did not represent the water level of the Ordovician limestone.

At Stage I, there was basically no cement slurry crossing between injection boreholes. As the grouting pressure was increased and the grouting entered into the subsequent stages, the cement slurry crossing, and water inrush became frequent. For example, cement mud crossing occurred between Z5-2′ and Z2-2, Zl-2 and Z5-2, Z3-2 and Z5-2. The slurry crossing indicated that the slurry movement changed from the initial vertical diffusion to lateral diffusion subsequently. As the fracture passages between boreholes were gradually filled, the cement slurry injected at each borehole overlapped with each other to form one continuous solid grout-plug.

During reinforcement grouting, no more cement mud crossing was observed between boreholes. However, the water inrushes at monitoring boreholes became more frequent, which suggested that the large fractures had been sealed and the grout mixture was pushed into the small and micro fractures. Under relatively high injection pressure, the residual water in small fractures was squeezed out and drained through the non-grouted boreholes. This artificially created flow field also produced the conditions for hydrodynamic grouting in small fractures.

4.6.3 Completion Criteria of Grouting

Injection pressure and pump delivery rate are commonly used standards in determining the completion of a grouting project. During the initial stage of grouting, the cement slurry had to overcome the maximum static water pressure to push out the water in voids and fractures. Based on the control radius of each borehole and past engineering experiences, the injection pressure used in the project was approximately 1.5 times the maximum hydrostatic pressure. The grouting process itself was a dynamic and iterative process. The grouting parameters were adjusted constantly to the dynamic geological conditions of the boreholes according to any additional geological and hydrogeological data that were obtained during the grouting. The completion criteria should also consider factors such as the length of injected section and cement mud performance parameters to be more objective.

In order to precisely describe the permeability of the injected section and to reflect actual grouting effect, we proposed in this project the use of water intake rate as the completion criteria. The water intake rate was defined as the amount of water intake along 1 m of grout section under 1 m hydraulic head pressure within 1 min. It was determined by conducting specifically designed water pressure tests at the end of grouting. Such water pressure tests are in principle similar to the Lugeon test that is often used at dam sites. The Lugeon test takes place in section of 5 m intervals. The water is injected into the section isolated by packers, which are mechanically, hydraulically or pneumatically expanded and pressed against the borehole walls. A Lugeon unit is defined as the amount of water received by rock mass within 1 m borehole length at a pressure of 10 bars in 1 min.

The water intake rate is closely related to the permeability of a rock mass. A very small water intake rate usually suggests that the rock mass is less permeable to water. Table 4.11 shows seven levels of rock permeability and their relationship to water intake.

The water absorption rate was a comprehensive index that took into account the following factors such as hydraulic pressure in grout section, pressure at the mouth of water-pressing boreholes, and the length of grout section. The result was not influenced by the cement mud performance. Such a completion criteria reflected more precisely the function of the grout-plug as a hydrologic barrier. Because the purpose of each grouting stage was different, the threshold values for completion at each stage were different.

Table 4.11 Relationship between permeability and water intake in rock masses

Rock description	Permeability K (cm/s)	Water intake q (L/min m m)	Characteristics of rock mass
Impermeable	$K < 10^{-6}$	$q < 0.001$	Competent rock with fractures of equivalent opening <0.025 mm
	$10^{-6} \leq K < 10^{-5}$	$0.001 \leq q < 0.01$	Rock mass with fractures of equivalent opening of 0.025–0.05 mm
	$10^{-5} \leq K < 10^{-4}$	$0.01 \leq q < 0.1$	Rock mass with fractures of equivalent opening of 0.05–0.1 mm
Permeable	$10^{-4} \leq K < 10^{-2}$	$0.1 \leq q < 1$	Rock mass with fracture of equivalent opening of 0.1–0.5 mm
	$10^{-2} \leq K < 100$	$q \geq 1$	Rock mass with fractures of equivalent opening of 0.5–2.5 mm
	$K \geq 100$		Rock mass with fractures of equivalent opening ≥2.5 mm

- Water intake threshold for the early grouted boreholes Z1-2, Z5-2, and Z2′-2 was 0.05 L/min m m. The relaxed threshold was set because the grouting at this stage was to fill large voids and fractures in the karst collapse column, to transform highly permeable rock into weakly permeable rocks, to ensure relatively stable borehole walls, and to reduce borehole cave-in accidents.
- Final water intake threshold for other primary boreholes was 0.005 L/min m m. At gradually increased grouting pressure, the medium-sized fractures were filled; the loose rocks in the karst collapse column were reinforced; and the grout section was transformed into less permeable rock mass.
- For the late grouted three primary boreholes Z3-2, Z4-2, and Z5′-2, the water intake threshold was 0.005 L/min m m. After the large fracture passages and voids were filled in the karst collapse column, the walls of these boreholes were relatively stable. The borehole collapses rarely occurred.

The water intake threshold for the secondary and tertiary grout boreholes was 0.001 L/min m m. After a great amount of cement mud was injected into the primary boreholes and the cement mud was sufficiently hydrated and consolidated, the small and micro fractures were further filled and consolidated through the subsequent hydrodynamic grouting and reinforcement grouting. This strict threshold would ensure that the entire grouting section of the karst collapse column became an impermeable or weakly permeable grout-plug.

4.6.4 Grout Intake Distribution

From August 13, 2006 to September 13, 2006, a trial injection was carried out in the upper section of borehole Z5-2, in which 3.5 million kg of cement was consumed. The formal injection of grout started on September 9, 2006 and ended on May 29, 2007. A total of 49,339 m^3 cement mud was used, including 41.6 million kg of #42.5 Portland cement, 8000 kg triethanolamine, 80,000 kg industrial salt and 1500 kg soluble silicate. Figure 4.32 shows the percentage distribution of injected cement in each borehole. The injected cement mixture at boreholes Zl and Z2′ in the north of the karst collapse column accounted for 24% and 30%, respectively, while the sum of the injected volumes at Z5 and Z5′ accounted for 26%. These boreholes were the primary injection boreholes of the project and indicated that the strata in the south and the east of the karst collapse column were fragmented and broken and consisted of voids and fractures. Coring of Z3 and Z3-J2 further confirmed that rocks at that part of the karst collapse column were fragmented, and the core recovery was extremely low. The injected volume at borehole Z4 in northwest of the karst collapse column accounted for only 6%, suggesting relatively competent rock masses at the location. In addition, Z4 was drilled after some of fractures may have been filled.

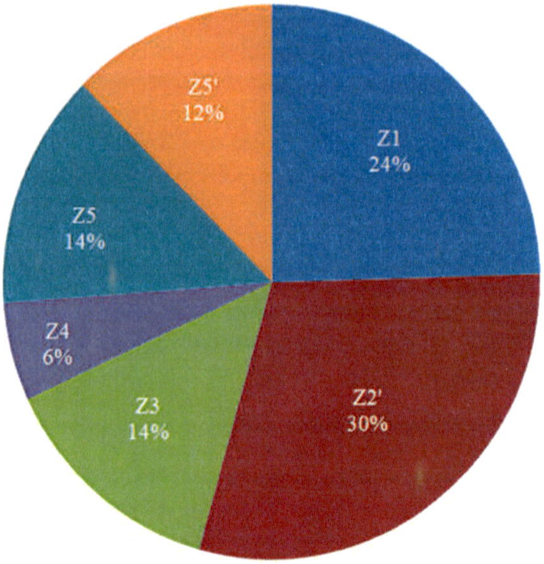

Fig. 4.32 Grout intake distribution among grouting holes

4.6.5 Evaluation of Plug Effectiveness

4.6.5.1 Borehole Coring

Z4 and Z3 were used to check the quality of the constructed grout-plug. Boreholes Z4-J-1 and Z3-J2 were cored from 610 m to the bottom of boreholes. The average core recovery at Z4-J1 was 37%. The borehole encountered normal formation from 610 to 657.4 m. Coring was relatively easy. Cement fills were observed in the core samples, as shown in Fig. 4.33.

4.6.5.2 Water-Injection Test

Water-injection or water-pressing test was conducted at depths of 610 m, 664 m and 742 m at Z4-J1 and Z3-J2, respectively. The water absorption rate calculated from the water-injection test was smaller than the pre-determined threshold, i.e., 0.001 L/min m m.

4.6.5.3 Water Release Test

The quality of the grout-plug was defined by the water-sealing effect of the plug, the recharge inflow of the karst collapse column, and the spatial hydraulic conductivity in the upper section of the plug. It was quantitatively evaluated by conducting an underground aquifer test in which 242 m^3 of water was drained. The drainage borehole was in the −480 m drift with a total depth of 341.6 m. Small-scale faults were intercepted at the depth of 289.6 m. The borehole intercepted the karst collapse column at 308 m, and a water inrush occurred at 310 m with the maximum inflow of 14.2 m^3/h. The inflow was stabilized at 11.8 m^3/h.

There were six surface observation boreholes Z3-J2, Z2, Z2′-1, Z3-1, Z4-3 and Z5′-2 (Fig. 4.34). Of them, Z3-J2 was the principal observation borehole and was drilled into the center of the upper section of the karst collapse column. The maximum drawdown at Z3-J2 was 36.3 m. The water level was recovered to the level of 16.4 m amsl. During the water drainage test, the water levels at Z2′-l, Z4-3, and Zl-1 did

Fig. 4.33 Core samples at select grout holes

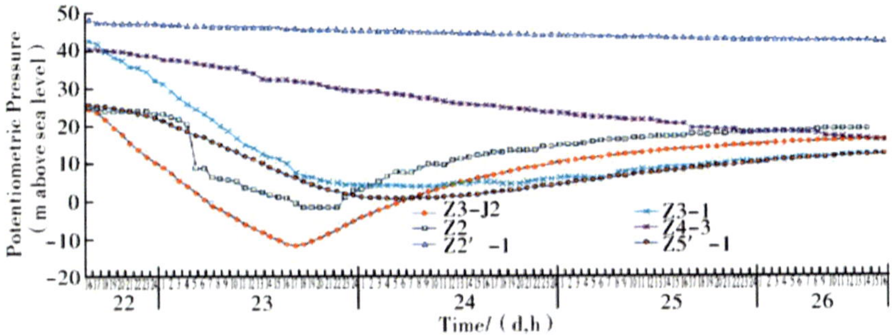

Fig. 4.34 Potentiometric pressure measurements at observation wells in response to water release test

not change, which suggested a lack of connection between these three boreholes and the drainage borehole. The water in these boreholes were confined and isolated with no hydraulic connection to either the Ordovician limestone or the voids in the upper part of the karst collapse column.

In summary, the results from the core samples, the water-injection test, and the water drainage test showed clearly that the constructed grout-plug had successfully cut off the hydraulic connection between the coal seam and the lower Ordovician limestone.

4.6.6 Summary

Design and construction of watertight plugs in underground mines is a complicated process, and the design equations available for plug length estimate can give results over a wide range. Aside from following best practices in mining engineering and civil engineering, the practitioners' knowledge and experience are critical in their design and implementation. In response to a water inrush in which pressurized groundwater in the underlying limestone suddenly flowed into the mine through a karst collapse column in Dongpang Mine, China, a vertical watertight plug was successfully constructed within the karst collapse column to restore the flooded mine to full operation. The grouting was conducted on the surface by drilling boreholes and pumping grout into the collapse column, while monitoring of the effectiveness of the plug was carried out underground by artesian flow tests through directionally drilled wells. The completed watertight plug is 73 m high at depths from 664 to 737 m below groundwater surface. Construction of such a large in-situ plug in a concealed collapse feature overcame several engineering challenges. A key component to the success of this project was that all the grout holes including primary, secondary, and tertiary ones were directed precisely to the designed locations. In addition, the technique, quality control, grouting stages, and completion criteria of each stage were tailored to the site-specific conditions. Core samples collected at boreholes and underground water-injection and water drainage tests verified effectiveness of the completed plug.

4.7 Utilization of Paleokarst Crust of Ordovician Limestone in Water Inrush Control in Sihe Coal Mine, Shanxi Province

4.7.1 Introduction to Paleokarst Crust

In China, karst develops and distributes widely in rocks that range in age from Archeozoic to Cenozoic but are predominantly Paleozoic. Carbonate rocks occupy an area of about 3.25 million km^2 of the country: of this bare karst is some 1.25 million km^2 and the rest is covered or buried karst. Groundwater in the karstified carbonate rocks is a valuable natural resource for local people. Unfortunately, many mineral deposits such as coal, iron, lead and zinc, gold, aluminum, and copper are located in between, or above, or below the karst aquifers. The majority of the well-known deposits with large quantities of water (pumping water over 1 m^3/s) are karst water-impregnated deposits.

One of the major impacts of the groundwater on the mining activities is the unpredictable occurrence of water inrush, in which a significant amount of water suddenly invades the underground working areas from the underlying aquifer under potentiometric pressure. Struggling against the water inrushes in the past few decades has led to multiple theories and techniques of evaluating the inrush risks and preventing the pressurized water from the limestone into underground mining areas. These theories emphasize the potentiometric pressure in the Middle Ordovician limestone and the aquifuge between the coal seams and the Middle Ordovician limestone. The term "aquifuge" is defined in this paper as an impermeable body of rock which contains no interconnected openings or interstices and therefore neither absorbs nor transmits water and it is interchangeable with the term "aquiclude." The commonly used water inrush coefficient is defined by:

$$\mathrm{T} = \mathrm{P}/\mathrm{M}$$

where T is the water inrush coefficient; P is the potentiometric pressure; and M is the thickness of aquifuge. The greater the water inrush coefficient is, the more likely occurs a water inrush. The potentiometric pressure is typically measured at monitoring wells in the Middle Ordovician limestone while the limestone has been treated as a unified aquifer system. Determination of the aquifuge thickness is less definitive but it is an equally important parameter.

With increase in mining depth of the Carboniferous-Permian coal seams, the potentiometric pressure of the Ordovician groundwater becomes higher, indicating a higher risk of water inrush. It is particularly important to study the heterogeneity of

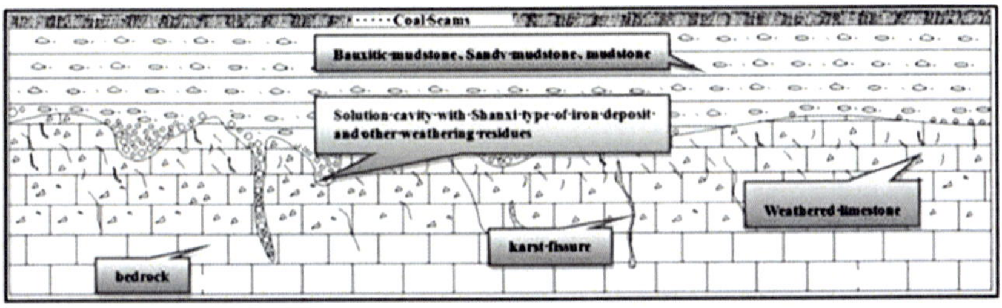

Fig. 4.35 Palaeokarst profile in Fengfeng Formation

karst development in the underlying Middle Ordovician limestone and determine any impermeable strata that may prevent the pressurized karst water from bursting into coalmines. Detailed analysis of the exploratory borehole data suggests presence of a paleokarst crust (hereinafter referred to as crust) at the top of Middle Ordovician, which is often referred to as the Fengfeng Formation (Fig. 4.35). The enormous variation in water-resisting capacity makes recognition and application of the crust less straightforward in coal mining.

4.7.2 *Characteristics of Paleokarst Crust at Sihe Mine*

Sihe Mine is in Shanxi Province, which is a large anthracite base with an annual production of 1200 million tons. The average thickness of the Fengfeng Formation is approximately 100 m. The distance between the lower coal seams and the Middle Ordovician limestone is approximately 20 m. The water pressure exerting on the coal seam floor ranges from 1 to 3 megapascal (MPa). The potential of water inrush has become a key factor restricting its sustainable production at the mine.

4.7.2.1 Lithology of Fengfeng Formation

Figure 4.36 shows the relationship between the coal seams and the underlying limestone formations. The Fengfeng Formation at Sihe Mine consists of cyclic deposition of pure limestone and argillaceous limestone with variable thicknesses. This cyclic deposition structure is common in the middle of the Fengfeng Formation with a thickness between 20 and 50 m. The top 20 m of the Fengfeng Formation is pure limestone which is the country rock for the paleokarst with any fractures filled with calcite and weathering residuum. This altered top section of the Fengfeng Formation provides a geologic barrier to groundwater flow.

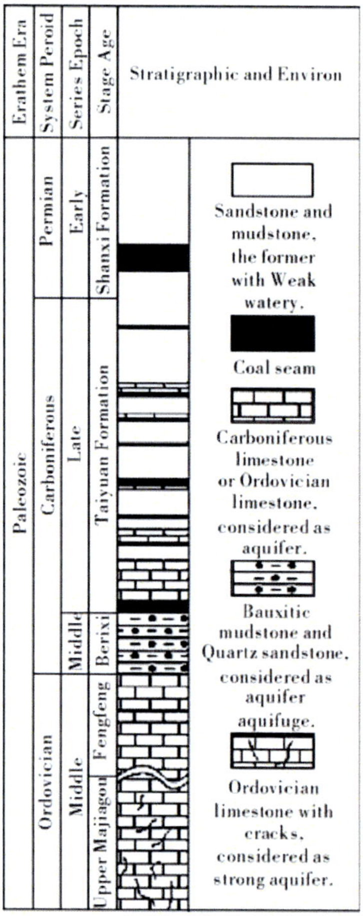

Fig. 4.36 Lithology in Sihe Coal Mine

4.7.2.2 Fractures in Fengfeng Formation

Three types of fractures were observed in the Fengfeng Formation: cross joints, inter-laminar joints, and micro-structural fractured fractures (Fig. 4.37). The upper part of the Fengfeng Formation mainly develops the micro-structural fractures with the joints being secondary. The lower part of the Fengfeng Formation is a water-bearing region and the plaeo-karstification and weathering is not as obvious as the top part. The main form of fractures is either oblique cross joints or inter-laminar joints.

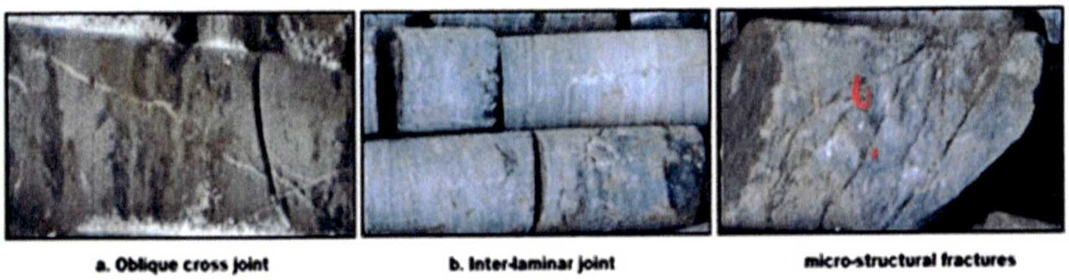

Fig. 4.37 Three different forms of fractures in Fengfeng Formation

Fig. 4.38 Cemented fractures in Fengfeng Formation

4.7.2.3 Filling Materials

Physical and chemical weathering over 150 Ma has produced a large amount of breccia cemented with calcite, the weathering residues, or argillaceous filler. As shown in Fig. 4.38, the high extent of cementation and filling in breccia or fractures makes the paleokarst crust in the Fengfeng Formation possessing strength and water-resisting capacity.

4.7.2.4 Water-Resisting Capacity

Pumping tests, flow-rate logging, chemical analyses of groundwater, and water-pressure tests were conducted to determine the water-resisting capacity and spatial distribution of the aquifuge in the Fengfeng Formation. The hydrogeological characteristics of the Fengfeng Formation were then compared with those of the underlying upper Majiagou Formation (Fig. 4.36), which is often considered as a strong aquifer.

The water-pressure tests were conducted in thirteen boreholes. The test results show that the top 35 m of the Fengfeng Formation has an extremely poor permeability ranging from 0.00002 to 0.00024 m/d, which indicates that this part of stratum has a certain water-resisting capacity. The water pressure in the Ordovician limestone gradually decreases in the underlying stratum. The thickness of this part is 8.6 m and the water pressure drop is 3.27 MPa. As a result, the water resistance coefficient is approximately 0.38 MPa/m.

4.7.3 Hydrogeogical Properties of Fengfeng Formation

4.7.3.1 Drilling Fluid Loss

Of seventeen Ordovician hydrogeological boreholes drilled in the Fengfeng Formation in Sihe Mine, fourteen boreholes lost drilling fluid at normal rates from 0.1 to 1 m^3/h and only three encountered significant loss of more than 1 m^3/h in the lower segment of the Fengfeng Formation. Of the sixteen boreholes drilled through the upper Majiagou Formation, eight boreholes' consumption of drilling fluid was at normal rates and significant or massive loss occurred in the other eight boreholes in the middle segment of the upper Majiagou Formation. The difference of drilling fluid loss between the Fengfeng Formation and upper Majiagou Formation suggests that the fractures in the Fengfeng Formation might have been filled and the formation is not as conducive to water flow and storage.

4.7.3.2 Pumping Test Results

Table 4.12 lists the thirteen boreholes in which pumping tests were conducted in both the Fengfeng Formation and the upper Majiagou limestone. The specific capacity of the Fengfeng Formation is between 0.0009 and 0.0058 L/s m. Those values are two

Table 4.12 Result of Ordovician limestone pumping tests

Borehole	Comparison of groundwater level (m)			Comparison of specific capacity		
	Fengfeng Formation	Majiagou Formation	Difference	Fengfeng Formation	Combination of Majiagou and Fengfeng Formations	Ratio of combined and Fengfeng Formation
SWY2	555.8	521.2	34.6	0.000942	0.000859	0.9
SC2	449.7	483.5	33.8	0.001	0.188	188
GZ	450.9	481.5	30.6	0.00588	0.0366	6.2
SWY1	482.4	493	10.6	0.0026	0.144	55.4
SB1001	511.4	488.79	22.61	0.000477	0.13947	292.4
SB 1002	546.54	508.94	37.6	0.00299	0.112379	37.6
SB 1003	544.92	513.54	31.38	0.004489	0.06624	14.8
SB 1004	495.46	478.75	16.71	0.004721	0.044552	9.4
SB 1005	466.84	477.75	10.91	0.0004	0.00052	1.3
SW2	497.7	513.1	15.4	0.00305	0.00324	1.1
SWY3	488.7	487.6	1.1	0.00478	0.594	124.3
SC3	489.2	487.1	2.1	0.00435	0.0268	6.2
SW1	512	488.4	23.6	0.00403	0.302	74.9

to three orders of magnitude smaller than those of the upper Majiagou Formation combined, which ranges from 0.0005 to 0.594 L/s m. In addition, the water level of the Fengfeng Formation is higher than that of the upper Majiagou Formation, with a difference ranging from 1.1 to 37.6 m. The tests have proved that the water yield of the Fengfeng Formation is much weaker than that of the upper Majiagou Formation.

4.7.3.3 Hydrogeochemical Test Results

Figure 4.39 shows the Piper diagrams based on the chemical analyses of the water samples collected in these thirteen boreholes at which the pumping tests were conducted. The main anion in the Fengfeng Formation is SO_4^{2-}, secondly Cl^-, while the content of HCO_3^- is little. The main cations in the Fengfeng Formation are Ca^{2+} and Mg^{2+}. The composition of mixed samples of the Fengfeng and Maijiagou Formations is quite different. The main cation is Ca^{2+}, and the main anion is SO_4^{2-}, followed in HCO_3^-. These observations indicate that the Fengfeng Formation is in a more reductive environment than the Majiagou Formation. The water in the Fengfeng Formation has poor circulation, which is evidenced by the lack of HCO_3^-, a common cation in good groundwater flow conditions.

4.7.3.4 Flow Rate Logging Results

Based on the flow rate logging results in six boreholes, no flow date could be obtained in the Fengfeng Formation because of its poor water yield. In comparison, flow rate data were readily obtained in the Majiagou Formation. In addition, more boreholes showed greater water flow rates in the middle of the Majiagou Formation than in the top zone.

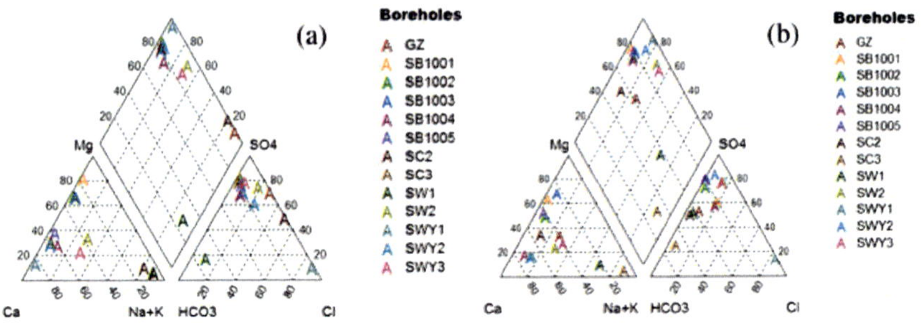

Fig. 4.39 Piper diagrams for groundwater in **a** Fengfeng Formation and **b** Majiagou Formation

4.7.4 Thickness of Aquifuge in Fengfeng Formation

In North China, the water production of an aquifer is often divided into four classes based on specific capacity:

- >5.0 L/s m—very strong
- 1.0–5 L/s m—strong
- 0.1–1.0 L/s m—moderate
- $\leq$0.1 L/s m—weak.

The specific capacity of the Fengfeng Formation is between 0.0009 and 0.0059 L/s m, which fall in the category of weak aquifer or relative water-resisting aquifuge. The thickness of relative aquifuge ranges from 35 to 70 m based on the summary statistics. This finding results in the thickness of aquifuge being much larger than that shown in Fig. 4.35 for the mining site. The ending result is that more areas are considered safe for mining.

4.7.5 Summary

Recognition of water-resisting ability of the upper Ordovician limestone is significant in evaluating water rush risks. Based on studies on the water-resisting capacity of the paleokarst crust in top part of the Middle Ordovician limestone, the paleokarst crust of 35–70 m thick is present and can be treated as an aquifuge in Sihe Mine. As a result of this study, more than 90 million tons of coal resources are considered to be not threatened by the underlying pressurized water. The heterogeneous nature of the karst aquifer provides us a new approach evaluating water inrushes when mining above the karst aquifer. Investigation of hydrogeological features of the Ordovician limestone requires multidisciplinary approaches. The water-resisting capacity of the paleokarst crust in the coal fields of North China may be variable in different areas. In some areas it may not exist, especially in areas with geologic structures. However, recognition of this potential significance of the paleokarst crust may increase the coal production or longevity of the operation of a mine.

Chapter 5
Prevention and Control of Mine Water Hazards from Overlying Aquifers

5.1 Water Control Technology for Overlying Thick-Bedded Sandstone Fissure Aquifer in Hujiahe Mine, Binchang, Shaanxi

5.1.1 Mine Background

Hujiahe Mine is in the Binchang Mining area of Huanglong Coalfield with a designed annual output of 5.0 Mt/a. The mine adopts vertical shaft and single level development and uses fully mechanized mining method. The roof is managed with natural caving. The Quaternary aquifer in Hujiahe Mine is not in close hydraulic connection with the lower Cretaceous aquifer. At the same time, the Anding Formation, Zhiluo Formation and Yan'an Formation are mostly composed of inter-bedded sand and mudstone. Because these formations produce little water, they have minimal effect on mining. The thickness of the sandstone aquifer of the Cretaceous Luohe Formation ranges from 200 to 300 m. The specific yield is 0.20 L/(s.m), whereas the average hydraulic conductivity is 0.198 m/d. The distance between the #4 coal roof and the bottom of the Luohe Formation aquifer varies between 100 and 190 m. The Luohe Formation is the main recharge aquifer to the mine.

5.1.2 Exploration and Prevention Techniques for Water Hazards Posed by the Overlying Thick Sandstone Fissure Aquifer

Comprehensive analysis of preliminary geological exploration and regional hydrogeological conditions concludes that the Luohe sandstone aquifer overlying #4 coal

S. Dong et al., *Methods and Techniques for Preventing and Mitigating Water Hazards in Mines*, Professional Practice in Earth Sciences,
https://doi.org/10.1007/978-3-030-67059-7_5

seam is the main threat to mine safety during the mining. This overlying aquifer is the focus of mine water control in Hujiahe Mine.

The exploration methods of thick sandstone aquifers are different from those of the commonly encountered thin aquifers. Because of the great thickness of the aquifer, the obvious heterogeneity, and the relatively good water richness, the combined ground and underground exploration methods are used. The detailed hydrogeological investigations take place on surface in combination with three-dimensional geophysical surveys. Methods used in the underground investigation are primarily electrical resistivity imaging.

5.1.2.1 Supplemental Hydrogeological Investigation

(1) The exploration drilling layout follows the following principles:

- Meet the requirements of hydrogeological monitoring. Based on the geological conditions revealed by the current production, adhere to the principle of using one borehole for multiple purposes. For example, a hydrogeological test holes can be used as a groundwater observation well.
- The hydrogeological characteristics of the thick sandstone aquifer need to be investigated to determine the hydraulic connection of different sections.
- Design the borehole layout in combination with ground geophysical prospecting results in the first mining area.
- The borehole layout also needs to meet the calculation requirements of mine water inflow and control the boundary conditions.
- Other factors to be considered in determining the borehole locations include terrain conditions, construction conditions, borehole protection requirements, and factors that may not affect coal mine safety production.

(2) Borehole layout and construction

Based on the principles listed above, the Hujiahe Mine conducted a supplemental hydrogeological exploration project in February 2012. The layout of the supplementary survey project is shown in Fig. 5.1, and the workload of the supplementary survey project is shown in Table 5.1. The project includes the following tasks:

- A total of nine surface hydrogeological boreholes were completed with a total footage of 6,470.33 m.
- Borehole geophysical logs were conducted in all boreholes and they included conventional geophysical logging and flow logging at discrete intervals.
- 21 single-well pumping tests were conducted.
- Three multi-well pumping tests were conducted.
- One large-scale inter-resistance pumping test was conducted in which multiple pumping wells were involved.
- One water injection test was conducted.

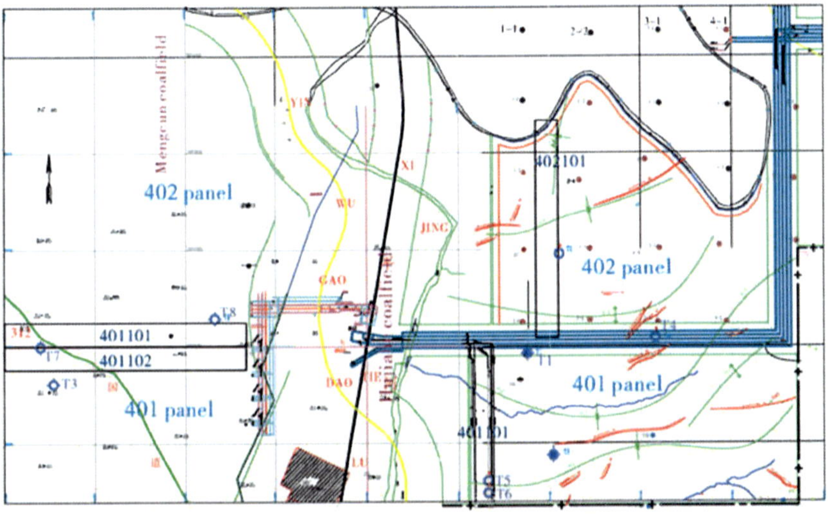

Fig. 5.1 Layout of Hujiahe Mine hydrogeological exploration project (not in scale)

- A hydrodynamic automatic monitoring system was established. The monitoring system consists of nine monitoring wells installed on the ground surface and the underground permanent drainage system.
- 106 water samples were collected for chemical analyses.
- 16 water samples were collected for isotope (δD (deuterium), δO^{18} (oxygen), and δT (tritium)) analyses.
- 60 groups of rock samples and 236 specimens were collected in eight boreholes for physical and mechanical properties testing.
- After the first mining face was recovered, the loss of drilling fluid was recorded and borehole televiwer was used in boreholes T5 and T6 to investigate and quantify the heights of the cave-in zone and fracture zone in the overlying formations.

(3) Results of hydrogeological exploration

The following results are achieved from the hydrogeological studies:

- Based on the lithological characteristics of the exploration boreholes, intervals between 80 and 100 m above the floor of Luohe Formation consist of coarse-grained aquifers such as conglomerate and medium-grained sandstone. These intervals are underlain by less permeable mudstone, sandy mudstone, or siltstone. The mudstone formations are up to tens of meters thick and can be used to divide the thick sandstone into the upper and lower aquifers.
- There are differences in water temperature between the upper and lower aquifers in the Luohe Formation. The water temperature in the upper aquifer of Luohe Formation is 19 °C, whereas the water temperature in the lower section is 21 °C. The difference is 3 °C.

Table 5.1 Summary of hydrogeological supplemental exploration activities in Hujiahe Mine

Item	Boreholes									Notes
Borehole number	T1	T2	T3	T4	T5	T6	T7	T8	T9	9
Location	Between 401 and 402 mining areas	402 mining area	Mengcun first mining area	401, 402, between mining areas	401 first mining face	401 first mining face	Mengcun first mining area	Mengcun first mining area	401 mining face	9 exploratory boreholes were drilled
Purpose	Groundwater long-term monitoring wells				Two-zone exploration boreholes		Groundwater long-term monitoring wells			The two zones are mining induced caved-in zone and fracture zone in the overlying formations
Final depth (m)	668	711.73	825.74	733	609.22	653.55	754.95	763.75	750.39	6,470.33 m in total
Final borehole formation	Yan'an Formation									
Geophysical logging (m)	668	698	823	725	605	450	750	753	740	6,212
Flow logging (m)	625	667	820	725	380	450	723	680	707	5,777
Single-hole pumping test (times)	3	3	2	3	–	1	3	3	3	21 times
Multi-well pumping test	When borehole T5 is used as the pumping hole borehole T6 is used as the observation hole. When borehole T6 is used as the pumping hole, borehole T5 is used as the observation hole. When borehole T3 is used as the pumping hole, borehole T7 is used as the observation hole. Three times in total									3 times

(continued)

Table 5.1 (continued)

Item	Boreholes									Notes
Borehole number	T1	T2	T3	T4	T5	T6	T7	T8	T9	9
Location	Between 401 and 402 mining areas	402 mining area	Mengcun first mining area	401, 402, between mining areas	401 first mining face	401 first mining face	Mengcun first mining area	Mengcun first mining area	401 mining face	9 exploratory boreholes were drilled
Mutual resistance pumping tests between boreholes	Boreholes T1 and T9 are used as main pumping holes, and boreholes T2, T4, T5, T6 and T9 are used as observation holes. One time									1 time
Cementing	1	1	1	1	2	2	1	1	1	11 times
Sealing	2	2	2	1	2	2	2	2	2	
Water samples for chemical analysis (group)	12	9	9	7	3	4	9	9	12	21 groups in reconnaissance, 74 groups in pumping tests, and 11 in groundwater inrush points, a total of 106 groups
Isotope analysis				1		3			1	
Rock mechanics test (group/specimen)	3/10	7/24		2/10	17/65	12/59	4/18	11/37	4/13	60 groups and 236 specimens

(continued)

Table 5.1 (continued)

Item	Boreholes									Notes
Borehole number	T1	T2	T3	T4	T5	T6	T7	T8	T9	9
Location	Between 401 and 402 mining areas	402 mining area	Mengcun first mining area	401, 402, between mining areas	401 first mining face	401 first mining face	Mengcun first mining area	Mengcun first mining area	401 mining face	9 exploratory boreholes were drilled
Status	Groundwater level gauging				Sealed		Groundwater level gauging			1 main station, 3 flow monitoring devices, 9 water level monitoring devices
Civil survey of borehole coordinates (times)	2	2	2	2	2	2	2	2	2	Placement and re-measurement, 18 times in total
Hydrogeological reconnaissance	Conducted on four transects from east to west and north and south; recorded more than 30 observation points; collected 21 sets of fully analyzed water compositions; and 12 sets of isotope water samples. Collected and reviewed the previous research results of the region and analyzed the hydrogeological data exposed by the mine. The reconnaissance results in a more comprehensive understanding of the Jinghe-Malian River secondary groundwater system where the Hujiahe mine field is located and has achieved the design purpose									
Height measurements in two zones	One method is based the observation of borehole T5 and T6 fluid losses. The observation of the borehole leakage fluid and water level are carried out before and after mining; The second method is based the digital imaging televiewer observations on boreholes T5 and T6 walls									The two zones refer to mining induced caved-in zone and fracture zone in the overlying formations

- The water quality is obviously different between the upper and lower aquifers of the Luohe Formation. The average pH in the lower section is 8.6, whereas the average pH in the upper section is 7.8. The Na^+ and Cl^- concentrations in the lower section is much higher than those in the upper section, while the HCO_3- concentration is lower. The lower section salinity and TDS are much higher than the upper section.
- The hydrogeological parameters of the upper and lower aquifer of the Luohe Formation are significantly different. Based on the results of the aquifer pumping tests in the upper section of the Luohe Formation, the hydraulic conductivity ranges from 0.25 to 1.35 m/d with an average of 0.64 m/d, and the specific yield ranges from 0.2746 to 0.4385 L/(s.m) with an average of 0.3621 L/(s.m). The water richness is medium. On the other hand, the hydraulic conductivity of the lower section of the Luohe Formation is 0.0730 m/d and the specific yield is 0.0228 L/(s.m). The water richness is weak.
- The results of borehole flow logging show that there is a total of 30 outflow sections of the nine boreholes that were drilled into the Luohe Formation. Of the 30 outflow sections, 23 are in the upper section of the Luohe Formation and seven are in the lower section of the Luohe Formation.
- Hydraulic connection between upper and lower sections of the aquifer of Luohe Formation is poor. Hydraulic connection tests were conducted between boreholes T5 and T6. When the lower section of borehole T6 was pumped, borehole T5, which is 126 m away from T6, was observed simultaneously. The water level did not decrease at T6 in response to pumping at T5, it instead increased by 0.58 m, indicating no hydraulic connection between the upper and lower aquifers.

The detailed hydrogeological investigations conclude that the sandstone of Luohe Formation in Hujiahe Mine can be divided into upper and lower water-bearing sections. The hydrogeological conditions of the upper and lower sections are different, and the hydraulic connection between these two sections is poor.

5.1.2.2 Integrated Geophysical Surveys

Comprehensive geophysical exploration was carried out in the first mining area of Hujiahe Mine. Transient electromagnetic method (TEM) and high-resolution reflection coefficient method (i.e., high-resolution electrical sounding method, or GF method for short) were used. The transient electromagnetic method was used mainly to determine the water richness of the Luohe sandstone, while the high-resolution reflection coefficient method was used to determine the degree of fissure development in the Luohe sandstone and the contact surface between the Luohe and Yijun Formation sandstones.

The three-dimensional seismic survey was applied on surface to an area of 20.38 km^2. The ground transient electromagnetic survey completed 46 transects with a total of 2256 TEM measurement points. High-quality and high-precision geophysical data were obtained for precise geological interpretations.

The comprehensive geophysical investigations identified 16 faults in the first mining area, including eight faults with displacements of more than 10 m and 7 faults with displacements of 5–10 m. The permeability of the faults was evaluated in combination with characteristics of the upper and lower segments. Based on the aquifer thickness changes in the Luohe Formation and Yijun Formation, five water-rich areas were delineated. The distance between the top of #4 coal seam and the bottom of the aquifer and its changes were analyzed and studied. The implementation of the geophysical project is of great practical significance for (1) ascertaining the occurrence of coal seams and the distribution of geological structures within the first mining area, (2) determining the water-bearing characteristics of the overlying strata; (3) delineating the water-rich anomalous areas; and (4) formulating scientific and rational development plan for the mine. The results provide an important basis for water inrush prediction and prevention and control of water hazards.

5.1.2.3 Underground Geophysical Prospecting

(1) Channel wave seismic exploration technology and electrical resistivity imaging

Subsequent to formation of the working face, the channel wave seismic survey and electrical resistivity imaging were used to verify the development of the geological structures in the overlying formation of the first mining face, including faults, distribution of river scouring zone, and water-bearing zones. The survey results show that there are no faults and collapse columns with displacements greater than 5 m in the working face, and there are a few faults with displacements between 3 and 5 m. The coal thickness of the entire working face is more than 22 m except a small area where the coal thickness is approximately 15 m because of river scouring. Three resistivity lows were identified in the roof of the working face (Fig. 5.2). These resistivity lows are interpreted to be partially water-containing in the overlying sandstone of the Zhiluo Formation. A vertical water-conducting fissure zone may be present in the overlying formation.

Hujiahe Mining Company installed 6 pre-dewatering boreholes prior to mining and 4 post-dewatering boreholes after mining in the three identified water-rich anomalies. The boreholes are terminated at the top of the Zhiluo Group sandstone with a cumulative footage of 780 m. At completion of the boreholes, the amount of water inflow is not significant, typical flow rates between 1 and 2 m^3/h and the maximum flow rate at 4 m^3/h. The geophysical anomalous areas were verified through the dewatering boreholes, and water in part of the roof formation was drained in advance to ensure the safe mining.

(2) Advanced dewatering in tunnel excavation

The advanced exploration and dewatering boreholes are drilled prior tunnel excavation for the underground working faces. The design of exploration and drainage is prepared by the department of geological survey and water prevention. A full-time exploration and dewatering team are responsible for the implementation. The mining

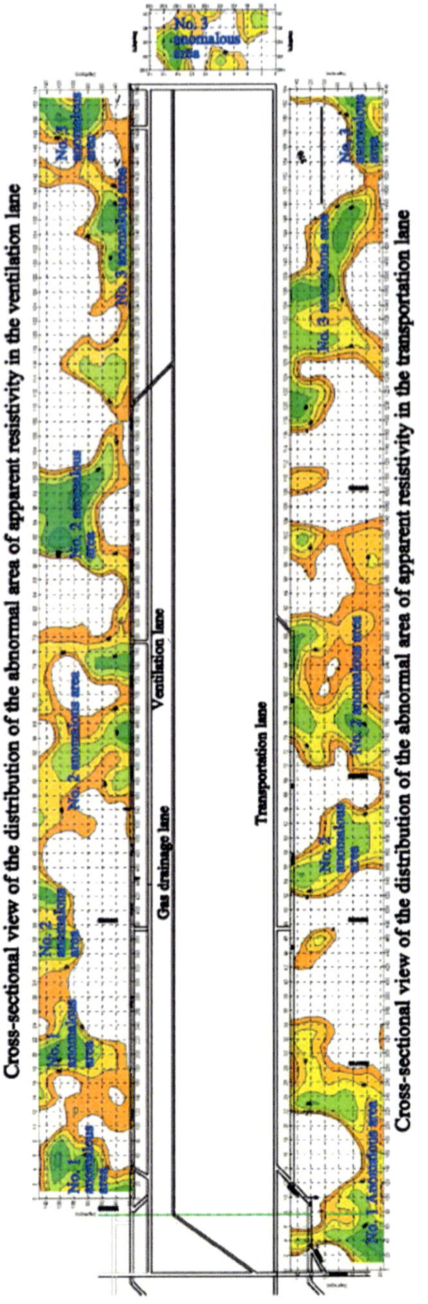

Fig. 5.2 Cross-sectional view of apparent resistivity anomaly in 401,101 working face

company adopts YDZ (B) type direct current method instrument, YHQ-G all-round borehole trajectory measuring instrument, and special drilling rig to perform the advanced detection and dewatering.

5.1.2.4 Detection of Damage Height in Overlying Formations

The overlying Cretaceous aquifer is the main water source that threatens Hujiahe Mine. When the coal seam is mined, the mining induced water-conducting fracture zone may extend to the Cretaceous aquifer. Water in the aquifer can suddenly flows into the mine, posing a threat to the safe mining. Therefore, the height of the mining induced fracture zone in the overlying formation is of great significance to prevention and control of water in the extensive sandstone aquifer.

(1) Layout of exploration engineering

Based on the ground conditions, geological conditions and requirements for damage height investigation in the overlying formations, two exploration boreholes T5 and T6 were drilled on the 401,101 working face to measure the damage height, which consists of the caved-in zone and fracture zone. Both boreholes are located approximately 30 m from the return air circulation lane of the 401,101 working face and 270 m and 150 m from the cut hole, respectively. The layout of the exploration boreholes is shown in Fig. 5.3.

(2) Introduction to exploration methods

The estimated height of the roof damage zone is based several pieces of evidence such as observation during the ground construction process, simple hydrogeological observation in the boreholes and the borehole visualization. A comprehensive analysis that integrates the change of water inflow during mining and the hydrogeological conditions of the mine is often carried out to make the final decision.

The content and technical requirements of exploration observation are shown in Table 5.2.

During the drilling process, the main indicators are bit drop, bit stuck, drilling speed change, and air suction phenomenon. The simple hydrogeological observations include drilling water level changes and drilling fluid loss observations. The method for observing the amount of drilling fluid loss is in accordance with fracture zone

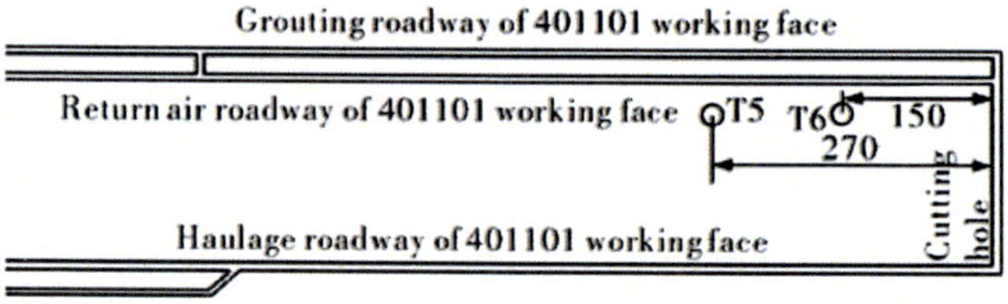

Fig. 5.3 Schematic diagram of exploration boreholes T5 and T6

Table 5.2 List of exploration and observation contents of roof diversion crack zone

Observation items	Observation content	Observation instruments and tools	Observation accuracy
Loss of drilling circulation fluid	The original water volume in the water source tank, the amount of water added during drilling, the remaining water volume, the measurement time, the number of drilling feet, the depth of the hole; record the drilling depth and time without returning water	Measuring ruler and drill pipe	• Depth error <0.15% • Ruler error <5 mm • Linear footage error <10 mm • Hydrogeological depth error <100 mm
Hydrogeological monitoring	After drilling a hole, observe the water level at a frequency of not less than 1 time per hour according to the hydrological telemetry device	Remotely operated monitoring device	
Borehole visual observation	Observation of borehole wall, lithology, fractures, water flow, and others	Downhole video camera or borehole televiewer	

observation guidance. Drilling fluid loss can be divided into loss of the circulation fluid, interruption of the fluid circulation, and atypical fluid circulation. The borehole visualization uses the borehole televiewer to observe the development height and connectivity of the cracks in the rock layer, the fracture of the rock mass and the distribution of the collapsed rock blocks.

(3) Determining the height of water-conductive fracture zone from simple hydrogeological observations

In investigating the two mining induced damage zones in the overlying formations, both the groundwater level change and the drilling fluid loss were observed in borehole T5. The results are shown in Figs. 5.4 and 5.5.

The drilling fluid loss in the upper part of borehole T5 fluctuated slightly, while a large amount of drilling fluid loss occurred at the depth of 424 m and below. No circulation fluid was returned at times, which suggests a total fluid loss. The loss of drilling fluid in the borehole was noticeable until completion of the drilling. The change of the internal groundwater level is consistent with the change of drilling fluid loss. The groundwater level fluctuation occurred at depth of approximately 424 m. The depth to groundwater increases from 50 m to more than 300 m when the drilling depth is greater than 424 m.

The simple hydrogeological observations of the boreholes suggest that the top of the water-conductive fracture zone is at a depth of approximately 424 m. Because

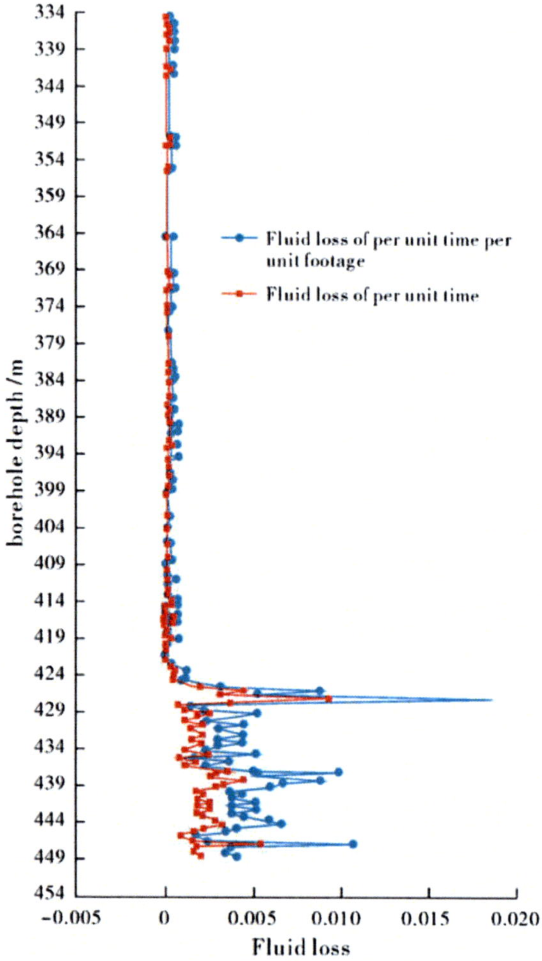

Fig. 5.4 Drilling fluid loss profile in T5

the depth to the coal seam is 644.3 m, the development height of the damage zone is approximately 220.3 m.

(4) Determination of height of the water-conductive fracture zone by borehole televiewer

Borehole televiewer technology was used to visualize the borehole walls. The results show that borehole T5 has fewer fractures and in smaller scales in the layers above 418.9 m deep. Fracture frequency and sizes are relatively greater below the depth of 418.9 m (Fig. 5.6a). Similar observations were made at borehole T6 with the threshold depth at 447.7 m (Fig. 5.6b). These observation results show that the top of the water-conductive fracture zone at borehole T5 is at a depth of 418.9 m, whereas the top of the water-conductive fracture zone at borehole T6 is at a depth of 447.7 m. The borehole

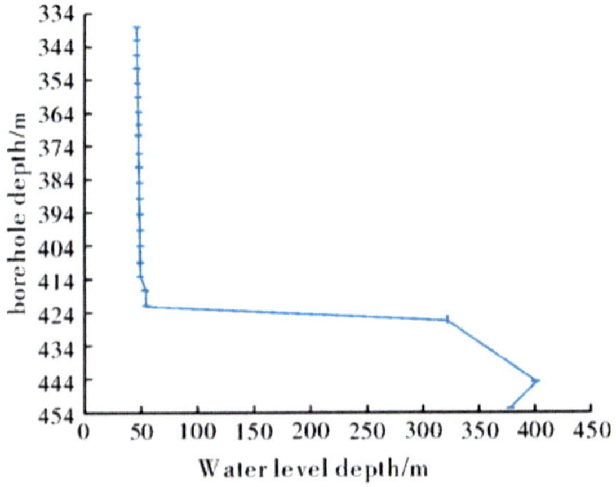

Fig. 5.5 Change in depth to groundwater in T5

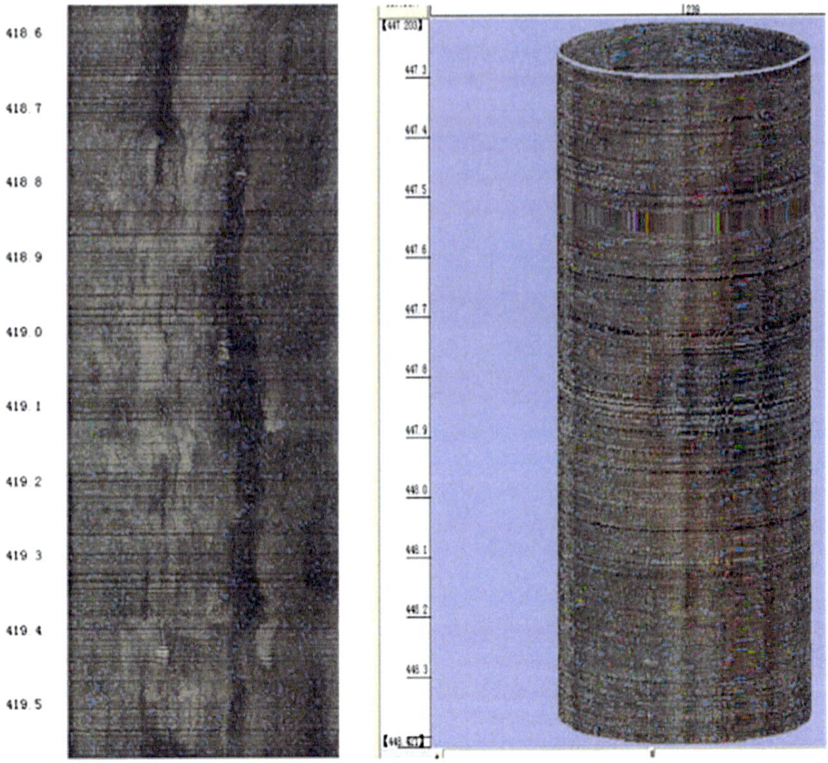

(a) (b)

Fig. 5.6 Borehole televiewer images

televiewer results are consistent with the simple hydrogeological observations, and the results are reliable. Based on the borehole televiewer results, the height of the water-conductive fracture zone at borehole T5 is approximately 225.4 m, which is 22.3 times the thickness of the coal seam. The development height of the water-conductive fracture zone at borehole T6 is 194.1 m, which is 16.04 times the thickness of the coal seam.

(5) Numerical simulation of mining induced destruction in overlying formations

FLAC3D was used to numerically simulate the development of the water conductive fracture zone under current mining conditions. In the modeling process, the actual measurements of pressures are used to evaluate the influence of pressures more objectively.

The initial stage of mining was assumed to advance 8 m each time until it reaches 80 m. After that the mining advances 16 m each time until it advances to 2 times the length of the working face. The results are shown in Fig. 5.7. The maximum height of the water-conductive fracture zone reached 171 m when the working face was at 320 m.

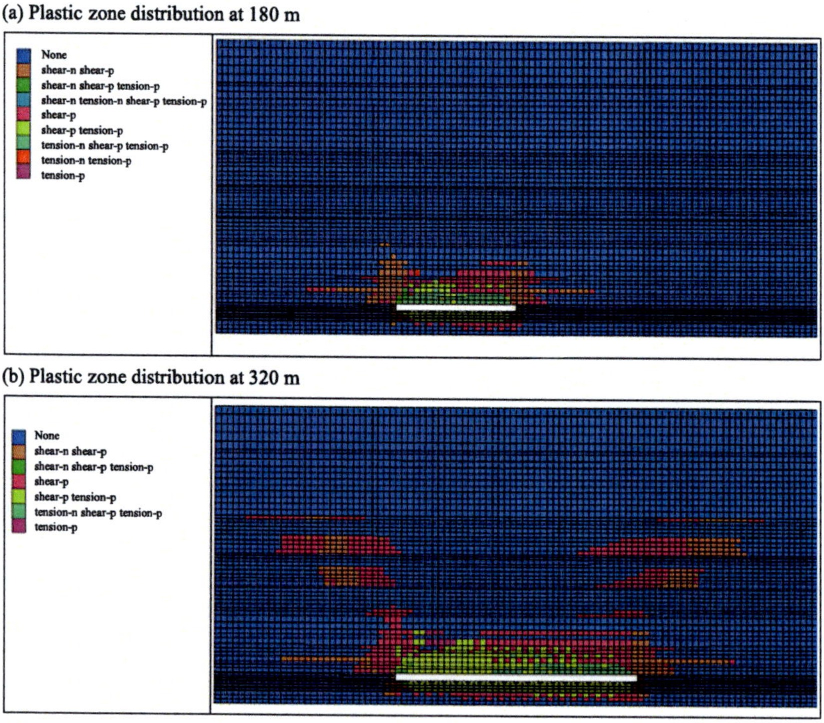

Fig. 5.7 Numerical simulation of development of water conductive fracture zone in the overlying formations

Fig. 5.8 Relationship between water inflow from working face and water level change

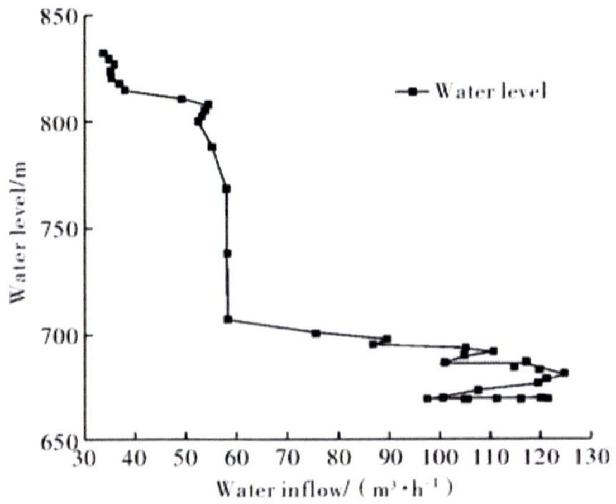

(6) Verification of the height of water conductive fracture zone

The geological conditions in the Binchang mining area provide conditions for the verification of the estimated height of the water-conductive fracture zone, and the water inrush risk on the working face will increase greatly if the water-conducting fracture zone reaches the Luohe Formation aquifer. The Luohe Formation aquifer in 401,101 working face of Hujiahe Mine is 189 m above the #4 coal seam roof. During the mining of the working face, the groundwater level of the aquifer in the Cretaceous Luohe Formation of the roof was monitored at borehole T6. Figure 5.8 shows the monitoring results.

At completion of mining of the working face, the groundwater level of the aquifer decreased significantly with the increase of water inflow from the mining panel. The drawdown was more than 150 m. Therefore, based on the change of groundwater level during mining, the height of the water-conductive fracture zone of the coal mining roof was more than 189 m.

Integration of the results from multiple methods and empirical data in surrounding mines, the ratio of the height of water-conductive fracture zone to the thickness of the coal seam is approximately 22.32.

The results from the geological and geophysical investigations, related aquifer testing and other observation indicate that the aquifer of the Luohe Formation is heterogeneous and anisotropic. It consists of two water-bearing zones, upper water-bearing zone and lower water-bearing zone. The geophysical survey was used to identify the possible distribution pattern of aquifers and anomalous water-rich areas. The height of the mining induced water-conductive fracture zone was comprehensively determined using several measurement methods and numerical simulations.

5.1.3 Exploration and Prevention Technologies of Water Hazards from Overlying Thick Sandstone Fissure Aquifers

The complex hydrogeological conditions of the thick sandstone fissure aquifers that overlie the coal seams make it necessary to implement comprehensive exploration and prevention technologies to control the water hazards. The specific measures in Hujiahe Mine are to focus on establishing and optimizing mining method of the working face, layout of working panel, and distribution of the mine drainage system. Gravity dewatering is the primary measure to prevent water hazards from the overlying Cretaceous formations. The secondary measure is to investigate and dewater the overlying Jurassic aquifer as needed.

5.1.3.1 Combination of Focused Investigation with Pre-drainage of Overlying Aquifers

The Jurassic aquifer is a direct water recharge source for mining in the mine. The aquifer is located within the potential range of the water-conductive fracture zone of the overlying rock in the coal seam. Groundwater in the aquifer directly enters the mine during underground working.

The Jurassic aquifer is generally characterized with deeply buried, poor water yield except for localized areas where the water yield may be large, unfavorable recharge conditions, slow groundwater movement, and low hydraulic conductivity. The main recharge forms from this aquifer include drips and seepage. The amount of water produced in the exploration boreholes drilled in the past is relatively small, mostly below 2 m^3/h. Overall it has little effect on the safety production of the mine. However, there is a possibility that localized water-rich and water-conducting fracture zones may pose water hazard threat.

The prevention of water hazards in the Jurassic aquifers is mainly concentrated on the localized fracture zones. Some examples are:

- Conduct geophysical surveys in anticline and fault development areas in front of roadways; arrange drilling and exploration investigations according to the distribution of geophysical anomalies; and dewater the anomalous areas if needed; and
- Apply underground audio frequency electric perspective technology and channel wave seismic exploration technology to detect the geophysical anomalies and geologic structures in the roof and internal formations. These anomalies may well be related to water-rich areas and can be verified by exploration boreholes. Dewatering may be conducted after verification to pre-drain the localized water-bearing area of the Jurassic aquifer.

5.1.3.2 Combination of Natural Gravity Drainage with Optimization of Mining Parameters

Combination of natural gravity drainage with optimization of mining parameters is proposed as an effective method to prevent and control water hazards from the overlying aquifer of the Luohe Formation. This approach is supported by the comprehensive analysis of the hydrogeological conditions of the overlying Cretaceous aquifer, the formation mechanism of mining induced fracture zone in the overlying rock, and the dynamic characteristics of the actual water inflows in the adjacent mines. This approach is also beneficial to the long-term production needs of the mine.

The optimization of mining parameters is to comprehensively determine the mining thickness, width of the working face, working face layout, and mining sequence according to the dynamic characteristics of the overlying aquifers so that the water recharge intensity into the working face is effectively reduce, and the long-term safe mining of the mine is ensured. The long-term groundwater monitoring at boreholes T5 and T6 installed in the first mining face and the supplemental hydrogeological investigation results conclude that the hydraulic connection between the upper and lower sections of the Luohe Formation aquifer is poor under natural conditions. When the mining induced fracture zone reaches the lower section of the Luohe Formation, the groundwater flows into the underground mining areas. As the lower section of the formation is drained, the hydraulic difference between the upper section and the lower section drives the groundwater to penetrate down into the mine. The groundwater recharge intensity of the working face is directly related to the extent of the overlying aquifer that is connected to the mining induced fracture zone. Therefore, proper control of the development of the water conductive fracture zone can effectively control the water recharge intensity into the working face.

The supplemental mine hydrogeological investigations and the empirical data of the surrounding mines revealed that there are differences in the water richness of the upper and lower sections of the Luohe Formation aquifer. Their hydrogeological parameters differ greatly. The water richness in the upper section of the aquifer is relatively strong, while the water richness in the lower section is relatively weak. The recharge conditions of the upper section of the aquifer are better. Through the actual measurements of the height of the water-conductive fracture zone and the simulation analysis, the height of the fracture zone is directly proportional to the thickness of the coal seam to be mined. Based on the height to thickness ratio, the allowable mining thickness of the coal seam can be estimated for each working face to prevent water hazards from occurring.

5.1.3.3 Improvement of Drainage Systems in Mine, Mining Area, and Working Face

Dewatering is a long-term task in Hujiahe Mine because of presence of the overlying thick and fractured sandstone aquifer. The mine established a complete dewatering system that encompasses working face, mining area, and the entire mine in response to the hydrogeological conditions. A more powerful drainage system was installed where the hydrogeological conditions are more complex.

A dedicated drainage lane was constructed in 401,101 working face. According to the irregular characteristics of Yunshun and Huishun grooves, four ring-shaped water retention silos, as shown in Fig. 5.9, were constructed. The volume of #1 water silo is 1100 m^3, whereas the volume of the other three silos is 500 m^3.

- The #1 silo is equipped with four BQW500-30*2-160 submersible pumps with rated flow of 500 m^3/h and head 60 m. Two submersible pumps are active, and the other two are standby. Two Φ315PE drainage pipes are connected to the No. 1 air return lane through one Φ559PE pipeline for drainage.
- The #2 water silo is equipped with two BQW500-30*2-160 submersible pumps, one active and one standby. The water is discharged to the No. 1 auxiliary transportation lane through one Φ315PE drainage pipeline.
- The #3 water silo is equipped with four BQW500-40*1-120 submersible pumps with flow rate of 500 m^3/h and head of 40 m. Two submersible pumps are active, and the other two are standby. Two Φ315PE drainage pipes are connected to No. 1 air return lane through one Φ559PE pipeline for drainage.
- The #4 water silo is equipped with four BQW500-40*1-120 submersible pumps with flow rate of 500 m^3/h and head of 40 m. Two submersible pumps are active, and the other two are standby. Two Φ315PE drainage pipes are connected to one Φ559PE pipe at the No. 1 return airway for drainage.

In addition, eight BQW500-30*2-160 submersible pumps are reserved on the ground surface. The designed maximum drainage capacity during mining of 401,101 working face is 1500 m^3/h, which meets and exceeds the predicted maximum water flow rate of 625 m^3/h.

The central mine drainage system, at the bottom of the mine shaft, is installed and put into use. The central water silo is composed of main and auxiliary water silos with a total volume of 7485 m^3. Five MD500-57*10 type pumps are installed in the central pump room. The pump motor is 1250 kw with rated flow of 485 m^3/h and the hydraulic lift of 568 m. They are all connected to three drainage pipes. The mine water is pumped to ground surface through the auxiliary well shaft via valve control and discharged to the water treatment plant. The drainage pipe is made of Φ325*14 seamless steel pipes. Three sets of pipes are installed with two active and

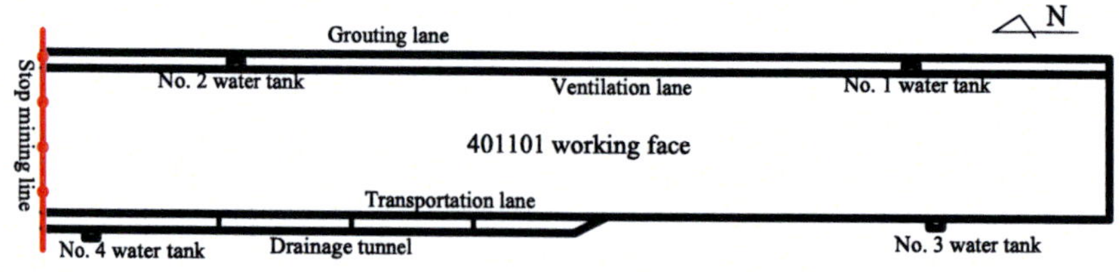

Fig. 5.9 Schematic diagram showing location of the water silos at 401,101 working face

Fig. 5.10 Hujiahe Mine central pump house

one standby. The designed maximum dewatering capacity is 970 m^3/h. Figure 5.10 shows the central pump house of Hujiahe Mine.

The mine established a robust underground drainage system in response to emergency. The mine drainage system consists of two BQ550-612/16-1600/W-S submersible pumps. The pump motor has power of 1600 kw. Two pumps and two pipelines can be in operation at the same time. The pipeline is made of two Φ325*14 seamless steel pipes and laid along the shaft of the return air shaft. The designed drainage capacity is 1100 m^3/h.

5.1.3.4 Combination of Lagged Dewatering and Optimized Dewatering

Under the circumstances where the amount of water in the goaf is large and the dewatering is challenging, the mine combines lagged drainage with the optimized drainage method to improve the drainage efficiency and eliminate the threat of water accumulation in the goaf.

The dewatering project is completed in grouting lane with 68 m in exaction to bottom, 280 m of drainage ditch, and the #13 communication lane. One Φ559 pipeline was laid from the #13 communication lane to the temporary water silo constructed in the grouting lane. Three BQW300-80-110 submersible pumps are installed. Two pumps are in operation, while the other one is standby. Two Φ315PE drainage pipelines are connected to the T-junction that is reserved for the pipe network in #2 communication lane.

After 401,101 working face passes the #1, #2, #3 water retention silos, the #1 and #2 retention silos for working face 401,102 is constructed and completed. Also completed is the #14 communication lane. Three Φ560 steel pipes lead back to the mine pool. Most of the old pond water in the 401,101 working face is currently

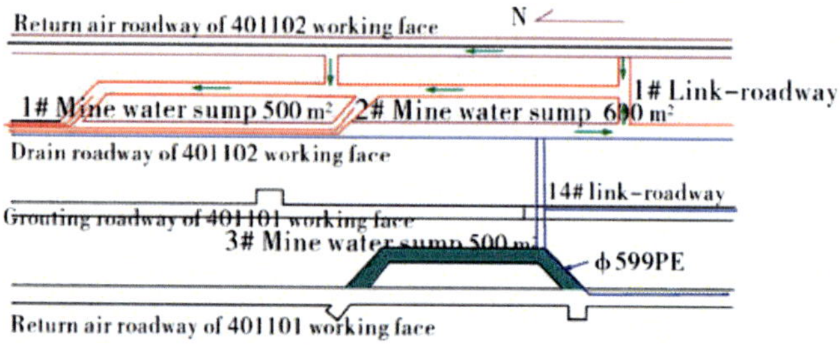

Fig. 5.11 Drainage layout at 401,101 and 401,102 working faces

introduced into 401,102 drainage lane sump through #14 contact lane for drainage. They are equipped with two BQW500-40*1-120 submersible pumps at rated flow rate of 500 m^3/h and hydraulic head of 40 m, one BQW800-35-110 submersible pump and one BQW600-35-110 submersible pump. Two are active, while the other two are backups. Three Φ315PE drainage pipelines are connected to No. 1 auxiliary transportation lane, No. 2 drainage pipeline of the auxiliary transportation lane, and No. 1 air return lane, as shown in Fig. 5.11.

Based on the actual elevations of the #4 water silo outlet to the No. 1 air return lane (No. 4 contact inclined lane) and the No. 1 auxiliary transportation lane, a 1 m retaining wall was constructed at the #4 water silo outlet. Then, two Φ325 plastic steel pipes were installed to extend to No. 1 auxiliary transportation lane. A pressure gauge and backflow pipe were installed at the exit position. The #3 sluice gate and two pressurized Φ325 plastic steel pipes were constructed to extend and connect to the Φ559PE pipeline of No. 1 air return lane. Two Φ325PE pipes were required to be laid on the joist, one joist every 4 m and bolt on the top of the roadway. Stranded steel wire was used to fasten the pipeline to enhance the firmness of the pipeline connection. After the #3 sluice gate was solidified, the outlet gate valve that connects No. 2 contact inclined laneway and two Φ325 plastic pipes was closed so that the old mine pool water passes through the Φ559PE pipeline of No.1 air return lane and the Φ559PE pipe of No. 1 Φ325 plastic steel pipe. The pipelines in the depot section of the auxiliary shaft discharge water to the permanent water storage. Figure 5.12 shows the pipelines and their connections.

5.1.3.5 Establishment of Hydrogeological Monitoring System

The hydrogeological conditions of the mine are investigated to understand the dynamic changes of the groundwater level. The aquifer recharge, groundwater flow characteristics, and discharge conditions are analyzed based on the data collected from monitoring of the groundwater level (or potentiometric pressure) and groundwater inflow of the roof aquifer. The roof aquifer is the main source that threatens the safe production of the mine. Hujiahe Mining Co., Ltd. established the mine hydrogeological dynamic automatic monitoring system to collect the data that provide

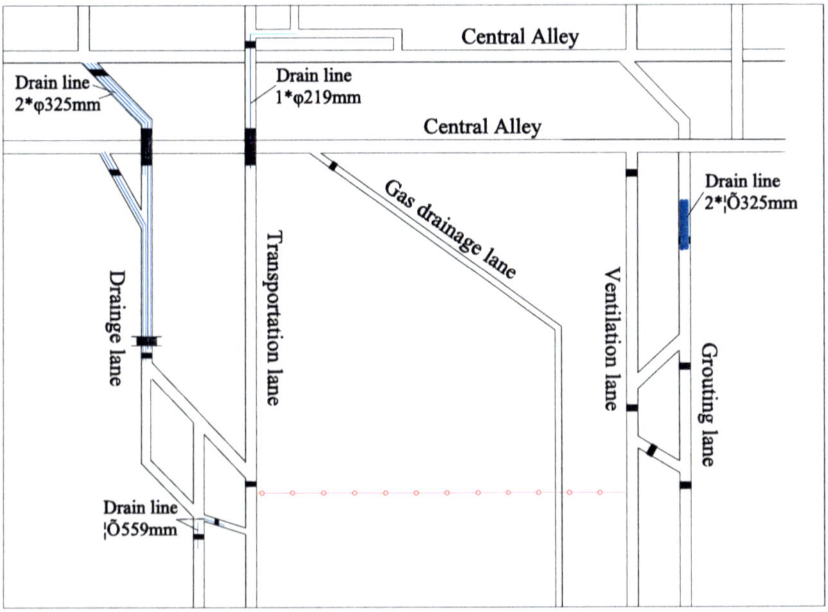

Fig. 5.12 Schematic drainage diagram of 401,101 working face

the technical basis and basic data for mine water prevention and control. The monitoring system uses the surface groundwater monitoring wells constructed in scientific research projects, the water supply wells in the industrial square, and the main underground drainage pipelines. Figure 5.13 shows the hydrogeological monitoring system including six sets of surface hydrogeological monitoring telemetry devices and three sets of flow monitoring devices for underground main drainage pipelines. The monitoring parameters are groundwater level and water inflow rate. The following objectives are accomplished through the monitoring system:

- Monitor the real-time changes of groundwater level in the aquifers of Cretaceous Luohe Formation and Yijun Formation.
- Generate real-time hydrogeological monitoring reports.
- Provide early warning on water hazards based on the upper limit, lower limit and change threshold of the hydrogeological monitoring value.

5.1.3.6 Technical Support to Roof Water Hazard Prevention

- According to the thickness of the aquifer and the distance of the aquifer above the coal seam, the mining thickness should be controlled to avoid inducing groundwater from the aquifer into the working face. The mining thickness of the coal seam of a single working face shall be formulated based on the change of the thickness of the overlying water-bearing formations. The mining height must be strictly

Fig. 5.13 Hydrological monitoring system (telemetry device and pipeline flow monitoring device)

controlled to avoid inducing water from the overlying formations, especially in areas where geological structures are present. Otherwise, mining is allowed only where fissures are not present. In the section where the roof is rich in water, the mining thickness shall be reduced. When the thickness of the top coal is too large, the mining height can be generally controlled according to the coal output during the mining process.

- Advance the mining process of the working face continuously, as fast as possible, to avoid long period stops in mining or slow mining progress.
- Adjust the mining sequence. It is recommended that the thicker section of the water barrier be mined first, such as the 401 panel, followed with the thinner section of the water barrier. Centralize mining in one panel to avoid the simultaneous mining of multiple panels so that drainage capacity is not stressed.
- Use down mining as much as possible in the design of working face roadways. Always install drainage ditches or special drainage channels in the roadway; or construct special drainage channels in low-lying places for centralized drainage according to the coal seam fluctuations in the panel. In the construction of underground drainage system, achieve self-flow of mine water inrush as much as possible to reduce the pressure on the temporary drainage system of the mine.
- Adopt measures such as reducing the width of the working face and exercising intermittent fully-mechanized caving to control the height of the mining induced fracture zone in the overlying formations and thus to reduce the groundwater recharge intensity of the roof aquifer.

- Establish a complete hydrogeological monitoring system for the mine, construct an automatic control system for mine water pumps and improve mine emergency rescue plans and emergency responses.

5.2 Prevention and Control Technology for Water Disaster from Bed-Separation Voids of Overlying Formations in Hongliu Coal Mine, Ningdong Coalfield

5.2.1 Mine Background

Hongliu Coal Mine is in the central and eastern regions of Ningxia Hui Autonomous Region. The mine is developed with both inclined and vertical shafts with a design production capacity of 8.00 Mt/a. The mine field is divided into three levels and eight sub-areas with a total of 35 mining area. The first level is divided into 12 mining area and mines #2, 3, and 4 coal seams. The second level is divided into 15 mining areas and mines #6, 10, and 12 coal seams. The third level are divided into 8 mining areas and mines #15, 16, 17, and 18 coal seams. The current mining coal seams are #2 and 3-1 coal seams. Table 5.3 summarizes the relationship between the atypical water inflows at the working face and the advanced distance.

The water gushing incidents in the working face result from the unique lithology and combination of the two coal roof formations, especially the geological characteristics of the interlayer mudstone. Figure 5.14 shows schematically the water inrush processes, which are described as follows:

- Mining of the working face leads to collapse of the direct roof of the coal seam. The overlying water-resisting layer of mudstone is bent and subsides due to the loss of support.

Table 5.3 Summary of relationship between atypical water inflow at working face and advanced distance

Atypical water inflow incident	Year. Month. Day	Advanced distance of ventilation lane (m)	Advanced distance of transportation lane (m)	Average advanced distance (m)	Water inflow prior (m^3/h)	Maximum water inflow (m^3/h)	Stable water inflow (m^3/h)	Volume of water inflow (m^3)
1	2009.11.03	39	54	46.5	15	212	135	31,800
2	2009.11.21	47.5	64.5	56	135	1,817	120	5,670
3	2010.03.03	100.7	124.3	112.5	90	793	108	55,700
4	2010.03.25	172	200	186	108	3,000	178	121,880

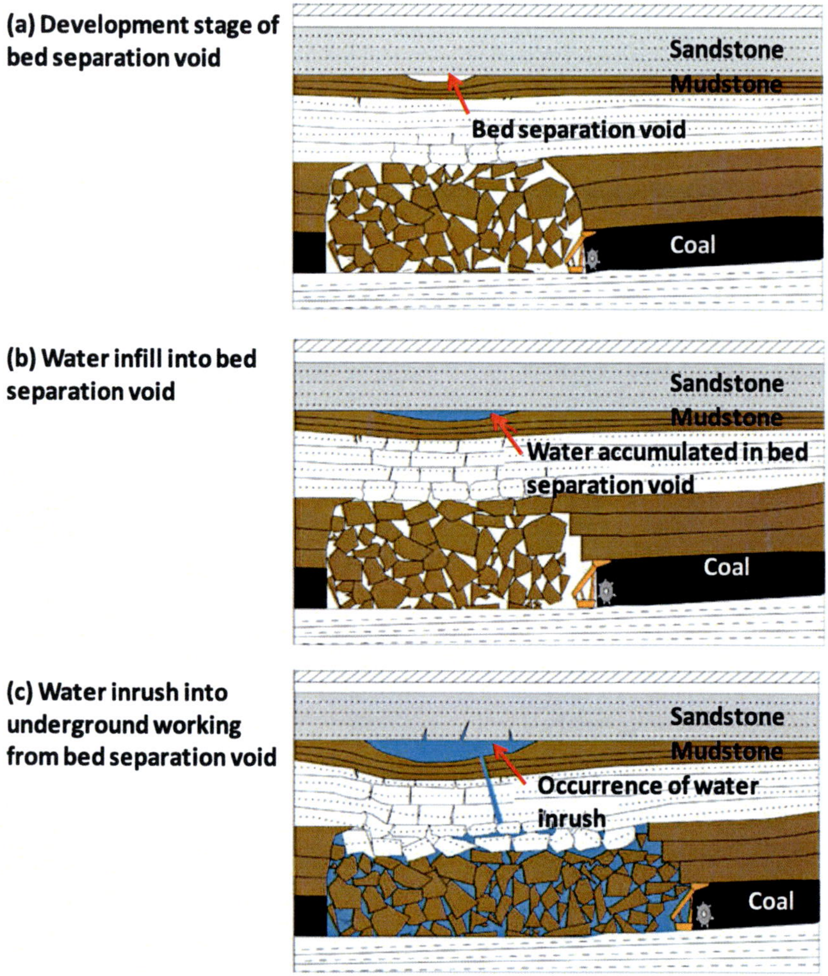

Fig. 5.14 Schematic diagram of water inrush from mining induced bed separation

- Because of the greater tensile strength of the mudstone layer the subsidence is large, whereas the inter-bedded sandstone subsides to a less extent. The subsidence difference in the formations results in bed partings in the contacts between sandstone and mudstone. The bed partings form a unique type of porosity for groundwater accumulation and storage in the overlying formations.
- As the mining impacted extent in the overlying formations increases, the water resistance capacity of the mudstone is weakened by its own weight, stress change, and potentiometric pressure. The bending of the mudstone continues until the displacement reached the tensile limit. The mudstone is then structurally changed and ruptured.

- The water in the bed separation void gushes into the underground workings.
- As the amount of water in the separated layers decreases, the mudstone aquifer expands under hydraulic action. Its fractures gradually close and become impermeable. The bed-separation void forms a self-contained water storage reservoir again.
- The same processes take place as the mining progresses, resulting in periodic water inrushes, as summarized in Table 5.3. The regeneration process of water resistance capacity in the mudstone plays a controlling role in the periodic water inrush from the water body that is accumulated and stored in the mining induced bed partings in the overlying layer.

Based on the above analysis, such a water inrush is characterized with hysteresis and periodicity and is often referred to as bed-separation water inrush.

5.2.2 Investigation and Mitigation of Bed-Separation Water Inrush

5.2.2.1 Overview of I020204 Working Face

I020204 working face is in the middle of Hongliu Mine. The #2 coal seam is the mining target. The working face has a buried depth between 298 and 481 m amsl. The coal thickness is relatively stable and varies from 4.82 to 4.98 m with an average of 4.89 m. The amount of coal reserves is approximately 1.578 million tons.

The geological and hydrogeological conditions are investigated with exposure of the working face and exploratory drilling of boreholes. The lithology directly overlying the coal seam of I020204 working face consists of off-white–grey siltstone and mudstone. The thickness of the siltstone and mudstone varies from 8.6 to 15.8 m with an average of 11.3 m. The indirect overlying formation is composed of coarse-grained sandstone of the lower Zhiluo Formation, and its thickness ranges from 38.2 to 58.1 m with an average of 50.9 m. The coarse-grained sandstone of the lower section of the Zhiluo Formation, also known as the Qilizhen sandstone, is the main recharge aquifer that affects mining of the working face. The Qilizhen sandstone is susceptible to occurrence of the bed-separation voids, which provide space for groundwater accumulation and storage during the mining process.

Experience with water control in #1121 working face indicates that the following effective approaches, if implemented in sequence, are effective in mitigating the potential groundwater threat:

- Drain the rock formation that directly overlies the coal seam.
- Drill dewatering boreholes in the sandstone and optimize their placements to prevent water inrush from the bed-separation voids.

5.2.2.2 Drainage of Overlying Formations for I020204 Working Face

The coal seam of I020204 working face is mined with a fully mechanized mining method. The groundwater drainage is accomplished by installing 89 dewatering boreholes from November 12, 2013 to February 3, 2014, including the following:

- 59 dewatering boreholes drilled from 11 drilling caissons within the air circulation lane. The largest water inflow occurs at F3-4. The maximum water inflow is 118 m^3/h, and the stable water inflow is 2.2 m^3/h.
- 14 dewatering boreholes drilled from 5 drilling caissons within the transportation lane. The largest water inflow occurs at J1-2. The maximum inflow is 12 m^3/h, and the stable inflow is 8 m^3/h.
- 16 dewatering boreholes drilled from 5 drilling caissons within the opening cut. The largest water inflow occurs at Q4-2. The maximum water inflow is 80 m^3/h, and the stable water inflow is 36 m^3/h. The flow rate at Q4-2 is also the largest after the water flow is stable.

As of February 3, 2014, the groundwater inflow from all boreholes was 317.5 m^3/h. The total groundwater volume was approximately 476,000 m^3. Figure 5.15 shows the layout of the dewatering boreholes for I020204 working face.

5.2.2.3 Hydrogeological Investigation of Bed-Separation Voids for I020204 Working Face

The locations of potential bed-separation voids are estimated with the site-specific conditions of I020204 working face and the mechanism analysis. The bed-separation voids occur more likely at distances of 30–40, 90–100, and 150–160 m from the open cut. Because the elevation of the machine lane is relatively low, there is a high probability of bed-separation occurrence at 15–60 m from the machine lane. Based on the above analysis, the areas with high risk bed-separation water inrush on I020204 working face are 15–60 m from the machine lane and 30–40, 90–100, and 150–160 m from the open cut.

A total of 24 exploratory boreholes are designed to be drilled in 3 drilling caissons. If the boreholes intercepted bed-separation water and the water flow in individual boreholes exceeds 10 m^3/h, additional boreholes are to be drilled to optimize dewatering efficiency. The design plan is shown in Fig. 5.16. Table 5.4 presents the drilling parameters.

The two boreholes that are 15 m away from the roadway are completed first. The boreholes further away from the roadway are then completed. The two boreholes furthest away from the roadway are completed in the last phase. The advance dewatering of the roof water and the investigation and mitigation of the bed-separation groundwater ensure that the bed-separation water inrush did not occur during mining of I020204 working face.

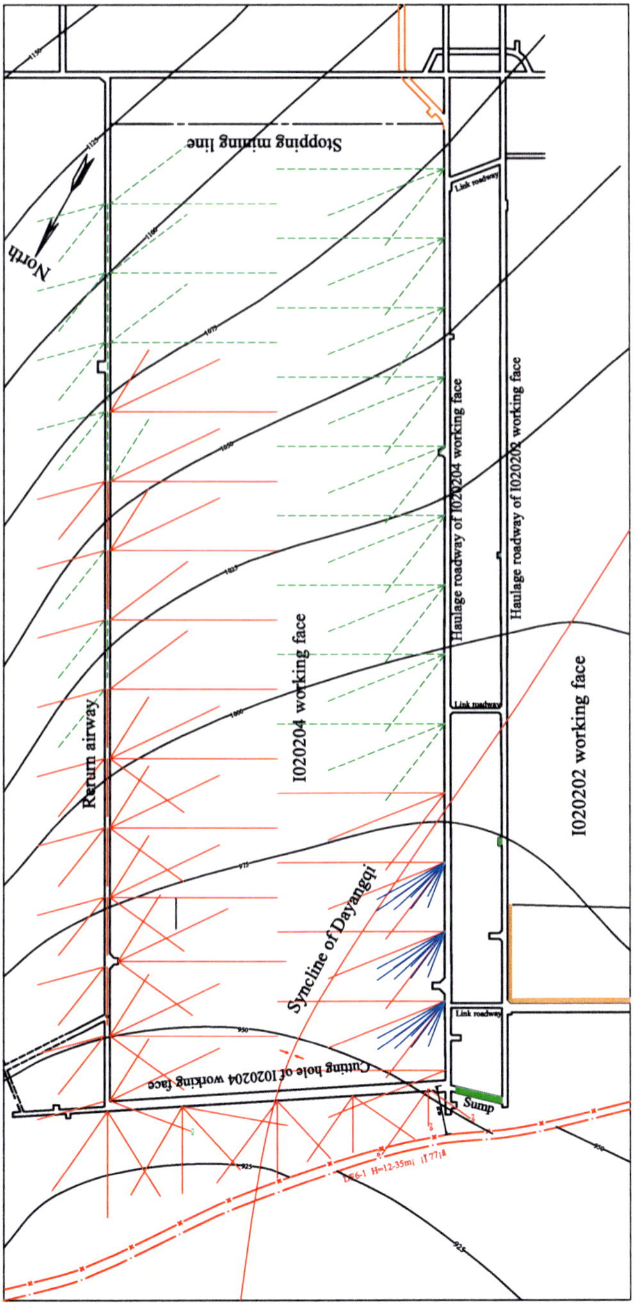

Fig. 5.15 Layout of dewatering boreholes for I020204 working face

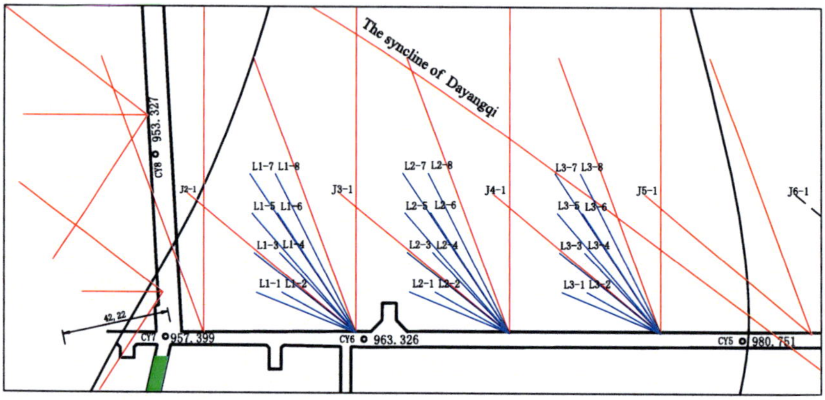

Fig. 5.16 Arrangement of exploratory boreholes for investigation and dewatering of bed-separation groundwater

5.2.3 Summary of Bed-Separation Groundwater Control

- The lowest point at which bed-separation water likely occurs during mining of the #2 coal seam in Hongliu Coal Mine is between 50 and 60 m above the floor of the coal seam. Based on theoretical analysis, numerical simulation, and in-situ verification, advance dewatering is the most effective approach. The best time, location and period are provided for the advance dewatering to avoid bed-separation water inrushes.
- Effectively drain the coarse sandstone aquifer that indirectly overlies the coal seam. The advance dewatering should be combined with the periodic collapse step of the working face and completed prior to the periodic collapses of the overlying formations.
- Such achievements are not only directly applied to the prevention and control of bed-separation water in all working faces of Hongliu Coal Mine that mine the #2 coal seam, but also applicable to other mines in Ningdong Yuanyanghu mine field, such as Shicaocun Coal Mine, Meihuajing Coal Mine, Qingshuiying Coal Mine, and Maiduishan Coal Mine. With a broad range of application and bright prospect, these achievements are significant in producing social and economic benefits.

Table 5.4 Borehole parameters for bed-separation investigation

Borehole number	Upward angle (°)	Azimuth angle (°)	Estimated depth (m)	Length of orifice (m)
L1-1	57	352	82	21
L1-2	63	358	79	21
L1-3	52	7	85	21
L1-4	57	16	83	21
L1-5	45	18	92	21
L1-6	49	26	89	21
L1-7	40	26	100	21
L1-8	43	33	98	21
L2-1	58	352	83	21
L2-2	64	359	80	21
L2-3	52	8	87	21
L2-4	57	16	84	21
L2-5	46	19	93	21
L2-6	50	26	90	21
L2-7	40	26	102	21
L2-8	43	33	98	21
L3-1	45	352	100	21
L3-2	45	359	101	21
L3-3	45	8	100	21
L3-4	45	16	101	21
L3-5	44	19	100	21
L3-6	45	26	100	21
L3-7	44	26	100	21
L3-8	45	33	100	21
Total			2,229	504

5.3 Prevention Technology on Water and Sand Inrush in Halagou Coal Mine, Shendong Coalfield

5.3.1 Mine Background

Halagou Coal Mine, a subsidiary of Shendong Coal Group, is in the northern Shaanxi Loess Plateau bordering the southern edge of the Maowusu Desert. Within the coalfield there is a perennial river, which has the water source at Halagou Guozhang springs. The 22,406 working face of the mine is the first fully mechanized mining panel. The working face is characterized with thin bedrock and thick unconsolidated water-bearing sand layer that overlies the coal seams. Water and sand inrush from the overlying formations threatens coal mining in the working face.

5.3.1.1 Overview of Working Face

(1) Neighboring conditions

The 22,406 working face is bordered by central air return lane to northwest, designed 22,407 working face to northeast, mined-out areas from the 22,609 and 22,610 working faces of Daliuta Coal Mine to southeast, and mined-out area from the 22,405 fully mechanized working face to southwest.

(2) Ground surface conditions and elevation

Ground surface of the working face, located in the middle of the western bank of Halagou Coal Mine, is characterized with large undulations. The topographic elevation is generally high on both sides and low in the middle. The lowest point is the bottom of the Halagou Gully with an elevation of 1,191 m. The terrain is covered with aeolian sand.

(3) Coal seam distribution

The average coal thickness in the mining area of the working face is 3.5 m. The coal seam structure is simple and belongs to the stable coal seam type. The coal seam is inclined to the southwest with axial orientation of NE-SW. The coal seam dip angle is between 1 and 3°. The floor elevation of the coal seam is generally higher in the incoming traffic lane than that in the outgoing traffic lane with the exception of an area from 896 m to 1709 m from the open cut where the elevation of outgoing traffic lane is higher than that of the incoming traffic lane. The average mining height is designed to be 3.5 m.

(4) Geology and geologic structures

At the working face the overlying bedrock is thinner at the bottom of the Halagou ditch with a thickness ranging from 21.5 to 98.3 m. At the same location, the thickness of the loose layer is 37.6 m, and depth to groundwater is 3.5 m. The fault F10 was exposed along the incoming traffic lane at 1395 m from the 22,406 opening cut. F10 is characterized with an azimuth of 53°, an inclination angle of 72°, a displacement of 7.3 m, and a width of 0.53 m. Fractures are developed alone the fault plane, and the rocks are fragmented. A small amount of water and sand is observed to percolate through the fault. The coal seam is underlain by siltstone, which affects the length of approximately 200 m in the working face.

5.3.1.2 Factors of Groundwater Recharge into Working Face

1. Recharge water sources

(1) Atmospheric precipitation

Atmospheric precipitation is the main source of groundwater and surface water. Atmospheric precipitation can directly or indirectly recharge the mine. According to the meteorological data collected at the Shenmu County Meteorological Station in Yulin City that was established in 1957, the average annual rainfall in the Halagou Coal Mine area is between 380 and 415 mm. The precipitation is concentrated in the rainy season from July to September, which accounts 50–70% of the annual amount. Observation data from the production mine (Jinjie Coal Mine) in the Shenfu mining area shows that the water inflow in the mine is closely related to the precipitation and the precipitation periods. The flow changes with the seasons and on average lags approximately one day after the precipitation. Based on the pattern of earth fissures that are induced from mining of #22 coal seam at 22,404 and 22,405 working faces, the ground fissures are connected to the underground working areas. There is a potential complete pathway between the precipitation and underground workings along which rainwater carried water and sand into the mine, especially in low-lying areas and in response to high intensity rain events. Therefore, atmospheric precipitation is the direct recharge source to the working face.

(2) Surface water

Halagou Gully is the main surface water source to 22,406 working face. Halagou Gully is a perennial river with flow varying seasonally at an average flow rate of 35 m^3/h. The bedrock overlying #22 coal seam is relatively thin along the Halagou Gully. The mining induced roof collapses result in the water-conductive fracture zone extending through the bedrock into the soil and sand layers. At the Halagou Gully, the fracture zone extends to the surface, becoming a flow path between the surface water and the working face. In the Halagou Gully section where the bedrock thickness is thinner, surface water, as well as the associated sediments, is another direct recharge source to the working face.

(3) Groundwater

The recharge aquifer in the working face mainly consists of the loose phreatic aquifer and the confined bedrock aquifer overlying #22 coal seam. The thickness of the overlying bedrock in the Halagou section of the 22,406 working face is less than the maximum height of the mining induced water-conductive fracture zone. Coal mining destructs the overlying rock layer in its full-thickness, and the mining induced water-conductive fracture zone extends through the bedrock to the soil layer or sand layer or the Haragou valley. The water in the loose sand layer and the confined water in bedrock fissures are direct water recharge sources to the working face.

2. Flow pathways

The main flow pathway in the 22,406 working face consists of the mining induced water-conductive fracture zone. The roof collapse in response to coal extraction result in water-conductive fractures that directly penetrate the bedrock and connect to the surface. Water inrush from overlying formations threatens coal mining in the working face. The challenge in prevention and control of this kind of water threat lies in selecting a suitable method to predict the water inflow. It is recommended that a pumping system with a capacity at 1.5 times the predicted water inflow be constructed to solve the water inrush issue. The 22,406 working face is currently facing the threat of water and sand inrushes. The prevention and control measures are based on the heights of mining induced caved-in zone and water-conductive fracture zone. It is therefore essential to estimate the heights of mining induced caved-in zone and water-conductive fracture zone.

(1) Calculation of caved-in zone height

① Method based on coefficient of crushing expansion

The coefficient of crushing expansion is taken to be the median value, i.e., $k = 1.2$, and then substituted into other relevant parameters, the caved-in zone height is:

$$H_{\text{caved-in}} = 17.5\,\text{m}$$

Subsidence may occur during the collapsing of the roof. Assuming a subsidence height of 0.5 m, we obtain the caved-in zone height:

$$H_{\text{caved-in}} = 15.0\,\text{m}$$

② Empirical method

Based on lithology of the overlying rock, the rock hardness is medium. Substitution of relevant parameters into the empirical equation leads to:

$$H_{\text{caved-in}} = 12.1\,\text{m}$$

From the perspective of disaster prevention and control, the maximum caved-in height of 17.5 m is selected as the most conservative estimate. Because the thinnest bedrock at the bottom of the Haragou ditch in the 22,406 working face is 21.5 m, the caved-in zone will not connect to either the sand aquifer or the surface water.

(2) Calculation of water-conductive fracture zone height

According to the calculation equation for the water-conductive fracture zone height, the height obtained under the geological conditions, H_w, ranges from 30.8 to 34.0 m. The calculated height range exceeds the 21.5 m bedrock at the bottom of the Halagou Gully. The mining induced water-conductive fractures act as pathways for groundwater flow. When the working face is advanced to the gully section, the overlying aquifers and surface water bodies and associated sands may gush into the working area through these channels. The threat of water and sand inrush is related to the hydraulic height of the water-bearing sand. If the hydraulic height of the sand aquifer exceeds the threshold hydraulic head designed for safety, the prevention and control measures are implemented with a focus on remediation of the mining induced water-conductive fractures.

3. Water recharge intensity

The water recharge intensity of the working face is related to various factors including thickness of the overlying bedrock, rock characteristics, aquifer thickness and water richness, water resistance of the aquiclude, rainfall intensity, mining method and mining intensity. In addition, it is also related to the spatial relation between the coal mining face and strong water yield portion of the overlying aquifer.

The Quaternary alluvial layers and the porous medium aquifer in the Salawusu Formation are widely distributed in the Halagou Gully. Many factors in this section are conducive to water recharge into the mining face such as thin bedrock, thin soil layer, strong phreatic sand aquifer, and accumulation of precipitation runoff. Therefore, the water recharge intensity is higher along the Halgou Gully.

5.3.2 Mechanism and Conditions of Water and Sand Inrush

5.3.2.1 Mechanism of Water and Sand Inrush

Water and sand inrush is a geological disaster that is caused by water and sand flowing into the mining face though mining induced water-conductive fractures under the combined action of groundwater dynamics and hydrostatic pressure. The flow sand may enter the working face continuously or episodically, resulting in paralysis of the dewatering system. The main feature is that the sand aquifer is weakly cemented under natural conditions and is integrated with groundwater. The water-bearing sand particles are typically stored in the fracture network. Under low hydrostatic pressures, no catastrophic water and sand breakthroughs occur, and a small amount of sand particles may flow into the working face. However, catastrophic water and sand inrush disasters occur when the hydrostatic pressure exceeds a critical value. The water and sand located in the water-conductive fracture network can be prevented by depressurizing the groundwater level. For the water and sand located in the caved-in zone where fracture apertures are wide and the moving distance is short, the water and sand potential energy can be directly converted into kinetic energy. Even under small hydrostatic pressures, water and sand inrush disasters can occur. Solely depressurizing the hydraulic head is typically not effective. Therefore, the threat degree of water and sand inrush is closely related to the height of the mining induced water-conductive fracture zone. Higher water-conductive fracture zones indicate greater threat to coal mining.

The flow sand breakthrough is a process that involves in accumulation and release of water and sand energy in the aquifer. When the mining induced fractures extend to the overlying sand aquifer, the bottom of the flow sand body is in a free-fall state under high hydraulic pressures. The natural stress distribution is destructed, and the new stress in the sand body is redistributed. A high-stress zone occurs at entrances of the flow pathways, whereas a low-stress zone is formed away from the pathways. The sand particles at the entrance points will first break out with gushing water under action of self-weight, rock stress, and hydraulic pressure. The groundwater level at the entrance points of the aquifer drops rapidly, resulting in a rapid increase in the hydraulic gradient. When the hydraulic gradient increases to a threshold at which sand particles start to mobilize, the sand particles, together with groundwater, begin moving along the flow pathways at high speeds to working faces, mined-out areas, or roadways. When the water inrush volume or the velocity of groundwater flow is large enough, the sand particles are transported to fill in the mined-out areas or roadways. Therefore, the degree of damage caused by water and sand inrushes is directly proportional to the hydrostatic pressure. Higher hydrostatic pressure means greater degree of hazard.

With decrease of the groundwater level of the aquifer the hydrostatic pressure is released. The water potential energy gradually decreases. When the water inrush speed is insufficient to carry away the sand particles, the eroded sand particles deposit and accumulate downgradient of the flow paths. The sand accumulation results in increased resistance for water and sand movement until the movement of water and sand stops. The potential for a pro-longed water and sand inrush is no longer present to bury the working face and roadway. However, if the vertical and lateral recharge sources are sufficient, the attenuation is small of the dynamic and hydrostatic pressure, the critical condition is still met in carrying sand particles to the working face or roadway. Under such a scenario, the water and sand inrushes continue and become aggravated over time. Therefore, for phreatic loose aquifers composed of aeolian sand, the main factors that affect the water and sand breakthrough and the degree of damage include the nature and extent of the mining induced fractures and sand particle size. Other factors are whether the hydrodynamic pressure formed under mining can mobilize the sand particles in the overlying fracture network and whether the gushing water can mobilize the sand particles that are accumulated in the working face.

5.3.2.2 Conditions for Water and Sand Inrush

Extensive mining practices have shown that the water hazard caused by the directional fractures that extend to the water-bearing sand layer generally is not catastrophic to the production of the working face. Increasing of the dewatering capacity can accomplish the prevention of water inrush disasters. However, when the caved-in zone spreads to the loosely cemented sand aquifer, the water and sand inrush disasters are significant. When the groundwater in the loosely cemented or semi-cemented sand aquifer is hydraulically connected to the overlying water-bearing layer, a dynamic complete pathway is established for the water and sand to transport into the working face. Consequently, the working face is flooded with water and sand and the working face and roadway are buried. Based on investigation of multiple cases of water and sand inrush accidents and the analysis of the water and sand inrush mechanism, the following four elements should be present when a water and sand inrush disaster occurs:

1. Material sources of groundwater and sand.

The occurrence of water and sand inrush accidents is closely related to the water yield of the sand aquifer and the composition of water and sand particles. The aquifer's water yield, lateral recharge of groundwater, and the recharge of atmospheric precipitation determine the magnitude of the hydrostatic pressure. For medium-strong aquifers, higher hydrostatic pressures lead to greater water potential energy, which in turn provides sufficient force to mobilize sand particles. Once the phreatic aquifer is converted into a direct recharge source, the water and sand inrush occurs in the working face or roadway. The amount of the water and sand depends on the aquifer characteristics and spatial composition of water and sand particles. The contact modes of bedrock and water and sand can be classified into the following three types:

(1) The formation directly overlying the bedrock consists of a sandy gravel layer aquifer, and the upper part is composed of a loess layer and a loose sand layer. In this case, the sand source of the water and sand inrush at the working face is the medium-fine sand sandwiched in the gravel layers. When the roof is fully cut through for the first time, the sand contained in the overlying formation directly enters the working face or mining. With the increase of water inrush time, the gravel layer becomes a filter layer for the sand particles, which has an effect of preventing further erosion. In addition, the reduction of water potential energy loses the erosion power of the sand particles.
(2) The formation directly overlying the bedrock is a loess layer, and the upper part is a loose sand layer aquifer. Such a structure has a good water resistance capacity because of the loess layer, which is characterized with a strong shear strength and a dense structure, especially when the loess layer is 5 m thick or greater. The passage of water and sand into the working face is blocked, and it is not easy to form an accident of water inrush and sand erosion. There are exceptions in which the bedrock overlying the coal seam is too thin, and the thickness of the loess layer is less than 5 m. Mining can cause "skylights" in the loess. The water and sand enter the working surface through the "skylights" under the action of hydrodynamic force, causing a catastrophic sand breakthrough accident.
(3) The sand aquifer directly overlies the bedrock. Under this scenario, mining of shallow coal seams can easily cause disasters of water and sand inrush. When the coal seam roof is severely eroded, the working face may have to be abolished. The case study is about this type of water–sand combination, i.e., the saturated water-bearing sand layer in direct contact with the bedrock.

2. Flow pathways for groundwater and sand

In mining of shallow coal seams underlain by water-bearing aeolian sand layers, the flow channels for water and sand can be divided into two types. The first type refers to the mining related channels, which consists of mining induced fractures and boreholes. The mining induced channel includes the caved-in zone and fracture zone in response to coal extraction. The borehole channel refers to unsealed or poorly sealed boreholes that can vary water and sand into underground working areas. The second type is composed of natural discontinuities such as faults, folds, and other geological structures. The inherent fracture channels and mining induced channels are all related to geological factors. The mining induced channels become complete pathways for water and sand and the mining area that is characterized with thin bedrock and thick loose sand layers. The second type of channels indicate that the channel connecting with the water-bearing sand layer has been formed before mining.

In the case of a certain water source, the width and height of the flow channel formed by mining activities determine the amount of gushing water at the working face, which in turn determines the instantaneous water flow rate and the intensity of water percolation and sand breakthrough. Mining of shallow coal seam results in roof breakage that inevitably extends to the overlying sand aquifer. The caved-in zone and fractures formed by roof collapse provide the passageways and places for water and sand. Of the mining induced destruction in the overlying formations, the caved-in zone channels are typically short and wide, and the buffer zone is thin or non-existence. The resistance along the way is low, resulting in a large amount of water and sand inrush with high flow velocity and high degree of damage. The preventive methods are most complicated. The mining induced fracture network consists of long and narrow fractures, which are well interconnected. There are more buffer zones during the process of water and sand movement toward the working face. The resistance along the pathways is higher. As a result, the amount and speed of water and sand inrush are limited. The prevention method is relatively simple. The subsidence zone consists of the longest mining induced fractures. During the process of water–sand erosion, the resistance along the way is large. Therefore, the amount of water–sand erosion and the flow velocity are relatively small. Reinforcing the support system may be adequate to prevent breakthrough of overlying formation, thus the prevention of water and sand inrush disasters.

The seepage flow in the natural discontinuities has insufficient hydrodynamic pressure to break through the argillaceous-cemented filling. Under mining disturbances, the hydrodynamic pressure is increased at the free-fall surface, and the increased pressure can cause water percolation and sand erosion. Under this scenario the effective preventive method for water and sand control is to grout or retain coal pillars.

3. Large hydrodynamic pressure

Hydrostatic pressure is present in the aquifer prior to coal mining. This pressure is proportional to the water richness of the aquifer. The aquifer contains significant gravitational potential energy with reference to the coal seam floor. When the sand aquifer is connected to the mining induced fractures, the aquifer water begins to flow, and the potential energy is immediately converted into kinetic energy. At this time, the sand layer not only is subjective to hydrofracturing by the hydrostatic pressure, but also undergoes the hydrodynamic pressure effect of the groundwater on the sand layer to overcome the resistance. The combination of the hydrostatic and hydrodynamic pressures mobilizes the sand-laten groundwater in a large range, resulting in a water and sand inrush incident.

4. Void to accommodate inflow of water and sand

When the sand aquifer is directly connected to the mining induced fractures, the sand particles in the aquifer move along the direction of the groundwater flow toward the entrance points under the combined action of hydrodynamic and hydrostatic pressures. If there is a large void below the discharge point, the sand particles will be driven toward the working face, mined-out area, and roadway. On the other hand, if the filled space is small, the eroded sand particles only accumulate below the sand breach, even if there is a large hydraulic gradient. The water and sand flow stops to prevent the further development. Therefore, the size of the mine-out voids determines the magnitude of a water and sand inrush and its progress.

In summary, occurrence of a water and sand inrush required four elements: water and sand source, pathway, driving force, and void to receive the water and sand. The interaction of these four elements is the basic condition that leads to water and sand breakthroughs. The water and sand inrush disaster occurs when all four elements are present in the working face.

5.3.3 *Prevention and Control Technology of Water and Sand Inrush*

The 22,406 working face is characterized with relatively thin bedrock over a trench and hidden gullies that result from undercutting erosion. The low-lying valley has collected a large amount of surface water and groundwater. In the section, the mining induced water-conductive fracture zone would directly connect to all overlying aquifers. In addition, there is a possibility that the caved-in zone is directly connected to the sand aquifer. Surface water and groundwater would carry a large amount of sediments into the working face, resulting in a water and sand inrush at the working face. The risk of sand and water inrush occurrence makes it necessary to implement prevention and control measures.

5.3.3.1 Prediction of Water Inflow into Working Face

In advancing 22,406 working face there will be a difference in the amount of water inflow with the different mining positions. The first water inflow peak at the working face occurs at the position of the first roof stress, which leads to the first collapse of the overlying layer. During the subsequent periodic collapses of the roof formation the water inflow is kept in a relatively stable range. When the working face advances to the Halagou Gully section, a second peak influx occurs at the working face. After the working face advances through the valley section, a third peak of water inflow occurs. The third inflow peak results mainly from the seasonal precipitation runoff that enters the working face along the fractures. The amount of water inflow is predicted for the various positions of the working face.

1. Water inflow at first collapse of the roof formation

Recovery of the 22,401–22,404 working face of the Halagou Coal Mine is completed, and the 22,405 working face is being recovered. This working face is no longer the first mining face. Therefore, the method of predicting the amount of water inflow from the secondary mining face can be used for calculation:

$$Q_{\text{initial pressure}} = \left[(2L_{\text{first time}} \times H \times W \times H)K_{\text{lateral}} \times J_{\text{lateral}} + S_{\text{first time}} \times H \times \mu/t\right]/24$$

where:

$Q_{\text{initial pressure}}$	Water inflow at 22,406 face at first collapse of the roof formation or occurrence of first stress on the roof, m^3/h.
$L_{\text{first time}}$	Length of the working face advancement when the fracture zone extends to the aquifer, 45 m.
W	Width of working face, 300 m.
H	Thickness of the aquifer affected by the fracture zone, 7.2 m.
K_{lateral}	Lateral hydraulic conductivity of the aquifer, 0.6327 m/d.
J_{lateral}	Lateral hydraulic gradient of the aquifer, 0.01.
S	Area of the working face when the fracture zone reaches the aquifer, 13,500 m^2.
μ	Average storativity of the aquifer within the influence range of the caved-in zone, 0.05.
t	Duration of the water inflow, 2 d.

Substitution of the input parameters into the above equation results in water inflow into the working face when the roof first collapsed:

$$Q_{\text{initial pressure}} = 102.0\,\text{m}^3/\text{h}$$

The 22,405 working face is adjacent to the 22,406 working face. The coal seam overburden structure, hydrogeological characteristics of the aquifer, and mining width of the working face are similar between these two working faces. According to the measured water inflow during the 22,405 working face mining process, the water influx at the working face was approximately 111.3 m^3/h when the roof of the 22,405 working face collapsed for the first time. The predicted water inflow for the 22,406 working face is comparable to the measured inflow at the 22,405 working face, which suggests that the water inflow prediction method and input parameters are suitable to the prediction of water inflow in different areas of the working face.

2. Normal water inflow into working face

The periodic collapses of the roof formation at the 22,406 working face occur at a distance between 12 and 18 m. The maximum value is taken as the conservative estimate from the perspective of water hazard prevention. Other parameters are the same as above, and the normal water inflow of the working face is:

$$Q_{\text{normal water inflow}} = 42.2\,\text{m}^3/\text{h}$$

3. Water inflow into working face over Halaguo Gully

(1) Water inflow forecast over the gully section

The flow rate at the Halagou Gully is 35 m^3/h, which is within the range of periodic collapses of the roof formation. The lateral hydraulic conductivity of the aquifer in the ditch section is 1.7424 m/d. Other parameters are the same as above. The available water inflow in the gully section is:

$$Q_{\text{water inflow over gully}} = 255.2\,\text{m}^3/\text{h}$$

(2) Forecast of water inflow during rainy season

During the rainy seasons, the surface runoff generated from torrential rains flows directly into the working face through the fractures. The amount of water at the working face can be estimated using the following equation:

$$Q_{\text{rain}} = \alpha \times \mu \times H \times F/24$$

In the above equation, the coefficient of precipitation infiltration is calculated at the conventional value of 0.5. The storativity of the working face after infiltration is assumed to be 0.8. The mined-out area in the gully section is 19,320 m^2. According to the rainfall data of the Shenmu County Meteorological Station, the 50-year-rain event is used for the calculation, i.e., the maximum daily rainfall is 135.2 mm, and the rainfall intensity is 7.04 mm/h. The amount of water influx into the working face through the collapse area is:

$$Q_{rain} = 322\,\mathrm{m^3/h}$$

Adding the amount of short-term inflow by the surrounding gully area to the above calculated flow rate lead to the total inflow:

$$Q_{rain} = 542.2\,\mathrm{m^3/h}$$

The calculated water inflow is the largest when mining the Gully section and during the rainy season. Therefore, when the 22,406 working face is mined, the water control measures include dewatering of the aquifer in the Gully section and treatment of the ground subsidence area during the rainy season.

5.3.3.2 Risk Analysis of Water and Sand Inrush

Based on geological reconnaissance, the mining area of the 22,406 working face is mostly covered with sand at the gully section. Portions of the valley have exposed bedrock. The hydrostatic pressure is significant in the water-bearing sand layer. When the mining induced fracture zone connects with the aquifer, the groundwater in the aquifer begins flowing. The hydrostatic pressure immediately transforms into hydrodynamic pressure. When the hydrodynamic pressure reaches a threshold value, a large amount of mixture of water and sand flows into the working face, causing the occurrence of water and sand inrush disaster. If the hydraulic head in the water-bearing sand layer is small, the sand particles cannot be mobilized during the conversion from hydrostatic pressure to hydrodynamic pressure. The water and sand inrush disaster will not occur on the working face. Therefore, to determine whether a water and sand inrush accident occurs on the working face, first of all, it is necessary to determine the safe hydraulic height of the water-bearing sand layer under which the water and sand inrush does not take place. Then, comparison of the measured water level of the aquifer against the safe hydraulic head determines whether there is a risk of water and sand inrush. Decrease of the groundwater level in the sand layer below the height of the safe head ensures elimination of one of the four necessary conditions for water and sand inrush occurrence.

1. Determination of water and sand inrush width

(1) Assume no rotation of the overhanging rock beam

The mining height is 3.5 m in the 22,406 working face of Halagou Coal Mine. The thickness of the direct roof, $\sum h$, is 4.7 m. The rock expansion coefficient after the mining collapse, Kp, is assumed to be 1.3. The rock collapse angle, α, is assumed to 80°. If we assume that the overhanging rock beam does not rotate, the channel width at which the water and sand inrush occurs is:

$$B_1 = 0.36\,\mathrm{m}$$

(2) Assume rotation of overhanging rock beam

Based on stress measurements at the 22,405 working face of Halagou Coal Mine, the distance in which the periodic roof collapses occur, l_1, is 18 m. The thickness of the overlying bedrock, h, is 21.5 m. The rotation angle of the rock beam, θ, is assumed to be 0.5°. The channel width of the water and sand inrush occurrence after rotation is:

$$B_2 = 0.16\,\mathrm{m}$$

When the overhanging rock beam turns to 0°54′9", it encounters the collapsed rock beam. Due to the squeezing of the two rock beams at the contact, the flow channels tend to be closed. The amount of water and sand flowing into the working surface would be reduced or even ceased.

(3) Determination of the radius of the circular section

According to the actual measurements, the length of the cross section of the gushing channel is l m. The equivalent maximum radius of the gushing channel is:

$$r_w = 0.34\,\mathrm{m}$$

2. Determination of radius of influence

The thickness of the loose aquifer on the working surface of 22,406 is 37.60 m. The hydraulic conductivity is 1.7424 m/d. The water level decrease by dewatering is assumed to be half of the thickness of the aquifer, i.e., 18.80 m. When the water inflow occurs, the radius of influence of the phreatic aquifer is:

$$R \geq 304.34\,\mathrm{m}$$

3. Determination of critical hydraulic gradient

The critical hydraulic gradient J_{cr}, is between 0.826 and 1.15, based on the Terzaghi equation. The critical hydraulic gradient measured in the experiments is typically greater than the theoretical values because of the influence of friction of the soil particles. In view of the small amount of soil in the sand layer, the calculated values for this mine varies from 1.076 to 1.40. From the perspective of disaster prevention and control, the minimum value is taken as the conservative hydraulic gradient.

4. Determination of critical hydraulic head

The critical hydraulic head is obtained by a comprehensive analysis:

$$H \geq 6.63\,\text{m}$$

For the 22,406 working face, the thickness of the loose aquifer sand layer where the bedrock in the Gully section is the thinnest is 37.6 m. The actual hydraulic head of the aquifer is much greater than the critical hydraulic head. Therefore, before mining of 22,406 working face, it is necessary to pre-drain the loose aquifer and ensure that the residual hydraulic head is lower than 6.63 m so that no large-scale water and sand inrushes occur.

5. Analysis on probability of water and sand inrush

The natural hydraulic head of the sand aquifer at the 22,406 working face is greater than the critical hydraulic head. There is a threat of water and sand inrush disaster at the working face. In addition, during the rainy season, the surface runoff will flow to the bottom of the gully where fracturing direct the water and sediments to the underground working areas, resulting in occurrence of water percolation and sand erosion disasters. To ensure the safety of mining through the gully section, the engineering measures include installation of dewatering system with sufficient capacity to avoid causing water inrush hazards and modification on the necessary conditions for water and sand inrush. These measures ensure that a large volume of water and sand does not gush into the working face. Water and sand inrush paralyze the dewatering system and causes secondary disasters at the working face.

5.3.3.3 Measures to Prevent Water and Sand Inrush

Because of the potential threat of water and sand inrush over the gully section during mining of the 22,406 working face and the possibility of surface runoff flowing into the underground working area in surface subsidence areas during the rainy season it is necessary to apply targeted water prevention and control technology and take engineering measures underground and on ground surface. The objective of these measures is to ensure that water and sand inrush will not occur during mining of 22,406 working face.

1. Underground groundwater control measures

Occurrence of a water and sand inrush requires four necessary conditions. Change of just one of the conditions may result in avoidance of water and sand inrush occurrence. Based on the site-specific conditions such as layout of working faces, characteristics of surface gully and underground drainage system, the method of dewatering the sand aquifer before mining is selected. Dewatering transforms the sand aquifer into a vadose zone. The hydraulic head of the aquifer is significantly lowered. The residual water pressure can no longer mobilize the sand particles. No water and sand inrush can occur during mining of the working face.

There are two methods to drill the dewatering borehole prior to mining. One method is to drill borehole underground to release the groundwater into the underground drainage system, while the other is install pump wells on surface. The aquifer overlying the coal seam in Haragou Coal Mine is relatively shallow. Although it is possible to install pump wells on surface to perform the dewatering, however, the site access is challenging and coordination with property owners is not an easy task. Therefore, the underground dewatering approach is selected to drain the aquifer, lower the water level of the aquifer, reduce the intensity of water inrush from the mining face, and alleviate drainage load. The purposes of drilling borehole to release groundwater are:

(1) Reduce the water content of the loose aquifer so that it does not have the driving force for water and sand inrush.
(2) Pre-drain the bedrock aquifer and loose sand aquifer overlying the #22 coal seam to reduce the intensity of water inrush.

The main threat to the 22,406 working face is the phreatic loose sand aquifer, which is the target formation of the exploration and water release. The radius of influence of each borehole is calculated by the following equation:

$$R = M \times S \times \sqrt{K}$$

where:

M Aquifer thickness, m
S Aquifer drawdown, m
K Hydraulic conductivity of the aquifer, m/d.

Because of the limited effect of a single borehole on the aquifer, the aquifer drawdown is assumed to be 1 m. The calculated radius of influence is 44.5 m. The borehole spacing is designed according to this radius of influence at 44.5 m. The borehole spacing can be 30 m to enhance the dewatering effectiveness.

2. Groundwater control measures on surface

The average flow rate in Halagou Gully is approximately 35 m^3/h. The flow is significantly larger during rainy seasons. A surface dam is constructed upstream of mining section of the gully to prevent surface water from flowing into the working face.

(1) Location of river dam and layout of conveyance pipes.

The river dam is located to avoid the surface collapse or subsidence areas that are induced by mining. The collapse or subsidence area is extrapolated from the mined-out area in reference to the characteristics of the overlying soil. The collapse range

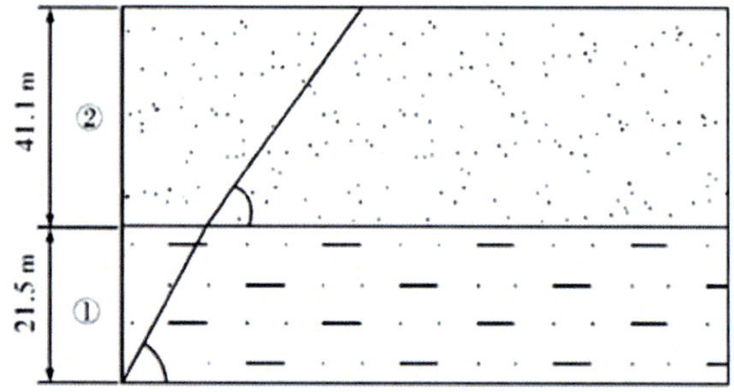

Fig. 5.17 Extrapolated ground subsidence range over 22,406 traffic lane

provide the essential data for the outlet of the dam and the conveyance pipelines. The ground collapse range above 22,406 traffic lane is shown in Fig. 5.17.

The ground subsidence range of 22,406 traffic lane is: cot80° × 23.73 + cot45° × 35.74 = 42.11 m. The positions of the dam and drainage outlet of the 22,406 working face is thus determined: 42.11 + 20 = 62.11 m. The following descriptions are provided to explain the calculation processes:

- In Fig. 5.17, ① represent bedrock, and ② denotes overburden. The weathered bedrock is considered as overburden.
- The thickness data of bedrock and overburden are from the boreholes in the Supplemental Hydrogeological Exploration Report of the Sipan District of Haragou Coal Mine.
- The bedrock movement angle and topsoil movement angle come from "Mining Engineering Design".
- The ground protection area includes the protected objects and surrounding enclosures. The protection level of the water retaining dam and the outlet of the drainage pipeline is class I. The width of the enclosure should be 20 m.

(2) Determination of surface runoff

According to the rainfall data of the Shenmu County Meteorological Station, the 50-year heavy rain in Shenmu County (maximum daily rainfall of 135.2 mm, rainfall intensity of 7.04 mm/h, August 1, 1977) is analyzed to determine the relationship between rainfall and runoff. Based on geomorphology and lithology, the infiltration rate is used to calculate the runoff depth using the infiltration runoff model.

According to the supplemental hydrogeology exploration, R_S (sand layer) = 0.69 mm and R_S (soil layer) = 2.50 mm. The surface runoff R in the Halagou Gully area is taken from the calculated value for the Haizigou area. The equation for calculating the river catchment is as follows:

$$Q_{catchment} = 1000 \times F \times R/T$$

where F is the catchment area in km^2. Considering sand layer water catchment area $F_{sand} = 0.626\ km^2$, loess layer catchment area $F_{soil} = 0.568\ km^2$, we obtain the following:

$$Q_{sand\ layer} = 431.9\,m^3/h$$

$$Q_{sand\ layer} = 1420.0\,m^3/h$$

Then the estimated total surface runoff is:
$Q_{surface\ runoff} = Q_{sand\ layer} + Q_{soil\ layer} + Q_{normal\ runoff}$
$= 431.9 + 1420.0 + 35.0 = 1886.9\,m^3/h$

The interception amount of plants is 20%-30%. Vegetation in the Haragou watershed is lush with a coverage rate of approximately 85%. The terrain is undulating with closed ponds. Therefore, the 0.70 empirical coefficient is used to modify the surface runoff:

$$Q_{surface\ runoff} = 1320.8\,m^3/h$$

(3) Determination of capacity of the conveyance pipeline

Based on the topographic map of Halagou Coal Mine and actual survey, the elevation difference along the conveyance pipeline is approximately 9 m. The hydraulic head difference between the two ends of the pipeline is thus designed to be 9 m. Three pipelines are designed with each being 708.8 m long. The hydraulic gradient is calculated as follows:

$$J = H/L = 9/708.8 = 0.0127$$

where:

J Hydraulic gradient
H Hydraulic head difference, m
L Pipeline length, m.

The double-wall corrugated polyethylene pipes (PE) are used with the roughness n between 0.008 and 0.009. Assuming the maximum roughness n of 0.009, we calculate the hydraulic radius:

$$R = D/4 = 0.3/4 = 0.075\,m$$

where:

R Hydraulic radius, m
D pipe diameter, 0.3 m.

The Chézy coefficient is:

$$C = R^{1/6}/n = 0.075^{1/6}/0.009 = 72.15$$

where:

C Chézy coefficient
R Hydraulic radius, m
n Roughness.

The water flow velocity in the pipeline is:

$$V = C \times (RJ)^{1/2} = 72.15 \times (0.075 \times 0.0127)^{1/2} = 2.23\,\text{m/s}$$

where:

V Water flow velocity in the pipeline, m/s.
C Chézy coefficient.
R Hydraulic radius, m
J Hydraulic gradient.

Therefore, the drainage capacity of the pipeline is:

$$Q = 3.14 \times (D/2)^2 \times V = 3.14 \times 0.15^2 \times 2.23 = 567.2\,\text{m}^3/\text{h}$$

The drainage capacity of three pipelines is: $3 \times 567.2 = 1{,}701.6$ m^3/h.

The calculated runoff generated from atmospheric rainfall upstream of the dam is 1,320.8 m^3/h. Three φ300 mm HDPE double-wall corrugated pipes can direct the runoff upstream of the dam to downstream. It is recommended to install four φ300mm HDPE double-wall corrugated pipes for safety insurance. Two grates are installed at the proper position of the nozzle to prevent debris carried by the flood from blocking the pipelines.

(4) Specifications and construction of the dam.

When the 22,406 fully mechanized mining face passes through the gully section, the water hazard prevention and control measures are to "intercept, block, and direct" the water flow. These measures are based on the following facts:

- They are consistent with the principles of "technically feasible, economically reasonable, and practically implementable".
- They comply with the relevant provisions of "Coal Mine Safety Regulations", "Coal Mine Water Prevention and Control Regulations".

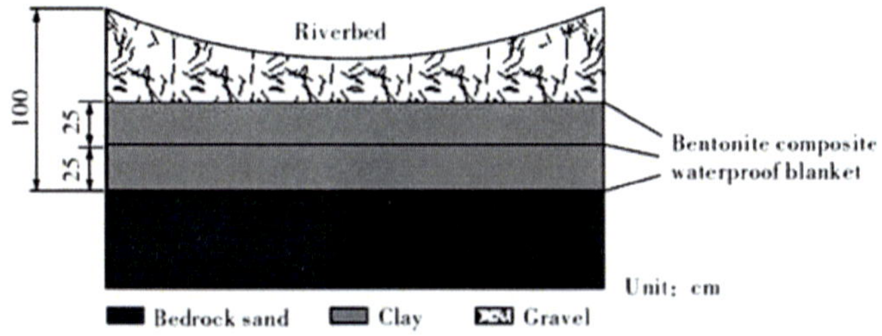

Fig. 5.18 Schematic diagram of river control project

- They are tailored to the site-specific conditions at the 22,406 working face on ground surface and underground.
- They are in line with the experience of surrounding mines.

Because the mining induced roof collapse is free-fall along shear planes, the three deformation zones are not obvious or cannot be fully developed, as shown in Fig. 5.18. In the gully sections where the overlying bedrock is thin, the caved-in zone can directly communicate with the loose aquifer and cause water and sand inrushes with water inrush as the primary form of hazard. To prevent water and sand inrush, it is necessary to control surface storm floods and surface runoff. A dam can be constructed upstream of the Halagou Gully outside the subsidence range caused by mining of the 22,406 face. The dam intercepts water in the Halagou Gully and surface runoff from heavy rains. The conveyance pipes are installed to divert the surface water accumulated in the dam from the area that is directly above the 22,406 working face. At completion of mining of the 22,406 working face, the riverbed should be paved immediately with high-quality clay. The clay would form an artificial impermeable layer to prevent the surface runoff from percolating into the underground during the storms.

5.3.3.4 Technical Measures Against Water and Sand Inrush

- Significant monitoring effort should be made on water inflow at the working face and water inflow points at both sides of the working face. Before the rainy season arrives, a full-time professional should be assigned to organize a dedicated team to build dams and dig ditches according to the topographic and geological conditions so that the surface runoff can be properly directed away from the area directly above the underground working face. No surface water should be accumulated in fracture zones. If conditions permit, the gully section should be mined after September of the year to avoid the impact of the rainy season flood.
- The capacity of the water silo constructed at the mining area should be designed to accommodate the normal water inflow for at least 4 h. During the rainy season, strictly implement the emergency management system during the rainy season. Based on the progress of the working face, detailed emergency response plans

should be formulated for water disaster control. Drills on emergency response to water disaster should be conducted according to the emergency plan.

- When the working face advances to the gully section, the advancement speed should be accelerated to pass the gully section quickly. The accelerated advancement would make the mining induced damage in the overlying formations lag the working face. The risk of surface water directly flowing into the working face along the subsidence cracks is minimized as much as possible. The change of this mining process helps reduce the difficulty of drainage and avoid impacting the coal quality.
- Water pumps, water pipes, valves, power distribution equipment and transmission lines for drainage must be regularly inspected and maintained. All equipment parts and accessories should be fully supplemented. All water pumps should be concurrently tested for drainage once. Any identified issues should be addressed in time.
- Set a sand pump at each end of the working face. The flow rate of the sand pump is no less than 50 m^3/h, and the hydraulic head is no less than 20 m. The outlet of the sand pump is directly connected to the 22,405 mined-out area. It may extend into the mined-out area further to ensure that the sand inlet has a certain sand storage space.
- Strengthen the organization and management of water prevention and control. Before mining, detailed water prevention and control measures and measures for disaster mitigation must be formulated. Responsible parties must be clarified. Professional workers should be assigned for duties, and a buddy system should be implemented. All underground worker should be properly trained on prevention and control of water safety.
- Detailed safety technical measures for preventing water and sand damage accidents should be formulated in the entire mine. Before passing through the gully section, an emergency drill should be carried out for water and sand disaster incident at the working face.

5.3.3.5 Performance Evaluation of Engineering Measures

Mining of the 22,406 working face of Halagou Coal Mine began on July 25, 2014. Before the mining, the mine constructed a dedicated drainage channel in the Halagou Gully area, which is designed to collect water from the loose layer, the weathering zone of bedrock and surface runoff. A total of 135 exploration and dewatering boreholes were drilled in the drainage channel. The cumulative drainage volume was 456,324 m^3, and the groundwater level of the sand aquifer dropped from the original 37.6 to 6.63 m. The drainage effect is obvious. In the process of underground dewatering, the surface above the 22,406 working face was impermeabilized by installing impermeable membrane to minimize infiltration of surface water. Construction of dam and ditch was conducted to direct surface runoff away from the mining area. In addition, a dewatering system with a total pump capacity of 750 m^3/h was installed in the working face. Mining of the working face was completed on August 27, 2014, and no water issues encountered. Implementation of these water and sand prevention and control measures achieved great economic and social benefits and saved engineering investment. Another alternative was to grout and solidify the loose sand layer, which costs approximately 8.5 million Yuan.

Chapter 6
Investigation and Prevention of Water Hazards from Old Mine Pools in Ordos

6.1 Background of Mining Area

Early mines in Ordos mainly used the "room-and-pillar" or "residual pillar" coal mining method, which is known for low resource recovery rate and poor safety and reliability. The mines were poorly equipped with extraction techniques. There was no distinction between mining and excavation. The mine production capacity was low, and the management level was poor. The room and pillar coal mining method resulted in not only waste of coal resources but also were susceptible to geological disasters and threat to mine production safety. The coal pillars left during mining are irregular. Some coal pillars gradually break down over time. The long-term roof support is not ensured. The ground surface is vulnerable to subsidence or collapse in response to room and pillar mining. Large-scale roof collapse accidents also occur and induce seismic shocks. In addition, the old mined-out areas abandoned from the early mining are poorly recorded or documented. These abandoned mined-out areas are voids for groundwater and methane accumulation and spontaneous combustion. They are the sources for mine disasters in the future coal mining and in the neighboring areas.

Decades of mining results in a wide distribution of the mined-out areas with various sizes. The exact of number of old mines is not available because they were scattered across the region. There were no designs, no drawings, and no records on these mine excavations. Consequently, no records on the nature and extent of the goafs are available. Some goafs have partially collapsed; some remained intact; and some goafs are known for spontaneous ignition. These goafs lead to or will lead to serious geological environmental problems in the area. In order to promote the harmonious development of the local economy, society and environment, ensure the safety of human beings and mines, and eliminate any concealed dangers of the mined-out area, it is essential to conduct the following investigations:

S. Dong et al., *Methods and Techniques for Preventing and Mitigating Water Hazards in Mines*, Professional Practice in Earth Sciences,
https://doi.org/10.1007/978-3-030-67059-7_6

- Conduct a comprehensive reconnaissance on distribution and occurrence of the mined-out areas.
- Understand conditions surrounding these mined-out areas.
- Ascertain the nature and extent of the mined-out area.
- Investigate roof collapse status, distribution of self-ignition area, distribution of water accumulation area, distribution of surface building distribution, and the occurrence and distribution of aeolian sand or loess.
- Identify any concealed safety hazards in the old mined-out areas and then take targeted comprehensive treatment plans according to the different conditions of the mined-out areas to further prevent, control and manage the hazards.

6.2 Technical Approaches

The objective of this project is to investigate the distribution and status of goafs through inventory of mined-out areas, geophysical survey, and drilling exploration. The type and severity of existing disasters are analyzed and evaluated. The comprehensive mitigation approach is proposed to prevent and control of goaf-related disasters. The results of this investigation provide scientific basis for the plan to promote governance. The technical approach includes the following:

- Data collection and analysis of the survey area
- Ground reconnaissance and surveys on the ground surface
- Defining areas for geophysical survey
- Work plan preparation for geophysical survey
- Measurement positioning
- Geophysical work plan implementation
- Data processing
- Prepare investigation report.

6.3 Geophysical Methods

Based on the geological environment and geophysical conditions of the survey area, several geophysical methods are used to facilitate the investigation. These geophysical methods include high-density electrical resistivity imaging, transient electromagnetic method, shallow seismic, EH4 magnetotelluric method, and controlled source CSAMT.

6.3.1 High-Density Electrical Resistivity Imaging

The high-density electrical resistivity imaging is a variation of the conventional resistivity method. In principle, it is the same as the conventional resistivity sounding method. It is a type of exploration method that is based on the electrical resistivity difference of rocks. Application of the geophysical method helps solve mineral resources, environment and engineering geological problems by observing and studying the distribution of artificially established electric field in the subsurface. Different configurations of electrodes can be used in this method. Field measurements required that all the electrodes (tens to hundreds) be placed on the measuring point. The computer program is used to control the electrode switch and to collect quickly and automatically the data. The data is then processed on a computer, and various graphical presentations about the distribution of the geoelectric cross section can be generated. Obviously, the application and development of high-density electrical resistivity imaging has made the electrical exploration more flexible and effective.

In comparison with the conventional resistivity sounding method, the high-density electrical resistivity imaging method has the following advantages:

- The electrode layout is completed at one time, which can reduce the failure and interference caused by repeated electrode placement.
- Scanning of the measurement of various electrode arrangements can be effectively performed. It is possible to obtain volumetric geophysical data for interpretation of geological structures in cross sections.
- Field data collection is automated or semi-automated, which not only has a fast collection speed (about 2 to 5 s per measurement point), but also avoids errors due to manual operations.
- The data can be pre-processed on site, and the profile curve shape can be displayed quickly. Post-processing can generate various maps to facilitate interpretations.
- The new method is characterized with low cost, high efficiency, volumetric data, convenient explanation, and significantly improved detection ability.

6.3.2 Transient Electromagnetic Method

Transient-electromagnetic method, also known as time domain electromagnetic method (TEM), belongs to the electromagnetic induction detection method. It follows the principle of electromagnetic induction, as shown in Fig. 6.1. The mechanism is that the conductive medium generates eddy current in response to electromagnetic field changes. An ungrounded loop or magnetic dipole is used to emit pulsed electromagnetic waves in the subsurface as the excitation field source, commonly referred to as primary field. According to Faraday's law of electromagnetic induction, the ground or the detection target will generate an induced eddy current field under the action of the excitation field. The eddy current field has spatial characteristics and temporal characteristics. Its magnitude is related to many factors, such as the

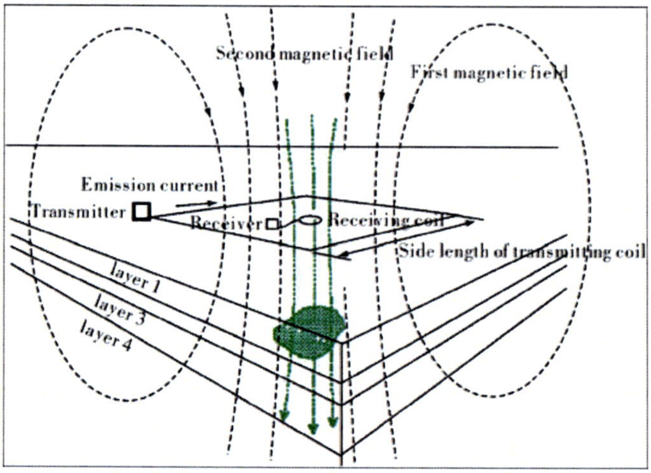

Fig. 6.1 Schematic diagram of TEM working principle

spatial and electrical characteristics of the target body and the characteristics of the excitation field. It gradually weakens until it disappears due to heat loss. Although people cannot directly measure the size of such eddy currents, they can use special instruments to observe the strength, spatial distribution, and temporal characteristics of the electromagnetic field, also known as secondary field, generated by this eddy current.

In principle, targets with large geometric scale, shallow depth, and better electrical conductivity generate strong secondary field signal with long the duration. Similarly, targets with smaller geometric scale, deep depth, and worse electrical conductivity generate weaker signal of the secondary field with short duration. By observing and studying the spatial distribution and temporal characteristics of the secondary field, it is possible to hypothesize and interpret the geometric and physical characteristics of the geological targets.

In comparison with other geophysical methods, the transient electromagnetic method has the following unique features and advantages:

- This method observes and studies a secondary field, which a pure anomaly field. There is no background interference from the primary field. This feature is different from the frequency domain electromagnetic method, which measures the integrated field. This unique feature simplified the anomaly interpretation.
- The TEM method is capable of penetrating high-resistance overburden layer. This capability is superior to conductive electrical detection method.
- The same point combination device (central loop or overlapping loop) can be used for measurements so that the best coupling with the detection target can be obtained. The anomaly response is strong, and the response form is simple.
- The topographic effect is insignificant. The field measurement operation is simple, and the working efficiency is high.

6.3.3 Shallow Seismic Method

Shallow seismic method uses the properties of seismic waves in propagating through underground rock formations in response to an artificial sound source such as hammering to investigate the depth, structure, and characteristics of the rock formations. The seismic waves are created at one point on the surface. The reflected waves from the interface of different elastic rock layers are recorded at several other points with geophones. The seismic records on wave propagation time, waveform, and amplitude are analyzed to study the spatial position and nature of the interfaces.

6.3.4 EH4 Magnetotelluric Method

In the electromagnetic exploration methods, the changes in both natural and artificial fields with frequency ranging from 0.01 to 104 are mainly used to study the internal structure of the earth. Electromagnetic fields of different frequencies have different penetration depths. Low frequency fields correspond to deep penetration. Therefore, the use of different frequencies in electromagnetic field changes can reflect the resistivity of formations at different depths. The penetration depth is proportional to the square root of the resistivity of the rock and inversely proportional to the square root of the frequency. Therefore, when the resistivity of the rock is constant, the high-frequency electromagnetic field measured on the surface reflects shallow geoelectric information, and the low-frequency electromagnetic field reflects the deeply buried formations.

However, the magnetotelluric sounding method uses natural electromagnetic field sources. The natural electromagnetic field is a rapidly changing electromagnetic field formed by the ionosphere in the 100 km sky under the influence of solar particle radiation and is interfered with industrial electricity leakage, ultra-long wave radio stations, atmospheric lightning phenomena and changes in the geomagnetic field. The characteristics of this electromagnetic field source are as follows:

- The natural electromagnetic field is unstable.
- Some frequency bands, primarily between 500 and 104 Hz, are characterized with strong interference and weak signals. These frequency intervals exactly reflect the subsurface formation within depth less than 200 m. Because of this characteristic of natural electromagnetic fields, the magnetotelluric method was mainly used to measure low-frequency signals to study the deep geological conditions of the earth prior to the 1990s. It was rarely used to study shallow geological conditions.

In order to make up for the shortcomings of weak signal strength in the frequency bands between 500 and 104 Hz, the United States GEOMTRICS and EMI (Electro-Magnetic Instruments) company jointly developed and produced the EH-4 magnetotelluric measurement system. The EH-4 system is mainly used for high-precision magnetotelluric measurements in the depth range from 10 to 1200 m. The EH-4 magnetotelluric system has the following characteristics:

- EH-4 is based on the principle of the magnetotelluric method. However, it uses two field sources in artificial electromagnetic field and natural electromagnetic field. The natural electromagnetic field source is used in the frequency range of 1 to 500 Hz, whereas the special artificial emission source is equipped in the 500 to 106 Hz frequency bands to compensate the natural sources and improve the accuracy of exploration.
- EH-4 not only has the stability of source controlled electric detection method, but also has the energy saving and portableness of non-source electromagnetic method.
- EH-4 can simultaneously receive and analyze the electric and magnetic fields in both X and Y directions and invert the X–Y conductivity tensor profile. These capabilities are particularly beneficial for investigating the two-dimensional structure.
- The EH-4 instrument is light, and the observation time is short. For example, it takes approximately 15 to 20 min to complete a sounding point at a depth of 1000 m, which allows continuous measurement of dense points (connected end to end) and continuous EMAP measurements.
- This method allows for real-time data processing and display, preliminary data interpretation, and intuitive imaging.

6.3.5 Control-Source Audio Magnetotelluric Method

The control-source audio magnetotelluric (CSAMT) method is a frequency domain electromagnetic method with a made=made source to generate the frequencies. Its principle is similar to the conventional magnetotelluric sounding (MT). Figure 6.2 shows the schematic diagram of the CSAMT method for operation. It is aimed to overcome the fact that the natural electromagnetic field signals are weak. A control source uses the transmission electric dipoles A and B spaced at 2–7 km and sends alternating currents of different frequencies to the subsurface. An alternating electromagnetic field is thus generated. The mutually perpendicular electric field signal Ex and the magnetic field signal Hy are measured at points far enough from the source. The Cagniard equation is then used to the apparent resistivity and impedance phase of the subsurface formations.

Because the CSAMT method is a geophysical method that uses active sources, the data collected in the field is of high quality and good repeatability. This method can be applicable to areas with relatively strong interference. The interpretation and processing methods are relatively simple and similar to those used in the MT method. The lateral resolution of the interpretation section is high. It has the advantages of not being shielded by the high-resistance layer. The CSAMT method has been widely used investigate mined-out areas or goafs.

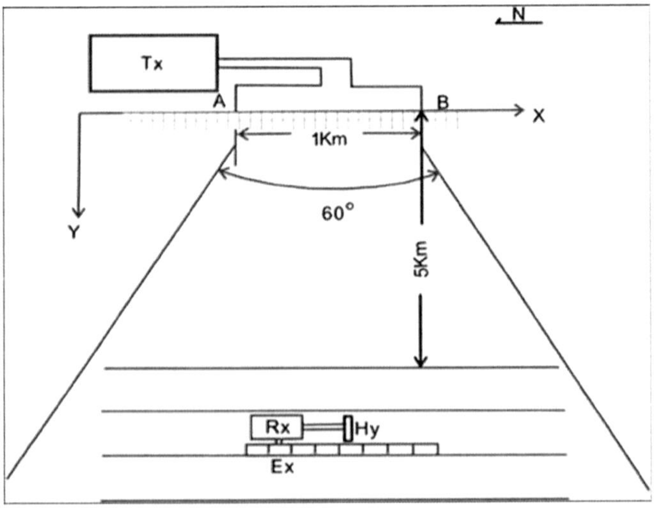

Fig. 6.2 Schematic diagram of CSAMT method

6.3.6 Magnetic Method

The magnetic method uses a magnetometer to measure changes in the magnetic field that is caused by differences in rock magnetism. The magnetic survey mainly uses relative measurement in the unit of nT (nanotesla). The principle is expressed by the following equations:

$$H = \frac{f}{Q_{m0}} = \frac{1}{4\pi\mu_0}\frac{Q_m}{\gamma^3}\gamma$$

$$B = \mu_0 H$$

where:

$\mu_0 = 4\pi \times 10^{-7} N/A^2$,

H/m or Henry/m. The SI unit of the magnetic charge is m N/A or Wb. The SI unit of magnetic induction is Wb/m^2 or N/(A m), which is called Tesla.

The measurement parameter is the geomagnetic field gradient value ΔT. The instrument is calibrated and inspected according to the specifications and instructions so that the performance of the instrument and various technical parameters and indicators meet the requirements. The measured data at each measuring point is corrected based on the measurement at the base station where the base value is collected and set. ΔT is the vertical component increment in the magnetic field between the measuring point and the base station. The ΔT value of the measuring point and the automatically generated ΔT curve are displayed on the screen of the instrument. The investigators can identify the ΔT mutation point and distortion point in time.

6.4 Achievements by Electrical and Magnetic Imaging

6.4.1 Geophysical Survey Layout

The electrical resistivity imaging and TEM method are used for the geophysical investigation. The spacing between electrical resistivity imaging transects is 10 m. The transects is arranged according to the topography and undulation and avoid crossing valleys if possible.

The TEM measurement network is 100 × 40 m, i.e., the line spacing is 100 m, and the data interval along the transect is 40 m. The high-density electrical resistivity imaging line is flexibly arranged according to the topography of the survey area, and the data interval is 10 m.

Based on the distribution characteristics of the measurement area, the distance between the measuring points of the magnetic method is designed to be 40 m or 50 m. The principle is to control the measurement area as much as possible. The data interval can be reduced to 20 m where the magnetic field shows greater changes.

The test results suggest that the TEM method uses a central loop configuration. The transmission coil frame is 120 × 120 m. The transmission frequency is 25 Hz. Figure 6.3 shows the layout of geophysical exploration in a mine goaf.

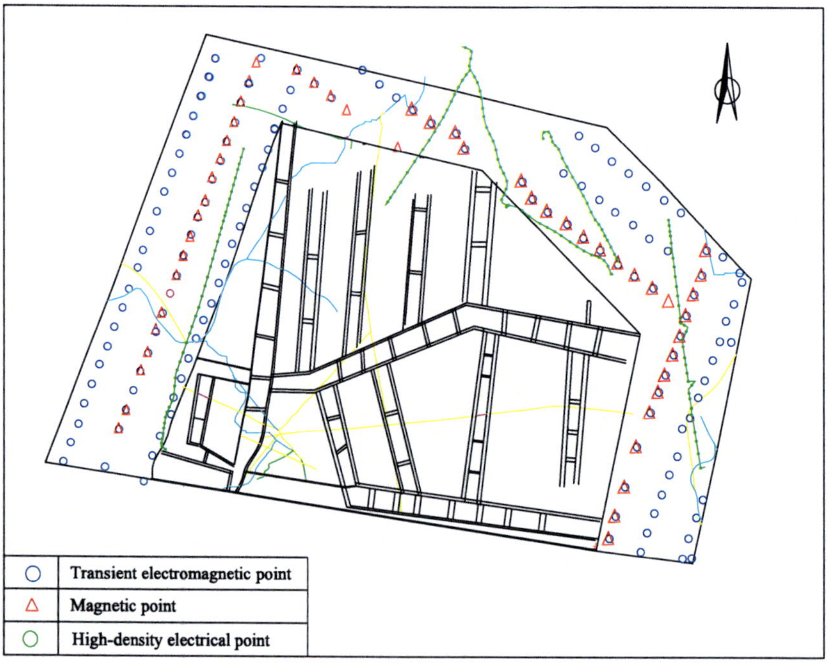

Fig. 6.3 Layout of geophysical survey in a mine goaf

6.4.2 Results of Electrical Resistivity Imaging Survey

Figure 6.4 shows a geoelectrical cross section from the high-density electrical resistivity imaging. The selected location of the survey line is directly above the two known goaf roadways. The tunnel depth is approximately 90 m. The terrain is relatively flat. There is no other interference. To obtain the largest possible section length, the joint measurements using A-MN and MN-B are adapted. Figure 6.4a, b are the cross-sectional views obtained by using different roll-along sections. The high-resistance anomalies of the two profiles accurately reflect the locations of the roadways, as labeled in blue rectangular lines. This correlation indicates that resistivity imaging with the design parameters can obtain reasonable interpretations of the subsurface features.

The first survey line of the electrical resistivity imaging is in the west of the survey area. The comprehensive profile is shown in Fig. 6.5. Three resistivity highs are identified in Fig. 6.4: anomaly #4 between the 8th and 15th measurement points, anomaly #3 from the 25th to 35th points, and anomaly #1 between the 47th and 59th points. Anomaly #4 has a greater amplitude, where the response in anomaly #3 is weaker. The response in anomaly 1 has a wider extent. The core parts of these three high-resistance anomalies are at elevation of approximately 1220 m amsl. This depth is where #6–2 coal seam is located. These three abnormal areas are interpreted as potential local mined-out areas or goafs.

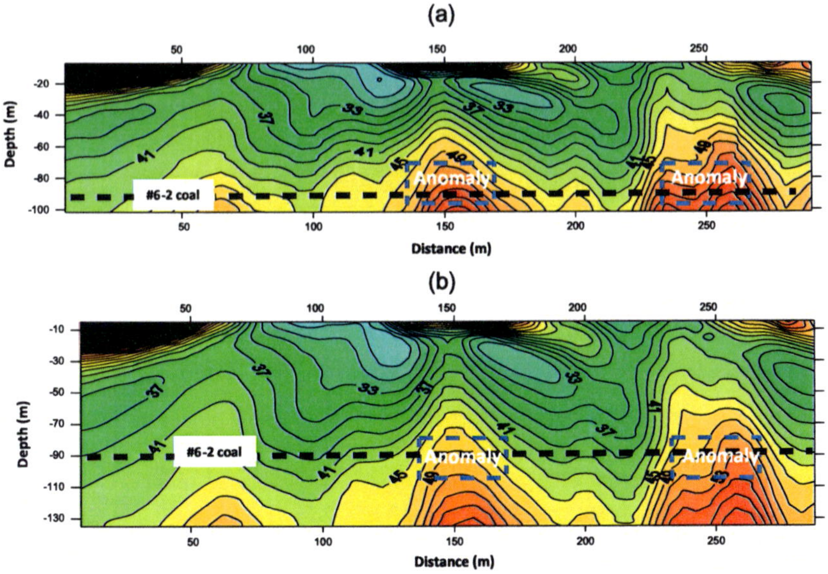

Fig. 6.4 Geoelectrical cross-sections from electrical resistivity imaging

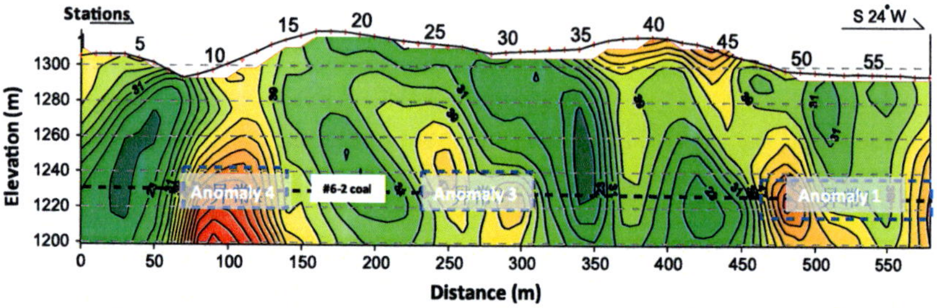

Fig. 6.5 Apparent resistivity cross section along first line

The high-density electrical resistivity imaging results in the first transect identified three resistivity highs where #6–2 coal seam is located. It is possible that these three anomalies are caused by mined-out areas or goafs. This interpretation is corroborated with other data available. There appears to be no or little water accumulation in the mined-out areas.

6.4.3 Results of Transient Electromagnetic Survey

The TEM method is carried out to test the different sizes of the transmission coils. Figure 6.6a, b, c show the cross-sectional diagrams obtained by launching the frame with side lengths of 30 m, 60 m, and 120 m, respectively. The apparent resistivity contours of the first two figures are seriously distorted. The high and low resistances are staggered, and the corresponding curve is severely distorted. Figure 6.6c reflects the profile of a normal formation. Comparison with the electrical resistivity imaging results indicates that the electrical resistivity distribution of the formation is basically the same from these two methods. The two known roadways are clearly reflected on the figure. The corroboration with the known features demonstrates the effectiveness of both the electrical resistivity method and TEM method for detecting mined-out areas.

Figure 6.7 shows the geoelectrical cross section along first line from the TEM survey. The apparent resistivity value has a low-to-high change trend in the detection range, which is consistent with the electrical resistivity distribution of the actual geological horizons. The coal-measure formation where #6–2 coal seam is located is on a relatively high resistivity horizon. The high-resistivity anomalies in this horizon should be the reflection of unfilled goaf areas. Anomalies of different scales and amplitudes are identified on the first, second and third lines For example, four resistivity highs are shown in Fig. 6.7: anomaly 2 between 3rd and seventh data points, anomaly #3 between data points 10 and 13, anomaly 4 between data points 14 and 15, and anomaly #5 between data points 20 and 24. The extent and amplitude of anomaly 2 and anomaly 3 are stronger than other anomalies, and they are likely

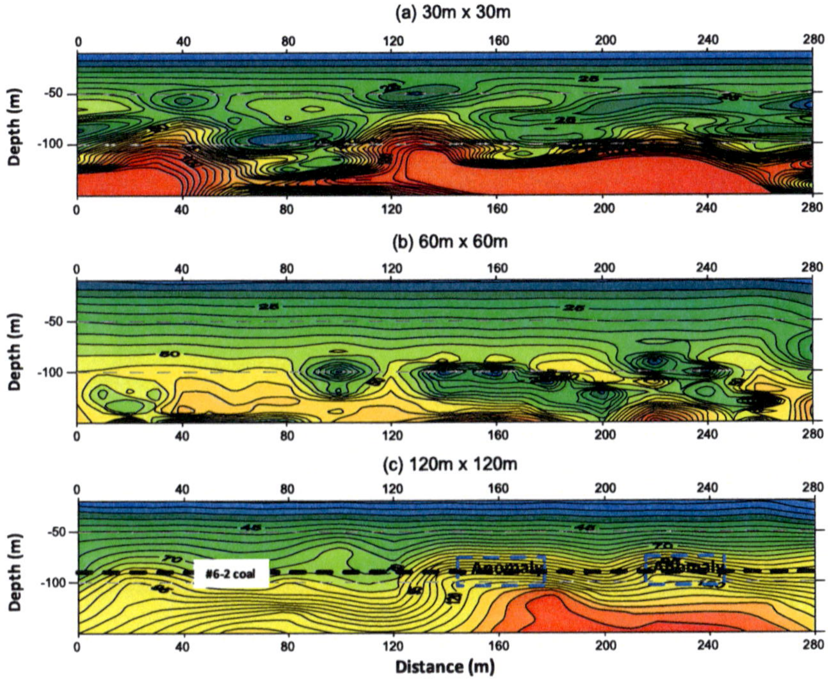

Fig. 6.6 Transient electromagnetic survey results for three transmission coil sizes

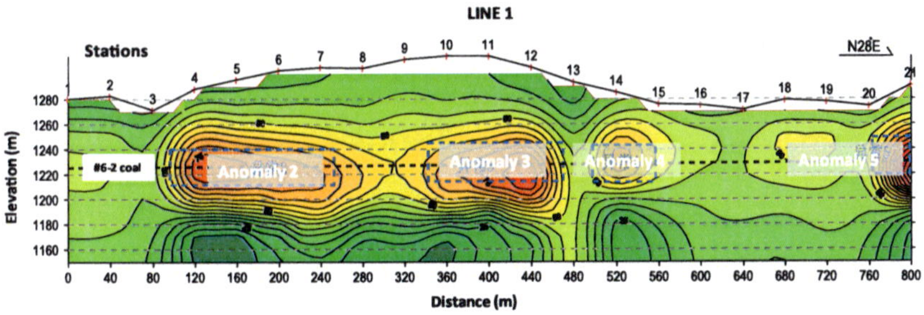

Fig. 6.7 TEM survey results along first line

related to goaf. The distribution form of anomaly #3 is unique and may be caused by two parallel tunnels.

The main roadway direction of the goafs is perpendicular to the direction of the survey lines. The high-resistivity anomalies on the first geophysical line is likely caused by mined-out areas or abandoned roadways. Additional geophysical survey can help determine the extent and direction of these anomalies.

6.4.4 Results of Magnetic Survey

The magnetic method is used in areas where terrain is open and flat, there is no human interference, and the magnetic field is free of changes. Four survey lines are completed with a total length of 2450 m. The average value of measurements is assumed to be the geomagnetic background value in the survey area. Its purpose is to make the average magnetic field value reflect the typical field in the area. The background field is low at 55,194 nT. This may be related to the thin coal seam in the survey area.

Figure 6.8 shows the measured magnetic profile along second line, which is in the west side of the survey area. The length of the survey line is approximately 960 m. The average total field value of the coal measures in the area is used as the background field. The residual magnetic values are mainly negative, varying between 100 and −300 nT (Fig. 6.8a). The residual magnetic field is relatively strong between 0 and 200 m, showing a strong negative anomaly. The residual magnetic field in remaining sections are small except for individual points with positive values. The red line is the residual magnetic field curve after the downward extension process. Its anomaly distribution characteristics show negative anomalies. Based on the analysis of the construction in this area, this type of characteristics is likely caused by oxidation at the outcrops of the coal seam.

Figure 6.9 shows the measured magnetic profile at measuring station 24. Station 24 and 26 are located on the north side of the survey area. The length of the profile is approximately 1000 m. The average total field value of the coal measures is calculated as the background field. The residual magnetic field along this survey line is mainly strong positive, varying from 300 to −200 nT (Fig. 6.9a). The section to the east of 640 m appears as a positive anomaly with a strong amplitude, and the west as a negative anomaly. The red line is the residual magnetic field curve after the downward extension process. Its anomaly distribution characteristics also mainly show positive trend. Based the ΔT curve and the downward extension distribution curve (Fig. 6.9b), it can be concluded that the gradient value of the survey line varies within the range between 12 and -8 nT. The gradient value of the entire survey line is dominated by

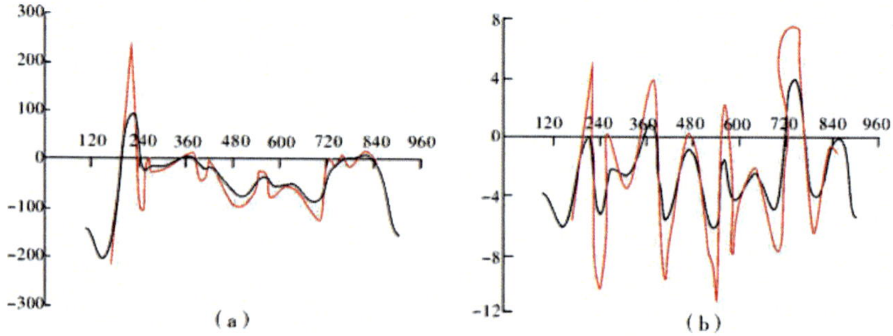

Fig. 6.8 Cross-sectional view of 2-wire magnetic survey

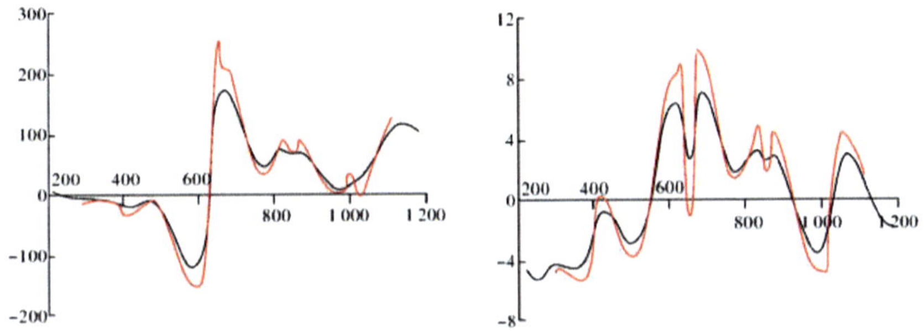

(a) (b)

Fig. 6.9 Sectional view of the 24-point magnetic method

positive and strong anomalies at intervals between 640 and 900 m and between 1040 and 1140 m.

Combination of the residual magnetic field and ΔT on this survey line indicates that that the overall performance of the magnetic field in this area is divided into two regions with the boundary at 640 m. To the east of the boundary is mainly positive anomaly, while to the west of the boundary is mainly negative anomaly. The residual magnetic field and ΔT in the sections between 640 and 900 m and between 1,040 and 1,140 m on the survey line are both positive anomalies, and the anomaly amplitudes are relatively strong. These anomalies are interpreted to be related self-ignition of the coal, which results in high temperature metamorphism in these sections.

The interpreted geophysical anomalies are corroborated with borehole exploration. The goafs identified from the geophysical prospecting are basically consistent with the actual conditions. Figure 6.10 shows the distribution of the goafs in this area.

6.5 Experience with Reconnaissance of Coal Mine Goafs in Ordos

Ordos City is rich in coal resources and has a long history of mining. The coal-bearing area accounts for approximately 70% of the city's land. There were more than 1,900 small coal mines in the 1970s and 1980s. The number of coal mines was reduced to 321 after concerted resource integration, rectification, closure, and technical transformation. The average single mine production capacity is more than 1.5 million tons per year. In 2012, the total raw coal production reached 590 million tons. Due to the disorderly mining of small coal mines in the past, old goafs of various sizes are widely distributed. The water and toxic and harmful gases accumulated in the old goafs poses a serious threat to normal production of current mines. The surface subsidence or collapse above the mined-out areas also affects the safety of ground buildings. Many disasters occurred in the city due to goafs. To fundamentally eliminate the hidden safety hazards of the old mined-out areas, the city learned lessons

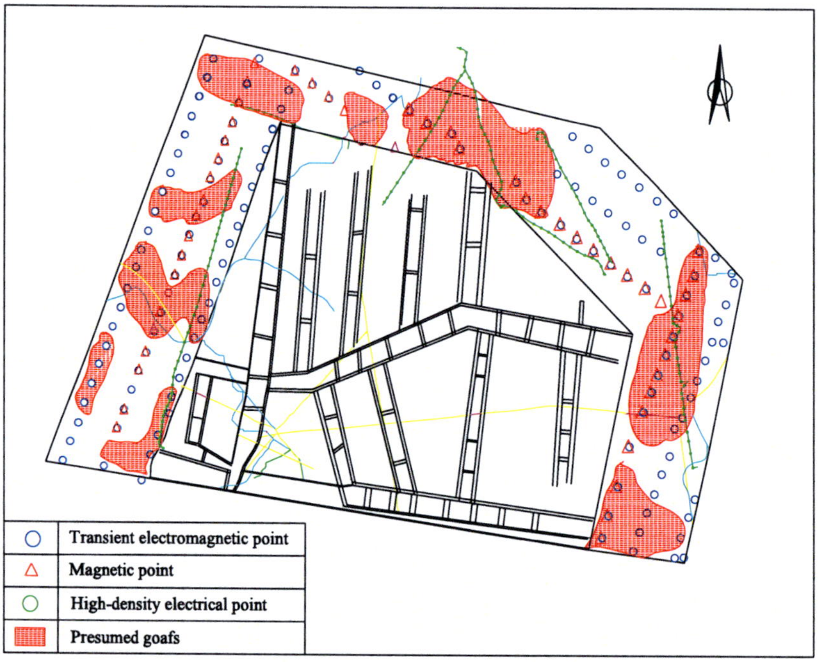

Fig. 6.10 Distribution of goafs in the geophysical survey area

from accidents and comprehensively surveyed the mined-out areas. The practices and experiences set examples in preventing the hazards associated with the mined-out areas.

6.5.1 Unified Organization and Implementation Led by Government

The Ordos municipal government is fully aware of the importance to prevention of coal mine accidents. The working concept consists of "advanced technology, high investment, and strict management". Professional trainings are implemented to constantly educate the technical foundation of coal mine safety production and improve the safety guarantee capacity of coal mines. The local government led the effort and financial funding to perform the unified survey of coal mine goafs in the city, with participation of coal mine enterprises. Since 2009, the city has invested a total of 88.6 million yuan in set-aside funds to determine the distribution of coal mine goafs, mined-out areas with groundwater accumulation, and formations with high yield of groundwater. The data provides technical guidance for coal mining safety.

6.5.2 Reliance on Technical Institutions to Improve Reconnaissance Effectiveness

In 2009, the municipal government commissioned a coal mine technical service institution to draft the report "Feasibility Study on the Distribution and Hidden Dangers of the Old Mined-out Areas in Ordos City". The report included data from 294 coal mines in the city (199 underground mines and 95 open pits). Based on the surveys and detailed technical discussions with mine technicians, five geophysical methods were used to delineate the extent and nature of the mined-out areas in 158 coal mines. The geophysical methods are high-density electrical resistivity imaging, transient electromagnetic method, shallow seismic, EH4 magnetotelluric method, and magnetic method. The survey area encompasses 122.61 km^2. The mined-out area was found to be 23.81 km^2. The reconnaissance data provides an important technical document for coal mine safety production.

In 2011, the municipal government again contracted technical service agencies to carry out surveys of the mined-out areas and water pooling areas during the 12th five-year plan period. Data at 211 coal mines in Ordos City was collected and catalogued. Detailed surveys were conducted in 113 coal mines, encompassing an area of 90.80 km^2. Six geophysical methods were used. The investigation identified the goaf area of 26.97 km^2, and water-accumulation goaf area of 3.96 km^2. The areas where groundwater is rick in the coal seam roof and floor is 7.39 km^2. The survey also delineated an area of 3.28 km^2 where the coal is burned. The results of the detailed survey provided the essential technical support for the development of the city's coal mine prevention and control plan.

6.5.3 Active Cooperation of Coal Mine Enterprises

The technical approach that is used by Ordos City for prevention and control of the goaf-related hazards can be summarized in three steps:

- Step 1: Conduct geophysical survey to determine anomalous areas
- Step 2: Conduct exploratory drilling to verify the anomalies
- Step 3: Perform comprehensive mitigation

The coal mining enterprises responded positively to the government's requests and provided technical data and information. Their cooperation is a key in completing the goaf reconnaissance. The city government shared the overall reconnaissance results of the old mined-out areas with the mine enterprises, who take the responsibility for conducting detailed investigations to determine their nature and extent. Drilling is used to verify any suspected areas. Comprehensive measures are taken to mitigate the mined-out areas that are susceptible to hazards. To prevent goaf-related disasters, the coal mining enterprises strictly follow the stipulation in which exploration must be conducted prior to excavation. Both geophysical and drilling methods are used to determine the nature and extent of the old mined-out areas. Mitigation is performed if the mined-out areas are susceptible to hazards.

6.5.4 Concerted Efforts from All Parties

The successful completion of the goaf reconnaissance results from close cooperation and concerted efforts of the municipal government, regulatory authorities, technical institutions, and coal mining enterprises.

- Firstly, the survey results make it possible to effectively prevent disaster accidents induced by goafs. The results have helped many coal mines to successfully discover unknown old mined-out areas in the active mining area. The mining efficiency is improved, and potential hazards such as water and gas inrushes from old mined-out areas and surface collapses are minimized or even eliminated.
- Secondly, coal mine safety supervision is more specific and targeted. The results of the census of the mined-out areas provide the most direct reference and basis for the regulation enforcement. When the coal mine is advanced or mined to the vicinity of the anomalous area discovered by the reconnaissance, the supervisory and inspection personnel takes the initiative to strengthen management, issue a warning of hidden dangers in advance, and urge the coal mine to take any necessary preventive measures.
- Thirdly, the company's self-safety awareness is significantly enhanced. Directed by the government's guidance and incentive, the coal mine enterprises continue to improve their safety investment. The enthusiasm for independent planning and research on safety-related projects increases significantly. Safety management becomes a conscious and proactive requirement for the enterprises.

Chapter 7
Technologies in Sealing Massive Karst Conduits in Restoration of a Flooded Open Pit Quarry in West Virginia, United States

7.1 Mine Background

This case study describes the design and construction of a multi-material grout curtain to mitigate a major water inrush through karstic conduits from a river to a nearby limestone quarry. A total of over 12,800 m^3 of slurry, low mobility grouts, and hot bitumen were injected to depths of over 60 m along a 460-m long curtain in several phases. The value of real time monitoring of all relevant hydrogeological as well as grouting parameters during the execution of such works is clearly demonstrated. This is believed to be the largest project of its type undertaken in North America, and details of this project can be found in publications by Bruce and Lolcama (Bruce et al. 2001; Lolcama 2005).

A large operational dolomitic limestone quarry is situated in West Virginia approximately 460 m from a regional river (Fig. 7.1). The geology of the quarry area consists primarily of the Ordovician limestone and dolomite. It is located on the eastern edge of the Valley and Ridge Province of the Appalachian and Mid-Atlantic regions. The geological features of the region strike primarily northeast southwest and are highly faulted with anticlines and synclines and fracturing of the bedding planes. Similar Palaeozoic geology stretches from New Jersey to Georgia and is observed in many areas to contain karstic features.

The Tomstown Formation was identified as the primary soluble carbonate lithology at the site. It consists of lenses and beds of dolomite and limestone and is exposed along the banks of the Shenandoah River adjacent to the quarry. Clay-filled karstic features are prominent in the walls of the quarry, while the adjacent property contains many large, deep sinkholes.

In April 1997, a major sudden inflow developed into the southwest corner of the quarry pit following production blasting activities and several abnormally severe precipitation events that caused flooding of the river and nearby sinkhole formation. The initial magnitude of the flow estimated at 132,500 L/min was far greater than the

S. Dong et al., *Methods and Techniques for Preventing and Mitigating Water Hazards in Mines*, Professional Practice in Earth Sciences,
https://doi.org/10.1007/978-3-030-67059-7_7

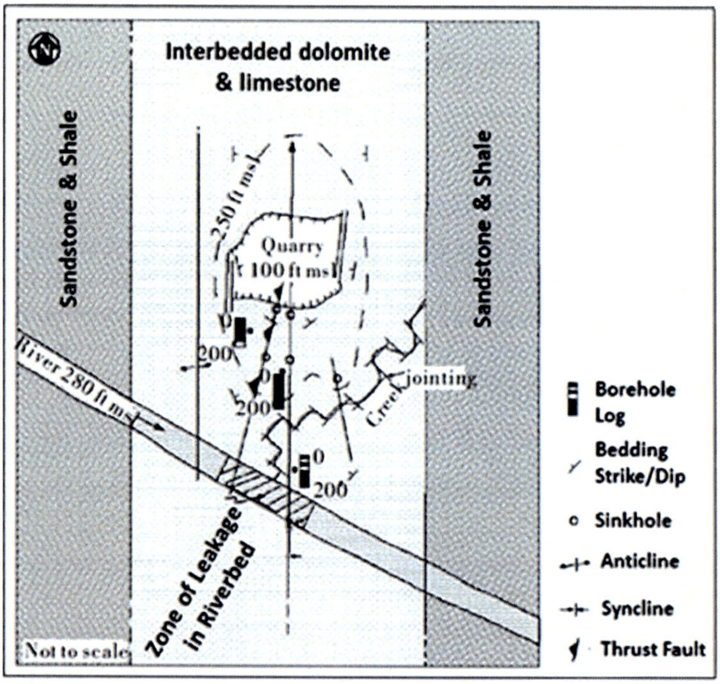

Fig. 7.1 Schematic map of mine site

capacity of the existing pit pumping facilities. Prior to the incident, the total flows from all sources into the pit were substantially less than 37,850 L/min.

The quarry inflow contained large amounts of red and brown sediment. This sediment was particularly evident after the initial flooding of the quarry, when the inflow was visible in the quarry as a plume up to 12 m in diameter. Such sediment washout indicated that the inflow was piping through erodible materials contained in karstic features. This observation again highlighted the Tomstown Formation as containing the most probable inflow conduits.

The new inflow posed a severe threat to both the current and future viability of the quarry, especially since the owner had to fulfill major aggregate supply contracts. A team of hydrogeologists, geophysicists, geologists, and geotechnical engineers was soon established to determine the source and path of the inflow, to define options for remediation, and to design and supervise the construction of the preferred option.

A vortex (or eddy) observed in the river (Fig. 7.2) appeared to be the point source of the flow. Initially, several attempts were made to construct a cofferdam with sandbags on and around the location of the vortex. Such actions resulted in immediate, yet very temporary, reductions of the new inflow to as little as 22,700 L/min. In each case, however, the flow conditions were re-established within 8–12 h as a new vortex re-formed, up strike in the riverbed. Of importance in evaluating such attempts was the discovery that the new vortex locations were all situated along an approximately

Fig. 7.2 Vortex (or eddy) observed in river during water inrush

straight line, extending out into the river and corresponding to the regional northeast to southwest strike of the formation.

In May 1997, pumping was discontinued, and the quarry water level was allowed to rise. This was done, in part, because the existing pumps were insufficient to maintain dewatering efforts. The partial flooding of the quarry would also assist later remedial operations by reducing the hydraulic gradient between the river and the inflow point. By the time remediation began in October 1997, the quarry water elevation had risen over 40 m, to approximate elevation 79 m above mean sea level (amsl), as compared to a river elevation of between 88 and 91 m. The total inflow was estimated at 83,300 L/min at this reduced hydraulic gradient. Dewatering was resumed in early December 1997, part way through remediation, to permit access to certain rock extraction areas in the pit.

7.2 Water Source and Pathway Investigations

The quarry mines the Cambrian-age Tomstown dolomite above the contact with the Antietam sandstone and quartzite. The Tomstown Formation has been intensively deformed into a plunging syncline fold structure. The massive flow conduit had formed within the lower Tomstown Formation, by karstification of a zone of intercalated limestone within the dolomite. The conduit pathway exactly follows the geologic strike of the limestone bedding, extending beneath the 122 m wide riverbed. The floodwater was entering the conduit through sinkholes in the bed of the river over an area of approximately 47 m^2. The conduit path length between river and the original flooding inflow is approximately 1,220 m. The original flooding inflow occurred at 53 m below river level. A chemical composition fingerprinting study was conducted. Comparison of the geochemical properties of the quarry inflow, the Shenandoah River and regional groundwater revealed a fingerprint of ions present both in the Shenandoah River and the quarry inflow. The presence of such ions, unique to the Shenandoah River and the inflow, therefore narrowed the possible primary inflow source to the Shenandoah River. Following successful construction of a grout curtain to cutoff the flowpath, three monitoring wells in the flow conduit pathway maintained a constant watch for signs of leakage of river water, signaling deterioration of the grout seal. During 1999 through 2001, these wells showed only a very small quantity of seepage occurring through the grout curtain.

Meanwhile, mining was continuing over 610 m away, strictly avoiding the original inflow location. The location of this mining was coincident with the conceptual pathway of the flow conduit that followed the beds of intercalated limestone through the limb of the local synclinal structure. During May 2001, a single artesian inflow of 1900 L/min with a surface water quality signature occurred, resulting in a suspension of mining in the area. The number and rate of artesian inflows increased rapidly to over 18,930 L/min over the following months. A site investigation program was initiated consisting of:

- Chemical fingerprinting to identify the source of the surface water leakage;
- Mapping of geologic structure adjacent to the artesian area; and
- Self-potential (SP) geophysical surveying.

The Shenandoah River was determined to be the source water. The downstream monitoring wells did not indicate significant leakage was occurring through the grout curtain. A deeply penetrating fault system was identified between the artesian area and the river, and the fault zone became the focus of exploration for a massive flow conduit. The fault was more than 915 m to the north of the previous flooding flow conduit. The SP geophysical survey, drilling, and water-quality profiling located a flow conduit with seepage of river water towards the quarry, but not the flooding flow conduit.

The downstream monitoring wells began to respond to substantial river flow through the grout curtain area after six months of artesian inflow of 18,930 L/min. Water levels were observed to rise rapidly and water temperature within the conduit zone changed rapidly. A river vortex appeared adjacent to the grout curtain. Self-potential surveying of the grout curtain identified a 46-m wide anomaly, 100 millivolts in magnitude. The anomaly was coincident with a massive void formation. Leakage in excess of 38,000 L/min was occurring through the massive void at depths of between 31 and 46 m below ground. Figure 7.3 shows the shape and magnitude of the SP anomaly. A new grouping of sinkholes was discovered on the opposite side of

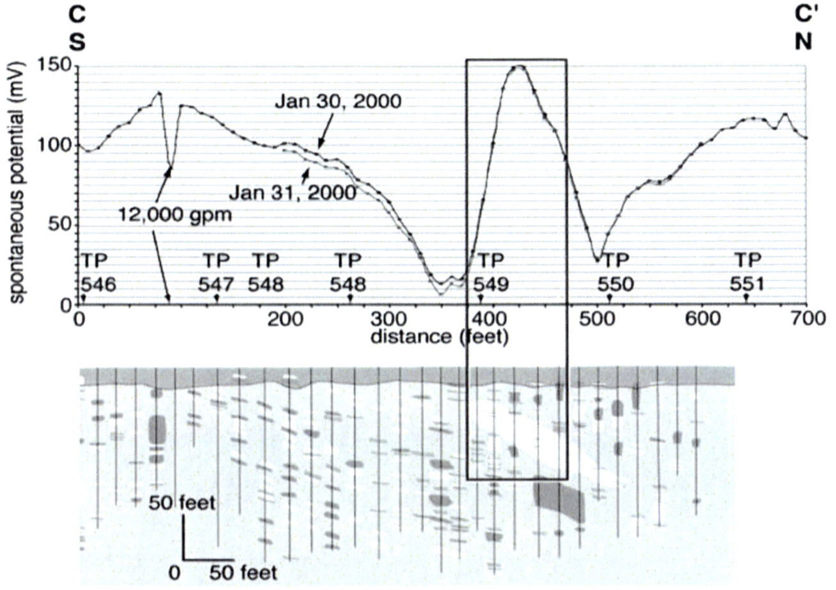

Fig. 7.3 Large magnitude SP response to conduit flow

the river from the grout curtain. The flooding inflow was being managed by pumping from a partially flooded pit.

Considering the evidence, it was interpreted that a new water inrush conduit had likely developed adjacent to the grouted feature. A conceptual flow pathway is shown in Fig. 7.4. The flow conduit system has a path length of well in excess of 1830 m and lies beneath the floor of the entire quarry pit, within the Fort Duncan dolomite member of the Tomstown formation. The flow conduits appear to be hydraulically disconnected from each other. Width of the flow conduit could be as wide as 244 m extending under the quarry floor.

7.3 Concept of Remediation Design

7.3.1 Selection of Cut off Methodology

The goal of the program was to reduce the total inflow into the quarry to an economically pumpable 30,300 L/min with the quarry completely dewatered. This would require reducing the flow from the river to below 11,400 L/min.

Three remediation alternatives were evaluated:

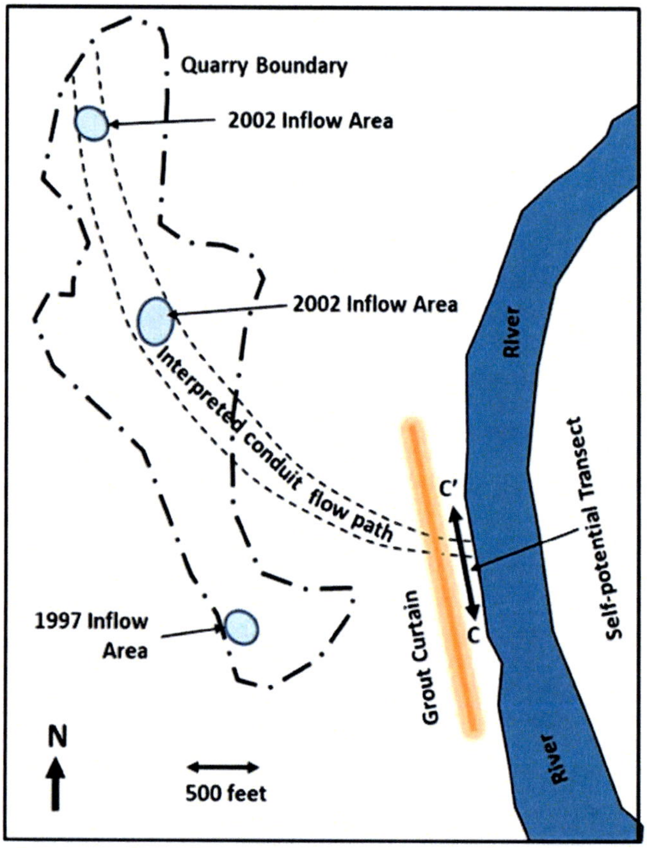

Fig. 7.4 Conceptual flow pathway of the karst conduit

- Alternative 1: Identify the specific solution cavities in the river and seal them there.
- Alternative 2: Construct an intercepting cut off at some appropriate location between river and quarry.
- Alternative 3: Treat the problem close to the quarry.

Alternative 1 was dismissed due to the demonstrated interconnection of the various entry points in the river, and the lengthy process needed to secure construction permits from the Federal and State environmental agencies. This option was also felt not to offer opportunities for a methodical, engineered approach.

Alternative 3 was also dismissed since the flow could bypass a near-quarry curtain and simply enter the quarry by exploiting different routes akin to the deltaic mouth of a major river. It would also be practically impossible to verify the extent and any deficiencies in a cut off. In addition, a cut off close to the quarry could hinder or be damaged by future mining activities.

Alternative 2 was preferred, on logistical, technical, and environmental grounds. It was decided to locate the cut off on a convenient road-side location approximately 15 m from the riverbank (Fig. 7.4).

The nature of the ground (containing incipient, potentially erodible karstic material) and the anticipated length and depth of a cut off, suggested initially either a diaphragm wall or a wall comprising large diameter overlapping piles. However, studies revealed that both were practically impossible, financially prohibitive, and overly time consuming. The decision was therefore taken to construct a grout curtain as the cut off, designed to both treat karstic features and permeate fissured rock in a preemptive fashion, and to create a permanent "plug" to the water inflow in the identified flow path as the active solution.

The installation began of vertical "exploratory holes" drilled in two parallel rows 3 m apart along the proposed grout curtain location. The holes in each row were 6 m apart and staggered to provide therefore one hole every 3 m. These holes were drilled initially for investigation and later used for grouting. The geological and hydrogeological data from these holes were interpreted to confirm the lateral and vertical boundaries of the treatment zones and the locations of the karst features. It was established that the curtain would conceptually extend laterally for at least 335 m and vertically as much as 67 m to ensure continuity with relatively competent, impermeable bedrock below and beyond the karsts feature delineated by the SP anomalies.

7.3.2 Selection of Grouting Concepts

The main challenges facing the construction of a successful grout curtain included the following:

- The groundwater flow rate was high and potentially through multiple conduits.
- Many of the karstic features were mud-filled, creating the possibility for erosion, piping, and so "blow out" after curtain placement when the hydraulic gradient acting across it began increasing; and the possibility of grout migration "upstream", into the river.

Several grouting related technologies were therefore studied to provide the curtain, in part or in whole jet grouting; polyurethane injection; low mobility grouting ("compaction" grouting); hot bitumen injection; accelerated cement based slurries; use of the multiple packer sleeve pipe (MPSP) system; and geotextile grout-filled bags. Each technique was assessed based on technical feasibility, likelihood of successful treatment of the inflow in both short and long terms, and cost, given the very severe geological and hydrogeological regimes to be accommodated. In summary, it was determined to first treat the peripheral zones with low mobility grout (LMG) and slurry grouts via the MPSP system (Bruce and Gallavresi 1988) using slurry grouts; and then treat the zones where water flowed, with hot bitumen from the downstream row of holes, backed up by slurry grouts simultaneously injected from the upstream row via the MPSP locations.

The use of grout-filled geotextile fabric bags (Bruce and Bell 1983) was considered to be potentially valuable for dealing with specific local features. The use of LMG for karstic remediation is often used (Byle 1997; Cadden et al. 2000) as are the details of slurry grouting. However, the following discussion relates to the less well-known technologies of hot bitumen and MPSP.

Hot bitumen grouting involves the injection of roofing grade bitumen when heated to a liquid state (over 200 °C). Hot bitumen is ideal for high velocity water flow conditions. As the fluid bitumen comes in contact with the colder, flowing water, it is rapidly cooled and so forms a very viscous mass with a semi-rigid skin. If the volume of water flow is large, the bitumen cools relatively quickly and forms a plug directly where the inflow is greatest. Although it is ideal for stopping flowing water, in-place hot bitumen grout remains slightly visco-plastic in a hardened state and is therefore susceptible to creep, and potentially to extrusion. Moreover, bitumen tends to shrink by as much as 10% as it cools and hardens, and so may permit residual flow paths around the edges of the plugs. Bitumen must therefore be injected simultaneously with cement-based slurries, which travel to these flow path areas to jointly create a permanent plug. Due to the relatively low cost associated with bitumen materials (approximately 35 times lower than polyurethane grout), and its suitability for treating high magnitude flows, bitumen was chosen as a primary technology for the construction of the grout curtain in high flow conditions. Such concepts are not new but have reached new prominence in fast flow applications in recent years (Bruce et al. 2001).

The MPSP system is applicable to highly variable collapsible and fractured formations such as encountered here. Geotextile bags are attached around plastic or steel sleeve pipes at specific locations. The bags are then inflated with cementitious grout after the pipes are placed in drill holes, and so act as packers that divide the annulus surrounding the sleeve pipe into specific sections or zones. In this manner, highly fractured or voided zones can be isolated from more competent zones. The formation grouting is conducted by means of a packer placed at successively higher levels within the sleeve pipe. The MPSP system was originally designed for permitting controlled treatment of badly fissured, unstable rock masses.

As a final general point, the grout injection strategy adopted involved conceptually channeling the inflow into the conduit in the center of the curtain. Therefore, the LMG and slurry grouting processes progressed simultaneously from the north and south ends of the curtain inwards, and were also conducted in the underlying peripheral karst zone all along the curtain, prior to then conducting the bitumen operation in the conduit flow zone, to hopefully stop the concentrated flow.

Table 7.1 Final low mobility grout (LMG) mix design

Ingredient	Ingredient quantity
Water	Approximately 190 L
Cement	160 kg
Fly ash	160 kg
Sand	950 kg
Anti-washout agent: UW450	3 L
Polypropylene fiber	1.4 kg
Maximum slump	<50 mm

7.3.3 Evolution of the Remediation Program

During the execution of the cut off, new geological data were continuously generated through the analysis of the drilling, grouting, and piezometer well data. In particular, the time-related performance of the curtain at its various stages of completion could be assessed and future directions of the program dictated. The successive phases of the work are summarized in Table 7.1. It is important to note that the need for, and nature of, each successive phase of treatment reflected a decision based on the consensus of all the stakeholders: owner, consultant, and contractor. For example, the bitumen injected in Phases II through V was used to seal especially large remnant flow paths discovered while conducting LMG or slurry operations. In contrast, the bitumen operations in Phases I and VI were conceived from the beginning as an integral part of the treatment program.

7.4 Execution of Mitigation

7.4.1 Drilling

Throughout the evolution of the project, numerous overburden/karst drilling systems were used together with various drilling rigs. Drilling to depths more than 15 m in karstic limestone invariably poses drilling problems, to which the keys are flexibility of response, appropriate methods and equipment, and experienced drillers. Systems ranged from pre-drilling with a down-the-hole hammer and driving 152-mm diameter temporary steel casings, through rotary drilling with end of casing flush, to rotary-percussive duplex drilling (Centrex system). The Centrex system, when powered by a rotary percussive diesel hydraulic track rig proved by far the most effective to depths greater than 43 m.

7.4.2 Grouting

7.4.2.1 Materials Mix Design and Delivery Systems

The LMG developed by Phase IV proved relatively inexpensive to produce and had optimal washout resistance and good pressure filtration characteristics (Table 7.1). A similar intensity of bench and field testing through a succession of mixes, finally produced the suite of slurry grouts shown in Table 7.2.

The quarry produced the LMG at its batch plant and it was trucked in 7 m^3 loads to the curtain. The grout was then injected with modified concrete pumps through 76-mm injection lines to steel casings as they were withdrawn from maximum depth.

Slurry grouts were also prepared in the batch plant and were delivered by truck to agitation tanks which fed centrifugal and Moyno pumps. The MPSP system comprised 76-mm-diameter PVC pipes with grouted geotextile bags at 7–9-m vertical intervals. As for the LMG, all major slurry grout injection parameters were recorded and displayed in real time.

The bitumen monitoring and injection plant was specially developed by the contractor for this project. Hot bitumen was supplied in 14.5 m^3 tanker trucks and pumped to a heated reservoir tank. The material was then drawn through the containerized control/pumping plant where pump rates were controlled, and all relevant injection data were displayed and recorded. The bitumen was pumped through insulated and heat traced field pipeline to the curtain injection points—specially developed steel "stingers" placed in the ground in a fashion not dissimilar to the slurry grout MPSPs. Prior to injecting the bitumen in any one hole, the field piping and stingers were flushed with hot oil (vegetable) to raise the temperature to a level that would prevent the bitumen prematurely plugging the lines prior to reaching the discharge ports in the ground.

Throughout the operation, modifications to this basic approach were continuously made to enhance control, responsiveness, and effectiveness leading to, for example,

Table 7.2 Mixture designations

Ingredient	A0	A1	A2	A3	A4
Water (L)	310	401	390	397	386
Bentonite* (L)	257	76	98	98	95
Cement (kg)	331	389	399	406	393
Fly ash (/kg)	290	340	349	355	393
Dispersant 2000B (ml)	1833	2159	2188	2247	2159
Marsh time (sec)	55	80–∞	120–∞	∞	∞
Bleed %	0	5–6	2–5	2	2
Specific gravity	1.53	1.55	1.6	1.63	1.64

*8% Bentonite Slurry (prehydrated 24 h)

the development of special "stingers" which could allow simultaneous injection of both bitumen and slurry into the same hole concurrently, and enhanced pumping rates.

7.4.2.2 Procedures

Best industry practice procedures were used for injecting and controlling the LMG and the slurry grouts. For the former, upstage end of casing injection was conducted in appropriate stages to preset stage refusal pressure criteria, although injection in nearer surface horizons in Phase IV, did lead to occasional surface heave. The slurry grouting was conducted via double packers placed into the MPSPs, with each stage being brought to proper refusal by varying the mix rheology, rate of injection and pressure. Routine quality assurance and quality control tests were conducted on all mixes, and all injection parameters were continuously recorded and analyzed.

Bitumen operations were conducted in carefully coordinated lateral and vertical patterns, with as many as four holes being "active" at any one time. Total injection rates for the bitumen often averaged 25 m^3/h depending on the ground response.

Criteria for terminating bitumen injection on any one hole included:

- The refusal of the adjacent upstream slurry grout holes.
- Attaining the theoretical volumes required to achieve the desired bitumen travel.
- The travel observed by monitoring temperatures in adjacent MPSPs with a well monitoring probe.
- The travel observed from leaks to or connection with surface.
- The data from the downstream monitoring wells. The pH and groundwater elevation readings in these wells were especially important sources of data.

7.5 Drilling and Grouting Quantities

The quantities of drilling and grouting conducted in each phase are summarized in Table 7.3.

7.6 Impact of Grouting Program on Quarry Inflow Characteristics

Throughout each phase of treatment, the following parameters were recorded and evaluated as a basis for real time direction of the grouting program, and as a guide to the design of the next phase:

- Ground water elevation, temperature, and pH of monitoring wells.

Table 7.3 Quantities of drilling and grouting conducted in each phase

Phase	Dates	Slurry grout		Low mobility grout		Bitumen	
		Number of grout holes	Volume (m^3)	Number of grout holes	Volume (m^3)	Number of grout holes	Volume (m^3)
I (Peripheral zone)	10/31–12/21/97	30	207	6	61	–	–
I (Conduit zone)	12/14–12/21/97	22	1652	–	–	23	1270
II	1/9/98–2/24/98	51	1968	–	–	1	138
III	2/25–3/21/98	12	667	–	–	2	184
IV	4/14–5/23/98	1	24	120	1436	2	115
V	6/11–7/18/98	9	146	21	151	8	254
VI (Cement)	8/20–11/1/98	26	484	–	–	–	–
VI-A	11/1–11/23/98	11	417	–	–	15	1580
VI-B	12/16–12/19/98	11	900	–	–	7	1183
Totals	10/31/97–12/19/98	173	6465	147	1649	58	4724

- Quarry water elevation and drawdown rates (and so a calculated total quarry inflow rate, due to flow from the river).
- Visual observation of the river "eddy" and the pit "boil".

A direct and successful intervention by grouting was detected by significant and quick changes in the piezometer data. Space prevents these data being described in detail: it may be reiterated however, that especially during the grouting operations, the assessment of these data in real time played a vital role in the tactical and strategic direction of the work in the field, when integrated with the relevant drilling and grouting data. Table 7.4 provides data on the measure of success—the river inflow into the quarry itself.

Three months later the upstream piezometers had remained constant, the differential hydraulic head was 41 m and the total flow from all sources was approximately 11,400 L/min, confirming the existence of an efficient, durable curtain, even in such challenging conditions. The quarry was restored to full production. Since then, the differential head has been increased to approximately 43 m, and the flow remains minimal from the river. It can be estimated from the flow/head data in Table 7.4 that although flow rates of over 75,700 L/min reestablished themselves soon after each of the six phases of grouting, this flow rate was measured at ever increasing hydraulic heads. Thus, although residual flow rates remained high until the last phase of treatment, the actual permeability of the overall karstic system was being, despite internal erosion, systematically reduced, thus allowing the quarry to lower the water level in the pit and continue with production activities.

Table 7.4 Summary of inflow calculated as coming from the river

Phase	Flow from river (gpm)		Water level difference (feet)		Notes
	Start	End	Start	End	
I	Approximately 40,000 reducing to 22,000	1000	70	30	Flow increased after treatment due to massive surface and subsurface clay erosion
		Rising back to 15,000 within 2 weeks of treatment ending		35	
II	Approx. 22,000	14,000		45	Erosion continuing
III	Approximately 20,000	15,000		60	Erosion continuing and flow redirected
IV	Approximately 20,000	2000–3000		90	Temporary success followed by "blowout"
		Rising to 17,000 in 48 h		95	
V	Approximately 25,000	9000–10,000		110	Erosion in curtain between Sta. 650 and 800
VI-A	Approximately 20,000	15,000		115	Major flow path still undetected
VI-B	Approximately 18,000	Effectively zero		122	Major void found 33 m down at Sta. 668 (took 400 m^3 grout)

1 gpm = 3.785 L/min; 1 feet = 0.3048 m

7.7 Summary

During the crisis of a water inrush in a mine, a thorough site investigation of the karst flow conduit must be performed before any remedial work is attempted. A plan for remediation which is based on an accurate understanding of the flow conditions, rather than a concept, obviously has a much greater chance of success. The considerable risk of attempting to seal a flow conduit is reduced with an accurate site model. This case history illustrates many important features, but three are particularly noteworthy. Firstly, from the technical viewpoint, it is an illustration of how contemporary grouting technology can be used, if correctly designed, implemented, analyzed, and controlled, to provide a successful result in even the most adverse conditions. Secondly, it conveys such a message that all sources of information must be studied before and during such an operation in order to gain the best possible "picture" of what is really happening in the ground and the incremental changes actually brought about the grouting itself and changes in the hydrogeological regime. Thirdly, and perhaps most importantly, this project illustrates the need for all stakeholders to partner fully and openly, and to provide mutual support at all times and in all aspects. In such circumstances, patience and trust are vital ingredients to successful teamwork in arduous and stressful conditions.

References

Bruce DA, Bell LA (1983) Cement bags bring solutions to pipe laying problems. Petroleum News, pp 40–41

Bruce DA, Gallavresi F (1988) The MPSP system: a new method of grouting difficult rock formations. ASCE Geotechnical Special Publication no. 14:97–114

Bruce DA, Traylor RP, Lolcama JL (2001) The sealing of a massive water flow conduit through karstic limestone. ASCE Geotechnical Special Publication no. 113:160–174

Byle MJ (1997) Limited mobility displacement grouting: when "compaction grout" is not compaction grout. In: Vipulanandan C (ed) Proceedings of sessions sponsored by the grouting committee of the geo-institute of the ASCE Geotechnical Special Publication no. 66, pp 32–42

Cadden AW, Bruce DA, Traylor R (2000) The value of limited mobility grouts in dam remediation, ASDSO, September 2000

Lolcama JL (2005) Case studies of massive flow conduits in karst limestone. ASCE Geotechnical Special Publication no. 144:57–65

Chapter 8
Environmental Impact Assessment in Hongliulin Coal Mine

8.1 Mine Background Setting

8.1.1 Geographical Location

Hongliulin mining area is bordered by the coal seam outcrop in the east, the Jinjie mine field and Liangshuijing mine field in the Yushen mining area in the south, the Erlintu exploration area in the west, and the Caragana tower and Zhangjiamao mine field in the north. The mining depth ranges from 1,160 to 990 m amsl. The planned production capacity is 12 Mt/a. The mining area is approximately 20 km long from east to west, and 8 km wide from north to south, encompassing an area of 142.55 km^2.

Geographically, Hongliulin Mine is in the north of Shenmu County, Yulin City, approximately 15 km away from Shenmu County. It is administratively under the jurisdiction of Majiata Township. The transportation is convenient, and coal export is fully guaranteed. The mine is 6 km from Shenmu Railway Station. The railways include Bao (tou)—Shen (Mu), Shen (Mu)—Suo (Suozhou), and Xi (An)—Bao (tou). The newly planned special railway line for coal transportation in the southern area of Shenfu Mining Area—Hongning Special Line is under construction. The highways include G210 National Highway Xi (An—Bao (Tou) highway approximately 60 km from the west side of the mining area, and the provincial highway S301 Fu (Gu)—Xin (Jie) is along the Kaokaowusu valley to the north east of the mine. It can reach Xi'an via Yulin and Yan'an in the south, Ordos City and Baotou in the north, and Shanxi Province via Fugu County in the east. The regional Yulin Airport has planes flying to Xi'an, Taiyuan, and Beijing.

S. Dong et al., *Methods and Techniques for Preventing and Mitigating Water Hazards in Mines*, Professional Practice in Earth Sciences,
https://doi.org/10.1007/978-3-030-67059-7_8

Table 8.1 Mineable and partially mineable coal resources/reserves (Mt)

Coal seam	Geological resources	Industrial resources/reserves	Designed recoverable reserves
2^{-2}	263.92	237.53	173.96
3^{-1}	253.85	238.77	183.40
4^{-2}	419.51	402.45	308.59
4^{-3}	160.61	128.00	104.74
4^{-4}	81.92	13.21	8.81
5^{-2}	836.15	811.53	581.37
Total	2015.96	1831.49	1360.87

8.1.2 Mining History

Hongliulin Mine started construction in June 2007. Except for the underground blasting material distribution chamber and the air return tunnel that are under construction, the mine construction is essentially completed. Hongliulin Mine consists of Lucaogou Coal Mine and Aobaogou Coal Mine.

The Lucaogou Coal Mine is in Lucaogou in the middle and western part of the mine field. It mined coal seams #4-2. This coal mine was destroyed and closed in 1986 due to noncompliance mining procedures.

Aobaogou Coal Mine is in Kentielinggou, southeast of the mine field. It is currently mining #5-2 coal seam with the room and pillar mining method.

8.1.3 Resources and Reserves

The mineable and partially mineable coal resources/reserves are shown in Table 8.1. Geological resources are estimated at 2015.96 Mt, and industrial resources/reserves are 1831.49 Mt. The designed recoverable reserves are 1360.87 Mt after deducting various coal pillars and mining losses. At the designed production capacity of the mine of 12.00 Mt/a, the mine service life is approximately 87 years.

8.2 Geoenvironmental Background

8.2.1 Physical Geography

Hongliulin Mine has a typical mid-temperate semi-arid continental monsoon climate. The climate features are: cold winters, windy springs, hot summers, cool autumns, hot and cold throughout the four seasons, large temperature differences between day and night, drought and less rain, large evaporation, and rainfall mostly concentrated in July, August and September. The frost-free period is short throughout the year, with freezing in early November and thawing in March the following year. The monsoon period is from early January to early May, mostly northwest wind.

The main meteorological parameters of the meteorological data of Shenmu County Meteorological Station are as follows: the extreme maximum temperature of 38.9 °C the extreme minimum temperature of −29.7 °C the annual average temperature of 8.8 °C the annual average rainfall of 436.6 mm; and the annual maximum rainfall of 553.1 mm. The maximum daily rainfall of 135.2 mm occurred on August 1, 1977. The annual average evaporation is 1774.1 mm. The annual average absolute humidity is 7.6 mbar. The average wind speed is 2.5 m/s. The maximum wind speed is 25 m/s that occurred on July 18, 1970. The maximum frozen soil depth is 146 cm that was measured in 1968.

The precipitation in Shenmu County decreases from south to north. The average annual rainfall in the area where the Hongliulin Mine is located ranges from 375 to 425 mm (Fig. 8.1).

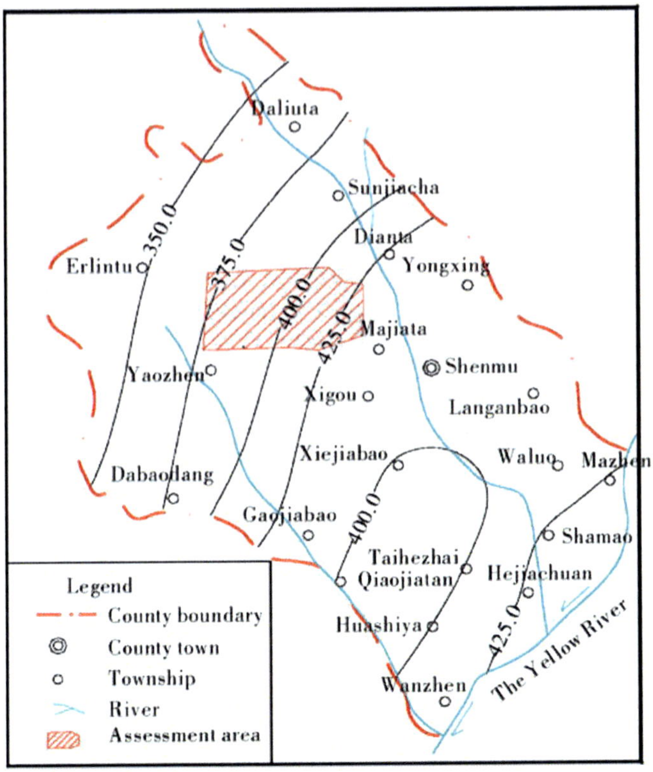

Fig. 8.1 Schematic diagram of annual rainfall distribution in Shenmu County

Precipitation in Shenmu County has obvious seasonality. Precipitation is mainly concentrated in July to September, accounting for 69% of the total. Geological disasters such as collapses and landslides also occurred from July to September, which coincided with the rainy season.

The Kuye River originates in Yijinhuoluoqi of Inner Mongolia flows from north to south at 5 km east of the mine into the Yellow River. According to the data from Shenmu County Hydrological Observatory, the river basin encompasses an area of 7298 km^2. The average flow is 17.40 m^3/s with a range from 0.02 to 13,800 m^3/s.

The east-west Majiata River near the southern boundary is a first-level tributary to Kuye River. It flows year-round and flows into Kuye River in north of Shenmu County. According to Changguan data, the flow of Majiata River is generally 0.50 m^3/s. The Lucaogou gully is the largest perennial tributary to Majiata River. The average flow is 0.14 m^3/s. The flows in other tributary gullies are seasonal.

8.2.2 *Topography*

Hongliulin Mine is located at the border of the Loess Plateau in northern Shaanxi and the Maowusu desert. The terrain is generally high in the west and low in the east. In the Tielinggou area, the elevation is approximately 1090.00 m amsl. The maximum elevation difference is 243.50 m. The average elevation is approximately 1200.00 m amsl. The landform types of Hongliulin mine field include aeolian sand landform, loess gully, alluvial landform, and sand-covered loess hills. Fengshatan in the west of the mine is relatively flat and open. The surface is covered by a loose layer. The loess gully and alluvial landforms are distributed in Lucaogou, Aobaogou and Maolutangou. The gullies are relatively developed on both sides of the rivers. The loess hills and gully areas are distributed in the southeast where the terrain is fragmented by alternating valleys and ridges.

8.2.3 *Stratum Lithology and Geological Structure*

8.2.3.1 Lithology

Figure 8.2 presents the lithology at Honglin Coal Mine. Most of the surface in the area is covered by the Quaternary aeolian sand and loess. The bedrock is mostly exposed in larger valleys. From old to new, the formations are:

- Mesozoic Triassic Upper Yongping Formation
- Lower Jurassic Fuxian Formation, Middle Yan'an Formation, Zhiluo Formation
- Cenozoic Neogene Pliocene Baode Formation, the Quaternary Middle Pleistocene Lishi Formation, the Upper Pleistocene Sarawusu Formation, Holocene aeolian

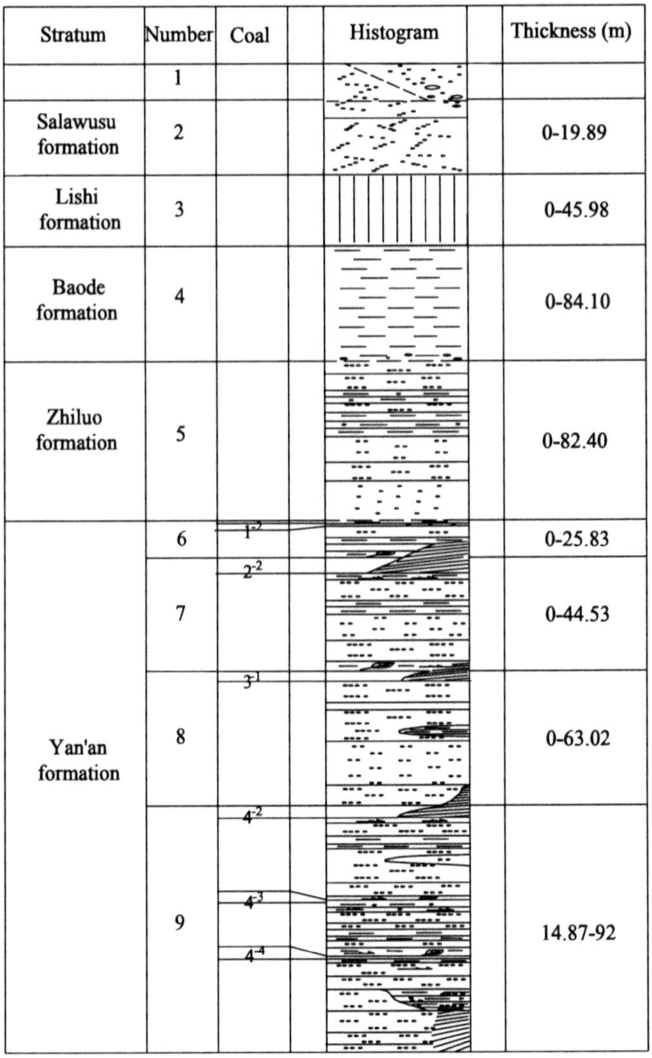

Fig. 8.2 Lithology at Hongliulin Coal Mine

sand and alluvium. The upper part of the Yan'an Formation is missing due to denudation in the east.

Upper Triassic Yongping Formation (T3y): This formation is the base bedrock of the coal-measure formations. The lithology consists of thick layered light gray-green, gray-green medium-fine-grained sandstone, which contains mica fragments, chlorite, quartz, and feldspar. The nature and degree of rounding are moderate with argillaceous or argillaceous cementation. There are cross-bedding and horizontal bedding with lenses of quartz gravel and pyrite nodules.

Lower Jurassic Fuxian Formation (J1f): This formation is in pseudo-conformity contact with the Yongping Formation. It is intercepted in individual boreholes. The thickness varies from 0 to 8.80 m. The lithology consists of gray-white medium-coarse quartz sandstone with poor sorting and roundness, and argillaceous cementation. Layering is not obvious.

Middle Jurassic Yan'an Formation (J2y): This formation is part of the Fuxian Formation and a coal-bearing stratum in the area. The thickness varies from 44.03 to 192.05 m with an average of 77.34 m. The stratum gradually thins from west to east due to denudation. Most of the formation is overlain by loose strata. Intermittent outcrops occur only in the valleys.

The stratigraphy of this group is composed of terrigenous clastic deposits. The lithology consists of light gray medium-fine feldspar sandstone, lithic feldspar sandstone, gray to black sandy mudstone, mudstone, and coal seams. There are small amounts of lenticular marl and pillow-shaped or spherical siderite.

Middle Jurassic Zhiluo Formation (J2z): This formation has a pseudo-conformity with the Yan'an Formation. Only part of the stratum remains in the western part of the minefield due to strong denudation in the later stage. The thickness varies from 0 to 25.83 m with an average of 10.13 m. The lithology consists of medium-coarse-grained feldspar sandstone with thick layered yellow-grey, yellow-green, and local purple variegated sandstone with calcium cementation. The sand grains are mediumly sorted, and sub-angular. Layering is not immediately obvious.

Neogene Baode Group (N2b): This group of formation is distributed in Liangmao loess hilly area. Its thickness ranges from 0 to 84.10 m with an average of 30.63 m. It is exposed on both sides of major valleys and the top of Liangmao plateau. The lithology consists of light brown-red clay and silty clay with multiple layers of calcareous nodules. The thickness of the nodules is generally 0.40 m. The thickness of the clay layer varies from 0.50 to 2.00 m. The clay is structurally interlayered and compact and has good plasticity. A layer of gravel with a thickness of 10–30 cm is present in part of the bottom of the section. The gravel composition is mostly hard and dense quartz sandstone or conglomerate with calcareous cementation.

Middle Pleistocene Lishi Formation (Q2L) of the Quaternary series (Q): The thickness is between 0 and 45.98 m with an average of 18.99 m. It is widely distributed in the area and exposed in the Liangmao valley area. The formation consists of brown, yellow to yellow-brown silty clay, partially mixed with gray-yellow sandy soil and paleosol layer. It has a uniform texture without apparent bedding. The upper part contains scattered calcareous nodules. The gravel diameter varies from 3 to 5 cm. The largest diameter is 10 cm. Scattered vertical joints are reported. It is in unconformity contact with the underlying strata.

Lower Pleistocene Sarawusu Formation (Q3s) of the Quaternary series (Q): The thickness is between 0 and 19.89 m. This formation is scattered in the Miaolvtan and Lucaogou valleys. The lithology consists of gray-yellow, gray-green, gray-brown and gray-black silt, fine and medium grained sands. There are bean-shaped calcium nodules at the bottom. It is in unconformity contact with the underlying strata.

Holocene system (Q4) of the Quaternary series (Q): This formation consists of aeolian layer and alluvial layer. The aeolian layer (Q4eol) is widely distributed in the west and east of the mine, covering other strata in the form of fixed dunes and semi-fixed dunes. It is in unconformity contact with the underlying strata. The lithology consists of light yellow, brown-yellow fine and silty sands. It is characterized with uniform texture, well sorted, and poor grounding degree. Its thickness varies from 0 to 40.19 m with an average of 9.58 m.

The alluvial layer (Q4al) is distributed in Lucao Gully, Maolvtan, Kentieling Gully. The lithology is mainly gray-yellow, gray-brown fine sand, and silty clay with a small amount of humus soil. Most of the bottom contains a gravel layer of approximately 7.00 m in thickness. The gravel diameter is between 3 and 4 cm. This layer is in unconformity contact with the underlying strata.

8.2.3.2 Coal-Bearing Strata

The coal-bearing rock series of the Yan'an Formation is composed of inland shallow lake-delta facies deposits with great changes in the lateral facies' environment. The vertical sequence structure is clear and characterized by minable coal seams. According to lithological characteristics, the coal-bearing strata are divided into five intermediate cycles from bottom upward. There are 5 coal groups, #1 through #5 from top to bottom. The main mineable coal seams are located on the top of the cycle rock section.

The first section of Yan'an Group (J2y1): This section is divided into two secondary sedimentary cycles from the bottom of Yan'an Formation to the top of #5-2 coal seam, containing coal seams #5-2, #5-3 and #5-4, respectively.

The lithology of the first cycle is gray-white medium-grained sandstone, followed by gray silt sandstone; locally it is gray siltstone with thin mudstone or thin coal seams, with parallel bedding, wavy bedding and horizontal bedding; top It is coal seam 5-3. The lithology gradually becomes thinner and the thickness becomes thinner to the east of this cycle.

The lithology in the middle and lower part of the second cycle is dominated by thick layered gray-white, gray medium-coarse sandstone, local fine-medium sandstone with parallel bedding; the upper part is gray siltstone, fine-grained sandstone, with horizontal bedding and Wavy bedding; the top is the #5-2 coal seam, and the local part of the #5-2 coal seam has #5-2 lower coal. This section has a thickness from 7.40 to 44.65 m with an average of 19.37 m.

The second stage of Yan'an Group (J2y2): The thickness varies from 14.87 to 92.00 m with an average of 65.73 m. This group is divided into three secondary sedimentary cycles, containing coal seams #4-4, #4-3, and #4-2, respectively.

The lithology in the lower part of each cycle is mainly gray-white and gray fine-grained sandstone. The upper part is composed of fine-grained sandstone and thick layered gray siltstone with thin layers of gray-black mudstone. Horizontal bedding and microwave-like bedding are reported. On the top of each sedimentary cycle are coal seams #4-4, #4-3, or #4-2.

The third section of Yan'an Group (J2y3): The thickness of this section varies from 0 to 63.02 m, and the average thickness is 36.14 m. The lithology of the lower part consists of thick layered gray-white medium-grained sandstone with argillaceous cementation without obvious bedding. The upper part is dominated by dark gray siltstone, fine-grained sandstone, and mudstone with thin coal seams interposed with wavy bedding. Due to later denudation, this section is preserved only in the west, with #3-1 coal seam at the top.

The fourth section of Yan'an Group (J1y4): The thickness varies from 0 to 44.54 m, and the average thickness is 30.03 m. The upper lithology is dominated by gray siltstone, fine-grained sandstone, and mudstone, whereas the lower lithology is dominated by gray-white fine-grained sandstone-medium-grained sandstone, with microwave-like bedding and horizontal bedding. Due to later denudation, the eastern part of the mine field is missing, and only the western part remains with #2-2 coal seam on the top.

The fifth section of Yan'an Group (J2y5): The lithology is mainly gray siltstone, mudstone, and thick layered gray-white fine-grained sandstone to medium-grained sandstone. Large-scale cross bedding and massive bedding are developed. Due to late denudation, the east is missing, and only the northwest corner remains. The thickness varies from 0 to 25.83 m, and the average thickness is 10.13 m. The top is #1-2 coal seam.

8.2.3.3 Geological Structure

The stratum of the mine is a gentle monoclinic structure dipping to the west. The dip angle is between 1° and 2° and accompanied by wide undulations. No faults and magmatic rocks are reported. The geologic structure is simple.

8.2.4 Aquifer and Aquiclude

8.2.4.1 Porous Medium Aquifers

- Quaternary Holocene alluvial (Q4al) porous medium phreatic aquifer: Mainly distributed in large valley terraces and floodplains. The lithology is composed of fine sand, medium-coarse sand, sub-sand, and gravel layers, with large pores, superior replenishment conditions, and large water yield. The average thickness is 7.00 m. The depth to groundwater is approximately 6.00 m. Springs are present, and their flow rate varies from 0.014 to 2.97 L/s.
- Quaternary loose sand porous medium phreatic aquifer (Q4eoL+ Q3S): This aquifer is composed of the fourth system aeolian sand and the fourth system upper Pleistocene Salawusu Formation. The aeolian sand is widely distributed on the surface. The lithology consists of silt-medium sand. The thickness varies from 0 to 40.19 m with an average of 9.58 m. It is permeable to water. The sand dunes

do not contain groundwater. The drainage area of the valley slope and low land is moderately rich in groundwater. The depth to groundwater varies from 2.16 to 4.06 m. Springs are present, and their flow rate ranges from 0.014 to 1.519 L/s. The largest wetland spring group occurs at the head of Wubangshigou Valley with a flow rate of 7.734 L/s.

The Salawusu Formation is unevenly distributed. They are distributed in small areas such as some valleys, low-lying lands, and sand dunes, and are absent in most of the sections. The water abundance is uneven. The water abundance is controlled by the topographical undulations. Where the recharge source is favorable groundwater is abundant. Springs occur and their flow rate ranges from 0.114 to 5.648 L/s. The water quality is HCO_3–Ca type. The total dissolved solids concentration varies from 0.264 to 0.307 g/L. The eastern loess hilly and gully area does not contain groundwater but is permeable to groundwater flow.

- Quaternary Middle Pleistocene Lishi Formation loess (Q2L) porous medium-fracture phreatic aquifer: Lishi loess is dominated by sandy soil with dispersed calcareous nodules. It is characterized with large porosity, loose structure, and well-developed joints. The thickness varies from 0 to 45.98 m, and the average thickness is 18.99 m. There is a loose gravel layer at the bottom of the valley terrace area. The spring flow rate ranges from 0.014 to 0.039 L/s. The water richness is extremely weak.

8.2.4.2 Neogene Pliocene Baode Formation (N2b) laterite aquiclude

The laterite aquiclude is distributed in the Liangmao hilly area of Huangtu with outcrops on both banks of the valley. Its thickness varies from 0 to 84.10 m with an average of 30.63 m. The upper part is brown-red clay and sandy loam with multiple layers of calcareous nodules. They are uniformly dense and have strong plasticity. It functions as an aquiclude in the area.

8.2.4.3 Middle Jurassic Zhiluo Formation (J2z) Bedrock Fracture Aquifer

The bedrock fracture aquifer is present in part of the stratum in the southwest part of the mine. The thickness varies from 0 to 82.14 m with an average of 42.52 m. The lithology is composed of gray-yellow-green and gray-white thick layered medium and coarse-grained sandstone, with thin layers of fine-grained sandstone intercalated locally.

The depth to the groundwater level varies from 16.02 to 22.90 m. The specific water yield ranges from 0.0255 to 0.0350 L/s·m. Based on pumping tests, the hydraulic conductivity is between 0.0756 and 0.111 m/d. The water abundance is weak. The water quality is HCO_3–Ca·Na type with total dissolved solids concentration ranging from 0.212 and 0.228 g/L.

8.2.4.4 Middle Jurassic Yan'an Formation (J2y) Aquifer

The Yan'an Formation is composed of medium-fine-grained sandstone, siltstone, mudstone, and coal seams. The aquifer consists of medium- and fine-grained sandstone, which is alternately stacked with the mudstone aquicludes. Fractures are not well developed. The water yield of each section is extremely poor. Based on the borehole pumping data, the main aquifers in this group include the following:

- Fracture aquifer overlying #4-2 coal seam: This aquifer is sporadically exposed in the valleys of Maolvtan, Kentielinggou, Lucaogou and Caojiepan. The lithology is composed of gray, gray-yellow, gray-green medium-fine-grained sandstone, mudstone, partially intercalated with siltstone and sandy mudstone. Severe weathering of the upper part results in fractures, which increases the aquifer's permeability and storage conditions. Its thickness varies from 11.62 to 158.88 m, and the average thickness is 55.01 m. The depth to groundwater ranges from 2.75 to 8.00 m in the valley area. The typical flow rate at the spring is between 0.027 and 2.970 L/s. The maximum flow rate of the spring group is 11.203 L/s. The lower part of this formation is not as permeable as the upper part. The average water inflow from boreholes is 0.155 L/s. The average specific water yield is 0.007841 L/s.m. The average hydraulic conductivity is 0.0152129 m/d. The overall water richness is weak. The water quality type is HCO_3–Ca with total dissolved solids concentration of 0.263 g/L.
- Confined fracture aquifer of #4-2–#5-2 coal seams: The formation of this aquifer outcrops only in Maolutan Gully and Kentieling Gully. The aquifer is deeply buried deep. The lithology is composed of light gray medium-fine grained sandstone and siltstone, intercalated with thin sandy mudstone and carbonaceous mudstone. The thickness varies from 20.88 to 96.20 m, and the average thickness is 65.38 m. Because joints and fractures are not developed the water yield is poor. According to the pumping test data at boreholes 6–4 and 6–8, the upper interval has better permeability than the lower part. The borehole water inflow rate varies from 0.091 to 0.114 L/s. The specific water yield ranges from 0.00129 to 0.00131 L/s.m. The hydraulic conductivity ranges from 0.00146 to 0.00131 m/d. The water quality type is HCO_3–Ca+Mg or Cl–K+Na. The total dissolved solids concentration ranges from 0.275 to 1.344 g/L.
- Phreatic fracture aquifer of burnt rock: The burnt rocks result from the spontaneous combustion of the thick coal seams at their exposed outcrops in the valley section. The spontaneous combustion process of the coal seam causes the upper rock to be baked and metamorphosed until it melted and produced numerous pores. The subsequent collapses result in fractures and cavities of up to 30 cm in dimension. The fracture frequency in individual locations is up to 15%. These fractures provide pathways for surface water and atmospheric precipitation to percolate into the subsurface. The thickness of burnt rock varies from place to place, depending on the thickness of the coal seam, the degree of spontaneous combustion, and the geomorphic location. It typically varies from 15 to 30 m with the maximum thickness of more than 40 m.

The water richness of burnt rocks varies greatly and is controlled by the replenishment conditions, the shape of the water-retaining floor and the topography and is closely related to the development of various burnt fractures.

The burnt metamorphic rocks of #5-2 coal seam are distributed in the eastern boundary of the minefield and are discharged by the gully cutting groundwater and become permeable rock formations.

The thickness of coal burnt metamorphic rock of #4-2 coal seam is thin. The rock is cut by valleys and does not have the condition of a good aquifer. Drilling of exploratory boreholes encountered loss of circulation fluid. A string of springs is present in the low-lying land of the valley drainage area. Their flow rate varies from 0.054 to 5.002 L/s. The water quality type is HCO_3–Ca or HCO_3–K+Na. The total dissolved solids concentration ranges from 0.247 to 0.357 g/L.

8.2.5 Groundwater Flow, Recharge, and Discharge

8.2.5.1 Quaternary Porous Medium Aquifers

The phreatic sand aquifers receive direct recharge from atmospheric precipitation, supplemented by condensation. The flow direction of the groundwater in this aquifer is controlled by the top surface of the laterite aquiclude. It flows from elevation highs to lows and discharges in valleys or low-lying places in the form of descending springs. Evaporation is another form of discharge. In the area where the laterite layer is missing, the porous medium aquifer recharges the underlying fracture aquifer composed of weathered bedrock.

8.2.5.2 Confined Porous Medium and Fracture Aquifers of Jurassic Clastic Rocks

This group of aquifers accepts regional lateral recharge and infiltration from the overlying aquifers. Where the bedrock outcrops the atmospheric precipitation directly recharges the rock layer along the fractures. The second recharge source is the infiltration of porous medium aquifer in the overlying sand layers through the permeable skylights where the laterite layer is absent. Generally, the groundwater moves along the bedrock surface from bedrock highs to lows. The groundwater emerges from the underground in the form of springs and discharges to surface water after being intercepted by the valleys.

8.2.5.3 Phreatic Fracture Aquifer of the Burnt Rock

The phreatic fracture aquifer of the burnt rock receives vertical recharge from the overlying porous medium phreatic aquifer and direct infiltration of atmospheric precipitation. Some sections receive the lateral recharge of the same rock layer and surface water. The groundwater flows toward the gully area along the aquicludes and discharges as springs.

8.2.6 Analysis of Groundwater Recharge Conditions in the Mine

8.2.6.1 Groundwater Recharge Source

Atmospheric precipitation: According to data from the Meteorological Bureau of Lushenmu County, the annual average precipitation in the mining area is 436.6 mm. The precipitation is concentrated in July through September, accounting for approximately 70% of the annual precipitation. Surveys of active mines in this area indicate that mine water inflow varies with season. The mine water inflow is directly affected by the atmospheric precipitation.

Surface water: The surface water in this area is distributed in Lucaogou, Maolutangou and Kentielinggou in the southern basin. The bedrock overlying the mineable coal seams is thin in the river valley area. The mining induced water-conductive fracture zone is expected to pass through the upper aquiclude (N2b) into the porous medium aquifer. The induced fractures allow the surface water to communicate with the underground mining areas. Under this scenario, the surface water is a direct recharge source to the mine. The surface water not only leads to a sudden increase in water inflow but also poses threat to water and sand inrush in the mine.

Groundwater: The groundwater that affects the future mine is the fracture aquifers of weathered bedrock and the Yan'an formation sandstone.

Old mine pools: The dewatering rate at Aobaogou Coal Mine varies from 74.40 to 102.30 m^3/d. Precautions should be taken to prevent mine water inrush from the old mine pools in the future coal mining.

8.2.6.2 Water Recharge Channel

The water recharge channels are mining induced caving zones and fracture zones, followed by primary structural fractures.

8.2.7 Geotechnical Conditions

According to the genesis of rock and soil, engineering geological characteristics and physical and mechanical properties, the rock (soil) in this area is divided into three types, seven rock groups and four structural types, as shown in Table 8.2.

Table 8.2 Engineering geological classification of rock and soil

Engineering geological classification	Rock formation	Saturated compressive strength (MPa)	Spatial distribution	Rock structure
Extremely weak (soil)	Loose sand group		Distributed in the western part of the mine, part of Liangmao and valley terraces	Loose structure
	Soil group		Widely spread throughout the region, including loess and laterite	
Weak	Burnt rock group		Distributed in strips along both sides of the major valleys	Fragmented structure
	Weathered rock group		Mainly rock formations on top of bedrock	
	Coal rock formation	10.1–19.7	All mineable and non-mineable coal seams	Layered structure
Semi-hard	Siltstone, mudstone, and interbedded rock group	12.6–30.0	Direct roof and direct floor of coal seam	
	Sandstone group	13.6–34.2	Basic top of coal seam and sandstone layer in the middle of each section	Block structure

8.2.7.1 Weak Rocks

This group of rocks includes burnt rock group, weathered rock group, and coal rock group.

- Burnt rock group: The original rock mass structure has undergone major changes. The fractures of the burnt rock mass are developed and open in a wedge shape. Massive and other burnt rock blocks together form a fragmented rock mass. There is no cohesive force between the rock blocks, and the rock mass is balanced mainly by the friction of the fracture surface and the mutual support between the rock blocks, and the stability is very poor. Therefore, the strength of a rock block can only represent its own mechanical strength and has no practical significance with the stability of the entire rock mass. The rock mass RQD $<10\%$, the rock quality is extremely poor, and the rock mass is broken.
- Weathered rock group: The weathering degree of the rock layer gradually weakens from top to bottom. The rock layer loses its original luster, and its strength

is reduced compared with the original rock. The rock structure of the strong weathering zone is destroyed, secondary fractures are developed, the bulk density decreases, and the porosity increases. The water content increases, and the strength decreases. Most rocks disintegrate or segregate along cracks in a short time when exposed to water. The RQD value is <30%. The rock quality is poor, and the rock mass integrity is poor.

- Coal rock formation: The saturated compressive strength of each main mineable coal seam is between 10.1 and 19.7 MPa with an average of 14.15 MPa. It is brittle but not tough and suitable for impact crushing.

8.2.7.2 Semi-hard Rocks

Siltstone, mudstone and interbedded rock group: This rock group is directly related to coal seam mining. It is the main rock group of coal-measure strata. It is composed of siltstone, mudstone, argillaceous siltstone, sandy mudstone, carbonaceous mudstone, and thin coal. It mostly occurs in the direct roof and floor of coal seams. The rock contains high clay minerals and organic matter. There are many horizontal beddings and small cross beddings, which are characterized by structural surfaces such as joints, fractures, and slippery surfaces. The saturated compressive strength varies from 12.6 to 30.0 MPa with an average of 20.13 MPa. The average RQD value is 58%, the rock quality is medium, and the rock mass is medium and complete.

Sandstone group: This rock group is dominated by medium- and fine-grained sandstone, followed by coarse-grained sandstone, which mostly forms the old roof or bottom of the coal seam. The primary structure surface has block bedding, trough bedding, and large plate-shaped cross bedding. The thickness of the single layer is large, and the structural surface is not developed. The compressive strength varies from 13.6 to 34.2 MPa with an average of 25.4 MPa. The average RQD value is 78%. The rock has decent stability in the region.

8.2.7.3 Rock (Soil) Structure Type and Rock Mass Quality

Rock (soil) structure type: According to the characteristics of the structural surface and structure of the rock-soil mass, the rock-soil mass in the area is divided into four major rock mass types: loose structure, fragmented structure, layered structure and block structure.

- Block structure: Mainly refers to the rock mass structure of sandstone. The rock mass is composed of thick layered sandstone, and the layer thickness of the rock mass is generally greater than 1.5 m. The lithology is single, the structural surface is less than the layered structure rock mass, and the bedding features are discontinuous staggered, wavy, and parallel bedding. The fractures are weakly developed, with good continuity, weak compression deformation, and rigid state. The mechanical properties of rocks are little affected by groundwater, and the integrity and stability of the rock mass are good. It is the rock mass structure with the best integrity and stability.

- Layered structure: It is a siltstone and mudstone (sandy mudstone) rock mass structure. The rock mass is mostly thin to medium-thick layered with some weak interlayers (such as mudstone, coal, carbonaceous mudstone, etc.). Its characteristic is that the rock mass is layered, and the structure is rectangular, and plate shaped. The flat bedding is developed and blocked by the soft and weak layer, the vertical cracks develop extremely unevenly, and the extension length is small. Water is easy to form above the weak layer, and the rock is easy to soften, disintegrate, and delaminate under the influence of groundwater, and it is easy to lose the original rock stress balance state. The main manifestations are delamination or instability along the sliding surface. When conditions are available, interlayer slippage and instability will occur. The direction and degree of slippage are controlled by the slope direction and stratum tendency of the ground, and the stability is generally poor.
- Fragmented structure: Composed of weakly weathered rock. The continuity of the rock mass is poor, the structural surface is rough, the rock blocks are of different sizes and shapes, and many invisible micro-cracks are often developed. Groundwater has a certain inducing effect on its stability. The distance between the structural planes of this kind of rock mass is generally less than 0.5 m, and they cut each other to form rock blocks of different sizes and shapes. The porosity of the rock block increases and the network of micro-weathered fractures is hidden, reflecting the changes in the internal material structure and mechanical properties of the rock blocks that make up the rock mass, which easily causes the roof pressure to increase and the initial failure period of the old roof to shorten. Especially when coal is mined in the lower part of loose aquifer or near weathered zone, the surrounding rock of shaft is easy to lose stability and damage, and the roof of working face or goaf is easy to cut and collapse in a large area.
- Loose structure: It is composed of loose soil and strong weathered rock mass. Loose soil particles are stacked in disorder, high compressibility, large deformation, poor compression and shear resistance, and rock and soil have no strength or very weak strength, which is the structure with the worst engineering geological conditions. The bulk structure has poor stability and is like porous medium, which is likely to cause many engineering geological problems.

In summary, the block structure of the rock mass in the area is characteristics of the coal-measure strata (Fuxian Formation, Yanchang Formation) and the middle and bottom strata of each section of the Yan'an Formation. The rock mass is complete and stable; the rock mass is layered. This structure is in the direct roof and floor of each coal seam, and the general stability is poor. The rock fragmentation structure is in the weathered zone of the top surface of the bedrock and the fractured zone, and the integrity and stability of the rock mass are poor. The loose structure is distributed in in the valleys and slopes on the surface and bedrock surface where the rock mass integrity and stability are the worst.

8.2.7.4 Rock Mass Quality

The rock mass quality is evaluated according to the rock mass integrity, rock mass quality coefficient method, and rock mass quality index method. The results are shown in Table 8.3. As presented in the table, results of the three quantitative evaluation methods are basically the same. The qualitative evaluation of various structural rock masses is consistent with the quantitative evaluation conclusions, indicating that the division of rock mass structure types reflects objectively the engineering geological characteristics of the rock mass in the mine.

8.2.7.5 Evaluation of the Roof and Floor Stability of the Main Coal Mining Seam

The roof of #4-2 coal seam is dominated by siltstone and fine-grained sandstone. The rock RQD value of mudstone is between 67 and 76% with an average of 71.30%. The RDQ of fine-grained sandstone varies from 57.30 to 72.80% with an average of 65.05%. The uniaxial saturated compressive strength of the rock is less than 30 MPa. This formation is classified as a weak rock with medium integrity of the rock mass and an unstable roof. The coal seam floor is dominated by mudstone, and the rock RQD value is between 28 and 78% with an average of 58.85%. The compressive strength is less than 30 MPa, which belongs to an unstable floor. The mudstone swells in contact with water and is prone to bottom heave.

The roof of #5-2 coal seam is dominated by fine-grained sandstone, and the rock RQD value of fine-grained sandstone is between 55 and 87% with an average of 73.14%. The uniaxial saturated compressive strength of the rock is less than 30 MPa. The roof formation is classified as a weak rock with medium rock integrity. It belongs to the unstable roof. The floor is composed of siltstone and fine-grained sandstone. The RQD value of siltstone is between 54.50 and 84.20% with an average of 66.85%, whereas the RQD value of fine-grained sandstone varies from 59.40 to 94.10% with an average of 80.51%. The uniaxial saturated compressive strength is less than 30 MPa. The rock mass integrity of the floor formation ranges from medium to relatively good. As a result, the floor stability is good.

8.2.8 Characteristics of Coal Seam

Except for coal seam #5-2 which is distributed in the entire area, the other coal seams including #2-2, #3-1, #4-2, #4-3 and #4-4 are locally distributed. Coal seams #5-2 and #4-2 are the primary mining coal seams. The main features are shown in Table 8.4.

Table 8.3 Quantitative evaluation results of rock mass quality grade

Rock structure	Saturated compressive strength (MPa)	Structural surface friction coefficient f (empirical)	Rock mass quality grade evaluation					
			RQD value		Rock mass coefficient method		Rock mass index method	
			RQD (%)	Rock integrity	Z value	Rock mass quality grade	M value	Rock mass
Unconsolidated loose structure				Loose, extremely broken				
Fragmented structure		0.30	<30	Poor				
Layered structure	20.13	0.40	58	Moderate	0.56	Typical	0.47	Moderate
Block structure	25.4	0.50	78	High	1.07	typical	0.73	Moderate

Table 8.4 Summary of mineable coal seams

Coal seam	Thickness (m)	Distance between coal seams (m)	Depth (m)	Consistence	Mineable	Recoverable area (km^2)	Notes
	Range (average)	Range (average)	Range (average)				
2^{-2}	3.20–8.68 (4.31)	24.68–30.95 (28.33)	0–175.72 (107.07)	Consistence	Yes	49.24	The structure of the coal seam is simple with localized gangue. There is a large area of spontaneous combustion in the east. The reserve area is in the west
3^{-1}	2.50–3.25 (2.91)	33.85–60.20 (42.48)	0–207.32 (113.57)	Consistence	Yes	68.52	The coal seam has a simple structure and generally does not contain gangue. The east is subject to self-combustion and erosion. The reserve area is in the west
4^{-2}	1.15–4.65 (3.16)		0–256.60 (120.51)	Consistence	Yes	99.61	The coal seam has a complex structure and generally does not contain gangue. The eastern part suffers from self-combustion and erosion. The reserved area is in the central and western part
		7.20–26.74 (18.70)					

(continued)

Table 8.4 (continued)

Coal seam	Thickness (m)	Distance between coal seams (m)	Depth (m)	Consistence	Mineable	Recoverable area (km^2)	Notes
	Range (average)	Range (average)	Range (average)				
4^{-3}	0.80–1.66 (1.29)	9.40–22.15 (12.26)	0–270.82 (152.25)	Relative consistence	Yes	84.78	The deposits are relatively stable
4^{-4}	0.80–1.13 (1.00)	16.83–43.25 (33.15)	0–193.49 (114.75)	Relative consistence	Yes	64.14	The deposits are relatively stable
5^{-2}	1.15–9.30 (5.87)		0–302.65 (156.97)	Consistence	Yes	138.23	The coal seam has a complex structure, generally containing 1–2 layers of gangue. Part of the eastern part suffered from spontaneous combustion

8.2.9 Other Human Engineering Activities in the Mine and its Vicinity

The existing engineering activities within the assessment area include Shuitugou Reservoir and supporting facilities, three small-scale coal mines (Chikaigou Coal Mine, Hewan Coal Mine, and Aobaogou Coal Mine), agricultural activities, and the construction of rural roads. Also present are 110 kV high-voltage transmission lines and other low-voltage transmission lines.

- Reservoir and supporting facilities: Shuitugou Reservoir was built in the 1990s and is located in Kentie Linggounao. It consists of two water storage areas. The first storage area is located in the downstream part, approximately 320 m long, 30 m wide, and water depth varying from 4.5 to 0.4 m with an average of 2.5 m. The maximum storage capacity is 15,000 m^3. Seasonal changes have little effect. It is used for nursing and agricultural irrigation. The second storage area is located in the upstream part, 180 m long, 40 m wide, and water depth varying from 3 to 1.5 m with an average of 2.5 m. The maximum storage capacity is 18,000 m^3.
- Small-scale coal mines: Lucaogou Coal Mine and Aobaogou Coal Mine are within the minefield of Hongliulin Mine. Other two small coal mines, Chikaigou Coal Mine and Hewan Coal Mine are in the vicinity of Hongliulin Mine. Lucaogou Coal Mine extracted #4-2 coal seam and was destroyed and closed in 1986 due to incomplete mining procedures. Chikaigou Coal Mine is located to the north of the boundary of Hongliulin Mine and mines #3-1 coal seam via adits. Hewan Coal Mine and Aobaogou Coal Mine are located in the southwest of the boundary of Hongliulin Mine. They are both developed with inclined shafts. #5-2 coal seam is exploited. Drilling and blasting are used underground, and room-pillar mining is the method. Table 8.5 summarizes the small-scale coal mine and mined-out areas.

Table 8.5 Summary of small-scale coal mines

Name	Area (km^2)	Coal seam	MIned-out area (km^2)
Hewan coal mine	4.5927	5^{-2}	1.0914
Aobaogou coal mine	2.1870	5^{-2}	0.3789
Chikaigou coal mine	2.3457	$3^{-1}, 4^{-2}, 5^{-2}$	0.1080

8.3 Geoenvironmental Impact Assessment

8.3.1 Evaluation Scope and Level

8.3.1.1 Evaluation Scope

Hongliulin mining area is approximately 20 km from east to west and 8 km from north to south, encompassing an area of 142.55 km^2. The influence of the goaf on the surface deformation boundary is calculated based on the thickness of the overlying bedrock of the main coal seam near the boundary of the mine, the thickness of the loose layer and the boundary angle in different lithologies. Other factors including construction projects and topography are considered to determine the assessment scope.

The thickness of the overlying bedrock of #5-2 coal seam near the mine boundary varies from 22 to 252 m. The thickness of the overlying unconsolidated layer is up to 128 m. The mining induced angle is between 70 and 75° in the bedrock and 40–45° in the unconsolidated layer. The calculated deformation area on surface extends 22–165 m beyond the boundary. In consideration of the topographic changes and the 20 m wide coal pillars along the mine boundary, the assessment range extends 150 m from the boundary of the mine. The industrial site and waste disposal area extend by 200 m beyond the boundary. Another 50 m is added to the assessment scope for a conservative evaluation on geological disasters. The total assessment area is 155.95 km^2.

8.3.1.2 Evaluation Classification

According to the "Guidelines for the Compilation of Mine Geological Environmental Protection and Governance and Restoration Programs" (DZ/T 223-2009), the geological environmental impact assessment level in mines depends on importance of the assessment area, the scale of mine production and construction, and the complexity of the mine geological environmental conditions (Table 8.6).

- Importance of the assessment area: Table 8.7 summarizes the importance of the assessment area. Villages in the assessment area are scattered, and the residents of the same village are also scattered; the roads are all rural roads, the buildings are villagers' houses, and there are no important traffic roads or construction facilities; the Shuitugou reservoir in the assessment area is used for breeding and agricultural irrigation; there is no nature Protected areas and tourist attractions; destruction of cultivated land, shrubland, natural grassland, etc. Therefore, the assessment area is determined to be an important area.
- The scale of production and construction of the mine: The designed production scale of Hongliulin Mine is 12.00 Mt/a, and the mining method is underground mining. The scale of the mine construction and production is large.

Table 8.6 Classification of geological environmental impact assessment

Importance of assessment area	Scale on mine production and construction	Complexity of geological environmental conditions		
		Complex	Medium	Simple
Important	Large	Level 1	Level 1	Level 1
	Medium	Level 1	Level 1	Level 1
	Small	Level 1	Level 1	Level 2
Relatively important	Large	Level 1	Level 1	Level 2
	Medium	Level 1	Level 2	Level 2
	Small	Level 1	Level 2	Level 3
Average	Large	Level 1	Level 2	Level 2
	Medium	Level 1	Level 2	Level 3
	Small	Level 2	Level 3	Level 3

Table 8.7 Importance of assessment area

Important	Relatively important	Average	Assessment result
• A concentrated residential area with more than 500 residents • There are expressways, first-class highways, railways, medium-sized and above water conservancy facilities, electric power facilities and other important construction facilities • The mining area is close to national nature conservation areas (including geological parks or scenic and historical sites) or important tourist attractions • There are important water supply sources • High likelihood to destroy cultivated land and farmland	• A concentrated residential area with 200–500 people • Distributed with secondary roads, small water conservancy, power facilities or other important construction facilities • Adjacent to provincial and county-level nature reserves or important tourist attractions • There are relatively important water supply sources • Likelihood to destroy woodland and grassland	• Residents scattered, and the population of concentrated residential areas is less than 200 • No important traffic arteries or construction facilities • Distant from nature conservation sites and tourist attractions at all levels • No important water supply sources • Less likelihood to destroy other types of land	• Average. The villages have small population and the villagers live scattered • Average. The roads are all country roads, and the buildings are village houses • Average. There are Lucaogougou Naohazi and Shuitugou Reservoir in the area • Average. Lucaogougou Minhaizi and Shuitougou Reservoirs are in the area and they are not water supply sources • Important. Cultivated land of greater than 50 acres is within the mining area

Note The importance level of the assessment area adopts the principle in which a higher level is used if only one metric meets the level requirement

- Complexity of the mine geological environment: The complexity of the geological environment in the assessment area depends on six factors, as presented in Table 8.8.

Recharge to the groundwater in this area is provided by atmospheric precipitation, which is small and concentrated. Except for the alluvial layer of the Lucaogou and other large valleys where the local aeolian sand has medium water richness in the Sarawusu Formation, the water richness of other aquifers is dramatically different but mostly weak or extremely weak aquifers. Because the laterite layer in the Baode Formation is not continuous, the hydraulic connection between the porous medium Quaternary aquifer and the underlying confined fracture aquifers is complicated. The direct recharge source to the mine is the confined fracture aquifer overlying each coal seam. The direct recharge includes Quaternary porous medium aquifer, surface water and fracture aquifer of burnt rock at the boundaries with the main gullies and burned areas. In general, complexity of the hydrogeological conditions is medium.

The overlying formation of the mineable coal seams consists of mudstone, siltstone, fine-grained sandstone, and medium-grained sandstone. The underlying formation of the coal seams is composed of mudstone and siltstone. The surrounding rock of the coal seam is layered formation and belongs to soft-semi-hard rock with medium stability. The ground surface is covered by aeolian sand, loess and laterite with scattered alluvial layer, bedrock and burnt rock. The scattered and fragmented structures are dominant, whereas the layered structure is limited. The engineering properties of rock and soil are medium, so are the stability of the surrounding rock of the coal seam and the foundation of the site and the engineering geological conditions.

The strata are controlled by a gentle monoclinic structure that dips to the west at angles between 1° and 2°. No faults are reported. Under the current conditions, the types of geological and environmental problems are few with little harms.

Based on the evaluation results on importance of the area, the scale of mine construction, and complexity of the mine geoenvironmental conditions, environmental impact assessment is conducted at Level 1.

8.3.2 Assessment of Background Conditions

Assessment of the background environmental conditions is conducted in four aspects, i.e., geological disaster, aquifer, landform and landscapes, and land resources.

8.3.2.1 Assessment of Geological Disasters

Geological disaster reconnaissance identified two cases of potential ground subsidence in Aobaogou Coal Mine and Lucaogou Coal Mine. Table 8.9 summarizes the characteristics of the potential subsidence.

Table 8.8 Grade of complexity degree of geological environment conditions of mines

Factor	Complex	Medium	Simple	Assessment result
1	• Main ore layer (body) is located below the groundwater level • Mine has complex water recharge conditions with many water sources • Good recharge and water yield conditions in aquifers, structural fracture zones, karst fracture development zones • Strong hydraulic connections to concentrated preferential flow zones or surface water • Threat of old mine pools is great • Normal water inflow rate is >10,000 m^3/d • Regional aquifer is vulnerable to mining and dewatering	• Main ore layer (body) is located near or below the groundwater level • Recharge boundary conditions are medium • Water recharge aquifer, structural fracture zone, karst fracture development zone and other water richness are medium • Concentrated preferential flow zone or surface water has a certain connection to mine • Old mine pool (kiln) water threat is moderate. • Normal water inflow rate is between 3000 and 10,000 m^3/d • Mining and dewatering are more likely to cause damage to the surrounding main water recharge aquifers	• Main ore layer (body) is located above the groundwater level • Recharge boundary conditions are simple • Water recharge aquifer, structural fracture zone, and karst fracture development zone have poor water richness • Poor hydraulic connections to regional strong aquifers, concentrated preferential flow zone, and surface water • Normal water inflow rate of the mine is <3000 m^3/d • It is unlikely that the surrounding main water recharge aquifer is damaged by mining and dewatering	Medium

(continued)

Table 8.8 (continued)

Factor	Complex	Medium	Simple	Assessment result
2	• The rock mass structure of the surrounding rock of the deposit is broken structure and loose structure • Weak or loose rock layers are developed, alteration zone and karst fracture are developed, rock weathering is strong • Thickness of bedrock weathered broken zone is >10 m • Stability of roof and floor of the coal seam and the surroundings of the deposit is poor • Foundation stability of the mine engineering site is poor	• Rock mass structure of the surrounding rock of the deposit is thin-thick layered structure • Altered zone and karst fracture are moderately developed, and the rock is moderately weathered • Roof and floor and the surroundings of the deposit have medium stability • Foundation of the mine engineering site has medium stability	• Rock mass structure of the surrounding rock of the deposit is dominated by a huge thick layered-blocky overall structure • Weak alteration, lack of karst fractures, weak rock weathering • Thickness of the bedrock weathered fracture zone is <5 m • Roof and floor of the coal seam and the surrounding of the ore deposit have good stability • Foundation of the mine engineering site has good stability	Medium

(continued)

Table 8.8 (continued)

Factor	Complex	Medium	Simple	Assessment result
3	• Geological structure is complex • Occurrence of the ore layer (body) and the surrounding rock of the ore deposit varies greatly • Fault structure is developed or there are active faults, and the water-conducting fault zone cuts the surrounding rock, overlying rock, and main aquifer of the ore layer • Strong water conductivity and has a great impact on the safety of underground mining	• Geological structure is relatively complex • Occurrence of the ore layer (body) and surrounding rock of the ore deposit changes greatly • Fracture structure is relatively developed, and the surrounding rock, overlying rock and main aquifer of the ore layer are cut • Water conductivity is poor • Greater impact on the safety of underground mining	• Geological structure is simple • Occurrence of the ore layer (body) and the surrounding rock of the ore deposit changes little • Fault structure is not developed • Faulted uncut ore layer is covered by the surrounding rock • Fracture zone has little impact on the safety of underground mining	Simple
4	• Under current conditions, mine geological disasters develop, or there are many types of mine geological environmental problems, which are harmful	• Under current conditions, there are many types of mine geological and environmental problems, which are relatively harmful	• Under current conditions, there are few types of mine geological environmental problems with little harm	Simple
5	• Area and volume of the mined-out area is large • Mining and residual mining are repeated many times • Mined-out area has not been effectively treated • Mining impact is strong	• Area and volume of the mined-out area is large • Repeated mining is less • The part of the mined-out area is treated • Mining impact is relatively strong	• Area and volume of the mined-out area are small • There is no repeated mining • Mined-out area is treated • Mining impact is limited	Simple

(continued)

Table 8.8 (continued)

Factor	Complex	Medium	Simple	Assessment result
6	• There are many types of geomorphic units, the micro-topography is complex, the topography is undulating, which is not conducive to natural drainage • Topography slope is generally greater than 350 • Relative height difference is large • Ground slope is basically the same as the rock surface slope	• There are many types of geomorphic units, the micro-topography morphology is more complex, the topography is moderately undulating, which is not conducive to natural drainage • Topographic slope is generally between 200 and 350 • Relative height difference is large • Ground slope and the rock formation slope are mostly oblique	• The type of geomorphic unit is single, the micro-topography is simple, the topography is undulating smoothly, which is conducive to natural drainage • Topographic slope is generally less than 200 • Relative height difference is small • Ground slope and the rock formation slope are mostly inverse	Simple—Medium

Note Adopt the conservative principle. A higher level is designated if only one of the first 6 factors satisfies the level requirements

Table 8.9 Ground subsidence in small-scale coal mines within environmental assessment area

Mine	Mined-out area (km^2)	Survey results	Hazard object	Hazard level	Note
Aobaogou coal mine	0.3789	There is no obvious deformation of the ground and no collapse	None	Low	The mined-out area is 0.1031 km^2 within the assessment area.
Lucaogou coal mine	No data	There is no obvious deformation of the ground and no collapse	None	Low	Closed in 1986

8.3.2.2 Assessment of the Aquifers

Hongliulin Mine is a newly built mine. The groundwater flow conditions are affected by other engineering and agricultural activities.

The mined area of Aobaogou Coal Mine is adjacent to the Hongliulin Mine. The maximum water inflow of Aobaogou Coal Mine is 102.30 m^3/d, and the normal water inflow is 74.40 m^3/d. Water is observed dripping from the roof of the coal seam. The Quaternary aquifer was not affected. There are no villagers living in the goaf areas of Aobaogou Coal Mine and Lucaogou Coal Mine. Mining of these small mines have not affected the water supply sources for the villagers.

Four groundwater monitoring wells were installed in Hejia Village, Hewan Village, Gangfangliang Village, and Hewan Coal Mine to monitor groundwater quality. Table 8.10 presents the results in the villages, and Table 8.11 summarizes the groundwater quality at Hewan Coal Mine.

The monitoring results showed that the total number of bacteria exceeded the standard in Hejia Village. The water quality indicators of other villages' wells meet the Class III standard in the "Groundwater Quality Standard" (GB/T14848-93). The groundwater quality in this area is good.

This groundwater samples at Hewan Coal Mine were taken from the underground mine water discharge point. All water quality parameters analyzed meet the primary drinking water standards as stipulated in "Drinking Water Source Water Quality Standard" (CJ3020).—1993).

8.3.2.3 Assessment of Landform and Landscape

The landform types of Hongliulin mine field include aeolian sand landforms, sand-covered loess hills, loess hills and river landforms. The aeolian landform is the primary one. The distribution of main landform types in the assessment area are summarized in Table 8.12.

Table 8.10 Results of background groundwater quality at villages

Monitoring location	pH	Fluoride mg/L	Volatile Phenol mg/L	Hardness mg/L	Permanganate mg/L	Ammonia mg/ L	Petroleum mg/L	Total number of bacteria /mL	Sulfate mg/L	Suspended solids mg/L
Gangfangliang Village	6.81	0.48	0.002	128	1.1	0.013	0.03	48	73	9
Hejia Village	6.54	0.93	0.002	108	0.8	0.013	0.03	114	39.4	8
Hewan Village	6.66	0.51	0.002	124	1.1	0.013	0.03	72	55	12

Table 8.11 Results of background groundwater quality at Hewan Coal Mine

Parameter	Results	Primary drinking water standards	Monitoring method
pH	6.71	6.5–8.5	Glass electrode method (GB6920-1986)
Fluoride	0.46	≤1.0	Ion chromatography
Volatile phenol	0.001	≤0.002	Spectrophotometry (GB7490-1987)
Chemical oxygen demand	16	/	COD instrument rapid determination method
Biochemical oxygen demand	2.6	/	Chemical dilution (GB7488-1987)
Ammonia	0.013	≤0.5	Spectrophotometry (GB7479-1987)
Suspended solids	63	/	Weight method (GB11901-1989)
Total number of bacteria	264	/	Plate (GB/T5750-1985)
Coliform	170	≤1000	Multi-tube fermentation (GB/5750-1985)
Hardness	168	≤350	EDTA titration (GB7477-1987)
Cyanide	0.002	≤0.05	Colorimetry (GB7486-1987)
Arsenic	0.004	≤0.05	Spectrophotometry (GBGB/T7485-1987)
Sulfate	71.6	<250	Weight method (GB/T11899-1989)
Total dissolved solids	480	≤1000	Weight method
Nitrate Nitrogen	1.396	≤10	Spectrophotometry (GB7479-1987)
Mercury	0.00005	≤0.001	Ion chromatography
Zinc	0.02L	≤1.0	Atomic absorption method (GB7475-1987)
Iron	0.03	≤0.3	Atomic absorption method (GB11911-1989)
Manganese	0.01	≤0.1	Atomic absorption method (GB11911-1989)
Copper	0.02	≤1.0	Atomic absorption method (GB7475-1987)
Lead	0.2	≤0.05	Flame atomic absorption method (GB/T7475-1989)
Cadmium	0.008	≤0.01	Dithizone spectrophotometry (GB7475-1987)

Note Except pH, coliforms, and total number of bacteria, all other parameters are in mg/L

Table 8.12 Landform types in the assessment area

Geomorphology		Area (km^2)	%
Aeolian landform	Mobile sand dunes	28.07	18.36
	Semi-mobile, semi-fixed dunes	13.64	8.92
	Fixed dunes	77.62	50.76
	Subtotal	119.33	78.04
Sand-covered loess hills	Mobile sand dunes	3.17	2.07
	Semi-mobile, semi-fixed dunes	0.94	0.61
	Fixed dunes	1.66	1.09
	Subtotal	5.77	3.77
Loess hills	Loess ridge and loess valley	23.74	15.52
Alluvial landform	River terraces	4.08	2.67
Total		152.92	100

The assessment area has the largest aeolian landform, encompassing 119.33 km^2 and accounting for 78.04% of the mining area. The sandy land includes mobile dunes (land), semi-mobile, semi-fixed dunes (land) and fixed dunes, with fixed dunes as the primary form. The fixed sand dunes (land) have the largest area, accounting for 50.76% of the assessment area. They are widely distributed around Shilatajie in the northeast, east of Lucaogou in the north, and Dongshuangliang and Dashuliang in the west. The mobile sand dunes account for 18.35% of the assessment area, which is distributed in the west of Lucaogou, Wulbengshigou and west of Maolvtan. The semi-fixed dunes have a relatively small area, accounting for 8.92% of the assessment area. They are mainly distributed on both sides of the Wulbongshi River in the south, to the west of Dashuliang, and the area between Maolvtan and Kentieling River in the middle.

Sand-covered loess hills including mobile sand dunes, semi-mobile and semi-fixed sand dunes, and fixed dunes encompasses an area of 5.77 km^2, accounting for 3.77% of the mining area. The mobile sand dunes are distributed in the east of Kentieling River and north of Caojiepan, accounting for 2.07% of the assessment area. The semi-mobile, semi-fixed sand dunes, mainly distributed to the west of Zhujiamiaogou, accounts for 0.61% of the assessment area. The fixed sand dunes, distributed around the Caojiepan area in the northeast, occupy 1.09% of the assessment area.

The loess hills consisting of loess ridges and loess valleys are distributed in the area around the Kentieling River in the southeast, covering an area of 23.74 km^2 and accounting for 15.33% of the assessment area.

The alluvial landform is represented by terraces and distributed along the Majiatagou and Lucaogou, covering an area of 4.08 km^2 and accounting for 2.67% of the assessment area.

There are no geologically significant landforms in Hongliulin Mine. The topography is not affected by mining activities.

Aobaogou Coal Mine and Lucaogou Coal Mine both adopt room and pillar mining. The ground deformation in the mined-out areas is not obvious. The topography and landform are not affected by their mining activities.

8.3.2.4 Assessment of Land Resources

The land use in the assessment area is divided into seven types, i.e., forest, cultivated land (including irrigated land and dry land), grassland, residence, transportation, water body, and water conservancy facility. Table 8.13 summarizes the land use statistics. The cultivated land encompasses 6.83 km^2, accounting for 4.47%. The grassland occupies 100.50 km^2, accounting for 65.72% of the project area. The forest is 13.65 km^2 or 8.92% of the project area. The water bodies are dominated by seasonal rivers. The water bodies and water conservancy facilities encompass 0.09 km^2 or 0.06% of assessment area. The residential land covers an area of 0.95 km^2, accounting for 0.62% of the project area. The transportation facilities take an area of 0.40 km^2 or 0.26% of the assessment area. The unused land, mainly desert, encompasses 30.51 km^2 or 19.95% of the assessment area. The main crops in the assessment area are corn, sorghum, millet, soybeans, and potatoes.

Both Aobaogou Coal Mine and Lucaogou Coal Mine adopt room and pillar mining. The ground deformation in the goaf is not obvious, and mining has basically no impact on land resources.

Table 8.13 Current land use types and statistics in the assessment area

Land use type		Area (km^2)	%
Farmland	Irrigated land	1.51	0.99
	Drought land	5.32	3.48
Woodland	Shrubland	13.65	8.92
Grassland	Natural grassland	100.50	65.72
Land for transportation	Roads	0.40	0.26
Land for waters and water conservancy facilities	Water body	0.09	0.06
Residential land	Rural homestead	0.95	0.62
Other unused land	Desert	30.50	19.95
Total		152.92	100

8.3.3 Soil Erosion Intensity

According to the three types of allowable erosion and regional characteristics in the "Soil Erosion Classification and Classification Standards" (SL190-1996) promulgated by the Ministry of Water Resources, the allowable soil erosion in this area is 500 t/km^2. The soil erosion in the assessment area is divided into two types: water erosion and wind erosion. The soil erosion intensity is represented by five grades: slight, mild, moderate, strong, and extreme. The assessment area is dominated by wind erosion, and water erosion is secondary. The statistical results of soil erosion intensity and area are shown in Table 8.14.

8.3.3.1 Wind Erosion

Wind erosion occurs in the west and north sand dune area totaling 122.42 km^2, accounting for 80.05% of the assessment area. The wind erosion intensity includes slight, mild, moderate, intense, and extreme erosion.

- The area of slight wind erosion is distributed around the Shilata boundary in the east and north of Wujialiang in the north, covering an area of 7.47 km^2 or 4.88% of the assessment area.
- The area of mild wind erosion is distributed in east of Lucaogou, covering an area of 1.94 km^2 or 1.27% of the assessment area.
- The area of moderate wind erosion is the largest and distributed around Dongshuangliang, Dashuliang, and Yangjiamu River in the west, west of Dajinghao, and

Table 8.14 Statistics of soil erosion

Type and intensity of soil erosion		Erosion modulus (t/km^2.A)	Area (km^2)	%
Water erosion	Slight hydraulic erosion	<500	4.22	2.76
	Mild hydraulic erosion	500–2500	1.66	1.09
	Moderate hydraulic erosion	2500–5000	23.77	15.54
	Strong hydraulic erosion	5000–8000	0.86	0.56
	Subtotal		30.5	19.95
Wind erosion	Slight wind erosion	<500	7.47	4.88
	Mild wind erosion	500–2500	1.94	1.27
	Moderate wind erosion	2500–5000	72.54	47.43
	Intense wind erosion	5000–8000	22.03	14.41
	Extremely intense wind erosion	8000–15000	18.44	12.06
	Subtotal		122.42	80.05

Shilatajie, Liujia Village, Xiliang Village, and Maolvtan in the east, encompassing 72.54 km^2 or 47.43% of the assessment area.

- The area of strong wind erosion is distributed around Kentielinggou, Maolvtan, Luccaogou in the south, and the area north of Wulbengshigou in the west, covering an area of 20.14 km^2 or 12.67% of the assessment area.
- The Extremely strong wind erosion occurs in the Lucaogou and Wulbengshigou areas in the west and the surrounding areas of Taosuhaizi in the southwest, covering an area of 22.03 km^2 or 14.41% of the assessment area.

8.3.3.2 Water Erosion

The area where water erosion dominants is 30.5 km^2, accounting for 19.95% of the assessment area. Water erosion is distributed in the loess valleys, loess ridges and river terraces in the east and south. The erosion intensity includes slight, light, moderate and intensive grades.

- The slight water erosion is distributed in the terraces along the Majiatagou and its tributaries Lucaogou, Wulbengshigou, and Kentielinggou, covering an area of 4.22 km^2 or 2.76% of the assessment area.
- The mild water erosion is distributed in the valleys along the Maolvtan, covering an area of 1.66 km^2 or 1.09% of the assessment area.
- The area of moderate water erosion is distributed in the north and east of Kentielinggou in the east, covering an area of 23.77 km^2 or 15.54% of the assessment area.
- The intense water erosion area is 0.86 km^2 or 0.56% of the assessment area.

8.3.4 *Vegetation and Coverage*

8.3.4.1 Vegetation Type

Table 8.15 summarizes the vegetation types in the assessment area. Artemisia and salix shrubs have the largest area, accounting for 77.20% of the assessment area.

Table 8.15 Vegetation type and area statistics in the assessment area

Vegetation types	Area (km^2)	%
Artemisia sphaerocephala and salix shrub	118.06	77.20
Artemisia sphaerocephala, caragana korshinskii shrub	3.16	2.07
Miscanthus grass	25.53	16.70
Agricultural vegetation	5.48	3.58
Not covered by vegetation (construction land)	0.69	0.45
Total	152.92	100

They are widely distributed in the surrounding areas of Lucaogou in the west, and the Shilata boundary in the east. There are also large areas around the Shilata boundary in the east. Miscanthus grasses, distributed around the Kentieling River in the east and the area to the east, and in the north around Qianhupan and Chikaigou, accounts for 16.7% of the area of the assessment area. The area of artemisia sphaerocephala and caragana korshinskii is only distributed to the north of Caojiepan in the northeast small, occupying 2.07% of the assessment area.

The agricultural vegetation is distributed in Majiatagou, Lucaogou, Maolvtan, and terraces along the Kentieling River. The main agricultural types are millet, grain, sorghum, corn, soybean, and potato. They have one-year life cycle. The area is 5.48 km^2 or 3.58% of the assessment area.

8.3.4.2 Vegetation Coverage

According to the percentage of vegetation coverage on the ground, the vegetation coverage in the assessment area is divided into four levels, i.e., medium-high coverage, medium coverage, low coverage, and very low coverage. There is no hierarchy for the agricultural vegetation. Table 8.16 summarizes the vegetation coverages.

- The medium-high coverage area is distributed in the north of Wujialiang in the north, around Dashuliang in the south and east of Lucaogou, covering an area of 16.70 km^2 or 10.92% of the assessment area.
- The medium coverage is the largest and distributed around the Shilata boundary in the east, to the east of Kentielinggou and Maolvtan, between the trenches and Qianhupan in the northwest, and around Dongshuangliang in the west and in the south, east of Yangjiamu River and west of Lanjia Village. The total area is 64.75 km^2, accounting for 42.34% of the assessment area.

Table 8.16 Vegetation coverage statistics in the assessment area

Coverage classification	Coverage percentage for classification (%)	Area (km^2)	% in mining area
Medium-high coverage	50–70	16.70	10.92
Medium coverage	30–50	64.75	42.34
Low coverage	10–30	26.56	17.37
Extremely low coverage	<10	38.75	25.34
Agricultural vegetation		5.48	3.58
Not covered by vegetation (construction land)		0.69	0.45
Total		152.92	100

- The area of low coverage is distributed around Caojiepan in the north, Maolvtan and Kentielinggou in the south, and Yangjia Village in the west. The coverage area is 26.56 km^2 or 17.37% of the assessment area.
- The very low coverage is distributed along the banks of Lucaogou in the west, around Taosuhaizi in the southwest, between Maolvtan and Kentieling River in the south and to the east. The total area is 38.75 km^2 or 25.34% of assessment area.

8.3.5 Summary

No significant geohazards are identified in the background assessment. The ground subsidence in the mined-out areas of Aobaogou Coal Mine and Lucaogou Coal Mine are potentially concealed hazards. Because no obvious ground deformation is reported, the degree of harm is small, thus the risk is small. The agricultural activities are the most important factors that affect the groundwater quantity, groundwater quality, landscape, and land resources.

8.4 Predictive Geoenvironmental Assessment

The mine geoenvironmental prediction assessment mainly focuses on the following four aspects in response to five-year mining in Liuhonglin Coal Mine:

- Geohazards caused or exacerbated by mining activities
- Impact or destruction on aquifers
- Impact or destruction on geologically significant landforms and landscapes
- Impact on land resources.

8.4.1 Predictive Assessment of Geological Disasters

8.4.1.1 Prediction of Surface Deformation

Mining can cause and aggravate existing geohazards such as ground subsidence, ground fractures, landslides, and collapses. These geohazards result in damages to buildings and casualties. The surface deformation predictions are conducted for the worst-case scenarios. The references include the observations of surface movement and deformation in Shendong mining area, the site-specific conditions of Hongliulin Mine's coal seam conditions and mining methods, and the regulations on buildings, water bodies, railways and main shafts and coal pillars in coal mining:

- Maximum subsidence (mm): $W_{cm} = M \times q \times \cos\alpha$
- Maximum slope (mm/m): $i_{cm} = W_{cm}/r$

- Maximum curvature (10^{-3}/m): $K_{cm} = 1.52 \times W_{cm}/r^2$
- Maximum horizontal movement (mm): $U_{cm} = b \times W_{cm}$
- Maximum horizontal deformation (mm/m): $\varepsilon_{cm} = 1.52 \times b \times i_{cm}$.

where

M Mining thickness of coal seam, m
H Burial depth of coal seam, m
A Dip angle of coal seam, 1–2°
r Radius of influence by mining, m
q Subsidence coefficient
b Horizontal movement coefficient.

The parameters related to surface deformation calculation are provided as follows:

- Subsidence coefficient of the initial mining $q = 0.62$
- Subsidence coefficient of repeated mining $q = 0.65$
- Horizontal movement coefficient $b = 0.31$
- Tangent of influence angle of initial mining $\text{tg}\beta = 2.0$
- Tangent of influence angle of repeated mining $\text{tg}\beta = 2.4$
- Radium of influence $\text{r} = H/\text{tg}\beta$.

Table 8.17 summarizes the mine-wide prediction results using the above equations and parameters in response to 5-year mining of the mineable coal seams in Hongliulin Mine. Table 8.18 presents the prediction results on ground subsidence for each working face.

For the six mineable coal seams, the superimposed deformation is as follows:

- Maximum settlement value is 12,877.54 mm
- Maximum slope is 133.93 mm/m
- Maximum curvature is 7.78×10^{-3}/m
- Maximum horizontal movement is 1969.66 mm
- Maximum deformation is 63.11 mm/m.

According to the mining plan, the five-year mining sequences are #4-2 coal seam in 24206 working face and #5-2 coal seam in 15201, 15202, 15203, 15204, 25201, 25202, and 25203 working faces.

Mine-wide surface deformation: Based on the coal seam distribution, buried depth, and mining thickness, the ratio of the thickness of the overlying bedrock to the mining thickness, or effective depth-to-thickness ratio, is used to analyze the surface deformation. The effective depth-to-thickness ratio of the coal seams in this area is between 2.4 and 29.1, with an average of 12.07. Because the effective depth-to-thickness ratios are less than 30, mining of the coal seams can cause ground subsidence discontinuously and develop geohazards.

The length of planned working faces ranges from 250 to 300 m. With an advancing length between 1700 and 4600 m, which is equivalent to 1.2–1.4 times the average mining depth, the ground surface is expected to be affected to such an extent that a mining induced subsidence basin will be formed.

Table 8.17 Summary of predicted mine-wide ground subsidence, slope, horizontal movement, and deformation

Coal seam	2^{-2}	3^{-1}	4^{-2}	4^{-3}	4^{-4}	5^{-2}	Cumulative
Depth (m)	0–175.72 107.07	0–207.32 113.57	0–256.60 120.51	0–270.82 152.25	0–193.49 114.75	0–302.65 156.97	
Thickness (m)	3.20–6.40 4.25	2.50–3.25 2.91	1.15–3.50 3.08	0.80–1.66 1.29	0.80–1.13 1.00	1.15–6.00 5.23	
Range and average of subsidence (mm)	1983.70–3,967.40 2636.15	1549.76–2112.18 1850.13	712.89–2274.65 1943.73	584.91–1078.84 834.72	519.92–734.39 632.89	747.39–3899.41 3368.72	991.85–12,877.54 5531.17
Slope (mm/m)	24.21–91.46 53.69	21.08–59.89 36.63	11.29–87.97 36.22	8.87–24.04 14.06	7.22–48.88 14.21	14.78–127.15 52.47	19.47–133.93 65.82
Curvature (10^{-3}/m)	0.41–5.57 1.86	0.37–2.93 1.19	0.27–5.93 1.13	0.13–1.47 0.42	0.14–5.33 0.62	0.22–6.30 1.38	0.53–7.78 1.99
Horizontal movement (mm)	614.95–1229.89 817.21	480.43–654.78 573.54	221.00–705.14 602.56	181.32–334.44 258.76	161.18–227.66 196.2	231.69–1208.82 1044.3	307.47–1969.66 1194.84
Horizontal deformation (mm/m)	11.41–43.09 25.30	9.93–28.22 17.26	5.32–41.45 17.07	4.18–11.33 6.62	3.40–23.03 6.7	6.97–59.92 24.72	9.17–63.11 31.01

Table 8.18 Summary of 5-year prediction of ground subsidence, slope, horizontal movement and deformation for mining faces

Mining panel	24206	15201	15202	15203	15204	25201	25202	25203
Average depth (m)	100	130	120	110	110	140	140	140
Thickness of coal seam (m)	2.8	6.8	6.8	6.8	6.8	6.8	6.8	6.8
Range and average of Subsidence (mm)	991.85–2169.6 1694.41	3719.43	3719.43	3719.43	3719.43	3719.43	3719.43	3719.43
Slope (mm/m)	19.47–34.75 29.64	44.76–94.38 69.19	41.84–94.38 60.48	41.84–86.04 59.88	51.76–86.04 66.66	51.2–55.37 53.29	51.2–60.48 55.84	45.1–52.38 48.74
Curvature (10^{-3}/m)	0.58–0.96 0.79	0.82–3.64 2.12	0.72–3.64 1.67	0.72–3.03 1.61	1.09–3.03 1.90	1.07–1.25 1.16	1.07–1.49 1.28	0.83–1.12 0.98
Horizontal movement (mm)	307.47–672.60 525.27	1153.02	1153.02	1153.02	1153.02	1153.02	1153.02	1153.02
Horizontal deformation (mm/m)	9.17–16.37 13.97	21.09–44.47 32.6	19.72–44.47 28.5	19.72–40.54 28.22	24.39–40.54 31.41	24.13–26.09 25.11	24.13–28.5 26.31	21.25–24.68 22.96

In the loess hilly and gully area, the valleys such as Wulbengshigou, Lucaogou, Maolutangou and Kentielinggou have deep entrenchments varying from 20 to 50 m. The slope ranges from 30° to 50°, while some are even upright. This area is composed of sandy, soil and rocky high slopes with poor stability. With coal mining, the formation and expansion of goafs, the surface will undergo various degrees of deformation. Ground fractures are formed in edge areas of the valleys, around coal pillars at the mine boundary, roadways, and ventilation shaft. The ground fractures lead to slope instability, collapse, and landslides. Stepped ground fractures and graben ground fractures and ground subsidence are formed in larger ridges. Such geohazards pose threat to Shuitougou Village, Kentieling Village, Maolutan Village, and Lucaogou Village. They also pose threat to the irrigation land in Kentielinggou, Maolutangou and Lucaogou. Developed in other loess areas are gullies with large reliefs and slopes and are prone to ground fractures, collapses, and landslides. The impacts on dry land and the normal farming of villagers are moderate.

Coal mining results in formation and expansion of the mined-out areas. A mining induced ground subsidence basin will be formed. The stepped and graben-type ground fractures will be formed in the outer marginal area of the basin along the coal pillars. The surface slope of the inner marginal area will increase. The center will consist of ground subsidence and ground fractures that tend to close or leave only traces. The damage to Caojiepan village will be large. Other places are relatively flat, the damage and threat are low.

The surface deformation is closely related to the advancing speed and distance of the underground mining face. Generally, when the advancing distance equals to 1/4H where H is the average mining depth in m, the mining effect will reach to the surface and cause surface deformation. The duration of surface movement (T in days) can be calculated in the following equation:

$$T = 2.5\,H$$

The estimated duration of surface movement after coal seam mining is as follows:

- 2^{-2}: 0.3–1.2 years
- 3^{-1}: 0.4–1.4 years
- 4^{-2}: 0.3–1.8 years
- 4^{-3}: 0.3–1.9 years
- 4^{-4}: 0.2–1.3 years
- 5^{-2}: 0.4–2.1 years

Surface deformation over working faces after five-year mining: Based on the mine plan, the following working faces will be completed in five years:

- #4-2 coal seam: 24206 working face.
- #5-2 coal seam: 15201, 15202, 15203, 15204, 25201, 25202, 25203.

Ground deformation will occur with the advancement of the mining face. The settlement center is basically the same as the mining center. The range of the moving basin expands outward on the plane beyond the goaf itself. The #4-2 coal seam

Table 8.19 Prediction of surface deformation at affected villages

Village	Depth to thickness ratio	Slope (mm/m)	Curvature (10^{-3}/m)	Horizontal deformation (mm/m)	Destruction Level	Household affected	Population affected
Shuitougou	18.1	66.3	1.5	31.2	IV	73	256
Kengtieling	14.9	80.4	2.4	37.9	IV	83	314
Maolvtan	16.4	73.1	2.6	34.4	IV	59	202
Lucaogou	12.9	98.5	4.1	46.4	IV	254	997
Caojiahuopan	16.5	113.9	3.2	53.7	IV	14	48
Total						483	1817

Note The number of damaged households and population are the data of the administrative village within the coal mining area

working face extends approximately 4–70 m, whereas #5-2 coal seam working face extends approximately 60–95 m. The estimated duration of surface movement after coal seam mining is as follows:

- #4-2 coal seam: 0.6 year
- #5-2 coal seam: 0.6-1.2 years.

Surface deformation in villages: There are sparsely populated and scattered villages within the mining area. The main villages are located in Kentielinggou, Maolutangou and Lucaogou. The villagers' houses are made of brick-concrete structure and brick-wood structure. There are also a few reinforced earth cave dwellings. The surface deformation resulting from the mining of coal seams will inevitably affect the lives and property safety of these villagers.

Table 8.19 summaries the deformation parameters at the main threatened villages and the potential hazard levels. The results show that all villages can be severely damaged in grade IV. The number of damaged households is 470, and the population is 1764. The potential risk is high.

The subsidence within the first 5-year mining is relatively minor. As a result, the potential risk and impact are small.

8.4.2 *Predictive Assessment of Aquifers*

8.4.2.1 Calculation of Three-Zone Height in Overlying Formations

Mining of the coal seams results in three zones in the overlying formations, i.e., caved-in zone, water-conductive fracture zone, and subsidence zone. Under certain conditions, the overlying aquifer and surface water will be affected because of the three zones.

The main aquifers in the assessment area include alluvial phreatic porous medium aquifers in gully terraces and floodplains, aeolian sand and Salawusu sand and porous medium-fracture aquifers in burnt rocks. The groundwater in these aquifers accumulates in low-lying places and discharges in the form of springs. They are the main recharge source to rivers and gullies. The primary surface water is Lucaogou River, and other tributaries have seasonal flows only.

The impact of overburden movement and deformation on the aquifer and surface water is controlled by the height of the caved-in zone and the water-conductive fracture zone. The subsidence zone consists of rocks with plastic deformation and their hydraulic properties are not affected. The impact is predicted and evaluated for coal seams #2-2, #3-1, #4-2, #4-3, #4-4, and #5-2. The calculation equations adopt the models recommended in the regulation on buildings, water bodies, railways and main shafts and coal pillar retention:

- Prediction of the height of the caved-in zone:

$$H_m = \frac{100\sum M}{4.7\sum M + 19} = 2.2$$

where:

H_m Caved-in height (m)
M Mining thickness of coal seam (m).

- Prediction of the height of the water-conductive zone:

$$\text{Mode 1}: H_{Li} = \frac{100\sum M}{1.6\sum M + 3.6} + 5.6$$

$$\text{Mode 2}: H_{LI} = 20\sqrt{\sum M + 10}$$

- Prediction of the height of the protective layer and water resistance coal and rock pillar:

$$H_m = 3\left(\frac{\sum M}{n}\right)$$

where:

H_b Height of protective layer
M Mining thickness of coal seam for each layer
n Number of mining layer.

$H_{sh} = H_{li} + H_b$

where:

Table 8.20 Prediction summary of caved-in zone and water-conductive fracture zone

Coal Seam	Mining thickness (m)		Height of water-conductive zone (m)		Height of caved-in zone (m)	Height of protective layer (m)	Height of coal and rock pillar (m)
			Mode 1	Mode 2			
2^{-2}	Maximum	6.00	51.05	58.99	14.91	10.50	69.49
	Minimum	3.50	43.64	47.42	12.07	9.00	56.42
	Average	3.86	44.70	49.07	12.48	10.29	59.36
3^{-1}	Maximum	3.18	42.20	45.67	11.57	9.54	55.21
	Minimum	2.50	38.49	41.62	10.33	7.50	49.12
	Average	2.86	40.54	43.80	11.01	8.58	52.38
4^{-2}	Maximum	3.50	43.64	47.42	12.07	10.50	57.92
	Minimum	1.15	26.74	31.45	6.91	3.45	34.90
	Average	3.01	40.82	44.44	11.14	9.22	53.66
4^{-3}	Maximum	1.66	32.13	35.77	8.39	4.98	40.75
	Minimum	0.90	23.46	28.97	6.07	2.70	31.67
	Average	1.23	27.43	32.03	7.11	3.68	35.71
4^{-4}	Maximum	1.10	26.12	30.98	6.75	3.30	34.28
	Minimum	0.80	21.99	27.89	5.71	2.40	30.29
	Average	0.99	24.72	29.92	6.39	2.98	32.90
5^{-2}	Maximum	6.00	51.05	58.99	14.91	18.00	76.99
	Minimum	1.15	26.74	31.45	6.91	3.45	34.90
	Average	5.27	48.57	55.45	14.00	15.81	71.26

H_{sh} Height of coal and rock pillar
H_{li} Height of the water-conductive zone
H_b Height of protective layer.

Table 8.20 presents the input parameters and calculation results.

8.4.2.2 Mine-Wide Prediction Analysis

#2-2 coal seam: The height of the water-conductive fracture resulting from mining of #2-2 coal seam varies from 47.42 to 58.99 m. The height of the water resistance coal and rock pillar ranges from 57.92 to 67.99 m. The depth to #2-2 coal seam is between 0 and 157.72 m with an average of 107.07 m. Comparison of the data suggests that the water-conductive fractures will pass through the bedrock to connect with the Quaternary porous medium aquifer or the surface in Lucaogou and vicinity.

#3-1 coal seam: The height of the water-conductive fracture resulting from mining of #3-1 coal seam varies from 41.62 to 45.67 m with an average of 43.8 m. The height of the water resistance coal pillar ranges from 49.12 to 55.21 m with an average of

52.38 m. The buried depth of #3-1 coal seam is between 0 and 207.32 m with an average of 113.57 m. Comparison of the data suggests that the water-conductive fractures will pass through the bedrock to connect with the Quaternary porous medium aquifer or the surface in Lucaogou and Wulbengshigou and their immediate vicinities.

The distance between #3-1 coal seam and the overlying #2-2 coal seam is between 24.68 and 30.95 m with an average of 28.33 m. Mining of #3-1 coal seam results in the water-conductive fractures extending to the 2-2 coal seam goaf.

#4-2 coal seam: The height of water-conductive fractures resulting from mining of #4-2 coal seam varies from 31.45 to 47.42 m with an average of 44.44 m. The height of water resistance coal pillars ranges from 34.90 to 57.92 m with an average of 52.16 m. The depth to #4-2 coal seam is between 0 and 256.6 m with an average of 120.51 m. Comparison of the data suggests that the water-conductive fractures will pass through the bedrock to connect with the Quaternary porous medium aquifer or the surface in Lucaogou, Wulbengshigou, Maolutangou and neighboring areas.

The distance between #4-2 coal seam and the overlying #3-1 coal seam varies from 33.85 to 60.20 m with an average of 42.48 m. Mining of #4-2 coal seam results in the water-conductive fractures extending to the 3-1 coal goaf.

#4-3 coal seam: The height of the water-conductive fracture resulting from mining of #4-3 coal seam varies from 28.97 to 35.77 m with an average of 32.02 m. The height of the water resistance coal pillar ranges from 31.67 to 40.75 m with an average of 35.7 m. Based on lithology of the mining area, the water-conductive fractures may terminate in the Baode Formation or the bedrock and do not extend to the Quaternary porous medium aquifer or the surface.

#4-3 coal seam is on average 18.7 m below #4-2 coal seam with a range from 7.20 to 26.74 m. Mining of #4-3 coal seam results in water-conductive fractures extending to the 4-2 coal goaf.

#4-4 coal seam: The height of the water-conductive fracture resulting from mining of #4-4 coal seam varies from 27.89 to 30.98 m with an average of 29.92 m. The height of the water resistance coal pillar ranges from 30.29 to 34.28 m with an average of 32.90 m. The water-conductive fractures may terminate in the Baode Formation or the bedrock and do not communicate with the loose layer of the Quaternary system or the surface.

#4-4 coal seam is on average 12.26 m below #4-3 coal seam with a range from 9.40 to 22.15 m. Mining of #4-4 coal seam results in water-conductive fractures extending to the #4-3 coal goaf.

#5-2 coal seam: The height of the water-conductive fracture resulting from mining of #5-2 coal seam varies from 31.45 to 58.99 m with an average of 55.45 m. The height of the water resistance coal pillar ranges from 34.9 to 76.99 m with an average of 71.25 m. The buried depth of #5-2 coal seam is between 0 and 256.6 m with an average of 120.51 m. Comparison of the data suggests that the water-conductive fractures will pass through the bedrock to connect with the Quaternary porous medium aquifer or the surface in Kentieling valley and Maolvtan valley and their adjacent areas.

The distance between #5-2 coal seam and the overlying #4-4 coal seam ranges from 16.83 to 43.25 m with an average of 33.15 m. Mining of #5-2 coal seam results in water-conductive fractures extending to the 4-4 coal goaf.

The prediction results indicate that that the water-conductive fractures resulting from mining of a single coal seam are relatively limited. However, Cumulative effects from mining of all coal seams are significant. Over 108 km^2 of groundwater and surface water will be affected. If no water control measures are taken, the overlying aquifers or surface water will become recharge source to the mine, causing not only water hazards but also loss of water resources.

Such analyses result in a green mining concept, i.e., water-conservation mining technology. This approach recommended in the environmental impact assessment is to take the Baode Group water resistance layer as the protection object during the mining stage. The height of water-conductive fractures is controlled with such measures as retaining waterproof rock pillar, limiting mining thickness, strip mining or reinforcing the Baode Formation.

8.4.2.3 Prediction Analysis for First Five-Year Mining

The first 5-year mining occurs in a weak water-rich area. The main aquifer is composed of the Yan'an Formation sandstone. The aquifer is extremely weak and does not provide meaningful water supply. The calculation results of the height of the three zones resulting from the first 5-year mining are summarized as follows:

24206 working face: The height of the water-conductive fracture resulting from the first 5-year mining of #4-2 coal seam in 24206 working face varies from 51.05 to 58.99 m. The height of the water resistance coal and rock pillar is 76.99 m. The water-conductive fractures pass through the Yan'an Formation into the Baode Formation.

25201, 25202, 25203 working faces: The height of the water-conductive fracture resulting from mining of #5-2 coal seam at 25201, 25202, and 25203 working faces varies from 51.05 to 58.99 m. The height of the water resistance coal and rock pillar is 76.99 m. Part of the water-conductive fractures pass through the Yan'an Formation and the Baode Formation to the surface in valleys.

15201, 15202, 15203, 15204 working faces: The height of the water-conductive fracture resulting from mining of #5-2 coal seam at 15201, 15202, 15203, and 15204 working faces varies from 51.05 to 58.99 m. The height of the water resistance coal and rock pillar is 76.99 m. Most of the water-conductive fractures pass through the Yan'an Formation and the Baode Formation to the surface in valleys.

The surface areas where the water-conductive fractures extend to surface depends on the mining induced fracture angles. The affected surface dimension is calculated by the following equation:

$$L = H \times \tan \alpha$$

where:

L Diameter of affected surface area, m

H Stratum thickness, m

A fracture angle, which assumes to be 75°.

Based on the equation, the affected range resulting from mining of 24206 working face is 10 m beyond the goaf, whereas the affected range resulting from mining of 15201, 15202, 15203, 15204, 25201, 25202, and 25203 working surface is 17.99 m beyond the goaf. The mining induced cone of depression in the aquifer is calculated by

$$R = 10 \times S_w \times \sqrt{K}$$

where:

R Radius of influence, m
S_w Drawdown in water level, m
K Hydraulic conductivity, m/d.

Based on site-specific data, the calculated radius of influence in 24206 working face is 10.8 m beyond the affected range, whereas the radius of influence in 15201, 15202, 15203, 15204, 25201, 25202, and 25203 working faces is 65.2 m beyond the affected range. These calculations suggest that the phreatic groundwater will be affected in the first 5-year mining area if no protection measures are taken. The water level of the Yan'an Formation will drop, and the spring water flow will decrease or stop. However, since there are no villages in the first five-year mining area, the impact is relatively minor.

Similar to the mine-wide analysis, the environmental impact assessment recommends implementing water-conservation mining technology in the first five-year mining. The height of water-conductive fractures is controlled with retaining water resistance rock pillar, limiting mining thickness, or using strip mining.

8.4.2.4 Groundwater Loss and Reuse

The mine water inflow is 9500 m^3/d including 8600 m^3/d from the overlying aquifer of the coal seams. The amount of water is considered to be a loss to the regional groundwater resource. The water pumped out of the mine is utilized for Shenmu County Power Plant as cooling water. If 330 days are assumed for coal production, the regional groundwater loss is 3.135 million m^3.

8.4.2.5 Predictive Assessment of Groundwater Quality

Coal mining results in changes in the original groundwater quality and mixture of different water sources. Tunneling and mining affect the water quality. Coal and cuttings increase the content of suspended solids and COD in the water. The mine water is continuously pumped to the surface as the mining progresses and reused after being qualified. The mine water will not affect the surface water. Even when a small amount of water infiltrates downwards, the filtering and purification effect of the underlying rock layer will prevent adverse effect on the water quality of the aquifers below the 5th coal seam. As a result, the impact on water quality is relatively minor. Moreover, the environmental impact assessment recommends that no polluted wastewater be discharged after the mine is put into operation.

8.4.3 Evaluation of Impact on Topography and Landscape

8.4.3.1 Impact on Topography and Landscape for Mine Site and Road Construction

The mine construction and mining will change to a large extent the original natural landscape in the direct implementation area. For example, the excavation and filling of the site will inevitably cause direct disturbance to the original surface form and vegetation. The waste rock and soil generated by the excavation are directly piled on the landform, which will modify the natural landscape. The permanent occupation of the land turns the original landscape into an industrial site.

During the road construction, activities such as filling and excavation, soil borrowing, and spoiling during the roadbed construction result in man-made landscapes such as bare slopes, borrow pits, and spoil grounds, which are inconsistent with the surrounding landscape. After the completion of the road, the original landscape will be separated, resulting in spatial discontinuity of the landscape ecosystem and impact on the original landscape.

At completion of the construction, site restoration effort is implemented. The abandoned rock and soil will be covered with vegetation, and the slope will be treated. The landscape of the project area will be more diversified and promote the development of a positive ecosystem. The final impact on the topography and landform is relatively light.

8.4.3.2 Impact on Topography and Landscape in the Subsidence Area

The landforms of Hongliulin Mine include aeolian sand landforms, sand-covered loess hills, loess hills and river landforms. The predictive evaluations indicate that the total mining induced ground subsidence varies from 0.99 to 12.88 m. The subsidence resulting from the first five-year mining of #4-2 coal seam varies from 0.99 to 2.17 m, whereas the subsidence resulting from the first five-year mining of the 5-2 coal seam is 3.72 m. Subsidence will have a certain impact on the ground elevation. The resulting surface fractures, especially large fractures, will change the integrity of the original landform. Surface subsidence will also cause changes in the surface slope, especially in the marginal area of subsidence. Because the surface subsidence takes a long time and form gradually, it will not change the overall topography of the area dramatically. The overall impact on the topography and landform is relatively insignificant.

Table 8.21 Land occupation by ground construction activities

No	Land use	Unit	Quantity	Land type
1	Industrial site	km^2	28.72	Grassland
2	Ventilation shaft site	km^2	0.65	
3	Gangue dump	km^2	7.35	
4	Ground blasting material storage	km^2	2.00	
5	Connection roads to industrial site	km^2	22.59	
6	Ventilation shaft road	km^2	5.75	
7	Gangue road	km^2	2.03	
8	Road for coal sale	km^2	1.82	
9	Roads for blasting material storage	km^2	1.79	
Total		km^2	71.4	

8.4.4 *Predictive Assessment of Land Resources*

8.4.4.1 Assessment of Mine Site and Road Construction

The newly built sites of Hongliulin Mine include industrial site, ventilation shaft, waste disposal site, and blasting material warehouse. New off-site roads include external connection roads to industrial sites, ventilation shaft, gangue disposal site, and blasting material warehouse. The area and type of land occupied by the site and off-site roads are shown in Table 8.21. The total area is 71.4 km^2, and the land type is natural grassland. The impact of site and road construction on land resources is relatively minor.

8.4.4.2 Assessment of Land Resources in Mining Areas

Land resources: The impact of the land affected within the evaluation area is predicted based on the prediction results of surface subsidence and the effective depth-to-thickness ratio of the coal seam. The predicted classification results are summarized in Table 8.22. The affected areas are mainly natural grassland, shrub forest land, unused land, and dry land.

Table 8.22 Land resource damage in mine mining area

Land type and use		Area (km^2)	Extent impact
Arable land	Irrigated land	1.51	Severe
	Dry land	5.32	Relatively severe
Woodland	Shrubland	13.65	Slightly impacted
Grassland	Natural grassland	100.50	Slightly impacted
Transportation	Road land	0.40	Relatively severe
Water bodies and water conservancy facilities	Water surface	0.09	Slightly impacted
Not used	Desert	30.50	Slightly impacted
Residential land	Rural homestead	0.95	Severe
Total		152.92	

Mining in this area results in geohazards such as ground subsidence and ground fracture. The impact of mining on land resources is determined to be light to severe. The impact is severe in irrigated land and rural homestead areas, relatively severe in dry land and roadways, and less severe in natural grassland, shrubland, and sandy areas.

Impact on land resources in first five-year mining is evaluated in analogy to the surface deformation degree of several large mines of Shendong Company. Tables 8.23, 8.24 and 8.25 present the results. The main affected areas are natural

Table 8.23 Land resources affected by mining of #4-2 coal seam

Land type	Affected area (km^2)				
	Lightly impacted	Relatively severe		Severe	
		≤7°	>7°	≤7°	>7°
Arable land	1.80	0.86	0.68	0	
Grassland	70.96	34.47	28.21	0	
Transportation	0.44	0.40	0.02	0	
Total	73.20	35.73	28.91	0	

Table 8.24 Land resources affected by mining of #5-2 coal seam in North #1 Panel

Land type	Affected area (km^2)				
	Lightly impacted	Relatively severe		Severe	
		≤7°	>7°	≤7°	>7°
Arable land	9.42	0	0	0	
Grassland	257.50	60.22	15.06	0	
Transportation	2.88	0.67	2.68	0	
Total	269.80	60.89	17.74	0	

Table 8.25 Land resources affected by mining of #5-2 coal seam in South #1 Panel

Land Type	Affected area (km^2)				
	Lightly impacted	Relatively severe		Severe	
		≤7°	>7°	≤7°	>7°
Arable land	12.19	0	3.07	0.49	3.73
Grassland	268.72	23.05	42.81	25.04	37.56
Transportation	1.85	0.28	0.20	0.39	0.21
Total	282.76	23.33	46.08	25.92	41.50

grassland, followed by cultivated land and less land for residential and transportation. According to the type of land and the degree of damage, the impact of mining on land resources is determined to be mild to severe.

Surface deformation caused by underground production will cause partial soil erosion in loess hilly and gully areas, resulting in partial loss of soil nutrients. Because the subsidence occurs continuously and slowly, the loss is limited. The impact of soil erosion on land resources is relatively minor.

Impact on vegetation: The impact vegetation occurs during the construction period and coal production. During the construction, vegetation is cleared for site leveling, construction of buildings, road construction and other engineering activities. Ground excavation completely destructed the surface vegetation. In the process of coal production, at the edge of the subsidence area, due to the influence of surface fracture and subsidence terraces, the surface soil is loosened and the water retaining capacity is reduced. Such soil conditions can cause trees to tilt and fall, which affects the growth of vegetation.

The vegetation in the assessment area is mainly artemisia sphaerocephala, salix sylvestris, and white chinensis. These species are common and widely distributed. The construction will cause local loss of the original vegetation but will not change the plant community in the assessment area. No plant species will disappear because of mining. Remediation of the subsided land and implementation of a reseeding program will help recover the plant community. The impact on the vegetation is minor.

Impact on soil erosion: Surface deformation, rock and soil disturbances caused by underground production change the soil structure, physical, and chemical properties, thereby increase the degree of soil erosion. In the loess hilly and gully area, the soil erosion will be intensive, causing loss of soil. The intensified decomposition of organic matter resulted in partial loss of soil nutrients. However, because the collapse occurred continuously and slowly, the loss caused was limited.

Summary: The effective depth-to-thickness ratio for the first five-years mining is less than 30. The surface deformation is extensive, resulting in strong potential of geohazards. In the loess hilly and gully area, mining will induce ground fractures, landslides, collapses, and ground subsidence. Although the ground subsidence is potential hazardous, there is no receptor for the threat. Therefore, adverse impact is small, and the risk is small.

The ground surface deformation resulting from mining of Hongliulin Mine causes damage to the buildings of each village and hazard level is grade IV. The risk to building damage ranges from high to medium. The impacts of site construction mining on land resource and landscape are overall insignificant, and any damages can be restored at completion of mining.

If no protective measures are taken, mining of the coal seams results in the water-conductive fractures connecting to the porous medium phreatic aquifers in the Lucaogou, Wulbengshigou, Kentielinggou, and Maolutangou areas. The groundwater level in some areas drops, and the water flow in Lucaogou and Wupengshigou will decrease within a certain period. The water-conservation mining technology is recommended as protective measures to avoid adverse impact on the groundwater resources. This approach uses the Baode Group water resistance layer as the protection object during the mining stage. The height of water-conductive fractures is controlled with measures such as retaining water resistance rock pillar, limiting mining thickness, using strip mining, or reinforcing the Baode Formation.

The geological environmental impact prediction and evaluation map was compiled based on the prediction and evaluation results (Table 8.26). The entire assessment area is divided into 10 regions with varying degrees of impact including four severely affected regions, two relatively severely affected regions, and four less affected regions.

Table 8.26 Summary of environmental impact assessment

Impact zone		Area (km^2)	Elements				
Classification	Hazard level		Geohazards	Water resource	Landscape	Land resource	Prevention difficulty
I_1	Severe	13.21	• Possibility of occurrence is high • Affect Lucaogou, Lanjia Village, Hejia Village, Wulbongshi, He Zeshang, Yangjia Village, Hejia Village, Gouzhang • Affect a population of more than 100 people • Affect Changjiagou Reservoir-Lucaogou diversion tunnel • Affect rural roads	• Water-conductive fractures will connect alluvial layer in Wupengshigou, surface in Lucaogou, the Salawusu Formation in the vicinity of the valley, and the surface aeolian sand aquifer • Decrease water flows in Lucaogou Wulbengshigou gullies • Cause the groundwater level decrease • Affect production and water supply for nearby residents	Large mining induced earth fracture will change the integrity of the original landform but will not change the overall topography of the area. The impact is relatively minor	Destroy shrubland, natural grassland, irrigated land, dry land, roads, sandy land, and rural homesteads	Difficult

(continued)

Table 8.26 (continued)

Impact zone		Area (km^2)	Elements				
Classification	Hazard level		Geohazards	Water resource	Landscape	Land resource	Prevention difficulty
I_2	Severe	1.55	• Possibility of occurrence is high. • Affect Maolutan Village with a population of more than 100 people. • Affect rural roads	• Water-conductive fractures will connect the alluvial layer of the Maolutan Gully, the Sarawusu Formation in the vicinity of the valley, and the surface aquifer of aeolian sand • Affect the domestic water consumption of residents			Difficult

(continued)

Table 8.26 (continued)

Impact zone		Area (km^2)	Elements				
Classification	Hazard level		Geohazards	Water resource	Landscape	Land resource	Prevention difficulty
I_3	Severe	4.17	• Possibility of occurrence is high • Affect Shuitougou, Majialiang, Zhangjiacun, Hejiacun, Kentieling Village • Affect a population of more than 100 people • Affect the water delivery channel of Shuitugou Reservoir • Affect rural roads	• Water-conductive fractures will connect the alluvial layer of Kentielinggou, the Salawusu Formation in adjacent areas of the valley, and the surface aquifer of aeolian sand • Results in the water level of Shuitugou Reservoir to drop • Reduce the water volume in Kentieling gully • Affect residents' water supply for production and living			Difficult

(continued)

Table 8.26 (continued)

Impact zone		Area (km^2)	Elements				
Classification	Hazard level		Geohazards	Water resource	Landscape	Land resource	Prevention difficulty
I_4	Severe	0.55	• Possibility of occurrence is high • Affect the residents of Caojiehopan Village • Affect rural roads	• Water-conductive fracture zone will connect to the burnt rock fracture aquifer in Caojiahopan Village • Cause groundwater level to drop • Affect water supply to residents			Difficult

(continued)

Table 8.26 (continued)

Impact zone		Area (km^2)	Elements				
Classification	Hazard level		Geohazards	Water resource	Landscape	Land resource	Prevention difficulty
II_1	Relatively severe	1.51	• Possibility of occurrence is likely • Affect dry land and rural roads	• Water-conductive fracture zone will connect to part of the seasonal flowing water and the aquifers of the Yan'an Group • Surface water flow will decrease • Groundwater level will drop • Affect the residents' production and water supply		Destroy shrubland, natural grassland, dry land, roads and sandy land. Mainly natural grassland	Relatively difficult

(continued)

Table 8.26 (continued)

Impact zone			Elements				
Classification	Hazard level	Area (km^2)	Geohazards	Water resource	Landscape	Land resource	Prevention difficulty
II_2	Relatively severe	25.47	• Possibility of occurrence is high • Affect rural roads	• Groundwater level in the Yan'an Formation will decrease • Water-conductive fracture zone will connect to some gullies • Affect domestic water supply of residents			Relatively difficult

(continued)

Table 8.26 (continued)

Impact zone		Area (km^2)	Elements				
Classification	Hazard level		Geohazards	Water resource	Landscape	Land resource	Prevention difficulty
III_1	Less severe	56.06	• Possibility of occurrence is high • Affect rural roads	• Water-conductive fracture zone will connect to the Yan'an Formation • Result in groundwater level drop		Destroy natural grassland, shrubland and dry land. Mainly natural grassland, occupying natural grassland	Easy
III_2	Less severe	17.03	Possibility of occurrence is small				Easy
III_3	Less severe	15.53	Possibility of occurrence is small				Easy
III_4	Less severe	17.84	Possibility of occurrence is small				Easy

Chapter 9
Impact of Open Pit Mining on Karst Springs

9.1 Introduction

Much of the mining and quarrying conducted in karst areas is for the rock itself. Some carbonate rocks contain valuable supplies of water, oil, and gas. They may weather to form bauxite deposits and host ore deposits such as lead, zinc, iron, and gold. Coal is often found within thick carbonate sequences, while diamond-bearing kimberlitic pipes may penetrate through them. To the mining industry, caves, sinkholes, and conduits are nuisance to their operations and lost revenue. Presence of karst often mean extra costs for stripping, lost drill steel during blast-hole drilling, reserves that disappear as caves are encountered, non-spec fines in the aggregate, flooded quarries, high pumping costs, and machinery damage and possible personal injury as caverns collapse or ground settles under mobile equipment. To the people living adjacent to quarries, mining means noise, dust, possible water loss and contamination. It is not uncommon that dry wells and springs, destruction of caves with endangered species, or structure damage caused by sinkhole formation lead to lawsuits. Many limestone quarries are located and engineered to attempt to reduce the karst-related problems to a minimum. Because each karst aquifer is unique, the exact inter-relationship between a mining operation and a karst aquifer can only be established by site-specific circumstances.

The geotechnical and environmental impacts are generally more significant when mining occurs below the groundwater level. Mining can create relatively abrupt and unnatural hydraulic gradients in the surrounding area by forming an open space beneath the natural piezometric surface. Water may slowly flow out of rock into the mine or it may flow into the mine at a rate of hundreds of liters per second if the rock is highly fractured. In many situations, it is necessary to pump the water out of the mine to create a workable condition. Disturbance of the natural water flow by dewatering can intensify the subsurface erosion processes and directly cause ground collapse. In 1949 dewatering in a limestone quarry in Hershey, Pennsylvania, caused more than 100 sinkholes in an area of approximately 2600 ha (Foose 1953). Mine dewatering

S. Dong et al., *Methods and Techniques for Preventing and Mitigating Water Hazards in Mines*, Professional Practice in Earth Sciences,
https://doi.org/10.1007/978-3-030-67059-7_9

produces a cone of depression in the water level contours. The shape of the cone of depression is often irregular and influenced by the hydrogeologic heterogeneity of the aquifer and its boundary conditions. If the cone of depression extends to include surface streams, it is not surprising that surface water loses some of its flow to the mine.

The amount of water flowing into a mine often experiences dynamic changes as the mining progresses. When mining intercepts large water-bearing conduits, the flow rates may increase substantially by an order of magnitude or more in a very short period of time. The flow may be discolored, new seepage entry and exit points may develop, and regional groundwater level may drop suddenly. If the pump capacity is less than the inflow rate, the mine may be completely flooded, and surface manifestations may occur in the form of depressions in embankments and sinkholes in overburden. Sudden, unexpected, significant, and obvious changes to the preexisting hydrogeological regimes characterize such a karst-related flow event. It may also trigger sinkholes within its influence area, and the additional sinkholes may bring more water into the mine, thus further enhancing the subsurface erosion processes. If the intercepted conduits carry majority of the groundwater, mining has the potential to completely change the groundwater flow direction. This occurred in a limestone quarry of Harmony, Minnesota, when quarrying progressed below the water table and intercepted the major conduits. Approximately 90% of the groundwater that formerly discharged from a large spring flows into the quarry (Green et al. 2003).

When groundwater is confined in conduits of a karst aquifer, mining may occur above the water body but below the potentiometric pressure. The competent and impervious limestone rock may act as a hydrologic barrier to temporarily prevent the groundwater from flowing into the mine. As the mining gets deeper, the hydrologic barrier becomes thinner. The pressurized groundwater may break through the barrier through hydrofracturing, and seemingly small modifications to the barrier may cause the inflow. Lolcama et al. (1999) reported such a flooding incident in a limestone quarry in West Virginia. Following a blast, an inflow of approximately 2524 L/s suddenly developed into the floor of the fully operational quarry. The water was traced to originate from a river approximately 460 m away with a head difference of approximately 50 m. The sudden inflow caused a rapid decline of the regional water level in the aquifer, and a water storage basin nearby the river drained rapidly into a sinkhole.

Mining may reactivate old filled karst features. Realization of this possibility is important in evaluating its environmental impacts because the results from traditional geological and hydrogeological investigations may provide misleading information. Boring may indicate clay-filled voids; pumping tests may suggest low hydraulic conductivities of the aquifer. While these findings represent the natural conditions prior to the mine operation, long-term and persistent seepage may eventually result in open channels through features filled with residual clay or other erosional or weathering materials. Abkemeier and Stephenson (2003) reported that a mine in Missouri that had been dry for more than 50 years intercepted a sediment-filled cave system below the water table. The dewatering of the quarry created a flow which eroded the fill, opening a subterranean channel to the Mississippi River. The catastrophic inflow exceeded the pumping capacity and flooded the quarry for over a year, while simultaneously triggering sinkhole collapses in the surrounding area, one of which caused a railroad wreck, diesel spill, and injuries to railroad personnel.

Mining can occur in unsaturated or saturated zone, although defining the peak groundwater level in a karst aquifer is challenging even with several years' groundwater monitoring data. In general, quarrying in the unsaturated zone may result in relatively local impacts such as increased runoff, reduction in temporary storage and potential destruction of hydrologically inactive caves. Exceptions may be those areas with thick unsaturated zone and well-developed epikarst, especially when the mine has a strong hydraulic connection with surface water. The case study presented in this chapter illustrates the methods and techniques that were used to investigate the relationship between a surface mine and a nearby spring as part of a comprehensive environmental impact study.

9.2 Background of Environmental Impact Study

Figure 9.1 shows the location of the study area and general geology. A marble quarry had been in operation from 1964 to 2010. The surface elevation ranged from 1,000 to 1,270 feet above mean sea level (amsl), whereas the bottom of the quarry was at an elevation of approximately 750 feet amsl with seven benches, each bench being approximately 40 feet high. The area in the vicinity of the quarry is heavily forested. To the south of the quarry, relatively steep hills and canyons exist. To the east and north, the topography becomes more moderate and is mantled with light forest and grassland. The environmental impact study was for an application to amend the use permit by expanding the working area of the 80-acre quarry into an adjacent area of approximately 17.1 acres (Fig. 9.1). The purpose of this study was to provide an evaluation of potential geologic, hydrologic, and hydrogeologic environmental impacts and constraints relevant to the proposed project. The proposed quarry expansion had the potential to cause environmental impacts with respect to geologic and hydrogeologic conditions.

The quarry was in an isolated body of marble surrounded by granitic, metamorphic, and sedimentary rocks. Dissolution of the marble rocks along fractures by percolating groundwater has created a localized karst aquifer system of interconnected subterranean channels and caverns that are fed by groundwater inflow from adjacent rock bodies and by percolating surface water and streamflow from above. Surface outflow from the karst aquifer system occurs at springs located in the local creek drainage on the southerly (downslope) side of the marble body. The principal spring, referred to as SP-1, is an important water supply for a nearby city. SP-1 is located approximately 750 feet south of the quarry at an approximate elevation of 588 feet amsl. The terrain at the spring slopes steeply down to the south. It is the outlet of a complex marble aquifer system and discharges from the southern end of the marble at the marble-granodiorite contact.

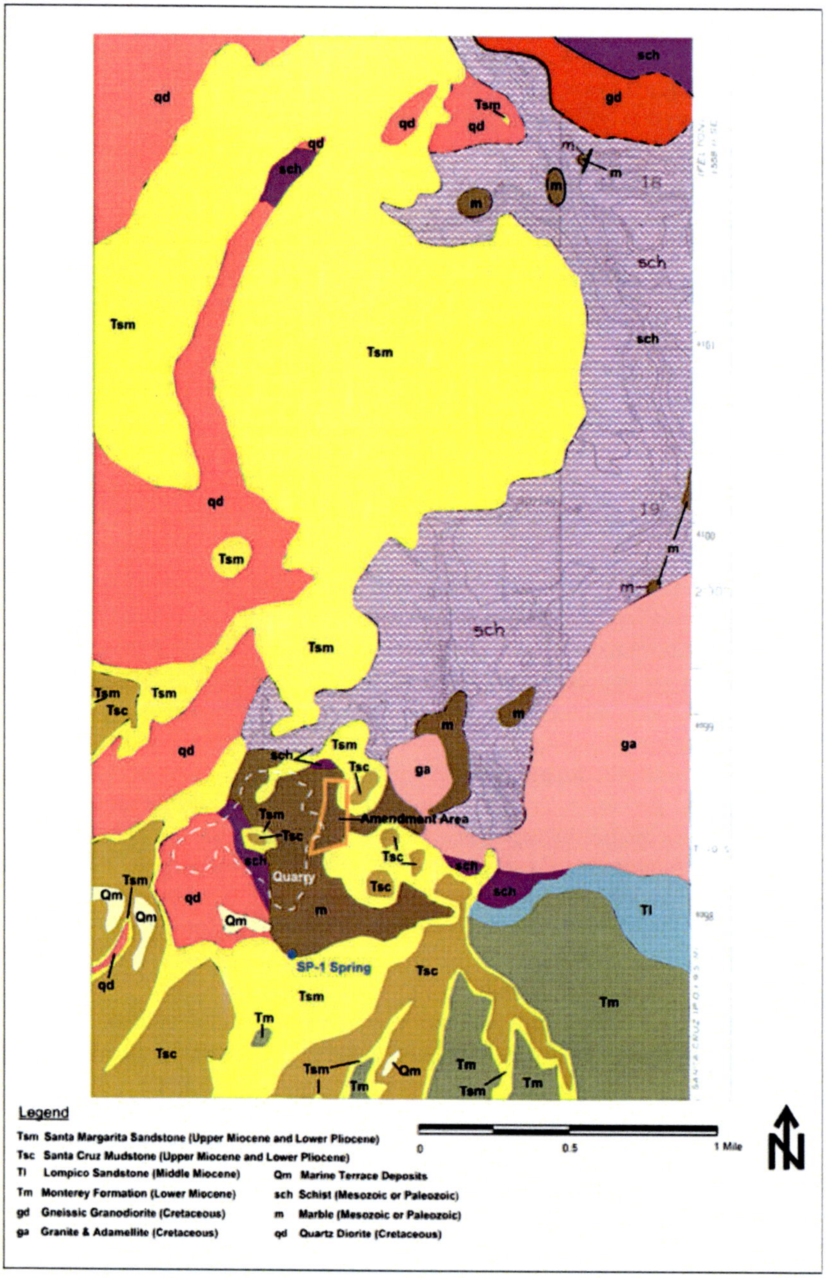

Fig. 9.1 Geology of study area

Environmental concerns associated with the proposed quarry expansion include:

- Quarry blasting could cause additional turbidity in the spring water.
- The expanded quarry activity could potentially alter flow paths in the karst aquifer and reduce flows to SP-1 Spring.
- The changes in surface or subsurface water flow could affect the stability of a landslide adjacent to the spring facility, potentially impacting the turbidity at the spring.
- Extensive grading to remove overburden from the marble deposit in the expansion area and placement of the overburden soils as fill in the existing quarry will expose a considerable area to accelerated erosion, potentially leading to increased turbidity.

9.3 Geologic Setting

The quarry property is situated on the western slope of the central Santa Cruz Mountains, part of the Coast Ranges physiographic province. The northwest-southeast structural grain of the Coast Ranges is controlled by a complex of Pleistocene- and Holocene-active faults within the San Andreas fault system. Southwest of the San Andreas fault, the Coast Ranges, including the Santa Cruz Mountains, are underlain by a large, northwest-trending, fault-bounded, elongate prism of granitic and metamorphic basement rocks. The marble being mined in the quarry is part of a metamorphic rock unit that also includes schist and quartzite. The granitic and metamorphic rock basement is overlain by a sequence of dominantly marine sedimentary rocks of Paleocene to Pliocene age and non-marine sediments of late Pliocene to Pleistocene age.

The study area is tectonically active. The Santa Cruz Mountains are cut by several active faults, of which the San Andreas is the most important. Along the coast, the ongoing tectonic activity is most evident in the formation of the series of uplifted marine terraces that sculpt the southwest slope of Ben Lomond Mountain. The Loma Prieta earthquake of 1989 and its aftershocks are recent reminders of the geologic unrest in the region.

California's broad system of strike-slip faulting has a long and complex history. Locally, the San Andreas, Zayante-Vergeles and San Gregorio faults and the Monterey Bay-Tularcitos fault zone present a seismic hazard to the subject project. These faults are associated with Holocene activity (movement in the last 11,000 years) and are therefore considered to be active. The most severe historical earthquakes to affect the project site are the 1906 San Francisco earthquake and the 1989 Loma Prieta earthquake, with Richter magnitudes of approximately 8.3 and 7.1, respectively.

9.3.1 Lithology

The lithology of the study area consists of Paleozoic, Mesozoic, Cretaceous, and Upper Miocene formations. The geologic units include three major lithologic groups: Mesozoic or Paleozoic metasedimentary rocks, Cretaceous igneous rocks, and an Upper Miocene sedimentary group (Brabb 1989). Sedimentary rocks (probably limestone and shale) of Permian age once formed the core of Ben Lomond Mountain. Before or during Cretaceous time (at least 65 million years ago) these rocks experienced folding, faulting and intrusion of igneous rocks (granodiorite, granite and related rocks), that metamorphosed the sedimentary rocks into marble and schist (Clark 1981).

During the subsequent geologic time periods (approximately 37 million years), these metamorphic rocks were folded further, uplifted, and then eroded. Sedimentary rocks including the Santa Margarita Sandstone of Miocene age were deposited on the eroded landscape (Clark 1981; Phillips 1990). Subsequent tectonic uplift of the coast range has further altered the landscape through canyon erosion, and erosional removal of the Santa Margarita Sandstone and other Tertiary formations from atop the marble body in some local areas. As a result of the complex geological processes, the sedimentary rocks make up the coastal terraces and canyon slopes to the south and form caps on some of the uplands to the north of the quarry (Wisser and Cox 1960). The metasedimentary rocks underlie the sedimentary sequence and are exposed in the quarry, hillsides, and canyons in the area.

The metasedimentary rocks are the oldest rocks exposed in the area. They are metamorphic rocks that include the marble that was being quarried at the Quarry and schist. The marble was originally deposited as limestone in a marine environment, and the schist represents fine-grained argillaceous sediments that were deposited at approximately the same time as the original limestone. The same intrusive events that metamorphosed the limestone to marble also changed the argillaceous materials to schist (Clark 1981). The marble is generally massive. Layered bedding is not readily apparent. The marble was most easily seen in the quarry area. Outcrops were also observed along local river valleys.

The schist is present over a large area to the northeast of the quarry and over a small area to the west of the quarry. Layers and lenses of schist are also present throughout the marble, as observed on the quarry walls. These schist stringers within the marble are most likely inter-bedded siltstone and mudstone within the original limestone, and they may help identify the orientation of the bedding in the marble. The overall dip of the inter-bedded schist units is to the southwest on the quarry walls, with a consistent strike to the northwest.

The Cretaceous igneous rocks are granitic rocks that have been described as granodiorite west and southwest of the quarry, and granite and adamellite northeast of the quarry (Brabb 1989). Intrusions of granitic rock are also evident in portions of the quarry walls. The intrusion of the igneous rocks provided the temperatures and pressures necessary for the metamorphosis to produce the marble and schist.

The igneous rocks form the basement rocks of the area, and the exposed marble and schist are the remnant pendants within or between the granite plutons.

Groundwater may occur within fractures in the granodiorite. Because the granodiorite appears to be producing only very limited quantities of water, it can act as a hydraulic barrier that confines most of the groundwater in the marble aquifer. The igneous rocks also help inhibit loss of water through deep flow.

The marble, schist, and granodiorite are overlain uncomfortably by the Upper Miocene Santa Margarita Sandstone, which consists of marine sandstone deposits. The Santa Margarita Sandstone has been described as thick-bedded to massive, fine to medium sand and thin-bedded conglomerates. It is gray to yellow, medium-grained, and generally well sorted in composition. It is friable to semi-consolidated. Its thickness is generally less than 50 feet but may be as much as 140 feet in some areas (Akers and Jackson 1977). The sandstone is overlain by minor outcrops of Santa Cruz Mudstone in portions of the area surrounding the quarry. This sandstone forms the erosional remnants on the tops of the nearby hills and northern uplands near the local resident communities. In addition to isolated sandstone outcrops surrounding the quarry, the sandstone also occurs south of the quarry near SP-1 Spring. Massive sandstone forms the lower slopes south of the quarry area. Further south of the quarry, the Santa Cruz Mudstone forms the high cliffs along the coast. These sedimentary formations strike northerly and dip gently to the west.

A significant landslide deposit exists in the immediate vicinity of SP-1 Spring. The spring itself appears to surface through the landslide deposits. This landslide area covers at least four to five acres. It includes debris from both the marble and Santa Margarita Sandstone. Intermittent seeps are present at the toe of the landslide. Significant shallow groundwater appears to be confined to the landslide deposits. Several seepage zones were identified in boreholes that were screened in the landslide deposit. Large-diameter boreholes at the toe of the slide indicated that groundwater does not appear to travel in significant quantities through the granitic bedrock, which acts as a boundary to water flow within the marble body. Monitoring data at several wells in the landslide deposit indicate that the water level increases, and conductivity decreases with precipitation events.

9.3.2 Geologic Structures

Geologic structure in the study area is a result of intrusion of granitic plutonic rocks at depth into metasedimentary country rock in Cretaceous time, followed by uplift and unroofing of the plutons and repeated cycles of sedimentary deposition and tectonic deformation throughout the Tertiary Period. Metamorphic rocks in the study area appear as a regularly bedded, but faulted, sequence of moderately to strongly metamorphosed sedimentary rock. The metamorphic rocks show a regular, approximate east–west strike with moderate northerly and southerly dips. However, bedded sections are discontinuous due to nearly ubiquitous faulting.

In areas with substantial marble, the metamorphic rocks are cut by orthogonal or near-orthogonal fractures, which are visible in the landscape as aligned valleys, swales, or notches in ridges. The fracture lineaments are often short and intersecting. In areas of karst, the fracture intersections are frequently marked by sinkholes. In some cases, these fractures are clearly faults with major displacement. In other cases, the fractures appear to have little displacement but have been etched into the landscape by preferential dissolution of the marble bedrock.

The marble, schist and granite rock in the quarry area have undergone extensive shearing and fracturing. The original bedding planes have been severely deformed and can hardly be recognized in the field. The quarry excavation reveals abundant fractures and joints at various orientations. The most common strike of the fractures is between N35°W and N50°W. These fractures dip vertically. They are often filled with clay.

Three faults were also inferred based on outcrop patterns before extensive excavation (Wisser and Cox 1960) and they are referred to as the north–south trending fault, the northwest-trending fault, and the northeast trending fault according to their orientations. The northeast-trending fault is coincident with the contact between the marble and schist. The northwest-trending fault is within the marble body. The fault oriented north south is roughly along the contact between the marble, schist and granodiorite and extends directly to SP-1 Spring. Aerial photographs also indicated a pronounced topographic lineament extending north-northwestwards from SP-1 Spring.

Figure 9.2 generally indicates that the surface topography and the paleo-bedrock elevation decrease from north to south. However, several large depressions are delineated based on the borehole data, and these depressions are within the marble body.

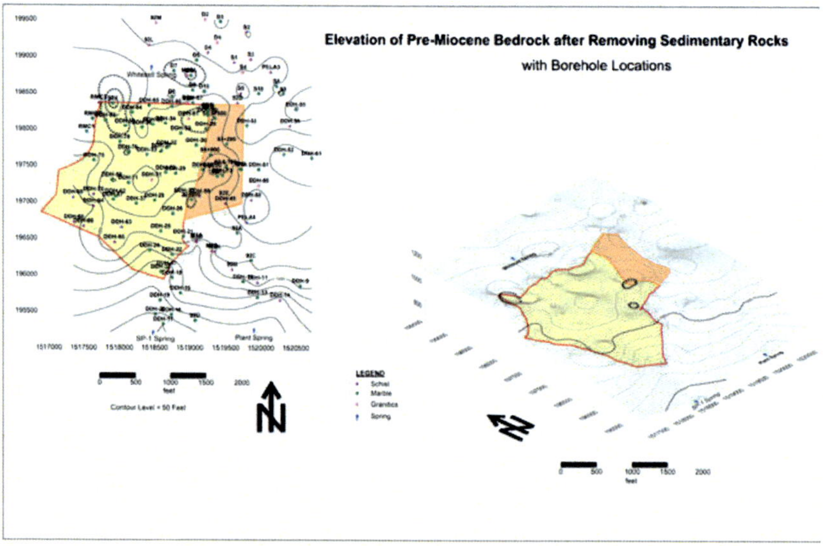

Fig. 9.2 Surface of pre-miocene bedrock

The most significant ones are along the contact between the marble and granitic rock to the north and northwest of the quarry. One is centered at M5A, and the other is centered at 92 J. These depressions appeared to be developed along fractures and joints.

Also evident is a valley trending from SP-1 Spring toward the north. This trend divides into two branches at the locations of DDH-23, DDH-67, and DDH-18. One trend continues in a northwest direction toward the depression centered at 92 J, and the other in a northeast direction, along which several depressions are aligned. These valleys are consistent with the inferred faults reported by Wisser and Cox (1960).

9.3.3 *Karst and its Influence on Groundwater Flow*

The study area has experienced episodes of subaerial erosion and submarine deposition throughout its geologic history, as indicated by the middle to upper Tertiary sedimentary sequence overlying the granitic and metamorphic basement. In the early to mid-Pleistocene, the study area was subject to sea level fluctuations of 300 to 400 feet every 100,000 years on average, caused by worldwide climatic variations and episodic glaciation in the higher latitudes. More recently, it has been elevated hundreds of feet above sea level and exposed to subaerial erosion for at least the last 500,000 years (mid Pleistocene to Holocene), based on estimated uplift rates for this section of the coastline. In other words, over the last 25 million years, the marble body in the study area has had an extraordinarily complex hydrogeologic history, characterized by dramatic fluctuations in climate and base level.

Groundwater flow in areas underlain by soluble rock, such as marble or limestone, is substantially different than groundwater flow in most other types of rock. In marble, the initial permeability may be quite low, but even slight downward flow over time will gradually dissolve the rock, forming solution channels through which water can flow more readily. In some cases, the solution channels enlarge to form caverns with underground ponds and streams. These solution cavities usually begin forming along bedding planes, fractures, or faults. In the quarry area, solution of the marble is strongly controlled by the fracture patterns. The influence of the fracture system on dissolution of marble bedrock and groundwater circulation is well documented by the alignment of sinkholes along major fractures (or faults) and particularly by the preferential location of prominent sinkholes (open or buried) at the intersection of two fractures.

The exposure afforded by the quarry walls provides a remarkable view of the hydrogeologic character of the marble over a vertical distance of 350 feet. Solution-widened fractures in the quarry walls are steeply dipping to vertical and commonly form continuous zones of solution channeling from the original ground surface through the quarry floor. The field reconnaissance of major karst features was conducted in accessible areas and the surface creeks. The main identified features included seven sinking streams, 31 springs, 27 sinkholes, and seven caves. The features relevant to the tracer tests are shown in Fig. 9.3.

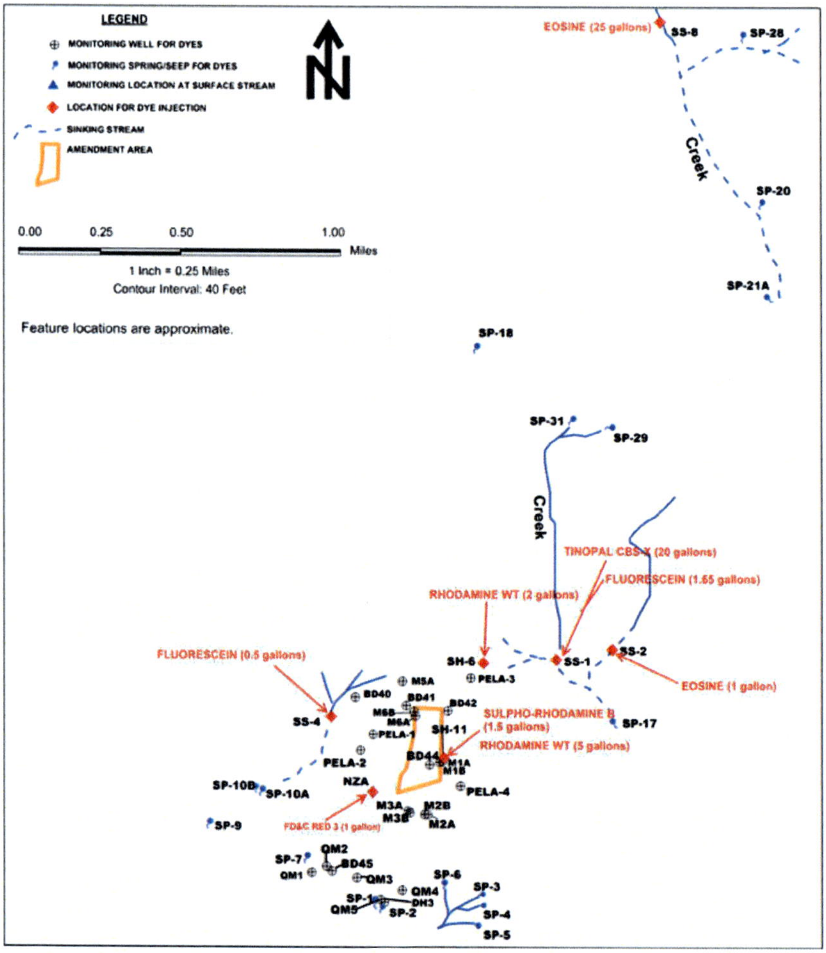

Fig. 9.3 Dye tracer injection and monitoring locations

9.4 Source Investigation for SP-1 Spring with Dye Tracing

The groundwater tracing was conducted to determine the water sources for SP-1 Spring and the preferential flow paths to the spring by introducing fluorescent dyes into sinking streams, sinkholes, and boreholes and monitoring their discharge at springs, seeps, streams, and wells. Information from this investigation has been critical in understanding the karst system at the site. It is also applicable to evaluating the potential impact of quarry operations on SP-1 Spring. The generalized capture zone for SP-1 Spring has been determined by combining the results from the tracing tests and other information related to the hydrogeology of the site. The delineation of the catchment basin for SP-1 Spring provides the information necessary to assist with the mitigation of the episodic turbidity problem at the spring. Groundwater-tracing data are also used to make recommendations for future quarrying operations to avoid adverse impacts on SP-1 Spring.

9.4.1 Previous Groundwater Tracing at Study Site

Five groundwater tracing tests had been conducted in the area prior to the current investigation. The details of each tracer test are summarized in the following section.

9.4.1.1 Tracer Test in DDH-33

On March 12, 1959, the first sugar tracer test was conducted in a diamond drill hole in the quarry area—DDH-33 (Wisser and Cox 1960). The linear distance between SP-1 Spring and DDH-33 is approximately 2450 feet. Groundwater elevation in this hole was reported at 768 feet amsl at the time of the tracing test. Five hundred pounds of sugar (300 gallons of sugar syrup in the concentration of 200,000 parts per million (ppm) were injected into DDH-33. SP-1 Spring and SP-3 Spring were monitored for sugar syrup for 14 days, and no sugar was detected at either of the monitored springs.

9.4.1.2 Tracer Test in DDH-26

On March 20, 1959, a second sugar tracer test was conducted in a diamond drill hole in the quarry area—DDH-26 (Wisser and Cox 1960). The linear distance between SP-1 Spring and DDH-26 is approximately 1670 feet. Groundwater elevation in this hole was reported at 810 feet amsl at the time of the tracing test. One thousand pounds of sugar (300 gallons of sugar syrup in the concentration of 400,000 ppm) were injected into DDH-26. Both SP-1 Spring and SP-3 Spring were monitored for sugar syrup for 14 days, and no sugar was detected at either of the monitored springs.

9.4.1.3 Tracer Test in SH-2

On May 13, 1959, a third tracer test was conducted in a sinkhole within SS-4 (most likely sinkhole SH-2) (Wisser and Cox 1960), which is located northwest of the quarry. Five hundred pounds of sugar (300 gallons of sugar syrup in the concentration of 200,000 ppm) were injected into the sinkhole. Both SP-1 Spring and SP-3 Spring were monitored for 82 h, and no sugar was detected at either of the monitored springs.

9.4.1.4 Tracer Test in DDH-19

On August 5, 1964, a fourth sugar tracer test was conducted in a diamond drill hole—DDH-19. DDH-19 is located approximately 450 feet north of SP-1 Spring. One thousand pounds of sugar (three hundred gallons of sugar syrup in the concentration of 400,000 ppm) were injected into this borehole. Only SP-1 Spring was monitored during the test. Sugar was first detected at SP-1 Spring 6.5 days after the sugar introduction and was consistently detected for a period of at least 26 days. The breakthrough curve at SP-1 Spring is very sporadic and does not represent a typical breakthrough curve. The highest concentration of sugar (10 ppm) occurred after a water sample at the injection borehole was collected and analyzed for the injected sugar.

9.4.1.5 Tracer Test in Surface Creek

A fifth tracer test was conducted in 1998 using bromide. The bromide tracer (amount unknown) was introduced into a sinkhole along a local Creek, which is most likely located at SS-1. This sinkhole was not observed in the present study, and it was likely filled up by river deposits. The total flow into the sinkhole was visually estimated at 75 to 150 gpm at the time of tracer injection. Both SP-1 Spring and SP-3 Spring were monitored over a period of 24 to 36 h, and no bromide was detected at either of the monitored springs.

9.4.2 Selection of Tracers

All chosen tracer substances should exhibit certain characteristics, and the selected tracers should not:

- Change the hydraulic characteristics of the traced water.
- Cause contamination of the water of the springs including SP-1 Spring beyond permissible standards.
- Lead to the emergence of toxic by-products following physical, chemical, or biological degradation.
- Affect tastes, odor, or color of the water at SP-1 Spring.

Other factors that were considered in selecting the tracers include conservative behavior, solubility in water, and ease of detection at low concentrations. Based on these criteria, fluorescent dyes were selected in the tracing study. Fluorescent dyes have been used as effective groundwater tracers for over 200 years, and they have been employed in most karst sites in the United States (U.S.) (Environmental Protection Agency [EPA] 1996). Fluorescent dyes are often the first choice to label groundwater for tracing studies in karst aquifers (Quinlan 1986). They are very soluble in water and can be detected at very low concentrations. Several of the dyes are sold under Drug and Cosmetic (D&C) names because they are used in such products. The fluorescent dyes are National Sanitation Foundation (NSF) certified for use in drinking water and they are often used to inspect water heating, cooling and processing systems for leaks.

There is extensive technical literature (Field et al. 1995; Smart and Laidlaw 1977; Smart 1984) demonstrating that fluorescent dyes present no health or environmental problems at concentrations five orders of magnitude or more above the method detection limit of modern protocols. A detailed discussion of the effects of the commonly used dyes on warm-blooded animals, fish and other water organisms, plants, and microorganisms is provided by Käss (1998). Based on the U.S. EPA approved methods, Field et al. (1995) concluded that none of the commonly used dyes presents a significant health or environmental problem if the dye concentrations are maintained below 1–2 mg/l for 24 h at the point of groundwater withdrawal or discharges.

Nonetheless, the commonly used fluorescent dyes have their maximum fluorescence in the visible wavelength range from 380 to 760 nm (nm). The psychological adversities that can occur due to a visible reemergence of dyes from SP-1 Spring were avoided in this study. The U. S. Geological Survey recommends that during tracing tests, the highest concentration of fluorescent dyes at the observation sites should be under 10 μg/l (Mull et al. 1988; Hubbard et al. 1982), which is far below the visible limit, to avoid such an aesthetic effect. In implementing the proposed tracer tests, the USGS guidelines were followed to minimize conspicuous coloration of the spring waters. Every effort was made to keep the concentrations of the fluorescent dyes less than 10 μg/l at SP-1 Spring while having a successful groundwater trace.

Six fluorescent dyes were used in this investigation, and their properties are listed in Table 9.1. The background conditions of these fluorescent dyes were tested at most of the monitoring locations prior to dye introduction. Although several of the dyes were detected at various locations, their concentrations are not unusually high and do not indicate that the water in the area was contaminated by any of the selected dyes.

9.4.3 Dye Injection

The dye injection was conducted in four separate phases with four dyes injected in the first phase, two dyes injected in the second phase, two dyes injected in the third phase, and one dye injected in the fourth phase. The initial dye tracing work plan included the first two phases. The third phase was added to the investigation based on the results from the first two phases. The fourth phase was added to the investigation because a new sinking stream was identified during low flow conditions. Within each phase involving more than one dye, the dyes were injected in a sequential manner to minimize their impacts on SP-1 Spring in case waters at more than one location have hydraulic connections to SP-1 Spring. This strategy would also help identify the presence of a particular dye in the dye analysis instrument spectrofluorophotometer.

Table 9.1 Selected fluorescent tracer dyes and their properties

Dye (Alternate name)	Color index name	Color index number	Chemical abstract service (CAS) number	Emission wavelengths[1] (nm)		Fluorescence intensity (%)	Adsorption tendency
				Eluted sample	Water sample		
Eosine (Eosine OJ, D&C Red 22)	Acid red 87	45,380	17,372–87–1	542	535	18	Low
Fluorescein (Uranine, Uranine C, Sodium Fluorescein, Fluorescein LT, Fluorescent Yellow/Green, D&C Yellow 8)	Acid yellow 73	45,350	518–47–8	516.5	508	100	Very low
Rhodamine WT (Fluorescent Red, Red Fluorescent Dye)	Acid red 388	None assigned	37,299–86–8	568.5	576	25	Moderate
Sulphorhodamine B (Sulfo Rhodamine B, Pontacyl Brilliant Pink B, Lissamine Red 4B, Kiton Rhodamine B, Acid Rhodamine B, Amido Rhodamine B, Fluoro Brilliant Pink)	Acid red 52	45,100	3520–42–1	576.5	582	30	Moderate
Tinopal CBS-X (Optical Brightener)	Fluorescent brighter 351		54,351–85–8	398	397	60	Moderate
FD&C Red #3	Food red 14	45,430	16,423–68–0	549	555	?	?

Sources (EPA 2002; Aley 1999; Crawford Hydrology Lab 2003)
?, No data available
1. Values are approximate only. Different instruments may yield slightly different results

The following sections discuss the locations and procedures for introducing the fluorescent dyes to trace the groundwater flow in the marble aquifer.

9.4.3.1 Locations for Dye Injection and Mass of Dye

The locations for the dye injection and the type of dye that was introduced at each location are shown in Fig. 9.3. Table 9.2 provides detailed information including the mass of dye that was used at each location. One monitoring well, two sinkholes, and four sinking streams were used for the dye injection. The same locations were used for the dye injection in the second and third phases. The fluorescent dyes used in this investigation are in liquid form, and their strengths (mass of dye [kg] per gallon of mixture) are also given in Table 9.2.

Determination of the necessary tracer mass is one of the most difficult aspects to estimate for a proposed tracer test (Field 2002). Although many estimation equations have been developed over the last century primarily for karstic terranes, the tracer mass released in most tracer studies is typically a guess utilizing the experience of the investigator (Alexander and Quinlan 1992). This is because each karst setting is unique, and a simple algebraic expression that appears to function adequately in one setting is not necessarily applicable to other settings. The equations are only useful to provide some general guidelines as to the amount of dyes to be used. Of these equations, three are particularly useful, as suggested by Worthington and

Table 9.2 Sequence of dye injection

Phase	Dye type	Location	Volume of dye mixture (gallon)	Mass of dye (kg)/Gallon mixture	Amount (kg)
I	FD&C Red 3	Borehole NZA	1.0	0.39462	0.39
	Fluorescein	Sink stream, SS-4	0.5	1.87	0.935
	Eosine	Sinking stream, SS-2	1.0	0.964	0.964
	Rhodamine WT	Sinkhole, SH-6	2.0	0.794	1.588
II	Sulphorhodamine B	Sinkhole, SH-11	1.5	0.20412	0.31
	Tinopal CBS-X	Sinking stream, SS-1	20.0	0.17962	3.59
III	Rhodamine WT	Sinkhole, SH-11	5	0.794	3.97
	Fluorescein	Sinking stream, SS-1	1.65	1.89	3.12
IV	Eosine	Sinking stream, SS-8	25	0.964	24.1

Smart (2003). These equations suggest that the amount of dye to be injected is closely related to the distance or travel time to the monitoring location, the discharge and the target dye concentration at the monitoring location. They were used for a rough estimation in this tracer design, and the amount of dyes estimated from these equations was modified based on the conditions of the specific site (sinking stream, sinkhole or borehole), estimated flow rate (if there is flow), type of overburden, type of dye, distance to SP-1 Spring, estimated discharge at SP-1 Spring, the maximum concentration to be detected at SP-1 Spring, and investigator's experience in similar settings.

The uncertainty of the dye adsorption by earth materials, the lack of adequate understanding of the groundwater flow (an answer to be obtained through dye tracing), dilution, dispersion and diffusion processes, and the vagaries of the weather (whether any rain is expected during the course of the tests) further complicate the accurate estimation of the mass of dye to be used. A safety factor is included in the mass estimation to account for the uncertainties of various factors and the complexities of the hydrogeological settings. Dye tracing may have to be repeated if a negative detection of dye is possibly caused by injection of an insufficient mass of dye.

9.4.3.2 Tracer Injection at Sinking Streams

Injection of fluorescent dyes into sinking streams is relatively simple. At SS-4, fluorescein dye was poured into the creek approximately 50 feet upstream from the main sinking point. The flow rate was approximately 5 gallons per minute (gpm). Several small dead-end side channels retained a small portion of the dye. The majority of the dye flowed with the water and sank into the subsurface. A swallow hole was visible at the bottom of the sinkhole and took most of the water into the subsurface. The dye injection lasted approximately 10 min.

Eosine dye was used at SS-2. The surface water gradually sank into the subsurface as it flowed through the marble section. No distinct sinking point was observed. The flow rate was approximately 80 gpm at the time of the dye injection. Eosine was poured approximately 20 feet upstream from the point where the valley was dry. The dye injection lasted approximately 8 min.

When Tinopal CBS-X was introduced at SS-1, a significant portion of the surface water was continuing to flow downstream of the marble section in the creek. The total flow was estimated to be approximately 700 gpm 50 feet upstream. The flow rate was approximately 300 gpm approximately 200 feet downstream. Two distinctive sinking points were observed at the time of injection and both are located on the western bank of the river at elevations that are a couple of inches higher than the riverbed. The major sinking point appears to be a cave entrance one foot in diameter. This small cave took approximately 50 gpm of water into the subsurface. It is conceivable that more water can flow into the cave when the water level in the creek is higher. Another sinking point is 10 feet south from the small cave. At least 10 gpm of water sank into the surface through the sand overburden. The continuous water flow into the subsurface at this location resulted in a small depression in the sand. Because of the flow conditions in the creek, Tinopal CBS-X was introduced into three different locations: 10 gallons of dye at the small cave entrance, 2.5 gallons of dye at the sinking point through sandy overburden, and 7.5 gallons of dye at the main stream. The total injection duration was 10 min.

The tracing was repeated during a low-flow condition when all the water in this creek sank underground at SS-1. The amount of 1.65 gallons of fluorescein dye was poured into the creek at the cave entrance just below the abandoned dam. The flow was estimated to be 15 gpm. No dye was visible beyond the sinking point in the creek.

Tracing at SS-8 took place after it was identified as a sinking stream during the low-flow condition. Twenty-five gallons of Eosine was poured at the waterfall 150 feet upstream from the sinking point. The total length of the sinking section was approximately 200 feet. The total stream flow, estimated approximately at 240 gpm, was lost over this section.

9.4.3.3 Tracer Injection at Sinkholes

Tracer injection at sinkholes is more complicated than that in sinking streams. Injection of the dyes into the sinkholes was facilitated by injection of water using a water truck. A swallow hole was visible at the bottom of sinkhole SH-6. Prior to the dye injection, approximately 2,500 gallons of water were used to pre-wet the sinkhole and the underground passages. Then two gallons of Rhodamine WT were injected at the bottom of the sinkhole in 5 min. The "primer" water was introduced into the sinkhole at a rate of approximately 50 gpm. The injected water and dye were pushed into the subsurface very quickly. An open channel was visible at the bottom of the sinkhole. After the dye injection, approximately 6,500 gallons of water were used to "chase" the dye and facilitate the injection process. The "chaser" water was injected into the sinkhole at a rate of approximately 100 gpm.

Sinkhole SH-11 was a backfilled sinkhole. In order to introduce the dye into this sinkhole, it was reopened by removing several tons of rock fragments and soil. The excavation stopped when a large void was exposed, and preliminary permeability tests indicated that the void could take at least 50 gpm of water. The void was formed between rock fragments, and it was not within the marble body. Similar to the injection process at SH-6, 2000 gallons of water were used to pre-wet the sinkhole and the underground passages; moreover, 4500 gallons of water were introduced one day earlier for the permeability test. One and a half pounds of Sulphorhodamine B was injected into this location in 5 min. After the dye injection, 7000 gallons of water were used to chase the dye and facilitate the injection process. The objective of using the chaser water is to prevent the tracer dye from getting stored in large dead-end voids and behind other obstructions.

The tracing at SH-11 was repeated using a different tracer in a larger quantity. After the sinkhole was reopened again, 3000 gallons of water were used to pre-wet the system. While the water continued to flow into the sinkhole, five gallons of

Rhodamine WT were injected. Immediately after the dye injection, 4000 gallons of water were used to chase the injected dye.

9.4.3.4 Tracer Injection at Monitoring Well NZA

Both primer and chaser water were used in tracer injection at the monitoring well NZA. NZA was drilled into the marble at the bottom of the quarry. The water injection rate was very small (<5 gpm) to avoid an excessive increase of the water level in the well. An excessive increase of the water level would create an unnatural condition. The water level was also monitored prior to the dye injection to ensure that the primer water had already moved into the aquifer. Two pounds of Red #3 were injected into the borehole via a garden hose. After the dye injection, a small amount of water was used to help push the dye into the aquifer. The total amount of water used to chase the dye was no more than 500 gallons.

9.4.4 Dye Monitoring

The main objective of the proposed groundwater tracing tests was to investigate the recharge sources for SP-1 Spring. However, SP-1 Spring was not the only location for dye monitoring in this study. An attempt was made to obtain as much information as possible to better understand the hydrogeological conditions in the quarry area, especially the flow paths to SP-1 Spring or SP-3 Spring. Therefore, the monitoring locations for the tracer tests included springs, monitoring wells, and locations at local creeks that are located downstream of the dye injection points. The monitoring effort is different at each location, depending upon its hydrogeological significance to the study.

9.4.4.1 Monitoring Locations

The monitoring locations were selected based on the results from the karst inventory and their elevations in relation to SP-1 Spring and the dye injection points. Table 9.3 lists the locations that were monitored during the tracer study. Forty-nine locations were identified for dye monitoring, including ten locations at local creeks, 13 springs (SP-1 through SP-10, SP-18, and SP-28), 25 monitoring wells, and one at the water tank in the quarry. The characteristics of the springs and creeks have been discussed in the karst inventory section. During the dye tracing, a request was made not use the water at the springs.

Table 9.3 Monitoring for fluorescent dyes

Monitoring Locations		Phases	Sampling effort
Springs/Seeps	SP-1	I, II, III, IV	The dye bugs were retrieved and replaced on a weekly basis. An automatic sampler was used to collect water samples. The sampling frequency varied from 30 min to 36 h, depending on the breakthrough process. Sampling frequency was much higher immediately after each dye injection
	SP-2	I, II	The dye bugs were retrieved and replaced on a weekly basis. The water samples were collected at the same time when their dye bugs were replaced
	SP-3	I, II, III, IV	The dye bugs were retrieved and replaced on a weekly basis. An automatic sampler was used to collect water samples. The sampling frequency varied from 1 to 48 h, depending on the dye breakthrough process. Sampling frequency was much higher immediately after each dye injection
	SP-4	I, II	The dye bugs were retrieved and replaced on a weekly basis. The water samples were collected at the same time when their dye bugs were replaced. Additional water samples were collected every 4 h for the first 10 days and then changed to daily until the end of their monitoring
	SP-5	I, II	The dye bugs were retrieved and replaced on a weekly basis. The water samples were collected at the same time when their dye bugs were replaced. Additional water samples were collected every 4 h for the first 10 days and then changed to daily until the end of their monitoring
	SP-6	I, II	The dye bugs were retrieved and replaced on a weekly basis. The water samples were collected at the same time when their dye bugs were replaced. Additional water samples were collected every 4 h for the first 10 days and then changed to daily until the end of their monitoring
	SP-7	I, II	The dye bugs were retrieved and replaced on a weekly basis. The water samples were collected at the same time when their dye bugs were replaced. Additional water samples were collected every 4 h for the first 10 days and then changed to daily until the end of their monitoring
	SP-8	I, II	The dye bugs were retrieved and replaced on a weekly basis. The water samples were collected at the same time when their dye bugs were replaced. Additional water samples were collected every 4 h for the first 10 days and then changed to daily until the end of their monitoring
	SP-9	I, II	The dye bugs were retrieved and replaced on a weekly basis. The water samples were collected at the same time when their dye bugs were replaced. Additional water samples were collected every 4 h for the first 10 days and then changed to daily until the end of their monitoring

(continued)

Table 9.3 (continued)

Monitoring Locations		Phases	Sampling effort
	SP-10A	I, II	The dye bugs were retrieved and replaced on a weekly basis. The water samples were collected at the same time when their dye bugs were replaced. Additional water samples were collected every 4 h for the first 10 days and then changed to daily until the end of their monitoring
	SP-10B	I, II	The dye bugs were retrieved and replaced on a weekly basis. The water samples were collected at the same time when their dye bugs were replaced. Additional water samples were collected every 4 h for the first 10 days and then changed to daily until the end of their monitoring
	SP-17	I, II	The dye bugs were retrieved and replaced on a weekly basis. The water samples were collected at the same time when their dye bugs were replaced
	SP-21	IV	The dye bugs were retrieved and replaced on a weekly basis. An automatic sampler was used collect water samplers at the confluence of the flows from SP-21 and SP-30. The sampling frequency varied from 24 to 48 h. After the automatic sampler was removed on December 8, 2004, the water samples were collected at the same time when their dye bugs were replaced
	SP-28	IV	The dye bugs were retrieved and replaced on a weekly basis. The water samples were collected at the same time when their dye bugs were replaced
	SP-29	IV	The dye bugs were retrieved and replaced on a weekly basis. The water samples were collected at the same time when their dye bugs were replaced
	SP-30	IV	The dye bugs were retrieved and replaced on a weekly basis. An automatic sampler was used collect water samplers at the confluence of the flows from SP-21 and SP-30. The sampling frequency varied from 24 to 48 h. After the automatic sampler was removed on December 8, 2004, the water samples were collected at the same time when their dye bugs were replaced
	SP-31	IV	The dye bugs were retrieved and replaced on a weekly basis. The water samples were collected at the same time when their dye bugs were replaced
	SP-18	IV	The dye bugs were retrieved and replaced on a weekly basis. The water samples were collected at the same time when their dye bugs were replaced
Creeks	SS-1	I, II, IV	The dye bugs were retrieved and replaced on a weekly basis. The water samples were collected at the same time when their dye bugs were replaced
	CR-2	I, II, IV	The dye bugs were retrieved and replaced on a weekly basis. The water samples were collected at the same time when their dye bugs were replaced

(continued)

Table 9.3 (continued)

Monitoring Locations		Phases	Sampling effort
	CR-4	I, II, IV	The dye bugs were retrieved and replaced on a weekly basis. The water samples were collected at the same time when their dye bugs were replaced
	CR-5	I, II	The dye bugs were retrieved and replaced on a weekly basis. The water samples were collected at the same time when their dye bugs were replaced
	CR-6	IV	The dye bugs were retrieved and replaced on a weekly basis. The water samples were collected at the same time when their dye bugs were replaced
Wells	QM-1	I, II	The dye bugs were retrieved and replaced on a weekly basis. The water samples were collected at the same time when their dye bugs were replaced
	QM-2	I, II	The dye bugs were retrieved and replaced on a weekly basis. The water samples were collected at the same time when their dye bugs were replaced
	QM-3	I, II	The dye bugs were retrieved and replaced on a weekly basis. The water samples were collected at the same time when their dye bugs were replaced
	QM-4	I, II	The dye bugs were retrieved and replaced on a weekly basis. The water samples were collected at the same time when their dye bugs were replaced
	QM-5	I, II	The dye bugs were retrieved and replaced on a weekly basis. The water samples were collected at the same time when their dye bugs were replaced
	M1A	I, II	The dye bugs were retrieved and replaced on a weekly basis. The water samples were collected at the same time when their dye bugs were replaced
	M1B	I, II, III, IV	The dye bugs were retrieved and replaced on a weekly basis. The water samples were collected at the same time when their dye bugs were replaced. In the third and fourth phases' dye tracing, water samples were collected daily
	M2A	I, II	The dye bugs were retrieved and replaced on a weekly basis. The water samples were collected at the same time when their dye bugs were replaced
	M2B	I, II	The dye bugs were retrieved and replaced on a weekly basis. The water samples were collected at the same time when their dye bugs were replaced
	M3A	I, II	The dye bugs were retrieved and replaced on a weekly basis. The water samples were collected at the same time when their dye bugs were replaced
	M3B	I, II	The dye bugs were retrieved and replaced on a weekly basis. The water samples were collected at the same time when their dye bugs were replaced

(continued)

Table 9.3 (continued)

Monitoring Locations		Phases	Sampling effort
	M5A	I, II, III, IV	The dye bugs were retrieved and replaced on a weekly basis. The water samples were collected at the same time when their dye bugs were replaced. In the third and fourth phases' dye tracing, water samples were collected daily
	M6A	I, II	The dye bugs were retrieved and replaced on a weekly basis. The water samples were collected at the same time when their dye bugs were replaced
	M6B	I, II	The dye bugs were retrieved and replaced on a weekly basis. The water samples were collected at the same time when their dye bugs were replaced
	BD-40	I, II	The dye bugs were retrieved and replaced on a weekly basis. The water samples were collected at the same time when their dye bugs were replaced
	BD-41	I, II	The dye bugs were retrieved and replaced on a weekly basis. The water samples were collected at the same time when their dye bugs were replaced
	BD-42	I, II	The dye bugs were retrieved and replaced on a weekly basis. The water samples were collected at the same time when their dye bugs were replaced
	BD-44	I, II	The dye bugs were retrieved and replaced on a weekly basis. The water samples were collected at the same time when their dye bugs were replaced
	BD-45	I, II	The dye bugs were retrieved and replaced on a weekly basis. The water samples were collected at the same time when their dye bugs were replaced
	DH-3	I, II	The dye bugs were retrieved and replaced on a weekly basis. The water samples were collected at the same time when their dye bugs were replaced
	NZA	I, II, III, IV	The dye bugs were retrieved and replaced on a weekly basis. The water samples were collected at the same time when their dye bugs were replaced. In the third and fourth phases' dye tracing, water samples were collected daily
	PELA-1	I, II	The dye bugs were retrieved and replaced on a weekly basis. The water samples were collected at the same time when their dye bugs were replaced
	PELA-2	I, II	The dye bugs were retrieved and replaced on a weekly basis. The water samples were collected at the same time when their dye bugs were replaced
	PELA-3	I, II	The dye bugs were retrieved and replaced on a weekly basis. The water samples were collected at the same time when their dye bugs were replaced
	PELA-4	I, II	The dye bugs were retrieved and replaced on a weekly basis. The water samples were collected at the same time when their dye bugs were replaced

(continued)

Table 9.3 (continued)

Monitoring Locations	Phases	Sampling effort
Water tank	I, II	The dye bugs were retrieved and replaced on a weekly basis. The water samples were collected at the same time when their dye bugs were replaced

9.4.4.2 Sampling Methods and Sampling Frequencies

Two different samples were collected at each location for dye analysis—grab samples of water and activated charcoal samples (dye receptors or "dye bugs"). The use of both sampling methods was to minimize the risk of misinterpretations and maximize the detection of any low concentrations of the dyes.

Water Samples

Grab samples were collected using 30 ml glass vials. Water samples at the creeks and springs were collected directly. At the springs, the sampling locations were chosen to be as close to the spring orifice as possible. At monitoring wells, the water samples were collected using disposable bailers. The sampling frequency was weekly at the monitoring wells. Two automatic water samplers were stationed at SP-1 Spring and SP-3 Spring to collect the water samples. A filter fluorometer was used to analyze the concentration of Fluorescein, Rhodamine WT, and Eosine to help select the water samples to be sent for laboratory analysis.

Dye Receptors

Packets of activated charcoal (commonly called dye bugs) were used in all the monitoring locations. The use of activated charcoal helps assure dye detection because the low dye concentrations found in water samples may not be measurable or sampling frequencies may not have been adequate. The ability of activated charcoal to continuously adsorb and concentrate fluorescent dyes provides a sound means for determining the occurrence of the fluorescent dyes but not a quantitative value of the concentration of the dye.

Two sets of dye bugs were placed at the springs and creeks to account for the possible loss of the bugs during their one-week deployment. The dye bugs at the springs and creeks were placed in such a way that the maximum amount of water would pass through them. The dye bugs in the monitoring wells were placed within their screen intervals. The dye concentration measured in the dye-eluant mixture cannot be related directly to its concentration in the water sample. However, bugs can accumulate and magnify low concentrations of dye passing through them over an extended period, thus effectively lowering the limit of detection.

Sampling Frequencies and Monitoring Duration

The dye bugs were retrieved and replaced on a weekly basis. The frequency of water sample collection was evaluated for each individual monitoring station, and the frequency varied as the groundwater flow system became more understood. Table 9.3 lists the sampling frequency and monitoring duration for each location.

9.4.5 *Dye Analyses and Results*

Collected dye receptors and selected water samples were analyzed analysis using a Shimazu RF-5301 PC Synchronous Scanning Spectrofluorophotometer. For the six dyes used in this study, a synchronous scan is performed where the excitation and emission monochromators are kept at a fixed wavelength separation during the scan. If the scan indicates positive results for the dyes, the concentration of the dye is then calculated based on the calibrations curves stored in the instrument or computer. If the emission spectra from two or more dyes overlay, then the spectra for each dye are separated by use of a non-linear curve-fitting computer software specifically designed for spectral separation (Crawford Hydrology Lab 2003).

Because both water samples and dye receptors were used for dye analysis, a positive detection is defined when either a typical dye breakthrough was established or when the tracer-dye concentration in two consecutive dye receptors exceeded ten times the background concentration at the same location, as recommended by Crawford Hydrology Lab (2003). If the background value is not detectable, then the quantitation limit for the dye was used for the background value (Table 9.4). Using ten times the background concentration ensures that the detection of the dye is positive, but it may also increase the false negative rate. Therefore, one more category was defined in this project—a likely positive detection. The likely positive detection was defined when the tracer-concentration in two consecutive dye receptors exceeded its background value by five times but less than ten times or when the breakthrough curve of a dye was not well defined due to low concentration of the dye.

Table 9.4 Quantitation limit of fluorescent dyes

Dyes	Quantitation limits (ug/l)	
	Dye receptor	Water sample
Eosine	0.005	0.01
Fluorescein	0.005	0.01
Rhodamine WT	0.005	0.01
Sulphorhodamine B	0.005	0.01
Tinopal CBS-X	0.1	0.1
FD&C Red 3	0.05	0.1

According to the criteria set above for positive detection, all four injected dyes in the first phase of tracing—Eosine, Fluorescein, Rhodamine WT and Red 3, were detected at SP-1 Spring or SP-3 Spring. Four monitoring wells—M1B, M5A, NZA, and QM5 had positive detections of at least one of the dyes. No dye was positively detected in other springs/seeps other than SP-1 Spring and SP-3 Spring.

The two dyes, Sulphorhodamine B injected at SH-11 and Tinopal CBS-X injected into sinking stream SS-1 during the second phase had not been detected at the monitoring station over the monitoring period. Tinopal CBS-X was detected at SP-17, which is in the same creek as SS-1 and downstream of the dye injection point. Because a portion of the water in the creek flowed to SP-17, this detection was not conclusive.

Of the two dyes injected in the third phase, the Fluorescein injected at SS-1 was detected at SP-1 Spring and SP-3 Spring. Two monitoring wells—M1B and M5A also had positive detection of the Fluorescein. The Rhodamine WT injected in SH-11 in the third phase, however, has not been detected at any monitoring stations.

The Eosine injected in SS-8 in the fourth phase was positively detected at SP-1 Spring and SP-3 Spring. The Eosine was also detected at springs SP-21 and SP-30 and monitoring well M1B.

The results of the groundwater tracing investigation are summarized in Table 9.5. The generalized routes of subsurface flow are depicted in Fig. 9.4.

9.4.5.1 Connection from Monitoring Well NZA to SP-1 Spring

Water at monitoring well NZA has a strong connection to SP-1 Spring. The Red 3 dye that was injected at NZA was detected in the water sample at SP-1 Spring 7 h after the dye injection, and its concentration reached a peak within 7.5 h, as shown in Fig. 9.5. After its peak, the concentration decreased quickly. The breakthrough curve is typical for karst aquifers and indicative of conduit flow in the marble aquifer beneath the quarry bottom. Well logs in NZA also show that this well intercepts a fracture system. The map distance between NZA and SP-1 Spring is approximately 800 feet. The groundwater flow rate based on the first arrival of the dye was estimated to be 110 feet/h, or 2,760 feet/day. Because of the tortuous flow path and the elevation difference between these two locations, this estimated water flow rate might represent the minimum flow rate. However, the dispersion process of the dye in the groundwater system further complicates the flow rate calculation. If the time to reach the peak concentration of the dye was used, the flow rate was approximately 107 feet/h, or 2,560 feet/day.

Although the dye concentration in water samples at SP-1 Spring decreased very quickly, the concentrations of Red 3 in the dye receptors were positively detected until June 25, 2004. Figure 9.6 shows the Red 3 concentrations in the dye receptors at SP-1 Spring. The dye receptors for the first two weeks had the most concentrated dye. It appears that a portion of the Red 3 was pushed into the aquifer system, but a significant amount of dye was retained in the well. The concentrations of Red 3

Table 9.5 Summary of dye tracing results

Phases	Monitoring locations	Tracer Insertion						
		Monitoring well, NZA	Sinking stream, SS-4	Sinking stream, SS-2	Sinkhole, SH-6	Sinkhole, SH-11	Sinking stream, SS-1	Sinking stream, SS-8
Phases I and II	SP-1	+7 h (1)	+66 days (2)	+ ? 29 days (1)	+3.3 day (1)	–	–	
	SP-3	–	–	+5.9 days (1)	–	–	–	
	SP-17	–	–	–	–	–	+7 days (2)	
	QM-5	+11 days (2)	–	–	–	–	–	
	M1B	–	–	+38 days (2)	–	–	–	
	M5A	–	–	–	+ ?	–	–	
	NZA	Injection point	+45 days (2)	–	+45 days (2)	–	–	
Phases III and IV	SP-1					–	+13 days (2)	+67 days (2)
	SP-3					–	+6 days (2)	+67 days (2)
	SP-21					–	–	+29 days (2)
	SP-30					–	–	+29 days (2)
	CR-2					–	–	+37 days (2)
	M1B					–	+21 days (2)	+37 days (2)
	M5A					–	+126 days (2)	–

\+ Positive detection
\+ ? Likely positive detection
– Negative detection
(1) The time is from dye insertion to first detection of the dye based on water samples
(2) The time is from dye insertion to first detection of the dye based on dye receptors

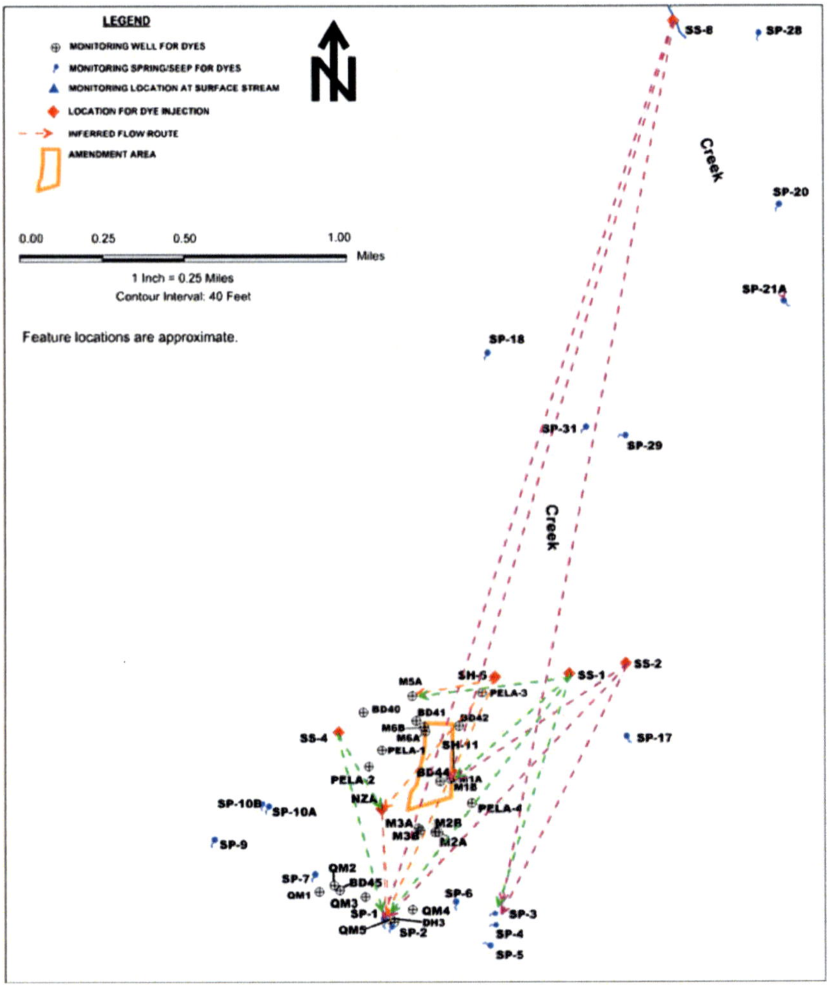

Fig. 9.4 Generalized groundwater flow routes based on tracer test results

in the dye receptors were still very high in NZA until the end of the monitoring of Red 3.

9.4.5.2 Connection from Monitoring Well NZA to Monitoring Well QM5

Red 3 was positively detected at QM5 in the dye receptors that were collected the second and third weeks after the dye injection. The dye concentration peaked 20 days after the dye injection. The concentrations are not as high at those at SP-1 Spring.

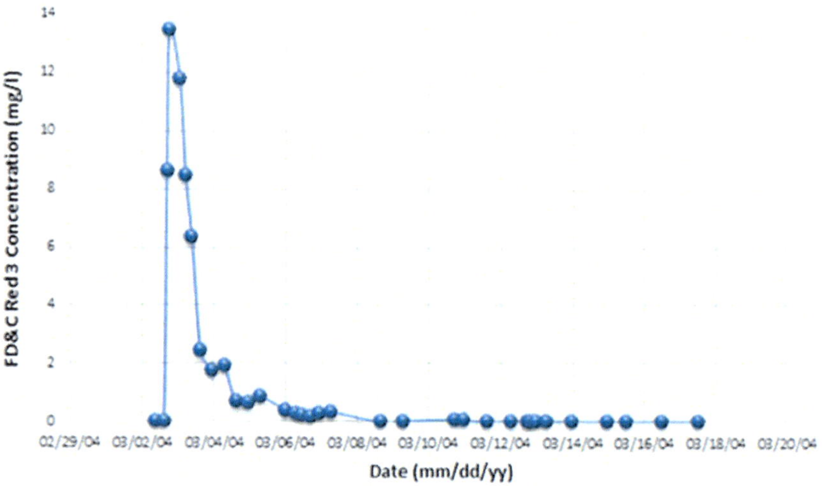

Fig. 9.5 Fluorescent dye red 3 breakthrough curve for water samples at SP-1 spring

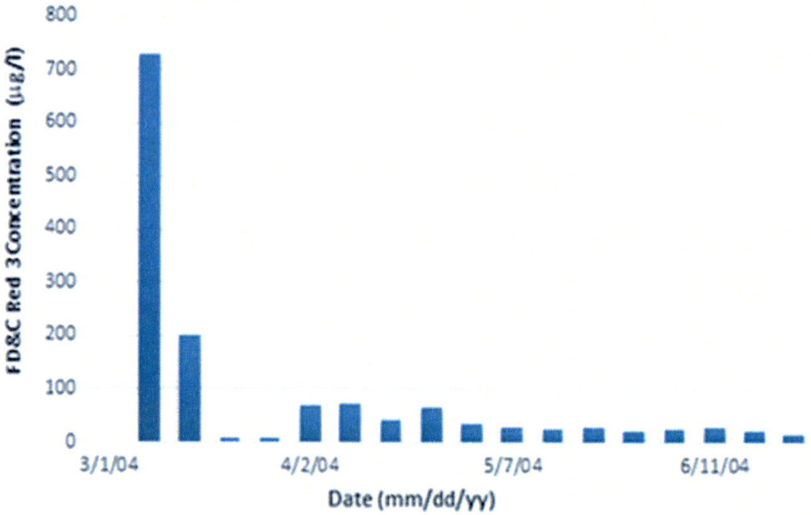

Fig. 9.6 Fluorescent dye red 3 laboratory results for dye receptors at SP-1 spring

QM5 is adjacent to SP-1 Spring but screened in the landslide deposits. It is unlikely that the Red 3-tagged water at NZA can flow to QM5 through a separate flow path that is different from that to SP-1 Spring. It is likely that the water at SP-1 Spring seeps through the landslide deposits into QM5. SP-1 Spring is hydraulically connected to the landslide deposit. Anecdotal data also support the hydraulic connection between SP-1 Spring and the landslide deposits. The travel time to QM5 is 11 days. The estimated flow rate is approximately 70 feet/day based on the first arrival of the dye. However, because the travel time was calculated from the results of dye receptors, the detection time could be earlier. The flow rate based on the time to the peak concentration is approximately 40 feet/day. Because of the combined flow route, this value is not meaningful.

9.4.5.3 Connection from Sinking Stream SS-4 to SP-1 Spring

Sinking stream SS-4 is 50 feet away from the quarry wall, where the water sinks into the surface through a sinkhole. However, no leakage of water was visible on the quarry wall. Approximately 1 kg of Fluorescein was introduced into the creek. The dye was detected at SP-1 Spring in the dye receptors collected 66 days after the dye injection. Since then the dye was detected in most of the dye receptors at SP-1 Spring until the end of the second phase monitoring (Fig. 9.7). The detected dye concentration is over ten times the background. The peak concentration occurred in the dye receptor collected on May 28, 2004, approximately 87 days after the injection. No breakthrough curve was established for this dye at SP-1 Spring because the dye concentration is below the quantitation limit for all the water samples. The map distance between the sinking point of SS-4 and SP-1 Spring is approximately 2900 feet. The average flow rate from SS-4 to SP-1 Spring was estimated to be 40 feet/day based on the first arrival of the dye. Because the water from SS-4 has to travel over 400 feet vertically to reach the saturated zone and the detection time was calculated from a weekly sampling frequency, the flow rate is not accurate and may represent the lower limit. The flow rate was estimated to be 30 feet/day based on the peak concentration in the dye receptors.

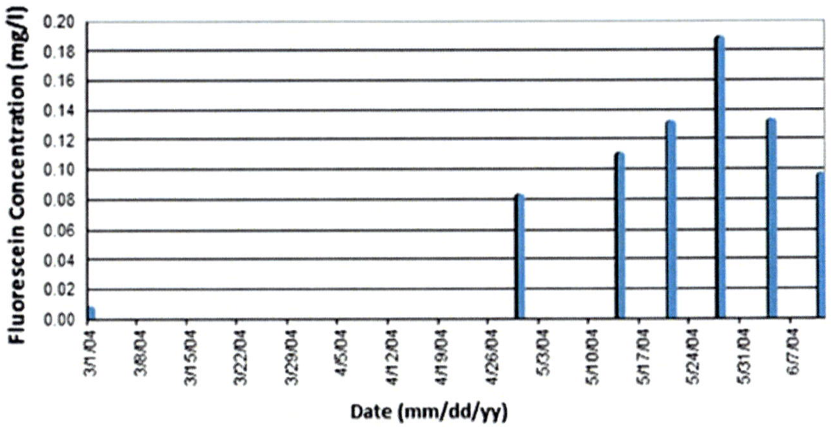

Fig. 9.7 Fluorescein dye laboratory results for dye receptors at monitoring well QM5

9.4.5.4 Connection from SS-4 to Monitoring Well NZA

Fluorescein was detected at NZA 45 days after the dye injection. The positive detection lasted three weeks. The peak concentration in the receptors occurred 52 days after the injection. Since the dye was detected at NZA before SP-1 Spring and the dye concentration at NZA appears to be higher than that at SP-1 Spring, it is possible that the water flows to NZA first and then moves to SP-1 Spring. However, it is also possible that the water takes a different route to NZA. The map distance between the sinking point of SS-4 and NZA is approximately 2,200 feet. The flow rate from SS-4 to NZA was approximately 50 feet/day based on the first arrival of the dye. Based on the peak concentration of the dye, the flow rate was estimated to be 40 feet/day.

9.4.5.5 Connection from Sinking Stream SS-2 to SP-3 Spring

A hydraulic connection was established between the sinking stream SS-2 and SP-3 Spring. Six days after Eosine dye was introduced into SS-2, it was detected at SP-3 Spring. As shown in the breakthrough curve (Fig. 9.8), the dye concentration increased very rapidly after its detection and reached its peak in 13 days after the dye injection. The presence of the dye is very persistent, and its concentration decreases gradually. The concentration in the dye receptors also indicates persistence of the dye at SP-3. The map distance between SS-2 and SP-3 Spring is approximately 4,200 feet. The flow rate was estimated to be 700 feet/day based on the first arrival of the dye. The flow rate was estimated to be 320 feet/day based on the time to reach the peak concentration.

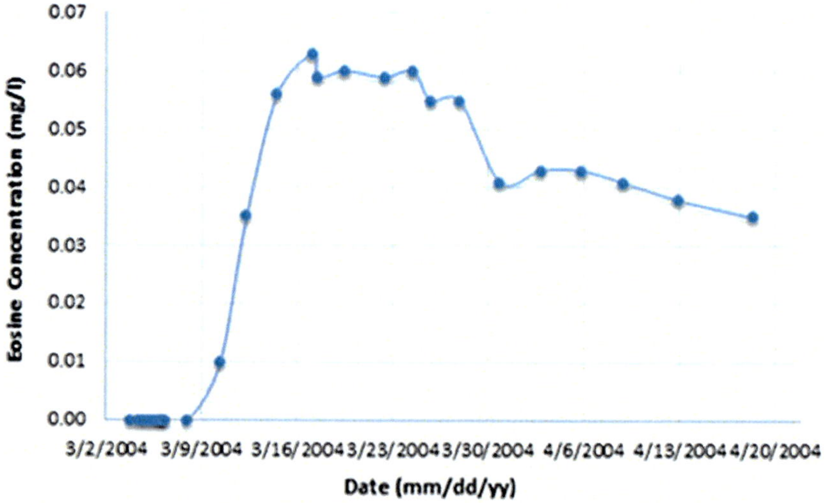

Fig. 9.8 Fluorescent dye eosine laboratory results for water samples at SP-3 spring

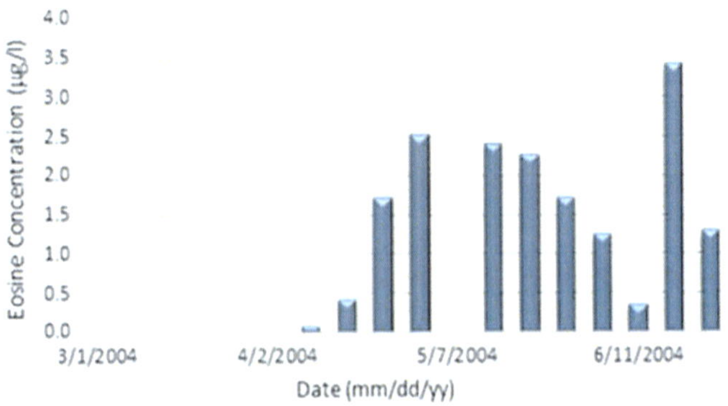

Fig. 9.9 Fluorescent dye eosine laboratory results for dye receptors at M1B

9.4.5.6 Connection from Sinking Stream SS-2 to Monitoring Well M1B

Eosine dye was detected 38 days after its injection at monitoring well M1B. The presence of the dye was persistent, and it was detected in most of the dye receptors that were collected after the first positive detection (Fig. 9.9). The peak concentration of the dye occurred on June 18, 2004, approximately 106 days after the dye injection. The map distance between the sinking point of SS-2 and M1B is approximately 3000 feet. Based on the first arrival of the dye, the flow rate was estimated to be 80 feet/day. Based on the time to reach peak concentration, however, the flow rate was estimated to be 30 feet/day.

9.4.5.7 Connection from Sinkhole SH-6 to SP-1 Spring

Rhodamine WT was injected into sinkhole SH-6. The dye was detected 3.3 days later in the water sample at SP-1 Spring. A dye breakthrough curve was established for the dye at SP-1 Spring (Fig. 9.10). The dye concentrations for all the samples were relatively low, but the curve is a typical breakthrough curve from a tracer test. The dye concentration increased relatively rapidly and reached its peak in approximately 10 days. The dye recession process was slow, and it lasted over a month before getting back to the non-detectable level. The dye was also detected in the dye receptors. The dye was not positively detected in the receptors until 16 days after the dye injection. The map distance between SH-6 and SP-1 Spring is 4100 feet. The flow rate from the Sinkhole Plain to SP-1 Spring could be 1240 feet/day based on the first arrival of the dye in the water samples. Based on the time to the peak concentration in the breakthrough curve, the flow rate was estimated to be 410 feet/day.

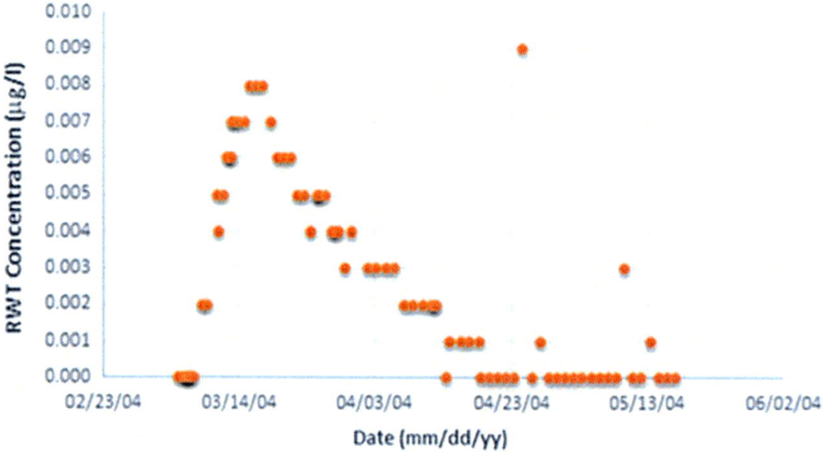

Fig. 9.10 Fluorescent dye Rhodamine WT laboratory results for water samples at SP-1 Spring

9.4.5.8 Connection from Sinkhole SH-6 to Monitoring Well NZA

Rhodamine WT was positively detected at NZA in the dye receptors 45 days after the dye injection. The peak concentration in the dye receptors also occurred 45 days after the dye injection. The dye concentrations were generally lower than those in SP-1 Spring. It seems that the water in the sinkhole area that flows to SP-1 Spring may not flow through the fracture system intercepted by NZA. It is possible that more than one fracture system is present in the saturated zone of the aquifer. The distance between SH-6 and NZA is approximately 3,250 feet. The estimated flow rate was 72 feet/day based on either the first arrival of the dye or the peak concentration of the dye.

9.4.5.9 Possible Connection from Sinkhole SH-6 to Monitoring Well M5A

The dye receptor collected at M5A on April 11, 2003, had a Rhodamine WT concentration of 0.3 μg/l, which is over 6 times the background value (0.047 μg/l). In most of the dye receptors collected in the following weeks, the dye concentration stayed at the same level. The hydraulic connection between M5A and SH-6 is probably valid. M5A is approximately 1300 feet west from SH-6. The flow rate based on the first arrival of the dye was estimated to be 30 feet/day. The flow rate based on the peak concentration of the dye was also approximately 30 feet/day.

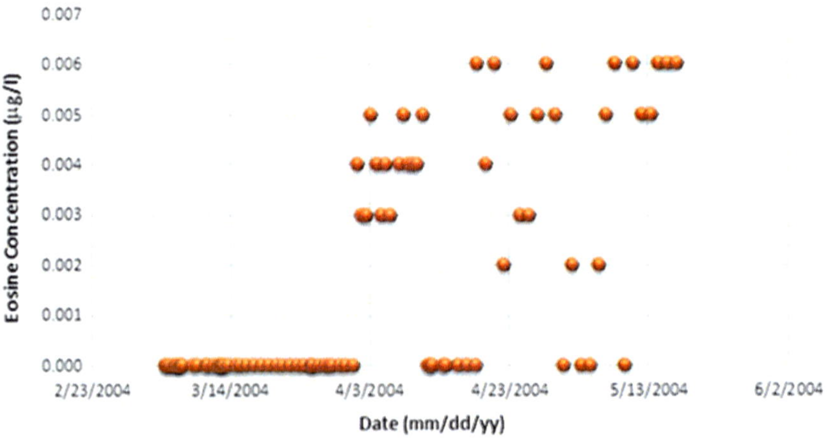

Fig. 9.11 Fluorescent dye eosine laboratory results for water samples at SP-1 spring

9.4.5.10 Possible Connection from Sinking Stream SS-2 to SP-1 Spring

Eosine was not positively detected in any of the dye receptors at SP-1 Spring. However, analysis from the water samples indicates an erratic dye breakthrough curve, as shown in Fig. 9.11. The Eosine dye had been below the detection limit until April 1, 2003 when its concentration increased to 0.004 μg/l. The concentration stayed at this level for approximately 9 days and then it dropped below the detection limit. A similar episode was repeated for subsequent water samples. The erratic nature of the curve is probably caused by the low concentration of the dye in the water samples. Although a definite hydraulic connection cannot be established between SS-2 and SP-1 Spring from the breakthrough curve, it is highly possible that water at SS-2 flows to SP-1 Spring. Well M1B in the saturated zone had already had a positive detection of the Eosine dye. The distance between SS-2 and SP-1 Spring is approximately 5400 feet. The estimated flow rate based on the first arrival of the dye from SS-2 to SP-1 Spring was 190 feet/day.

9.4.5.11 Connection from Sinking Stream SS-1 to SP-1 Spring

This connection between sinking steam SS-1 and SP-1 Spring was confirmed by injecting Fluorescein dye at SS-1 in the third phase of the dye tracing. Fluorescein was positively detected in the receptors 13 days after the dye injection. The maximum concentration of fluorescein in the dye receptors occurred three weeks after the dye injection. A small amount of Fluorescein was still detectable until January 7, 2005. The Fluorescein was also detected in the water samples. The breakthrough curve is shown in Fig. 9.12. The first water sample in which the Fluorescein dye was detected was collected at 11:00 p.m., August 24, 2004, 6.5 days after the dye injection. The

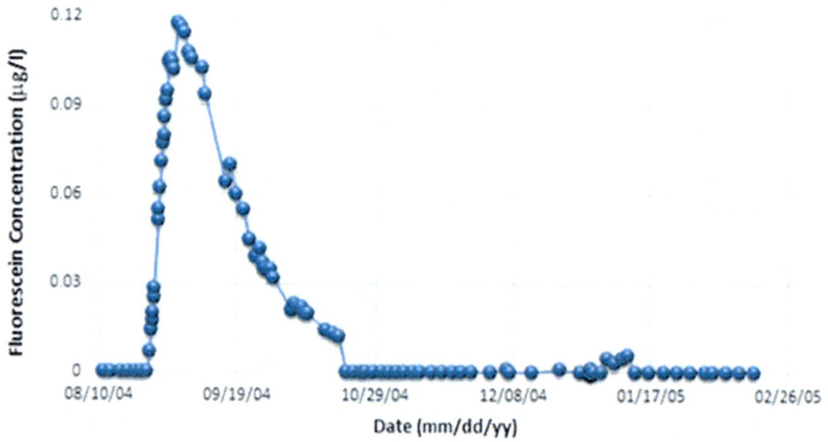

Fig. 9.12 Fluorescent dye Fluorescein laboratory results for water samples at SP-1 Spring

peak concentration occurred on September 3, 2004, i.e. 16 days after the dye injection. After the peak concentration, the Fluorescein concentration dropped gradually to the background level. The breakthrough curve is typical of karst aquifers in which conduit/fracture flow dominates.

The map distance between where the dye was injected at SS-1 and SP-1 Spring is approximately 4900 feet. The breakthrough curve established from the water samples was used to estimate the approximate flow rate for the connection. The flow rate based on the first arrival of the dye was estimated to be 754 feet/day. The flow rate based on the time to reach peak concentration in the water samples was estimated to be 310 feet/day.

9.4.5.12 Connection from Sinking Stream SS-1 to SP-3 Spring

The connection between sinking stream SS-1 and SP-3 Spring was also confirmed by the injection of Fluorescein dye at SS-1 in the third phase of the dye tracing. Figure 9.13 shows the laboratory results for water samples. The first water sample in which the Fluorescein dye was detected was collected on August 22, 2004, 4.125 days after the dye injection. The peak concentration occurred on August 26, 2004, 7.83 days after the dye injection. After the peak concentration, the Fluorescein concentration dropped gradually to the background level. The breakthrough curve is typical of karst aquifers in which conduit/fracture flow dominates. The Fluorescein was detected in the dye receptors also.

The map distance between where the dye was injected at SS-1 and SP-3 Spring is approximately 3,700 feet. The breakthrough curve established from the water samples was used to estimate the approximate flow rate for the connection. The flow rate of the groundwater based on the first arrival of the dye was estimated to be

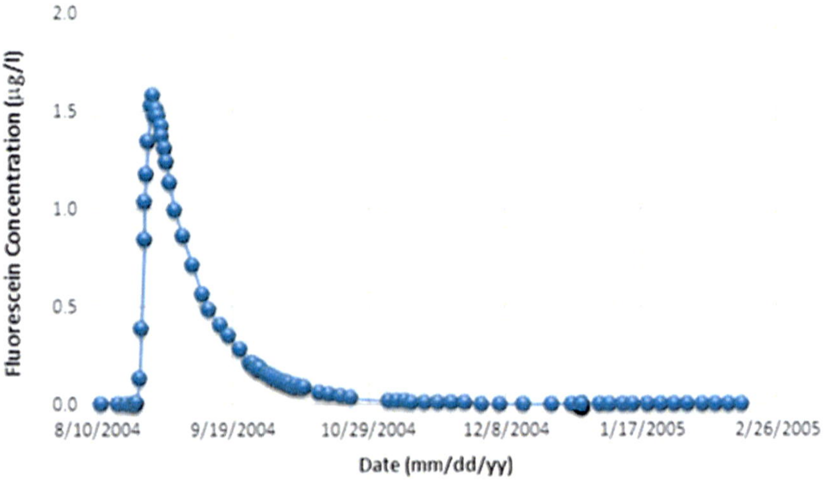

Fig. 9.13 Fluorescent dye fluorescein laboratory results for water samples at SP-3 spring

900 feet/day. The flow rate based on the time to reach the peak concentration was estimated to be 470 feet/day.

9.4.5.13 Connection from Sinking Stream SS-1 to Monitoring Well M1B

The connection between sinking stream SS-1 and monitoring well M1B was also detected from the injection of Fluorescein dye at SS-1 in the third phase of the dye tracing. Figure 9.14 shows the breakthrough curve for the dye receptors. The

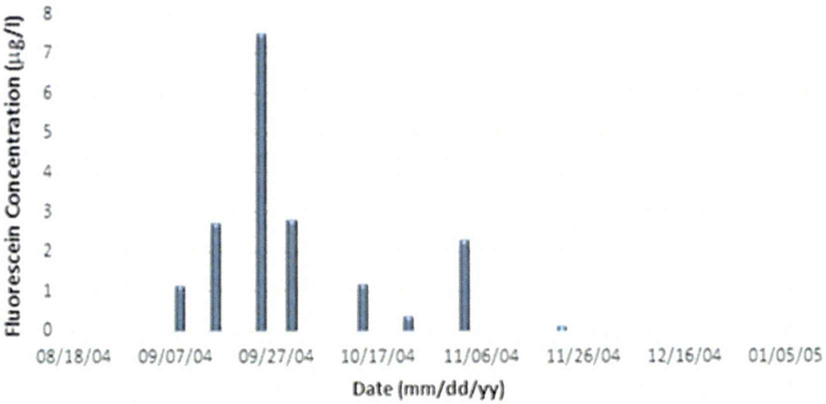

Fig. 9.14 Fluorescent dye Fluorescein laboratory results for dye receptors at M1B

Fluorescein dye began to show up in the dye receptors collected 21 days after the injection. The maximum concentration occurred 37 days after the dye injection. The dye was detectable in the dye receptors until November 3, 2004. The map distance between SS-1 to M1B is approximately 2,500 feet. Based on the first arrival of the dye, the flow rate for this connection was estimated to be 120 feet/day. Based on the time to reach the peak concentration in the dye receptors, the flow rate was estimated to be 70 feet/day.

9.4.5.14 Connection from Sinking Stream SS-1 to M5A

The connection between SS-1 and monitoring well M5A was also proven by the injection of Fluorescein dye at SS-1 in the third phase of the dye trace. Figure 9.15 shows the breakthrough curve for the dye receptors. The Fluorescein dye began to show up in the dye receptors collected 18 weeks, 126 days after the injection. The dye concentration was relatively small but more than 10 times the quantitation limit of 0.005 μg/l. The maximum concentration occurred in the dye receptor collected on January 8, 2005. The map distance from SS-1 to M5A is approximately 3,400 feet. The flow rate for this connection was estimated to be 30 feet/day if the time was based on the first arrival of the dye. If the time to reach the peak concentration of the dye was used in the calculation, the flow rate was estimated to be 20 feet/day. Because Fluorescein dye was also used at sinking stream SS-4 on March 3, there is a slight possibility that the Fluorescein detected at M5A came from SS-4. If this is the case, the travel time from SS-4 to M5A was approximately 293 days, indicative of a very slow rate of the groundwater flow. Based on all the other data, the former interpretation is much more probable.

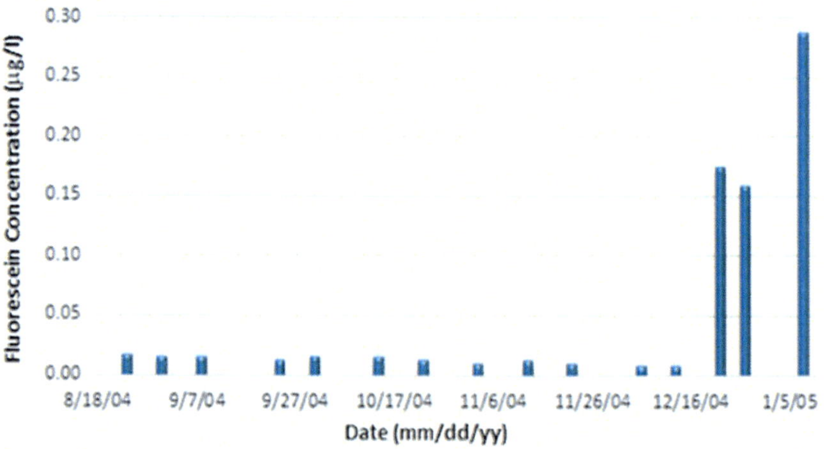

Fig. 9.15 Fluorescent dye fluorescein laboratory results for dye receptors at M5A

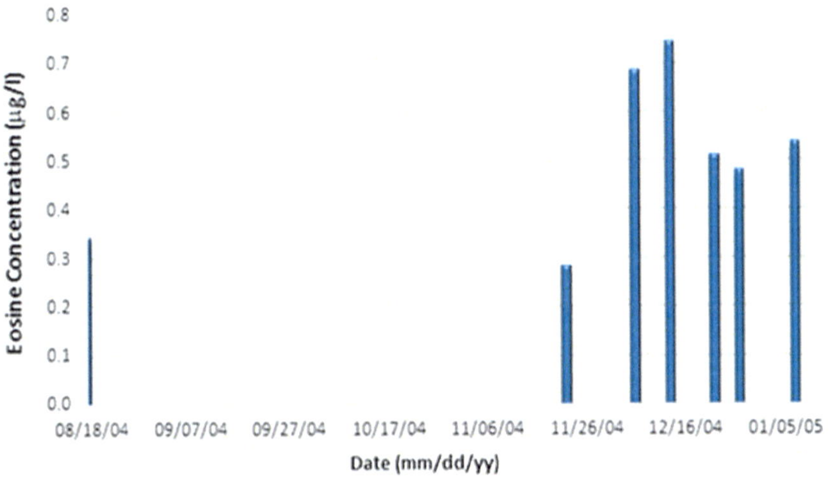

Fig. 9.16 Fluorescent dye eosine laboratory results for dye receptors at SP-1 spring

9.4.5.15 Connection from Sinking Stream SS-8 to SP-1 Spring

The connection between sinking stream SS-8 and SP-1 Spring was detected by the injection of Eosine dye at SS-8 in the fourth phase of the dye tracing. Figure 9.16 shows the laboratory results for the dye receptors. The Eosine dye began to show up in the dye receptors collected on November 22, 2004, 67 days after the injection. The dye concentration is relatively small but more than 10 times the quantitation limit of 0.005 μg/l. The Eosine dye has been detectable in all the dye receptors collected after November 22, 2004. The maximum concentration occurred in the dye receptor collected on December 13, 2004. The map distance between SS-8 to SP-1 Spring is approximately 2.75 miles, or 14,520 feet. The flow rate for this connection was estimated to be 220 feet/day if the time was based on the first arrival of the dye. If the time to reach the peak concentration of the dye was used in the calculation, the flow rate was estimated to be 160 feet/day. A breakthrough curve was not established at SP-1 Spring because the concentrations of the Eosine dye in all the water samples were below the quantitation limit of 0.01 μγ/l.

9.4.5.16 Connection from Sinking Stream SS-8 to SP-3 Spring

The connection between sinking stream SS-8 and SP-3 Spring was also detected by the injection of Eosine dye at SS-8 in the fourth phase of the dye tracing. Figure 9.17 shows the laboratory results for the dye receptors. The Eosine dye began to show up in the dye receptors collected on November 22, 2004, i.e. 67 days after the injection. The dye concentration in the dye receptor at SP-3 Spring is generally one magnitude larger than the concentration of the receptor collected at the same time as SP-1 Spring.

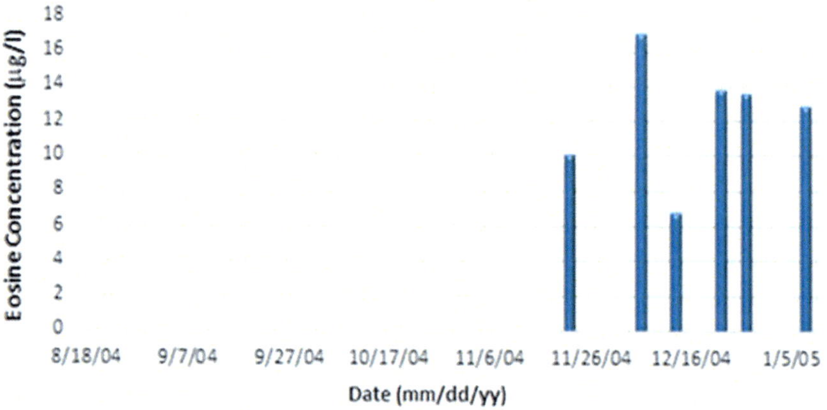

Fig. 9.17 Fluorescent dye eosine laboratory results for dye receptors at SP-3 spring

The Eosine dye has been detectable in all the dye receptors collected after November 22, 2004. The maximum concentration occurred in the dye receptor collected on December 6, 2004. The map distance from SS-8 to SP-3 Spring is approximately 2.63 miles, or 13,860 feet. The flow rate for this connection was estimated to be 210 feet/day if the time was based on the first arrival of the dye. If the time to reach the peak concentration of the dye was used in the calculation, the flow rate was estimated to be 170 feet/day. A breakthrough curve was not established at SP-3 Spring because the concentrations of the Eosine dye in all the water samples were below the quantitation limit of 0.01 μγ/l.

9.4.5.17 Connection from Sinking Stream SS-8 to Spring SP-21

The connection between sinking stream SS-8 and spring SP-21 was also established by the injection of Eosine dye at SS-8 in the fourth phase of the dye tracing. Figure 9.18 shows the laboratory results for the dye receptors. The Eosine dye began to show up in the dye receptors collected on October 15, 2004, approximately one month after the injection (29 days). The Eosine dye has been detectable in all the dye receptors collected since then, with concentrations often being over 100 μg/l. The maximum concentration of 443 μg/l occurred in the dye receptor collected on November 13, 2004. The map distance from SS-8 to SP-21 is approximately 1.13 miles, or 6,000 feet. If the time was based on the first arrival of the dye, the flow rate for this connection was estimated to be 210 feet/day, which is the same as the flow rate calculated for the flow from SS-8 to SP-3 Spring. If the time to reach the peak concentration of the dye was used in the calculation, the flow rate was estimated to be 100 feet/day.

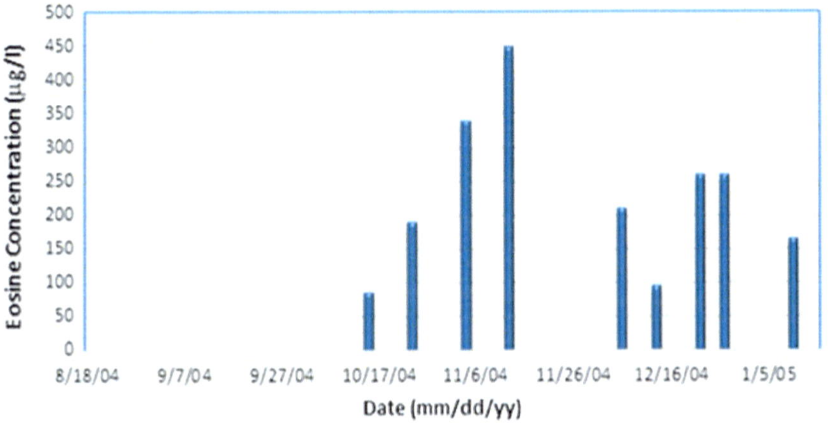

Fig. 9.18 Fluorescent dye Eosine laboratory results for dye receptors at spring SP-21

9.4.5.18 Connection from Sinking Stream SS-8 to Monitoring Well M1B

The connection between sinking stream SS-8 and M1B was also established by the injection of Eosine dye at SS-8 in the fourth phase of the dye tracing. Figure 9.19 shows the laboratory results for the dye receptors. The Eosine dye began to show up in the dye receptors collected on October 23, 2004, approximately 37 days after the injection. The Eosine concentrations in the receptors are less than 1 μg/l and the dye has been consistently detectable in the dye receptors collected since then. The maximum concentration occurred in the dye receptor collected on November 3, 2004. The map distance between SS-8 to M1B is approximately 2.3 miles, or 12,100 feet.

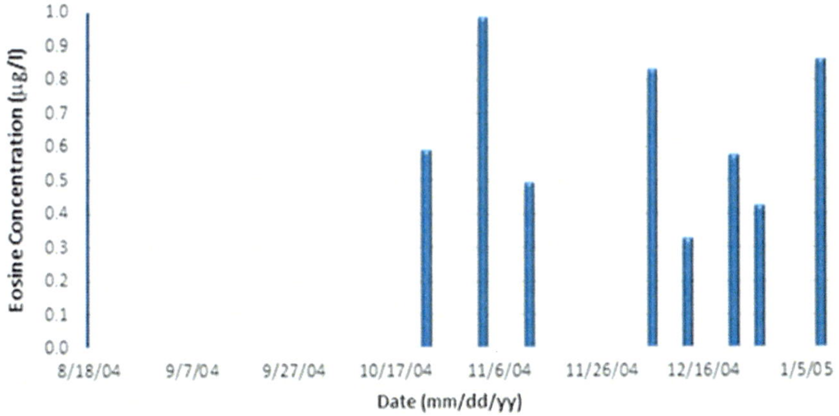

Fig. 9.19 Fluorescent dye eosine laboratory results for dye receptors at monitoring well M1B

If the time was based on the first arrival of the dye, the flow rate for this connection was estimated to be 330 feet/day. If the time to reach the peak concentration of the dye was used in the calculation, the flow rate was estimated to be 250 feet/day.

9.4.6 Summary of Dye Tracing

Table 9.5 summarizes the hydraulic connections established by the dye tracing and the estimated flow rates from the dye injection point to the monitoring point. The estimated flow rates were calculated from the tracer travel times (first dye detection and peak concentration) and the map distance between the two points. The flow rates can only be considered as an approximation. In addition, the flow rates vary with time. They can significantly increase in response to rain events. Nevertheless, these traces provide the essential information on the extent to which the various sinkholes, sinking points, or injection points are connected to SP-1 Spring or SP-3 Spring.

- Based on the tracing tests conducted in this study, SP-1 Spring is hydraulically connected to sinking streams SS-8, SS-1, SS-2, and SS-4, sinkhole SH-6, and monitoring well NZA.
- SP-3 Spring is hydraulically connected to sinking streams SS-8, SS-1, and SS-2.
- Monitoring well M1B is hydraulically connected to sinking streams SS-8, SS-1, and SS-2,
- Monitoring well NZA is hydraulically connected to sinkhole SH-6 and sinking stream SS-4.
- Monitoring well M5A is hydraulically connected to sinkhole SH-6 and sinking streams SS-1 and SS-4.

9.5 Conceptual Site Model for SP-1 Spring

In non-karst terranes there is a clean distinction between surface water and groundwater, and between surface drainage basins and groundwater aquifers. Drainage basins, which determine the catchment and runoff characteristics of surface water are defined by drainage divides set by local topography. Surface water is often only loosely coupled to groundwater system because of the slow rate of infiltration. These conceptual distinctions between basins and aquifers are blurred in karst areas like the one under study because of the integrated system of solution-enlarged fractures or conduits that carry water through the subsurface. Localized recharge to the conduit system occurs where surface streams sink underground and where sinkholes collect overland flow and direct it to the subsurface. The discharge of water from the conduit system often takes place at one or several large springs that often form the headwaters of a perennial surface stream. In karst there is a much more intimate relationship between groundwater and local surface water than in other types of aquifers.

9.5.1 Recharge Sources for SP-1 Spring

The boundaries of the delineated catchment basin for SP-1 Spring are determined from the data collected during this tracer study, the understanding of the local geology, and surface drainage patterns. The boundaries in the non-karst areas are based on the topographic divides. The SP-1 Spring catchment basin cannot be absolutely separated from the SP-3 Spring catchment basin because they share the water from SS-1, SS-2, and SS-8. These two catchment basins are hydraulically connected.

The data collected in this tracer study clearly indicate that SP-1 Spring receives recharges from sinking streams SS-8, SS-1, SS-2, and SS-4. Part of the water that sinks at SS-8 recharges SP-1 Spring, and part of the water re-emerges downstream at SP-21. During high-flow seasons, the surface waters flow past the sinking points, and only a percentage of the flow in each creek recharges the groundwater. During low-flow conditions, most of the water that sinks at SS-1 and SS-2 recharges SP-1 Spring. SP-1 Spring also receives water from SS-4, where most of the water sinks underground through a sinkhole.

Because of the uncertainties in the headwaters for the sinking streams, the total recharge area for SP-1 Spring and SP-3 Spring cannot be precisely computed. If the surface drainage areas of these sinking creeks are included in the catchment basin of SP-1 Spring, the total catchment area is approximately 3 mile^2, or 2,000 acres. This is not the catchment area solely for SP-1 Spring. A portion of the water in the catchment basin flows to SP-3 Spring and Springs SP-21. In addition, during high-flow seasons, a portion of the water in the surface creeks flows through the catchment basin and continues downstream. Therefore, SP-1 Spring gets its recharge from the delineated catchment area; however, not all the water in the basin flows to SP-1 Spring. The delineated catchment basin is an "open" basin, rather than a "closed" one.

Like any other karst aquifer, the marble aquifer includes input from surface catchments on non-karstic rocks such as the Santa Margarita Sandstone, granitic rocks, and schist. These non-karstic, but hydrologically connected rocks are referred to as "borderland." Surface runoff from the borderlands drains through surface streams that flow onto the karst area, where some sink underground at the margin of the marble or as they flow over the marble outcrop. These are the sinking stream inputs to the karst groundwater system. Little runoff is present on the karst land, which is typical in well-developed karst terranes. Overland flow during storms disappears into sinkholes to enter the groundwater system as internal runoff.

Depending on where the water comes from, the recharge sources to SP-1 Spring can be divided into two categories—allogenic recharge and autogenic recharge. Allogenic recharge derives its water from a neighboring non-karst borderland, whereas auotogenic recharge derives its water from precipitation falling on the karst land. Table 9.6 lists the various recharge sources to SP-1 Spring.

Based on the geologic map by Brabb (1989), the autogenic recharge area is approximately 200 acres and the allogenic recharge area is approximately 1800 acres. Thus, the recharge area for SP-1 Spring consists of approximately 10% karst land and 90% borderland. The existing quarry and the amendment area cover approximately 106

Table 9.6 Recharge sources for SP-1 Spring

Recharge sources		Descriptions
Allogenic recharge	Sinking streams	The groundwater tracing investigation has identified four such sinking streams that are hydraulically connected to SP-1Spring and/or SP-3 Spring. These four sinking streams are SS-8, SS-1, SS-2, and SS-4. The headwaters of these sinking streams have not been determined but likely from granitic rocks and/or on the Santa Margarita Sandstone southwestern side of Ben Lomond Mountain
	Sinkhole	Many sinkholes have been reported in the study area. The most significant sinkholes are on a sinkhole plain where dye tracing in one of the sinkholes (SH-6) indicates that this area is hydraulically connected to SP-1 Spring. The estimated flow rate is over 1,000 feet/day based on the first arrival of the injected dye at SP-1 Spring, whereas the estimated flow rate is approximately 410 feet/day based on the time at the peak concentration of the dye at SP-1 Spring. The sinkhole plain is developed over the marble but at the contact with Santa Margarita Sandstone. It collects water from the local surface areas and the discharges from springs. Because the main water sources are from the non-karstic borderland, the recharge at the sinkhole plain is considered to be allogenic
	Inter-aquifer transfer	Infiltration in the borderlands can produce aquifers in the non-karstic rocks. The water from these aquifers can move into the marble in the subsurface. No evidence is available for this inter-aquifer transfer; however, the marble is embedded in the granitic rock and is in contact with Santa Margarita Sandstone. The non-karstic aquifers located north and northeast have the potential to recharge the marble
	Landslide deposits	The tracer test at NZA and previous observations indicate that the landslide deposits and SP-1 Spring are hydraulically connected. When the water level in the landslide deposits is higher than that at SP-1 Spring, it is conceivable that SP-1 Spring gets recharge from the landslide

(continued)

Table 9.6 (continued)

Recharge sources		Descriptions
Autogenic recharge	Internal drainage through sinkholes	The dye tracing at sinkhole SH-11 does not indicate that this area is actively connected to SP-1 Spring. It is likely that the injected and the associated water are temporarily ponded in the unsaturated zone. Although the sinkhole has been reported to carry the overland runoff into the subsurface, it does not seem to have a drain that can efficiently carry the water to the saturated zone and then to SP-1 Spring. Water ponding at the bottom of a sinkhole is not uncommon (White 1988), especially in the study area where the saturated zone is underlain by over 300 feet of rocks that have fractures and voids
	Diffuse infiltration through soils and Santa Margarita sandstone	Rainwater may penetrate the soil on the karst surface as a diffuse infiltration. Observations on the quarry walls indicate that the soil–bedrock contact is irregular, and the marble consists of fractures that are filled with soil. Some soil-filled fractures have stains of water flow. The bedrock itself is relatively impermeable, and the open pathways available to carry water to the groundwater are limited to fractures. As a result, there is developed a temporary perched water body in what is called epikarst zone. The residence time in epikarst storage is typically weeks (Gunn, 1986; Williams, 1985), sometimes decades (Even et al. 1986) The epikarst zone, sometimes called the subcutaneous layer, is the dissolutionally weathered upper portion of the bedrock in carbonate landscapes. Epikarst zones store water seasonally or after major precipitation periods. Some sinkholes are discrete water input points into the epikarstic zone On the karst land, Santa Margarita Sandstone caps a small area. Exploratory boreholes indicate the presence of isolated water in the sandstone. Significant recharge to the marble from the caprocks does not seem to occur on an annual basis
	Quarry	The dye tracing at monitoring well NZA indicates that the water below the quarry floor is hydraulically connected to SP-1 Spring. Marble is exposed at the bottom of the quarry and the quarry floor collects rainwater. Standing water was observed on the quarry floor after rains. The marble itself has very low permeability, as indicated by the rudimentary aquifer tests in monitoring wells. If there is any water flowing into the saturated zone, the recharge is through fractures. Fine-grained "filtering" materials were placed on the quarry floor in 1998 to reduce the infiltration rate

acres in the karst land, which is approximately 5% of the total catchment area. The vast majority of the recharge for SP-1 Spring is allogenic.

The flow of SP-1 Spring has been measured intermittently since 1921. The annual average discharges for 1921, 1922, and 1931 ranged from 0.8 million gallons per day (mgd) to 1.2 mgd. The flow at SP-1 Spring for the period between 1961 and 1965 varied from 1 mgd to 1.34 mgd. These measurements were taken prior to 1968 when the quarry was opened. From 1988 to 1997, the discharge data are based on the production volumes at SP-1 Spring and the annual average flow varied from 0.76 mgd to 1.31 mgd, although not all the flow was measured. The time frames in which valid discharge data are available for 1999, 2000, 2002, 2003, and 2004, and their annual average flows were estimated to be 2.01 mgd, 1.07 mgd, 1.13 mgd, 1.41 mgd, and 1.53 mgd, respectively. Based on the aforementioned data, the annual average flow ranges at SP-1 Spring have been comparable before and after the quarry was in operation.

9.5.2 Groundwater Pathways to SP-1 Spring

In the marble, water is present in both an unsaturated zone and a saturated zone, as indicated by the groundwater monitoring data and isotope data and as supported by the groundwater tracing investigation. SP-1 Spring drains the water from the saturated zone that is characterized by fracture/conduit flow. The unsaturated zone is highly fractured and well developed. Many voids (including caves) are present. Isolated water levels exist, as measured in the monitoring wells that are screened in the unsaturated zone. Because of the great thickness of the unsaturated zone and the high porosity value, it has a great capacity for storing water. Following infiltration through topsoil and internal runoff through sinkholes on the karst land, water can be collected and temporarily stored in the unsaturated zone, until it subsequently recharges the saturated zone. A small proportion of water recharged at sinking stream SS-1 and the sinkhole plain may flow into the unsaturated zone, as indicated by the possible hydraulic connections between SS-1 and M5A and SH-6 and M5A. However, these connections are relatively poor, with the flow velocities ranging from 20 to 30 feet/day.

The data collected during this tracer study do not indicate a strong hydraulic connection between the unsaturated zone and the saturated zone. Two tracing tests through sinkhole SH-11 did not yield any positive detection of the dyes at SP-1 Spring and SP-3 Spring. Table 9.7 lists the main characteristics of the karst water system and their supporting evidence.

Like any other karst aquifer, the groundwater flow in the saturated zone of the marble ranges from conduit flow to diffuse flow. The flow is convergent to SP-1 Spring. The flow is concentrated into larger conduits closer to the spring. Although it is impossible to pinpoint the exact locations of the active conduits, the dye tracing data indicate that they may be located at, or near, NZA and M1B. The active conduits often lie very close to the local drainage level as observed in many active karst systems

Table 9.7 Main characteristics of the karst water system

Karst water system	Main characteristics	Supporting data
Saturated zone	Conduit/fracture flow. The saturated zone is heterogeneous and consists of multiple flow paths with different flow velocities. The groundwater flow is convergent toward SP-1 Spring through major conduits that are fed by water from small fractures	• A fracture zone was observed in monitoring well NZA • The tracer test at Monitoring Well NZA clearly demonstrated the karst feature of the saturated zone • The dye tracing at SH-6, SS-1, and SS-8 indicated a quick flow to SP-1 Spring • Relatively slow flow was observed from the tracing test at sinking stream SS-4 • Two monitoring wells in the saturated zone of the marble in the existing quarry and amendment area showed positive dye detection • The estimated flow rates range from 30 feet/day to 2,760 feet/day
	Strong hydraulic connection to allogenic recharge	• The water from SS-4 sinks vertically to the saturated zone • The tracing test at SS-4 is hydraulically connected to NZA and SP-1 Spring • The dye tracing at SS-2 is hydraulically connected to M1B (well in saturated zone) • The dye tracing at SH-6 is hydraulically connected to NZA • The dye tracing at SS-1 is hydraulically connected to M1B • The dye tracing at SS-8 is hydraulically connected to M1B
Unsaturated zone	High storage capacity	• Frequently encountered voids in boreholes • Presence of shallow caves • Thick unsaturated zone • Consistent presence of water over years in several monitoring wells such as BD40, BD41, BD44, M3A, and M6B
	Weak hydraulic connection to saturated zone	• Different water levels between wells in saturated zone and those in unsaturated zone. Water level in unsaturated zone is consistently higher than that in saturated zone • Different isotope signatures between the water samples in saturated zone and those in unsaturated zone. The water samples in unsaturated zone appear to be impacted by evaporation • Different conductivity values between the water samples in saturated zone and unsaturated zone. The water in unsaturated zone tends to have longer residence time • The tracing tests at SH-11 had a negative detection of the dye at SP-1 Spring and the monitoring locations. M1A and M1B are only a few tens of feet away from SH-11, no dye was detected in either of them

(White 1993). Vertical shafts or solution-widened fractures often develop along the borders between karst land and the borderland. The majority of the water draining from surface catchments, such as the karst of the Mammoth Cave area of Kentucky or of the Cumberland Mountains of Tennessee and northern Alabama often takes a direct route through the unsaturated zone to the underlying conduit system (White, 1988; Smart and Campbell 2003). The continuous uplift of the coastal areas may further enhance the development of the vertical shafts in the study area. At sinking stream SS-4, for example, the surface water sinks into the ground, and no water is visible on the adjacent quarry wall that is over 300 feet tall and only 50 feet away from the sinking point. At sinking stream SS-1, preliminary exploration of a cave indicated the possibility of a deeper passageway.

The geologic structure and the tracing tests suggest quick flow from SS-8, SS-1, and sinkhole SH-6 to SP-1 Spring. During recharge events, water can flow into the conduit system and quickly move to SP-1 Spring. During the tracing tests, the majority of the water in SS_8 arrived at SP-1 Spring in approximately 88 days; the majority of the water in SS-1 arrived at SP-1 in approximately 20 days, while the majority of the water in the SH-6 area arrived at SP-1 Spring in approximately 10 days. The quickest water flow was detected between monitoring well NZA and SP-1 Spring, where most of the water arrived at SP-1 Spring in approximately 7.5 h. Unlike the sinking creeks, NZA is only a monitoring well in the groundwater flow paths, and it is not a source of water for SP-1 Spring.

The approximately 3-month travel time from sinking stream SS-8 to SP-1 Spring may correspond with the observed lags in the seasonal fluctuations of SP-1 Spring, as noted by Todd (1963). In his report, Todd has the following description of the lag: "December is the maximum rainfall month in the area, but the peak spring flow is in March. Similarly, August is the minimum rainfall month before the increase in the fall months, yet November is the month for minimum discharge."

However, SS-8 is only one of the recharge sources for SP-1 Spring. While the recharge from SS-8 explains the seasonal fluctuation of water flow, other sources such as sinking stream SS-1 and the sinkhole SH-6 area, may explain the daily fluctuation of the water flow.

The karst aquifer also includes input from SS-4, SS-2, non-karstic aquifers in the borderland, small fractures around the conduits, and the unsaturated zone. These inputs occur through tight fractures in the low permeability rock in which the conduits are dissolved. They get recharged during the wet season. During dry periods, the water in these fractures gradually flows toward the conduits because the conduits provide pathways of the least resistance. This often forms a groundwater trough over the conduits. The recharge from the diffuse inputs helps sustain the flow at SP-1 Spring during dry seasons.

A conceptual diagram is presented in Fig. 9.20 to schematically illustrate the groundwater flow components and the complicated arrangement of flow paths brought on by the interplay between surface water and groundwater. This schematic diagram is not drawn to a specific scale, and the vertical profile is exaggerated. The lateral extent could range from several hundred feet to several miles. Because of the lack of information in the area between SS-8 and SP-1 Spring, this diagram is likely

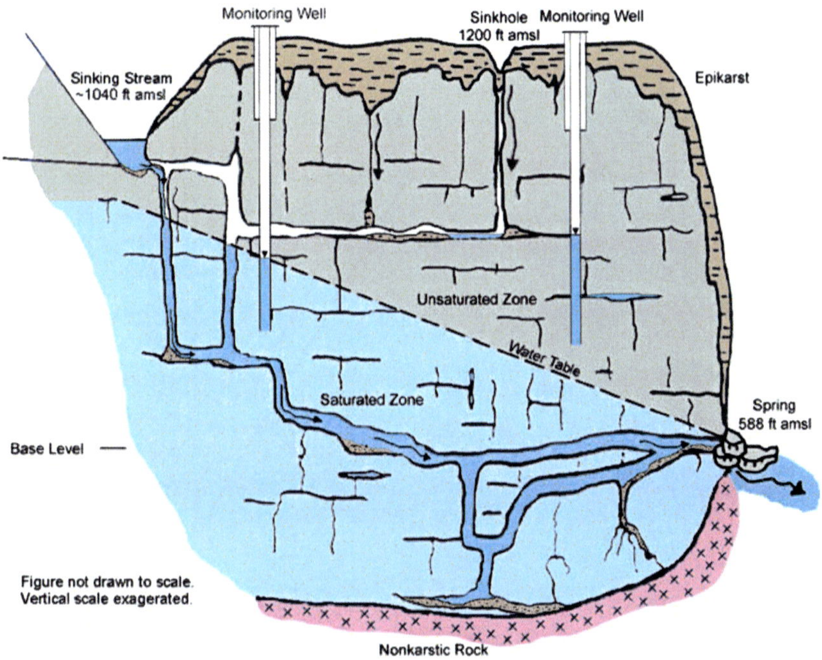

Fig. 9.20 Schematic conceptual site model for recharge source and flow paths to SP-1 spring

more representative of the flow conditions from SS-1 and SS-2 to SP-1 Spring. The water table fluctuates with the seasons.

9.6 Geochemical Characteristics at SP-1 Spring

9.6.1 Ionic Water Types

Figure 9.21 are trilinear plot of mineral analyses for approximately 20 different sources of water in the study area. The figure highlights two groups of samples that are representative of the marble and sandstone aquifers. Most area waters are of a calcium-bicarbonate type, including most streamflow. However, water associated with the marble aquifer is of a strongly calcium-bicarbonate type, with calcium comprising typically 75 percent or more of the major cations and sulfate comprising < 15 percent of the major anions (as milliequivalents per liter [meq/L]). The concentration of calcium in marble-aquifer groundwater is typically 6 to 13 times greater than the concentration of magnesium (as mg/L).

Water associated with the Santa Margarita Sandstone tends to be of a sodium-chloride or mixed type (e.g. springs along sinking streams SS-1 and SS-4). Water

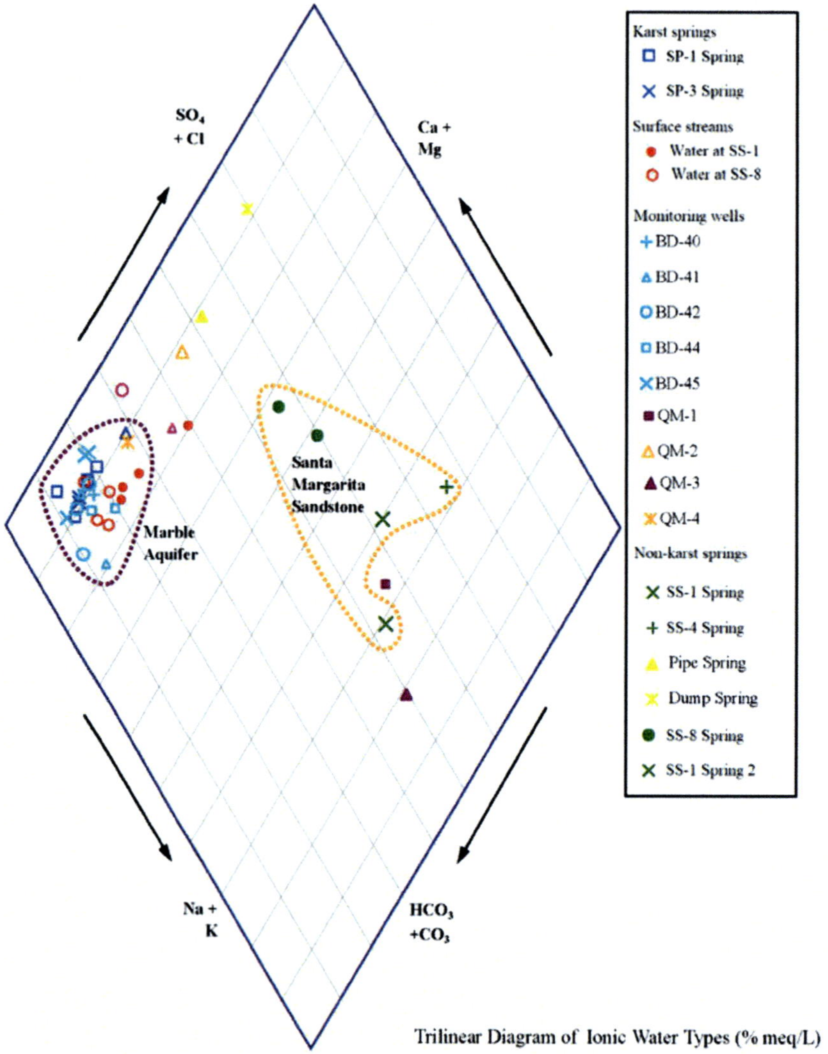

Fig. 9.21 Trilinear diagram of ionic water types

that has been in contact with stockpiled quarry waste tends to be of a calcium-sulfate type (Pipe and Dump springs, possibly QM-2,). Wells completed in granitic rocks near the downgradient boundary of the marble aquifer have a mixed-sodium water quality (QM-1 & -3).

9.6.2 Relationship Between Spring Discharge, Turbidity, Specific Conductance, and Temperature

9.6.2.1 Relationship Between Spring Discharge, Specific Conductance, and Temperature

The discharge, turbidity, specific conductance, and temperature have been monitored for many years at SP-1 Spring. Figure 9.22 shows their correlations for the data collected from 1998 to 2005. SP-1 Spring discharge ranged between 850 and 1100 gpm about 75 percent of the time. Water temperature provides a useful indicator of groundwater movement. The temperature of groundwater residing long-term within the aquifer typically approaches that of the mean annual air temperature (i.e. about 17.6 °C for the study area). Groundwater recharged by winter rainfall and runoff will reflect relatively cool temperatures for some period. Conversely, slow migrating groundwater will mimic seasonal air temperatures when approaching the ground surface, such as groundwater emerging from a seep. Figure 9.22d shows the correlation between SP-1 Spring discharge and water temperature during 2004–05. Spring flows below about 1100 gpm had little correlation with temperature whereas higher flows were generally greater than 16 °C, indicating that long-term groundwater storage contributes substantially to these flows.

Figure 9.22e shows the correlation between SP-1 Spring temperature and specific conductance. Flows with the lowest specific conductance (<400 microsiemens/centimeter [μS/cm]) are coldest (<15 °C) whereas flows with the highest specific conductance (>450 μS/cm) are warmest (>16 °C). Cold temperatures and relatively low mineral content reflect captured winter streamflow, whereas relatively warm water (approaching mean-annual air temperature) and higher mineral content reflect long-term groundwater storage. SP-1 Spring's discharge of relatively warm, mineralized water during peak flow conditions indicates a direct and substantial contribution from groundwater storage.

The water's specific conductance (i.e., electrical conductivity at 25 °C, expressed in μS/cm) is proportional to its concentration of total dissolved minerals. The recommended drinking water standard for specific conductance is 900 μS/cm, with enforceable limits set at 1,600 and 2,200 μS/cm.

Groundwater and surface water upgradient and cross gradient from the quarry marble aquifer tend to be low in dissolved minerals (<200 μS/cm). Water in contact with the marble becomes more mineralized and generally ranges between 300 and 1,000 μS/cm. Along major fracture zones east of the quarry, the dissolved mineral content tends to be relatively low in deep wells compared to nearby shallower wells (e.g., M6A vs. M6B, M1B vs. BD-44, M2A vs. M3B). Because dissolved mineral content usually increases with depth, this observation indicates that these deep wells encounter zones influenced by relatively rapid groundwater flow through deep karst conduits that drain stream swallow holes.

The average specific conductance of SP-1 Spring is approximately 480 μS/cm, which is intermediate between groundwater derived from captured streamflow

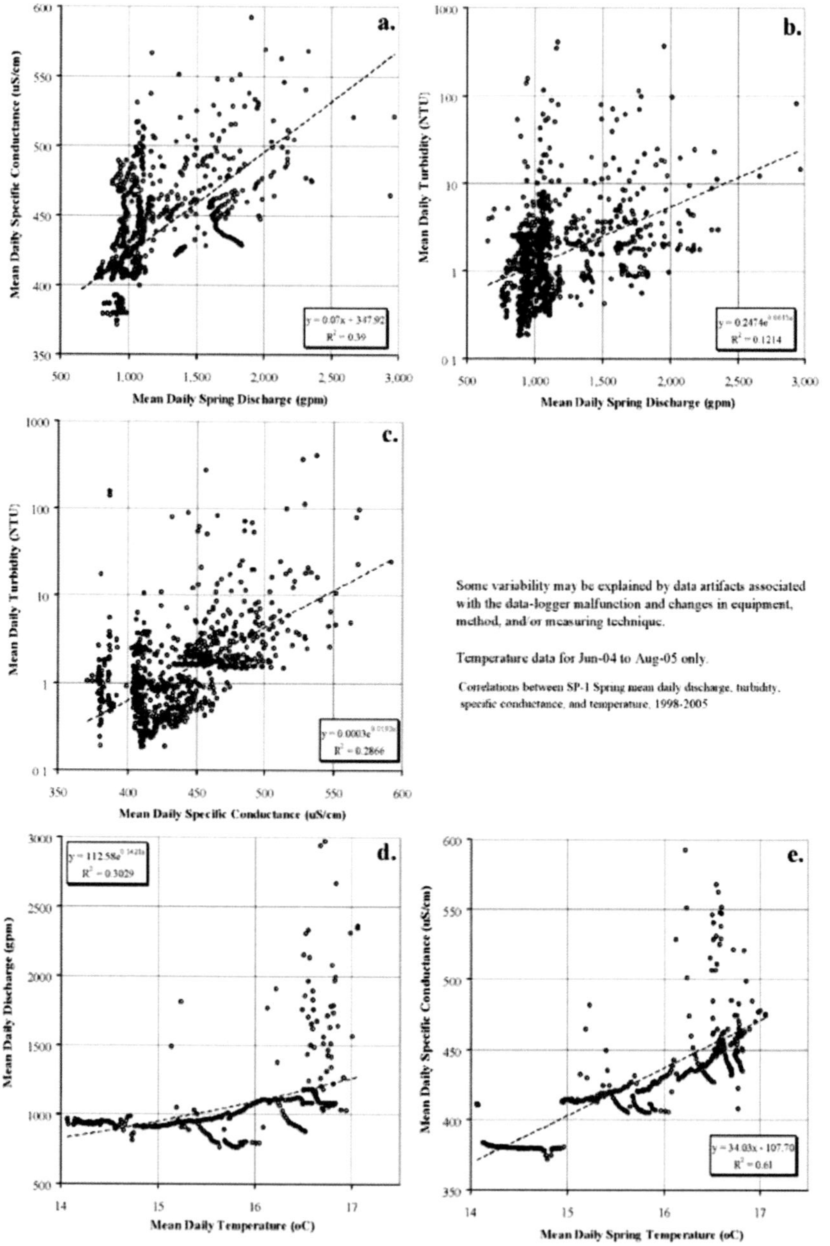

Fig. 9.22 Scattered plots between discharge, turbidity, specific conductance, and temperature at SP-1 spring **a** Relationship between specific conductance and spring discharge; **b** Relationship between turbidity and spring discharge; **c** Relationship between specific conductance and turbidity; **d** Relationship between temperature and spring discharge; Relationship between specific conductance and spring temperature

(generally 300–400 μS/cm) and more mineralized groundwater migrating relatively slowly from other sources of recharge. If it is assumed that upgradient groundwater has an average specific conductance of 600 μS/cm, then a roughly 50:50 mix of captured streamflow and upgradient groundwater is needed to produce the average specific conductance of SP-1 Spring.

The specific conductance of SP-1 Spring and discharge quickly rise to greater than 600 μS/cm during storm periods and gradually decline after each storm and throughout the dry season. This gradual recession is also apparent in the mean daily specific conductance duration curve. The scatter plot in Fig. 9.22a illustrates the direct, albeit rough, correlation between the spring's specific conductance and discharge. This trend is the opposite of that typically exhibited by streamflow, in which specific conductance gradually rises during the dry season as the groundwater contribution proportionally increases, and then quickly decreases as a result of dilution from precipitation and runoff during the wet season. The rapid pressurization of the karst aquifer by low-mineral winter recharge and captured streamflow causes the release of proportionally more high-mineral groundwater to SP-1 Spring. The rise in spring flow specific conductance usually peaks one-half to three days after a storm begins. This observation demonstrates that a large volume of dynamic groundwater storage exists and contributes substantially during periods of both high and low discharge.

The mean daily specific conductance and water temperature of SP-1 Spring are directly correlated (Fig. 9.22e). Flows with the lowest specific conductance (<400 μS/cm) are coldest (<15 °C) whereas flows with the highest specific conductance (>450 μS/cm) are warmest (>16 °C). Cold temperatures and relatively low mineral content reflect captured winter streamflow, whereas relatively warm water (approaching mean-annual air temperature) and higher mineral content reflect long-term groundwater storage. SP-1 Spring's discharge of relatively warm, mineralized water during peak flow conditions indicates a direct and substantial contribution from groundwater storage.

9.6.2.2 Relationship Between Spring Discharge and Turbidity

The mean daily turbidity correlates poorly with mean daily flow (Fig. 9.22b). For example, turbidities > 10 Nephelometric Turbidity Unit (NTU) have occurred on days with mean daily flows anywhere between 900 and 3,000 gpm (Fig. 9.22b.). The lower limit of the turbidity-versus-flow envelope generally increases with increasing flow. However, the highest recorded turbidities have not occurred at the highest flows, but instead are most associated with flows between 900 and 2000 gpm. In streams, turbidity is more highly correlated with flow, including maximum flows. SP-1 Spring turbidities are slightly better correlated with specific conductance (Fig. 9.22c). The poor correlation between SP-1 Spring peak flows and peak turbidities indicates that peak turbidities tend to occur early during the spring's stormflow hydrograph.

Springs with a small watershed may show a direct correlation between discharge and turbidity. For springs with large watersheds such as SP-1 Spring, a direct correlation between discharge and turbidity does not usually exist for two reasons—the phase difference between the discharge hydrograph and the turbidigraph and a phenomenon called turbidity hysteresis. The lack of a relationship between discharge and turbidity is characteristic of many large karst springs (Valdes et al. 2005; Fournie et al. 2007).

Figure 9.23 shows a clockwise or positive loop of the hysteresis curves for a storm event. This loop generally occurs during an increase of water discharge, which is indicative of sediment resuspension by pressure transfer. The shape of these hysteresis curves varies significantly between storms, suggesting that the complex relationship between discharge and turbidity can be further complicated by the antecedent groundwater flow conditions and the characteristics of each rainfall.

Clearly, the turbidity is not directly correlated to the spring flow. The discharge hydrographs are normally distributed over a much longer time scale than the suspended sediment response. The peak concentration of suspended solids tends to precede peak spring flow and have a much shorter duration, a relationship that has been observed in other karst springs (Mahler and Lynch 1999; Townsend and Macfarlane 2003).

Hysteresis curves are often used to describe sediment transport in surface streams (Ongley 1996). Recently, they have been used to describe the characteristics of suspended solids at karst springs (Valdes et al. 2005; Fournie et al. 2007). The following explain the occurrence of turbidity hysteresis at SP-1 Spring:

- The most intensive transport processes in the system occur during the pressure transfer caused by a flood pulse. The flood pulse causes a highly unsteady flow regime, which causes resuspension and transport of deposited sediments. The increased intensity in the rising limb is caused by the flood pulse overtaking the baseflow. It has been demonstrated that turbulence intensities are generally larger in the rising limb of a hydrograph than in the falling limb, and the turbulence intensity becomes larger as the unsteadiness of the hydrograph in-creases (Sutter and Verhoeven 2001). These findings suggest strongly that sediment transport is the strongest in the rising stage when the flow is the most unsteady. It is the flow velocity and the acceleration of the flow that provide the turbulent intensity to resuspend the particles in the water.
- Rapid depletion of sediments that can be resuspended occurs for the storm events. The most mobile sediments are those recently deposited. Compacted sediments, even of fine grain size, require stronger basal shear stresses to mobilize than uncompacted sediments. Larger storms are required to resuspend the compacted or larger sediments temporarily stored in the system. Thus, the turbidity does not increase as the discharge in-creases, once the more mobile sediments have been entrained.

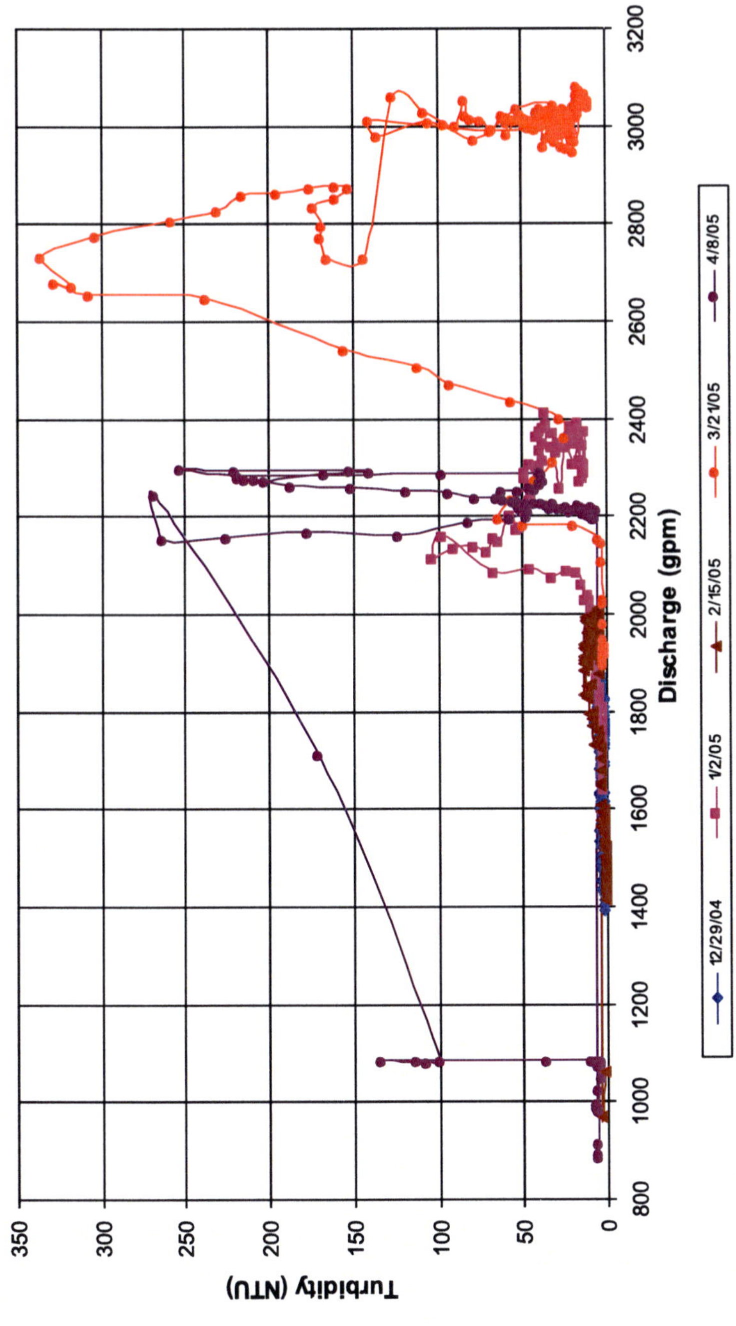

Fig. 9.23 Turbidity hysteresis at SP-1 spring for select storm events

9.6.3 Stable Isotopes

All groundwater has a "fingerprint" of naturally occurring isotopes that provides information about the origin of the water (Drever 1997). Waters that have the same sources tend to have similar isotopic signatures. Analysis of the stable isotopic ratios was designed to determine which waters have a similar isotopic signature to that at SP-1 Spring. A stable isotope survey was conducted at 39 quarry-area wells, springs, seeps, and streams in February 2003. Accumulated precipitation was 20 inches for the season to date, greater than 2 inches during the preceding 10 days, and about 1 inch during the sampling period.

Concentrations of the heavier isotopes of hydrogen and oxygen become depleted in water derived from evaporation. Thus, atmospheric moisture is depleted of the heavier isotopes relative to the ocean from which it evaporated. The isotopic composition of precipitation is further affected by the temperature and elevation of where it forms and falls. When precipitation accumulates on land, it becomes relatively enriched with the heavier isotopes as the lighter isotopes are disproportionately lost to evaporation. Once water percolates below ground it experiences no further evaporation and its isotopic ratios of hydrogen and oxygen remain fairly constant. Thus, groundwater at different locations, depths, and points of discharge can be grouped into similar origins and flow paths based on the similarity of these ratios, although various chemical processes can further alter the isotope ratios below ground. Figure 9.24 is a plot of the ratios of oxygen-18 to oxygen-16 ($\delta^{18}O$) versus the ratios of deuterium to hydrogen (δD). The corresponding values of specific conductance are posted beside each point. The isotope-ratio pairs are grouped within five separate envelopes based on their ratio values, specific conductance, and contextual hydrogeology.

- Group 1 consists of springs and seeps emerging from the Santa Margarita Sandstone (-5.4 to -5.8 $\delta^{18}O$, -32 to -34 δD). These have very low specific conductance values of less than 200 μS/cm. The envelope encompassing group 2 (-5.6 to -5.7 $\delta^{18}O$, -33 to -34 δD) nearly overlies group 1, however these samples are from monitoring wells completed in marble and have moderate to high values of specific conductance (350–900 μS/cm). The isotopic similarity of these groups, along with hydrogeologic interpretation, suggests that group 1 groundwater represents the source of group 2 groundwater, i.e., water recharged into the Santa Margarita Sandstone eventually migrates into the underlying marble aquifer.
- Group 3 consists of streamflow in the sinking streams SS-8 and SS-1 watersheds (-5.7 to -6.2 $\delta^{18}O$, -34 to -37 δD). Given the time of year, very low specific conductance, and approximately 3 inches of precipitation immediately prior to and during the survey, these samples are probably more representative of seasonal runoff than baseflow. Stream samples taken at other times would probably have substantially different isotopic signatures. However, these samples do represent the seasonal flows most available for capture by the karst sinkholes.
- Group 4 (-5.7 to -6.0 $\delta^{18}O$, -34 to -35 δD) consists of groundwater with moderately low specific conductance (~200–400 μS/cm) from relatively deep monitoring wells located along major fracture zones. This group overlies one end

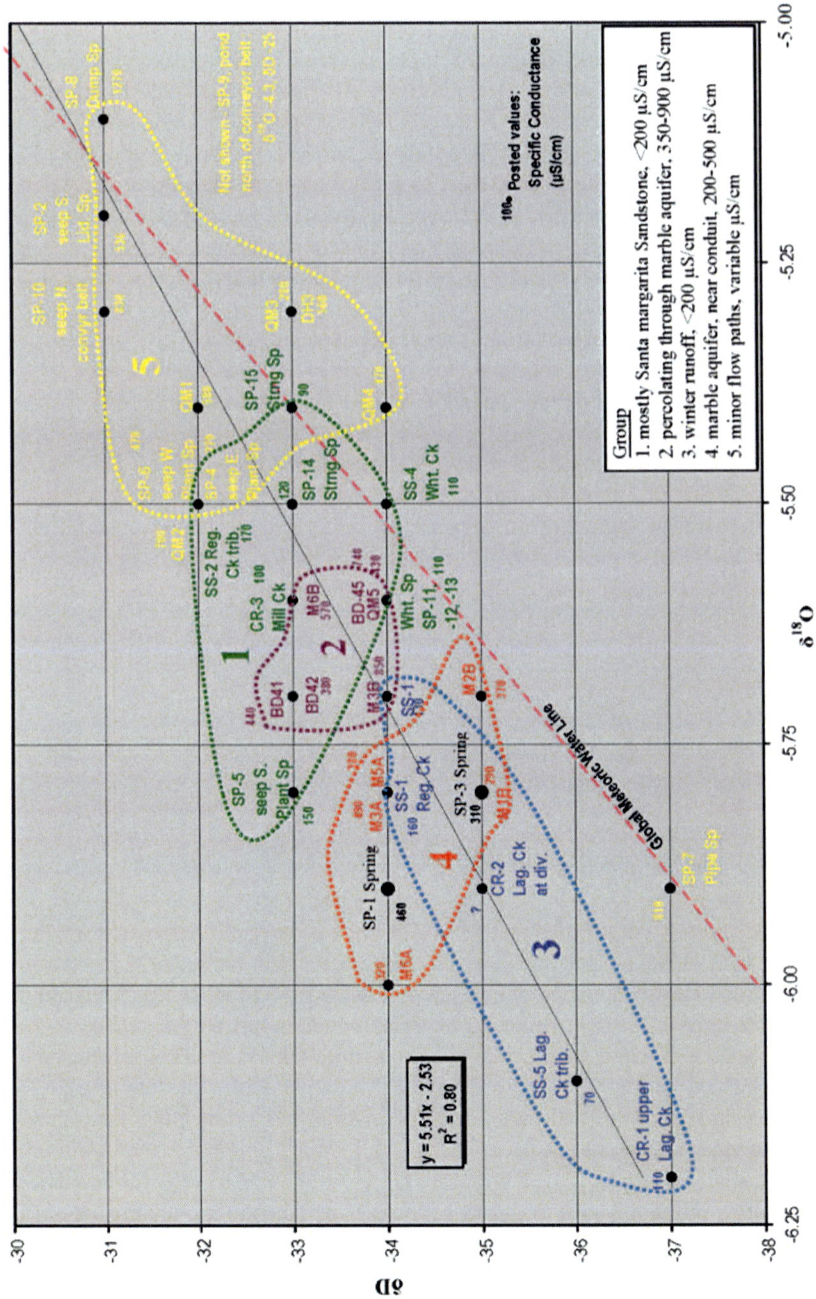

Fig. 9.24 Results of stable isotopes

of the group-3 envelope, consistent with the assumption that this groundwater is recharged predominantly by stream swallow holes. This group probably receives some leakage from shallower zones, thus explaining its location toward groups 1 and 2 relative to the center of group 3. Also, the rapid arrival of tracers to SP-1 Spring suggests that none of these monitoring wells directly encounter the highest permeability pathways.

- SP-1 Spring (-5.9 $\delta^{18}O$, -34 δD) plots midway between groups 1–2 and groups 3–4 in terms of δD, whereas in terms of $\delta^{18}O$ the spring plots approximately midway between groups 1–2 and the more depleted (i.e., more negative) portion of group 3. These data indicate that both sandstone recharge and sinkhole stream capture contribute roughly equally to the spring. The fact that the spring's $\delta^{18}O$ is skewed toward the depleted end of group 3 is reasonable given that the greatest volume of sinking recharge is from isotopically depleted stormflow. SP-3 Spring (−5.8 $\delta^{18}O$, −35 δD) plots within group 4 in terms of both δD and $\delta^{18}O$. This and its moderately low specific conductance suggest that its source is mostly from captured streamflow.
- Group 5 is isotopically enriched relative to the other groups (−5.1 to −5.5 $\delta^{18}O$, −31 to −34 δD, except for two outliers). This group includes all the non-calcium-bicarbonate waters, including those that appear influenced by quarry waste and/or migrate along fairly minor flow paths separate from the karst groundwater system (i.e. conceptual path *D*). This group partially overlies group 1 but is distinguished by generally higher values of specific conductance. A sample of highly enriched pond water (SS-9) suggests that the water had resided in the pond for an extended period of time and experienced considerable evaporation.

9.7 Turbidity and Sediment

9.7.1 Transport and Sources of Sediments in SP-1 Spring

Non-karstic aquifers do not often carry a sediment load because the pores through which the water flows are too small, and the flow velocities are too slow for sediment transport. The porous medium acts as an effective filter, allowing infiltrating groundwater to pass but rejecting solid material, which is left on the surface as a weathered residuum. Unlike a porous medium aquifer, a karst aquifer is characterized by enhanced permeability that permits high flow velocities capable of transporting suspended and bedload sediments. Calculations have shown that the transport of suspended sediments can occur in fracture openings as small as 1 cm (White 1988). Consideration of the critical boundary shear for bedload transport shows that flow velocities in the range of fractions of a meter per second are required to set the sediments into motion. Quantitative studies of sediment transport in conduits are sparse, but anecdotal evidence indicates that most sediment moves during flood events (White 1993). Extreme flood flows may strip a conduit system of its sediment filling. New material is deposited during flood recessions as in surface channels after floods.

Both SP-1 Spring and SP-3 Spring are karst springs. Although the flow paths to SP-3 Spring are different from those to SP-1 Spring, its water has turbidity spikes in response to precipitation events. The discussion of turbidity in SP-1 Spring and its close association with rainfall events dates back at least to 1956 (Todd 1963). Historic documents reporting turbidity spikes in response to rainfall events indicate a wide range in the turbidity levels for storms of similar magnitudes. Because of the complicated processes of sediment transport in the karst aquifer, the magnitudes of the turbidity response and the lag time vary significantly from one storm to another.

Like in any other well-developed karst aquifer, the transport of sediments is episodic and is very sensitive to the flow regime. The most significant sediment transport occurs primarily during storm flow and the rate of transport can vary greatly over the year, depending on storm intensity and overall flow rates. As storm flow recedes, the sediments are deposited in the conduit system and at the spring mouths. Both depositional and erosional processes occur in the conduits. Storm flow provides a means of flushing out the accumulated materials at springs. The presence of mobile sediments in karst has been recognized for several decades (Bretz 1964). It is well documented that water quality at karst springs can be impacted due to the increase in turbidity during storm events (Atteia and Kozel 1997; McCarthy and Schevenell 1998; Schevenell and McCarthy 2002; Mahler and Lynch 1999). Transport of sediment particles is an essential part of the functioning of the karst aquifer. In well-developed aquifers where surface streams sink underground, the insoluble materials carried by the sinking surface streams are carried through the conduit system and washed out through the springs. If this did not happen, the conduit system would eventually become clogged with sediment and drainage would be forced to the surface. A karst aquifer only continues to function if the flow is sufficient to transport the sediment and prevent clogging.

Similar to the water recharge sources for SP-1 Spring, the sediment sources can also be divided into allogenic and autogenic categories. Weathered products from the erosion of the borderland enter the aquifer through the sinking streams and the Sinkhole Plain. When these streams flow underground through the karst, so do the sediments. Carbonate rocks also contain a small percentage of insoluble material, and these materials accumulate as soils on the karst surface and as weathered residuum created in the subsurface by the solutional enlargement of the conduit system. According to the results from the tracing tests and the local geology, the following sources may provide natural sediments to SP-1 Spring.

- Sinkholes.
- Sinking streams SS-1, SS-2, SS-4, and SS-8.
- Landslide adjacent to SP-1 Spring.
- Weathered residuum on the surface and in the subsurface in the karst land.
- Quarry floor.

This study indicates that the sinkhole plain and sinking stream SS-1 are probably the most important sources supplying the sediments into the karst system. The sinkhole plain is active and sandy sediments are present. The runoff from the local springs or from rainfall can carry significant amount of sediments into the subsurface. At the sinking section of SS-1, sediments can be observed being transported into the subsurface through fractures in the marble. Small funnel-shaped depressions are present in the unconsolidated materials along the western riverbank, indicating downward transport of the sediments. The sediments can be stored in the conduit/fractures system temporarily and their further transport relies on the intensity of rainfall. Similar sediments have been observed in caves and some of the boreholes.

Sinking stream SS-8 is well connected to SP-1 Spring and can provide sediments to the karst system. Because the sinking section of SS-8 is approximately 2.7 miles away from SP-1 Spring, its contribution of sediments to SP-1 Spring is not instantaneous.

Because the spring box at SP-1 Spring receives water from the marble and the landslide deposits, a flow regime change in the deposits may move sediments in the slide deposits into the spring box. A local flow regime change can also mobilize the sediments that are stored at the spring mouth.

Because of the poor hydraulic connection between the unsaturated zone and the saturated zone, the sediment contribution from the autogenic source is limited. It is important to realize that most of the autogenic sediments are from natural dissolution processes, regardless of the existence of the quarry.

The autogenic source also includes the particulates present on the quarry floor. The sediments on the quarry floor may contribute to the turbidity at SP-1 Spring only if open fractures are present and when water flow carries the sediments into the subsurface. When no open fractures are present, the marble block is not permeable to sediments. The sediments on the existing quarry floor are angular and poorly sorted, whereas the sediments discharged at SP-1 Spring are rounded and well sorted. The X-ray diffraction analysis of suspended sediment samples at SP-1 Spring reported the average composition of the sediments: 32.7% quartz, 28.7% feldspars, 3.7% mica, 10% calcite, 1.5% hematite, 20.8% clay, and 2.7% unaccounted materials. Because the sediments on the quarry floor are predominantly calcite, the mineralogical composition does not point to a single source.

9.7.1.1 Transport and Sources of Sediments in SP-1 Spring

Turbidity represents the cloudiness of water caused by the presence of fine suspended matter. It can be measured by the presence of total suspended solids in water, which is reported in parts per million (ppm), or determined by measurements of light diffraction, which is reported in NTU. According to the National Primary Drinking Water Regulations (www.epa.gov/safewater), a Maximum Contaminant Level Goal is not established for turbidity. The Maximum Contaminant Level for turbidity value is 5 NTU. Turbidity has no health effects but can interfere with disinfection and provide a medium for microbial growth.

Elevated turbidity is a main concern at SP-1 Spring because it is a water supply source. Figure 9.25 shows the base-level turbidity data at SP-1 Spring from 1960 to 2004. The base level is defined a condition when no rainfall occurred during the month. These data demonstrate the turbidity variation when the precipitation impacts are excluded. A quality control chart is developed to demonstrate that no long-term

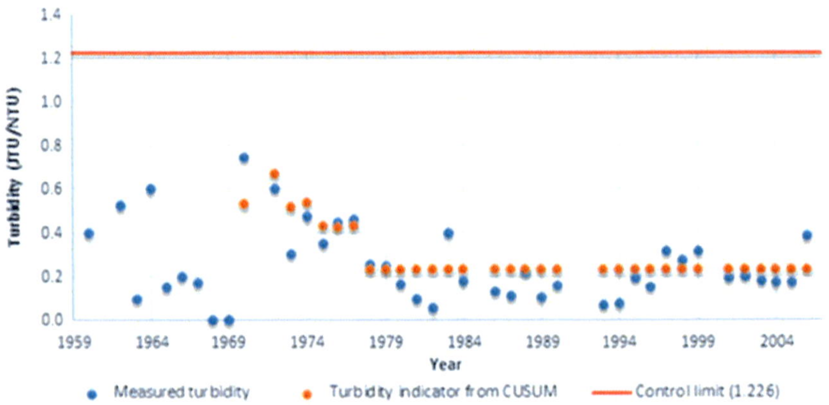

Fig. 9.25 Long-term base-level turbidity changes

degradation of base-level water quality due to turbidity. The pre-quarry samples collected from 1960 to 1969 are used to estimate the control limits.

Quality control chart method is a viable way to analyze the dataset to statistically determine if the quarry operations have changed the turbidity levels at SP-1 Spring. Because quarry operations may cause both large and small changes in the turbidity values, the Combined Shewhart-CUSUM control charts are used in the analysis. Combined Shewhart-CUSUM control charts are the statistical procedures directly recommended by U.S. EPA (1989,1992) and American Society for Testing and Materials (ASTM) (1998). As implied by its name, a combined Shewhart-CUSUM control chart is a mixture of two statistical schemes: Shewhart control scheme and CUSUM control scheme. The Shewhart control scheme is better in quickly detecting a large shift in the mean; whereas, the CUSUM scheme is usually faster in detecting a small change in the mean that persists.

Design of a combined control chart requires the determination of the following five parameters:

$\overline{X}$ Estimated mean of turbidity from pre-quarry samples from 1960 to 1969. The calculated annual average mean is 0.269 NTU.

S Estimated standard deviation of turbidity from pre-quarry samples from 1960 to 1969. The calculated standard deviation is 0.214 NTU.

h The value against which the cumulative sum will be compared.

k A parameter related to the displacement that should be quickly detected.

SCL The upper Shewhart limit, which is the number of standard deviation units for an immediate release.

For turbidity measurement x_i after the quarry began its operations, the standardized difference z_i is calculated by (EPA 1989):

$$z_i = \frac{x_i - \overline{X}}{S}$$

And, the cumulative sum Y_i is calculated by:

$$Y_i = \max\left[0, (z_i - k) + Y_{i-1}\right]$$

A significant turbidity change is declared when either $Y_i \geq h$ or $z_i \geq SCL$. EPA (1989) recommends using $SCL = 4.5$, $k = 1$, and $h = 5$ to allow a displacement of two standard deviations to be detected quickly (EPA 1992). For easy application, ASTM (1998) suggested the use of $h = SCL = 4.5$, which is slightly more robust in detecting any turbidity changes. For this application, we use $h = SCL = 4.5$.

The red horizontal line in Fig. 9.25 is the control limit. The control chart suggests that there are no statistically significant changes in turbidity due to quarry operations.

9.8 Statistical Analysis of Turbidity Responses to Quarry Blasting

Blasting is used to break-up the rock using ammonium nitrate as part of the operations at the quarry. Thirty to fifty-foot holes are drilled on a working bench and the charge is set in the blast holes. When an explosive is detonated, most of the energy from a properly designed blast works to displace rock from the quarry face. The remaining energy is released as vibrations through and along the surface of the earth and through the air. The blast-induced vibrations have been monitored as ground acceleration at various locations in the quarry and in its vicinity. The correlation between quarry blasting and turbidity spikes at SP-1 Spring is statistically analyzed to provide a detailed quantitative study on the possible impacts of quarry blasting on the turbidity at SP-1 Spring in combination with rainfall and antecedent soil conditions.

9.8.1 Data Collection and Validation

Valid statistical analysis relies on the availability and validity of the data to be analyzed. The purpose of this statistical analysis requires careful validation of three sets of data—blasting data in the quarry, precipitation data in the local area, and turbidity data at SP-1 Spring. Although precipitation and turbidity data are available for over decades, only those collected during the periods when the quarry blasting was monitored were analyzed during this study. Figure 9.26 shows one-month data as an example of data available for the analysis.

Because the data were collected by various organizations, different techniques, and different frequencies, professional judgment was applied to validation of the data. Figure 9.27 shows the general data selection processes. The precipitation data and turbidity data were converted to hourly intervals for the statistical analysis. The hourly turbidity is the average value if more than one reading is available within one hour, whereas the hourly precipitation is the cumulative value within each hour.

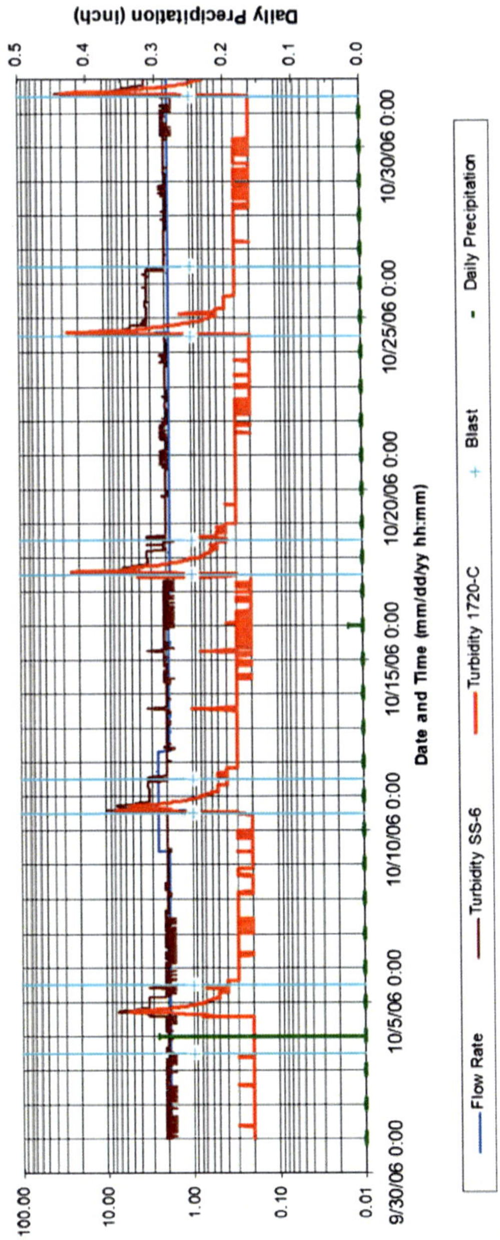

Fig. 9.26 Example of data available for statistical analyses

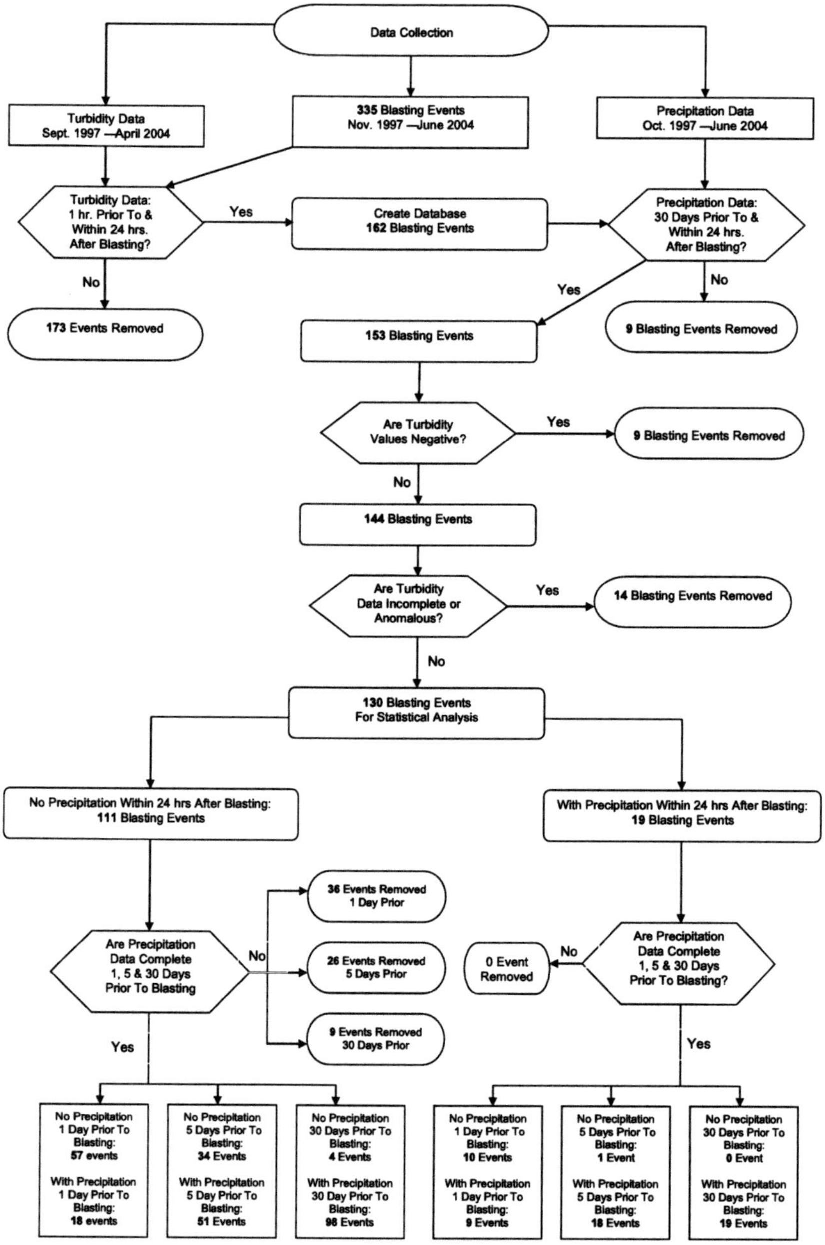

Fig. 9.27 Flowchart for blasting and turbidity data selection

Table 9.8 Blasting events used for statistical analysis for each scenario

		Precipitation 1 day prior to blasting		Precipitation 5 days prior to blasting		Precipitation 30 days prior to blasting	
		No	Yes	No	Yes	No	Yes
Precipitation 24 h After Blasting	No	57	18	34	51	4	98
	Yes	10	9	1	18	0	19

From November 1997 to June 2004, 335 blasting events were recorded in the quarry. Of these, 173 blasting events did not have corresponding turbidity data. The availability of precipitation data and the validation processes further eliminated 32 events. Therefore, 130 events were considered to consist of valid data, and they were used for the current study. Based on the precipitation data prior to and after each blast, the blast events were further divided into different categories. 111 blasting events occurred when there was no precipitation within 24 h after the blasting, while 19 events occurred when there were precipitation activities within 24 h after the blasting.

The antecedent soil moisture condition was considered by the cumulative precipitation activities prior to each blasting event. The precipitation 1 day, 5 days, and 30 days prior to the blasting event was considered. Table 9.8 provides 12 scenarios in which the statistical analyses are performed, and the number of events used for the respective analysis.

From a statistical perspective, reliable results rely on a large number of blasting events. When there is only one blasting event, such as the scenario in which there were precipitation activities within 24 h after the blasting but there was no precipitation 5 days prior to the blasting event, the results cannot be statistically analyzed. Because none of the blasting events occurred when there was a precipitation activity within 24 h after the blast, but there was no precipitation 30-days prior to the blast, no statistical analysis can be performed for this scenario either. The first measured turbidity change at SP-1 Spring tends to occur within a few hours; therefore, the turbidity values measured within 24 h after each blasting event were used for the analysis.

9.8.2 Statistical Analysis

The main challenge in the statistical analysis is to separate the impact of blasting from the impact of precipitation. When precipitation occurs prior to blasting, the water at SP-1 Spring is already affected by the rainfall, which will overshadow the possible impact of blasting. When precipitation occurs after blasting, the impact from the blasting is masked by the impact from precipitation. In addition, the turbidity at

SP-1 Spring is affected by other factors, such as activities in the catchment basin areas and the transport characteristics of the solid particles in the marble aquifer.

To evaluate the impact of blasting on turbidity at SP-1 Spring, the turbidity values within 24 h after the blasting were compared with the hourly average turbidity one-hour prior to blasting. The turbidity change was calculated for each of the 24 h after each blasting event by subtracting the hourly turbidity value one-hour prior to blasting from the hourly turbidity values after blasting. The turbidity change was used for the statistical analysis.

Statistically, the turbidity change would be zero or close to zero if the blasting has no impact on the turbidity at SP-1 Spring, provided that the impact of precipitation can be eliminated. It can also be reasonably assumed that the turbidity change would be greater than zero if the blasting has an impact on the turbidity at the spring. The magnitude of the turbidity change indicates the extent to which the blasting affects the turbidity at SP-1 Spring. Because none of the blasting events occurred when there was precipitation activity 24 h after the blasting but there was no precipitation 30-day prior to the blasting and only one blasting event is available for the scenario in which there were precipitation activities within 24 h after the blasting but no precipitation activities 5 days prior to the blasting event, only 10 scenarios of the 12 listed in Table 9.8 were actually analyzed.

Table 9.9 provides the analysis results. For each scenario, the average, median, mode, minimum, maximum, and standard deviation of the turbidity change were calculated for each of the 24 h after the blasting. Because the turbidity data were collected by different organizations and different sensors, the turbidity values may not be absolutely comparable throughout the monitoring period and the average value is probably the most representative of the turbidity change. Table 9.10 lists the main characteristics of the average turbidity change for the 10 scenarios analyzed. The data for the scenario in which only one blasting event is available are also presented in the table, although they do not have statistical meaning.

9.8.3 *Impact of Blasting on Turbidity at SP-1 Spring*

The impact of blasting on the turbidity at SP-1 Spring can only be evaluated when the effects of other factors can be precluded, i.e., when the impact of precipitation on the turbidity at SP-1 Spring is minimal prior to and after blasting. The scenario in which there was no precipitation 30 days prior to blasting and no precipitation within 24 h after blasting, may best fit this criterion. The average hourly turbidity prior to blasting is 1.44 NTU, which may represent the background value when SP-1 Spring is not affected by either precipitation or blasting. As shown in Table 9.10, the turbidity change has a positive response in the first hour after blasting and the response lasts approximately 18 h. The magnitude of the response is less than 0.5 NTU, with a range from 0.03 to 0.33 NTU. The maximum response occurs 5–6 h after blasting. For any particular blasting event, the turbidity change could be up to 1.27 NTU, which also occurred 5–6 h after blasting.

Table 9.9 Results of statistical analysis

Scenarios	Time after blast (h)	Statistical results					
		Average (NTU)	Median (NTU)	Mode (NTU)	Minimum (NTU)	Maximum (NTU)	Standard deviation
• No precipitation within 24 h after blasting • No precipitation 1 day prior to blasting • Number of Blasting Event: 57	0–1	0.11	0.02	0	−0.35	2.03	0.36
	1–2	0.36	0.05	0	−0.55	5.17	0.96
	2–3	1.3	0.1	0	−0.6	36.62	5.05
	3–4	0.84	0.1	0	−2.7	7.45	1.84
	4–5	1	0.15	0	−2.86	16.19	2.47
	5–6	1.5	0.15	0	−2.7	35.04	5.29
	6–7	1.27	0.1	0	−2.7	24.48	4.46
	7–8	1.09	0.1	0	−3.01	27.22	4.1
	8–9	0.74	0.08	0	−3.01	15.82	2.57
	9–10	0.54	0.08	0	−2.86	9.89	1.72
	10–11	0.37	0.08	0	−2.7	6.69	1.24
	11–12	0.28	0.08	0	−2.7	5.17	1.03
	12–13	0.22	0.05	0	−2.7	4.57	0.88
	13–14	0.2	0.02	0	−1.43	3.96	0.75
	14–15	0.14	0	0	−2.7	3.35	0.72
	15–16	0.12	0	0	−2.86	2.74	0.67
	16–17	0.11	0	0	−2.7	2.74	0.67
	17–18	0.1	0	0	−2.7	2.74	0.69
	18–19	0.07	−0.02	0	−2.7	3.42	0.77
	19–20	0.09	0	0	−2.7	2.74	0.69
	20–21	0.06	−0.03	0	−2.7	2.25	0.65
	21–22	0.07	−0.05	0	−2.7	2.58	0.71
	22–23	−0.04	0	0	−2.86	2.08	0.66
	23–24	−0.07	0	0	−3.78	2	0.72
• No precipitation within 24 h after blasting • With precipitation 1 day prior to blasting • Number of Blasting Event: 18	0–1	0.15	0	0	−35.91	48	14.9
	1–2	−2.95	0	0	−38.74	0.63	9.25
	2–3	−1.95	0.16	0.16	−43.16	5	10.5
	3–4	−2.49	0.74		−42.91	3.94	10.62
	4–5	−2.84	0.25	0.05	−42.24	4.13	10.63
	5–6	−3.22	0.14	0	−41.66	6.56	11.32
	6–7	−3.66	0.1	0	−41.91	3.04	11.26
	7–8	−3.94	0.13	1.37	−41.99	2.73	11.26
	8–9	−4.16	0.08		−41.83	2.05	11.3
	9–10	−4.18	0.09	0	−40.91	1.52	11.11
	10–11	−4.32	0.06	0	−40.74	1.52	11.13
	11–12	−4.37	0.03	0	−40.58	1.37	11.09

(continued)

Table 9.9 (continued)

Scenarios	Time after blast (h)	Statistical results					
		Average (NTU)	Median (NTU)	Mode (NTU)	Minimum (NTU)	Maximum (NTU)	Standard deviation
	12–13	−4.61	0	0	−39.99	1.44	11.37
	13–14	−4.58	0	0	−39.41	1.15	11.08
	14–15	−4.72	−0.02	−0.02	−39.08	0.88	10.93
	15–16	−4.74	−0.01	0	−39.66	0.9	11.08
	16–17	−4.8	−0.05	0	−39.91	0.45	11.06
	17–18	−4.79	−0.08	0	−39.08	0.7	10.89
	18–19	−4.88	−0.08	0	−40.16	0.08	11.11
	19–20	−4.92	−0.08	0	−40.91	0.88	11.3
	20–21	−4.98	−0.16	−0.08	−40.66	1.25	11.34
	21–22	−4.57	−0.1	0	−40.33	0.53	11.24
	22–23	−4.82	−0.13	0	−40.91	0.08	11.43
	23–24	−4.16	−0.1	0	−40.33	0.13	10.13
• No precipitation within 24 h after blasting • No precipitation 5 days prior to blasting • Number of Blasting Event: 34	0–1	0.08	0.03	0	−0.35	0.8	0.22
	1–2	0.18	0.07	0	−0.55	1.37	0.37
	2–3	0.28	0.09	0	−0.6	1.5	0.51
	3–4	0.53	0.08	0.08	−0.53	5.78	1.12
	4–5	0.86	0.09	0	−0.55	16.19	2.78
	5–6	1.39	0.09	0	−0.6	35.04	5.97
	6–7	0.91	0.1	0	−0.63	23.41	4
	7–8	0.58	0.03	0	−0.55	14.82	2.55
	8–9	0.47	0.05	0	−0.48	10.79	1.86
	9–10	0.38	0.06	0	−0.58	7.45	1.32
	10–11	0.26	0.04	0	−0.65	5.17	0.93
	11–12	0.17	0.05	0.13	−0.65	4.25	0.77
	12–13	0.15	0.03	0	−0.48	2.81	0.55
	13–14	0.11	0	0	−0.5	2.73	0.52
	14–15	0.08	0	0	−0.55	1.9	0.41
	15–16	0.08	0	0	−0.43	1.52	0.35
	16–17	0.05	0	0	−0.65	1.52	0.38
	17–18	0.03	0	0	−0.5	1.52	0.32
	18–19	−0.03	−0.03	0	−0.88	1.52	0.35
	19–20	0.03	0	0	−0.48	1.52	0.35
	20–21	0	−0.03	0	−0.53	1.52	0.32
	21–22	0.09	−0.03	0	−0.55	2.58	0.55
	22–23	−0.06	0	0	−2.18	1.52	0.49
	23–24	−0.1	0	0	−3.78	0.84	0.7

(continued)

Table 9.9 (continued)

Scenarios	Time after blast (h)	Statistical results					
		Average (NTU)	Median (NTU)	Mode (NTU)	Minimum (NTU)	Maximum (NTU)	Standard deviation
• No precipitation within 24 h after blasting • With precipitation 5 days prior to blasting • Number of Blasting Event: 51	0–1	0.15	0	0	−35.91	48	8.7
	1–2	−0.78	0.02	0	−38.74	5.17	5.72
	2–3	0.62	0.16	0	−43.16	36.62	8.3
	3–4	−0.28	0.16	0	−42.91	7.45	6.67
	4–5	−0.28	0.16	0	−42.24	7.63	6.7
	5–6	−0.03	0.23	0	−41.66	19.92	7.68
	6–7	−0.22	0.16	0	−41.91	24.48	7.93
	7–8	−0.37	0.2	0.13	−41.99	27.22	8.06
	8–9	−0.84	0.16	0	−41.83	15.82	7.42
	9–10	−1.03	0.1	0	−40.91	9.89	7.06
	10–11	−1.2	0.1	0	−40.74	6.69	6.99
	11–12	−1.3	0.02	0	−40.58	5.17	6.94
	12–13	−1.45	0	0	−39.99	4.57	7.1
	13–14	−1.42	0.02	0	−39.41	3.96	6.93
	14–15	−1.59	0	0	−39.08	3.35	6.85
	15–16	−1.67	0	0	−39.66	2.74	6.91
	16–17	−1.61	0	0	−39.91	2.74	6.9
	17–18	−1.66	−0.02	0	−39.08	2.74	6.81
	18–19	−1.71	−0.05	0	−40.16	3.42	6.96
	19–20	−1.63	0	0	−40.91	2.74	7.07
	20–21	−1.78	−0.08	0	−40.66	2.25	7.08
	21–22	−1.69	−0.08	0	−40.33	2.25	6.94
	22–23	−1.72	0	0	−40.91	2.08	7.09
	23–24	−1.51	−0.05	0	−40.33	2	6.25
• No precipitation within 24 h after blasting • No precipitation 30 days prior to blasting • Number of Blasting Event: 4	0–1	0.03	0	0	−0.19	0.31	0.21
	1–2	0.16	0.16		0	0.31	0.15
	2–3	0.09	0.16		−0.31	0.38	0.32
	3–4	0.09	0.06		−0.06	0.31	0.16
	4–5	0.11	0.16		−0.19	0.31	0.25
	5–6	0.33	0.28		−0.5	1.27	0.73
	6–7	0.22	0.22		−0.19	0.64	0.34
	7–8	0.11	0.16		−0.25	0.38	0.29
	8–9	0.09	0.13		−0.19	0.31	0.23
	9–10	0.12	0.13		−0.06	0.31	0.18

(continued)

Table 9.9 (continued)

Scenarios	Time after blast (h)	Statistical results					
		Average (NTU)	Median (NTU)	Mode (NTU)	Minimum (NTU)	Maximum (NTU)	Standard deviation
	10–11	0.05	0.06		−0.06	0.16	0.1
	11–12	0.04	0		−0.16	0.31	0.22
	12–13	0.11	0.16		−0.19	0.31	0.25
	13–14	0.08	0.13		−0.25	0.31	0.26
	14–15	0.14	0.16		−0.13	0.38	0.24
	15–16	0.1	0.12		−0.16	0.31	0.25
	16–17	0.03	0.13		−0.44	0.31	0.34
	17–18	0.03	0.06		−0.31	0.31	0.26
	18–19	−0.22	−0.16		−0.88	0.31	0.51
	19–20	0.02	0	0	−0.16	0.25	0.17
	20–21	0.17	0.16	0	0	0.38	0.2
	21–22	0.12	0.08		−0.06	0.38	0.19
	22–23	0.03	0	0	−0.25	0.38	0.26
	23–24	−0.02	0.09		−0.56	0.31	0.39
• No precipitation within 24 h after blasting • With precipitation data points 30 days prior to blasting • Number of Blasting Event: 98	0–1	0.08	0	0	−35.91	48	6.25
	1–2	−0.34	0.03	0	−38.74	5.17	4.16
	2–3	0.8	0.08	0	−43.16	36.62	6.79
	3–4	0.88	0.1	0.05	−42.91	72.19	8.79
	4–5	0.58	0.1	0	−42.24	32.31	6.13
	5–6	0.81	0.11	0	−41.66	35.04	7.29
	6–7	0.44	0.09	0	−41.91	24.48	6.43
	7–8	0.25	0.06	0	−41.99	27.22	6.32
	8–9	−0.1	0.06	0	−41.83	15.82	5.67
	9–10	−0.23	0.06	0	−40.91	15.25	5.41
	10–11	−0.37	0.07	0	−40.74	12.69	5.28
	11–12	−0.62	0	0	−40.58	5.17	5.06
	12–13	−0.62	0.02	0	−39.99	5.94	5.22
	13–14	−0.69	0	0	−39.41	3.96	5.05
	14–15	−0.74	0	0	−39.08	7.19	5.06
	15–16	−0.82	0	0	−39.66	4.88	5.07
	16–17	−0.77	0	0	−39.91	3.88	5.06
	17–18	−0.54	−0.01	0	−39.08	30.41	5.89
	18–19	−0.86	−0.03	0	−40.16	4.06	5.1
	19–20	−0.84	0	0	−40.91	3.75	5.17

(continued)

Table 9.9 (continued)

Scenarios	Time after blast (h)	Statistical results					
		Average (NTU)	Median (NTU)	Mode (NTU)	Minimum (NTU)	Maximum (NTU)	Standard deviation
	20–21	−0.95	−0.06	0	−40.66	3.69	5.18
	21–22	−0.88	−0.06	0	−40.33	2.75	5.1
	22–23	−0.95	−0.03	0	−40.91	2.94	5.18
	23–24	−0.79	−0.03	0	−40.33	4.25	4.6
• With precipitation within 24 h after blasting • All precipitation data points prior to blasting or 30 days prior to blasting • Number of Blasting Event: 19	0–1	−0.78	0.08	0	−16.38	1.14	3.84
	1–2	−0.86	0.03	0	−16.7	7.61	5.05
	2–3	0.43	0.16	0	−16.6	8.7	4.68
	3–4	0.58	0.3	0	−16.63	7.52	4.88
	4–5	0.69	0.25		−16.65	9.65	5.04
	5–6	0.59	0.08	0	−16.55	10.34	5.18
	6–7	0.34	0.13	0	−16.58	11.4	5.39
	7–8	−0.12	0.05	0	−16.53	6.38	4.76
	8–9	−0.45	0	0	−16.78	4.41	4.48
	9–10	−0.71	0.1	0	−16.4	3.56	4.74
	10–11	−0.76	0.08	0	−16.48	3.19	4.39
	11–12	−0.91	0	0	−16.55	3.25	4.3
	12–13	−1.26	0	0	−16.5	3.56	5.16
	13–14	−0.9	0.05	0	−16.6	3.69	4.43
	14–15	−1.04	0.03	0	−16.48	3.44	4.62
	15–16	0.29	0	0	−16.68	28.63	8.19
	16–17	2.75	0	0	−16.7	79.44	19.12
	17–18	14.67	0	0	−16.6	306.75	70.88
	18–19	4.32	0	0	−16.6	94.38	22.68
	19–20	7.65	0	0	−16.7	107.03	29
	20–21	9.52	0	0.1	−16.7	150.2	36.84
	21–22	7.41	0.05	−0.08	−16.75	122.11	29.4
	22–23	4.17	0.03	0	−16.6	86.11	20.77
	23–24	3.84	0.08	0	−16.45	78.86	19.18
• With precipitation within 24 h after blasting • No precipitation 1 day prior to blasting • Number of Blasting Event: 10	0–1	−1.48	0.07	0	−16.38	0.61	5.24
	1–2	−0.77	0	0	−16.7	7.61	6.08
	2–3	0.1	0.34	0	−16.6	8.7	6.45
	3–4	0.33	0.93		−16.63	7.52	6.44
	4–5	0.75	0.82		−16.65	9.65	6.89
	5–6	0.47	0.23		−16.55	10.34	6.89
	6–7	−0.17	0.09		−16.58	7.98	6.33

(continued)

Table 9.9 (continued)

Scenarios	Time after blast (h)	Statistical results					
		Average (NTU)	Median (NTU)	Mode (NTU)	Minimum (NTU)	Maximum (NTU)	Standard deviation
	7–8	−0.47	0.01		−16.53	5.78	6.03
	8–9	−0.78	0	0	−16.78	4.41	5.87
	9–10	−0.85	0.08		−16.4	3.56	5.66
	10–11	−1.04	−0.01		−16.48	3.19	5.57
	11–12	−1.29	0	0	−16.55	3.25	5.51
	12–13	−1.29	0	0	−16.5	3.56	5.51
	13–14	−1.3	0.03	0	−16.6	3.69	5.55
	14–15	−1.34	0.01		−16.48	3.44	5.52
	15–16	1.19	0	0	−16.68	28.63	11
	16–17	6.11	0.01	0	−16.7	79.44	26.3
	17–18	28.83	−0.01	0	−16.6	306.75	97.79
	18–19	9.3	0	0	−16.6	94.38	30.87
	19–20	15.54	0	0	−16.7	107.03	39.01
	20–21	19.04	−0.03		−16.7	150.2	49.86
	21–22	14.74	−0.04	−0.08	−16.75	122.11	39.83
	22–23	8.8	0.01		−16.6	86.11	28.2
	23–24	8.11	0.04	0	−16.45	78.86	26.04
• With precipitation within 24 h after blasting • With precipitation 1 day prior to blasting • Number of Blasting Event: 9	0–1	−0.01	0.08	0.13	−2.5	1.14	1.04
	1–2	−0.96	0.08		−11.44	1.45	3.97
	2–3	0.79	0.08		−0.44	3.78	1.47
	3–4	0.85	0.08		−3	5.86	2.61
	4–5	0.63	0.15		−2.81	3.7	1.97
	5–6	0.73	0.05	0	−3.31	6.48	2.62
	6–7	0.9	0.13		−5.56	11.4	4.41
	7–8	0.26	0.05	0	−6	6.38	3.12
	8–9	−0.08	0.05	0	−6.06	3.13	2.48
	9–10	−0.55	0.13		−10.44	2.7	3.81
	10–11	−0.45	0.13		−7.94	1.65	2.86
	11–12	−0.5	0.06		−7.44	1.23	2.65
	12–13	−1.22	0.1		−14.63	1.55	5.07
	13–14	−0.47	0.15		−8.31	1.6	3.02
	14–15	−0.71	0.13		−10.31	1.55	3.66
	15–16	−0.71	0.1	0	−10.06	1.45	3.57
	16–17	−0.98	0	0	−11	1.53	3.8

(continued)

Table 9.9 (continued)

Scenarios	Time after blast (h)	Statistical results					
		Average (NTU)	Median (NTU)	Mode (NTU)	Minimum (NTU)	Maximum (NTU)	Standard deviation
	17–18	−1.06	0	0	−11.81	1.43	4.07
	18–19	−1.21	0.05		−12.94	1.58	4.43
	19–20	−1.12	0		−11.88	1.43	4.07
	20–21	−1.07	0.05		−11.94	1.55	4.12
	21–22	−0.74	0.13		−11.69	2.36	4.19
	22–23	−0.96	0.06		−12.69	1.48	4.44
	23–24	−0.91	0.08		−11.81	1.52	4.13
• With precipitation within 24 h after blasting • With precipitation 5 days prior to blasting • Number of Blasting Event: 18	0–1	−0.83	0.07	0	−16.38	1.14	3.95
	1–2	−0.91	0.05	0	−16.7	7.61	5.19
	2–3	0.32	0.12	0	−16.6	8.7	4.79
	3–4	0.51	0.26	0	−16.63	7.52	5.01
	4–5	0.41	0.2		−16.65	9.65	5.03
	5–6	0.28	0.06	0	−16.55	10.34	5.15
	6–7	0.17	0.13	0	−16.58	11.4	5.49
	7–8	−0.27	0.03	0	−16.53	6.38	4.86
	8–9	−0.56	0	0	−16.78	4.41	4.59
	9–10	−0.85	0.09	0	−16.4	3.56	4.84
	10–11	−0.85	0.07	0	−16.48	3.19	4.5
	11–12	−0.97	0.03	−0.13	−16.55	3.25	4.42
	12–13	−1.33	0.04	0	−16.5	3.56	5.3
	13–14	−0.95	0.08	0	−16.6	3.69	4.56
	14–15	−1.1	0.08		−16.48	3.44	4.74
	15–16	0.31	0.03	0	−16.68	28.63	8.43
	16–17	2.9	0.01	0	−16.7	79.44	19.66
	17–18	15.49	0	0	−16.6	306.75	72.84
	18–19	4.57	0.03	0	−16.6	94.38	23.31
	19–20	8.08	0	0	−16.7	107.03	29.78
	20–21	10.05	0.03	0.1	−16.7	150.2	37.83
	21–22	7.83	0.09	−0.08	−16.75	122.11	30.19
	22–23	4.42	0.04	0	−16.6	86.11	21.34
	23–24	4.05	0.08	0	−16.45	78.86	19.72

Table 9.10 Characteristics of response in average turbidity for each scenario

Scenarios		Time to response[1] (Hours after blasting)	Maximum response (NTU)	Time to maximum response (Hours after blasting)	Duration of response (Hours)
No precipitation within 24 h after blasting	No precipitation 1 day prior to blasting	0–1	1.5	5–6	22
	No precipitation 5 days prior to blasting	0–1	1.39	5–6	18
	With precipitation 5 days prior to blasting	–[2]	–	–	–
	No precipitation 30 days prior to blasting	0–1	0.33	5–6	18
	With precipitation 30 days prior to blasting	2–3	0.88	3–4	6
With precipitation within 24 h after blasting	No precipitation 1 day prior to blasting	2–3[3]	28.83	17–18	22
	With precipitation 1 day prior to blasting	2–3	0.90	6–7	6
	No precipitation 5 days prior to blasting	0–1	6.08	5–6	17
	With precipitation 5 days prior to blasting	2–3[4]	10.05	20–21	22
	With precipitation 30 days prior to blasting	2–3[5]	14.67	17–18	22

Notes

(1) The response is considered to be positive when two consecutive average turbidity changes are greater than 0.

(2) The average turbidity change is greater than 0 at the first hour. Because the turbidity changes in the remaining 23 h are negative, the positive point was considered to be anomalous.

(3) The average turbidity change is greater than 0 in two separate time periods that is divided by negative changes. The duration is the response time to the end of the 24 h after the blasting.

(4) The average turbidity change is greater than 0 in two separate time periods that is divided by negative changes. The duration is the response time to the end of the 24 h after the blasting.

(5) The average turbidity change is greater than 0 in two separate time periods that is divided by negative changes. The duration is the response time to the end of the 24 h after the blasting.

Because only 4 events are available for this scenario, the statistical results may not represent a full range of conditions, such as the quantity of dynamite in the blasts, the bench level of the blasting, and the number of blast holes. The scenario, in which there were no precipitation activities 5 days prior to blasting and no precipitation within 24 h after blasting, was also used for the blasting impact analysis. The precipitation impact that occurred 6 days prior to blasting could be reasonably considered to be minimal. In this scenario 34 blasting events are available. The average hourly turbidity prior to blasting is 1.58 NTU, which is slightly higher than the previous value of 1.44 NTU. The turbidity change has a positive response in the first hour after blasting and the response lasts approximately 18 h. The magnitude of the response is less than 2 NTU, with a range from 0.03 to 1.39 NTU. The maximum response occurs 5–6 h after blasting. For any particular blasting event out of the 34 events, the turbidity change could be up to 35.04 NTU, which also occurred 5–6 h after blasting.

The scenario, in which there was no precipitation 1 day prior to blasting and no precipitation within 24 h after blasting, may provide the upper limits of the blasting impact on turbidity at SP-1 Spring, because the turbidity values may be impacted by both precipitation 2 days prior to blasting and the blasting itself. In this scenario 57 blasting events are available. The average hourly turbidity prior to blasting is 1.68 NTU. The turbidity change has a positive response in the first hour after blasting and the response lasts approximately 22 h. The magnitude of the average response is also less than 2 NTU, with a range from 0.06 to 1.50 NTU. The maximum response occurs 5–6 h after blasting. For any particular blasting event out of the 57 events, the turbidity change could be up to 36.62 NTU, which occurred 2–3 h after blasting.

The quantitative analysis indicates that the impact of blasting in the quarry on the turbidity at SP-1 Spring is likely no more than 2 NTU. The impact can occur in the first hour after blasting and reach a maximum 5–6 h later. The duration of the impact is approximately 22 h. These statistical conclusions are based on the data available in each scenario and they may not be applicable to any one particular blasting event because other factors may also impact the turbidity values at SP-1 Spring.

9.8.4 Impact of Precipitation on Turbidity at SP-1 Spring

The impact of precipitation on turbidity at SP-1 Spring is obvious when the precipitation occurred either prior to blasting or after blasting. In the scenario in which precipitation occurred within 24 h after blasting and no precipitation occurred 1 day prior to blasting, the turbidity change is 28.83 NTU based on 10 blasting events. Of the 10 blasting events, the turbidity change could reach up to 306.75 NTU in one particular event.

When precipitation occurred within 24 h after and 1 day prior to blasting, the turbidity change is relatively small, up to 0.9 NTU based on 9 blasting events. This is because the rainfall prior to blasting had already elevated the turbidity at SP-1 Spring.

When there was no precipitation within 24 h after blasting but rainfall occurred 1 day prior to blasting, 23 of the 24 turbidity changes are less than 0 based on 18 blasting events. The elevated turbidity caused by the rainfall 1 day prior to blasting clearly masked any impact caused by blasting. The average hourly turbidity prior to blasting is 4.18 NTU. The range of the turbidity change is from −4.98 to −2.49 NTU.

The masking effect is also present in the scenario, in which there was no precipitation within 24 h after blasting but rainfall occurred 5 days prior to blasting, albeit the effect is subdued. There are 51 blasting events that fall into this scenario. All the turbidity changes calculated for this scenario are less than 0, with the minimum being −2.00 NTU.

The masking effect of rainfall on turbidity at SP-1 Spring is further reduced when there was no precipitation within 24 h after blasting but rainfall occurred 30 days prior to blasting. There are 98 blasting events available for this scenario. The turbidity changes include values less than 0 and greater than 0. The positive values occurred 2–8 h after blasting. It seems that both precipitation 30 days prior to blasting and the blasting itself have some impacts on the turbidity at SP-1 Spring. When the precipitation 30 days prior to blasting has more significant impact than blasting, the turbidity changes are less than 0. Otherwise, the blasting impact would overcome the masking effect of precipitation on the turbidity and the turbidity changes would be positive. Overall, the impact of precipitation 30 days prior to the blasting is minimal but seems to be comparable in magnitude to the effect of blasting on the turbidity at SP-1 Spring. Obviously if these data were broken down into finer time increments, one could determine at what point the effects of blasting exceed the residual masking effect of the precipitation.

In order to further study the impact of rainfall on the turbidity at SP-1 Spring without the interference from blasting, the rainfall events that occurred 48 h prior to and after the blasting events were selected. Forty-six rainfall events are available from November 1997 to June 2004. Table 9.11 shows the selected events and their turbidity information. The average turbidity values range from 0.5 to 129.32 NTU, with an average of 14.12 NTU.

The quantitative analysis indicates that precipitation is a major factor that affects the turbidity at SP-1 Spring. Precipitation, either prior to or after blasting, impacts the turbidity changes. It should be pointed out that the turbidity at SP-1 Spring can be affected by many other factors such as logging, construction, and clearing in the spring catchment basin, and the transport characteristics of the sediments in the marble aquifer, above and beyond the two factors analyzed in this study. In addition, the turbidity values used in this study are hourly averages. More frequent measurements in turbidity values may show higher change in magnitude, and less frequent measurement may show lower change in magnitude. For example, if the analysis were performed using the turbidity values every 15 min, the turbidity change might be higher than the results obtained from the hourly averages; however turbidity spikes that only lasted minutes would have only minor impact. On the other hand, the turbidity change would be smaller if daily averages were used.

Based on the statistical analysis from the data available, the preliminary conclusions can be obtained as follows:

- Precipitation is the most important factor that impacts turbidity at SP-1 Spring. Unconsolidated particles are transported mainly during flood flow episodes in response to rainfall events. When water velocities decrease following a storm event, the sediments in suspension may settle out and temporarily be stored in the aquifer. Very small particles, mainly clay or silt, are transported in suspension in the water in between precipitation events. Some small particles may be rolled

Table 9.11 Rainfall events and their impacts on turbidity at SP-1 spring

Rainfall information				Turbidity data at SP-1 spring			
Beginning time (m/d/y hh:mm)	Ending time (m/d/y hh:mm)	Total rainfall (inch)	Average intensity (inch/h)	Background (NTU)	Minimum (NTU)	Maximum (NTU)	Average (NTU)
1/6/1998 20:00	1/20/1998 1:00	9.72	0.0307	2	0.5	204	81.24
1/26/1998 14:00	2/10/1998 20:00	12.74	0.0348	204	9.25	204	129.32
2/13/1998 4:00	3/3/1998 3:00	5.46	0.0134	9.42	8.58	204	50.76
3/11/1998 15:00	3/13/1998 15:00	0.49	0.0204	15.33	15.33	29.33	19.94
11/23/1998 13:00	11/30/1998 19:00	4.27	0.0245	0.5	0.2	3.28	0.5
1/15/1999 2:00	1/26/1999 14:00	15.36	0.0557	0.73	0.45	162.15	18.63
1/30/1999 23:00	2/1/1999 9:00	1.19	0.035	1.6	1.5	1.9	1.69
2/6/1999 10:00	2/10/1999 11:00	9.6	0.099	1.77	1.57	176.37	15
2/13/1999 17:00	2/13/1999 23:00	0.35	0.0583	3.25	1.65	3.28	2.45
2/16/1999 5:00	2/22/1999 6:00	4.4	0.0303	1.9	1.6	63.53	9.62
2/24/1999 23:00	2/26/1999 14:00	0.79	0.0203	1.75	1.58	1.85	1.74
3/1/1999 1:00	3/3/1999 6:00	0.7	0.0132	1.8	1.75	2.1	1.98
3/8/1999 13:00	3/11/1999 8:00	1.52	0.0227	1.68	1.73	16.88	4.1
3/14/1999 5:00	3/25/1999 6:00	4.87	0.0184	1.85	1.5	4.28	1.88
3/30/1999 20:00	3/31/1999 9:00	0.52	0.04	2.1	1.98	2.05	2.03
4/5/1999 8:00	4/12/1999 8:00	4.04	0.024	2	1.8	52.2	4.59
5/2/1999 19:00	5/3/1999 15:00	0.28	0.014	0.95	0.85	1.45	0.92
5/30/1999 8:00	6/3/1999 10:00	0.53	0.0054	0.78	0.73	1.18	0.84
9/8/1999 17:00	9/9/1999 4:00	0.18	0.0164	0.95	0.93	1.3	1.01

(continued)

Table 9.11 (continued)

Rainfall information				Turbidity data at SP-1 spring			
Beginning time (m/d/y hh:mm)	Ending time (m/d/y hh:mm)	Total rainfall (inch)	Average intensity (inch/h)	Background (NTU)	Minimum (NTU)	Maximum (NTU)	Average (NTU)
1/1/2000 14:00	1/1/2000 22:00	0.06	0.0075	0.68	0.63	0.7	0.68
1/11/2000 4:00	1/26/2000 3:00	5.15	0.0143	0.6	0.6	978.4	84.06
1/30/2000 0:00	2/1/2000 0:00	1.55	0.0323	4.78	3.35	4.85	3.89
12/13/2001 20:00	12/14/2001 14:00	0.89	0.0494	5.25	5.13	9.81	6.04
12/16/2001 23:00	12/17/2001 10:00	1.15	0.1045	2.94	0.94	35.5	5.56
12/21/2001 16:00	12/23/2001 5:00	0.89	0.0241	21.88	9.63	46.63	17.45
12/27/2001 16:00	1/3/2002 3:00	7.51	0.0485	1.94	0.13	1000.56	118.46
1/21/2002 14:00	1/22/2002 0:00	0.15	0.015	0.94	0.94	3.25	2.03
1/26/2002 4:00	1/29/2002 12:00	1.1	0.0138	4.75	0.94	5.5	2.62
2/16/2002 16:00	2/20/2002 2:00	2.04	0.0249	5.56	1.75	114.38	9.41
3/6/2002 0:00	3/7/2002 15:00	1.06	0.0272	2.5	2.06	13.38	3.49
3/10/2002 0:00	3/10/2002 8:00	0.52	0.065	2.56	1.94	6	2.87
3/15/2002 16:00	3/17/2002 16:00	0.61	0.0127	2.56	2.06	8.69	2.83
4/4/2002 8:00	4/5/2002 15:00	0.07	0.0023	2.38	2.13	3	2.55
4/22/2002 17:00	4/26/2002 10:00	0.23	0.0026	2.56	2.5	2.69	2.6
5/15/2002 1:00	5/16/2002 22:00	0.83	0.0184	2.75	2.25	2.75	2.52
11/14/2003 14:00	11/18/2003 1:00	0.81	0.0098	1.6	1.6	1.6	1.6
11/30/2003 3:00	12/7/2003 8:00	4.29	0.0248	1.6	0.23	4.33	1.39
12/19/2003 15:00	1/2/2004 20:00	12.42	0.0365	0.31	0.08	190.11	5.01

(continued)

Table 9.11 (continued)

Rainfall information				Turbidity data at SP-1 spring			
Beginning time (m/d/y hh:mm)	Ending time (m/d/y hh:mm)	Total rainfall (inch)	Average intensity (inch/h)	Background (NTU)	Minimum (NTU)	Maximum (NTU)	Average (NTU)
1/6/2004 21:00	1/10/2004 4:00	0.7	0.0089	2.81	1.6	11.94	4.98
1/13/2004 3:00	1/15/2004 4:00	0.06	0.0012	1.6	1.52	1.6	1.58
1/19/2004 2:00	1/21/2004 2:00	0.03	0.0006	1.6	1.6	1.6	1.6
1/24/2004 0:00	1/24/2004 10:00	0.68	0.068	1.6	1.45	3.58	1.75
1/26/2004 18:00	2/4/2004 5:00	3.46	0.0171	1.6	1.29	96.08	8.57
2/6/2004 20:00	2/6/2004 23:00	0.05	0.0167	1.6	1.52	1.6	1.57
2/16/2004 0:00	2/27/2004 14:00	4.99	0.0179	1.68	1.45	81.41	4.44
4/19/2004 2:00	4/21/2004 11:00	0.75	0.0132	1.6	1.6	1.75	1.66

along the bottom of the conduits by the water flowing over the particles. Precipitation can cause an elevation in turbidity of 14.12 NTU on average, with a range from 0.5 to 129.32 NTU.

- The blasting activities in the quarry have some impacts to SP-1 Spring. Blasting has the potential to elevate the turbidity at SP-1 Spring by less than 2 NTU on an average basis. Its maximum response tends to occur 5–6 h after blasting, and the impact can last up to 22 h.

9.9 Conclusions

The study area has very complex geology and hydrogeology and has a well-developed karst terrane characterized by sinkholes, sinking streams, caves and large springs in marble. The subsurface in the karst land consists of two distinctive zones—the deeper saturated zone (karst aquifer) and the overlying unsaturated zone (perched aquifer). The unsaturated zone has a high capacity for storing water temporarily, and it has a poor hydraulic connection with the saturated zone. The saturated water level in the saturated zone at the projected bottom of the quarry in the amendment area is below 741 feet amsl, according to the water levels measured in M1B in 2003 and 2004. The unsaturated zone can be over 300 feet thick.

The karst aquifer that feeds SP-1 Spring is well developed. Water flows through discrete solution-widened fractures/conduits in the saturated zone at estimated flow rates from 20 feet/day to 2560 feet/day. More than one conduit system may exist, and the conduits tend to be larger as they are closer to SP-1 Spring. Monitoring wells M1B and NZA likely intercepted, or are very close to, the conduits in the aquifer.

The karst aquifer receives recharge from allogenic as well as autogenic sources. Allogenic recharge is from the borderlands, including sinking streams SS-8, SS-1, SS-2, and SS-4, sinkhole plain, and the landslide deposits. The autogenic recharge is from the karst land including the existing quarry and the amendment area. The total delineated catchment area for SP-1 Spring is approximately 2000 acres. Allogenic recharge accounts for approximately 90% of the total catchment area; whereas autogenic recharge accounts for approximately 10% of the total catchment area. The recharge sources for SP-1 Spring are predominantly allogenic.

Turbidity increase in response to rainfall events is typical of well-developed karst springs and SP-1 Spring is no exception. The turbidity response of SP-1 Spring to rainfall is predominantly caused by the multiple allogeneic recharge sources, especially recharge from sinking streams SS-1 and SS-2, the sinkhole plain, and the landslide deposits. Transport of sediment particles through conduit systems and discharge of the sediments at springs are essential functions of a karst aquifer. In well-developed aquifers like the one under study, where surface streams sink underground, the insoluble materials carried by the sinking surface streams are carried through the conduit system and washed out through the springs. If the flow is not sufficient to remove sediments from conduits or fractures, the karst aquifer will eventually become plugged.

The discharge of SP-1 Spring in response to rainfall is predominantly caused by allogenic recharge. The characteristics of the response may depend on the sources that dominate a particular recharge event. It takes approximately one to two weeks for the majority of water recharged at sinkhole SH-11, sinking streams SS-1 and SS-2 to arrive at SP-1 Spring, whereas it takes approximately 3 months for the majority of water recharged at sinking streams SS-8 and SS-4 to reach SP-1 Spring.

Multivariate statistical analysis further substantiates the fact that precipitation is the most important factor impacting the turbidity at SP-1 Spring. Rainfall events can raise the turbidity level of 14.12 NTU on average, with a range from 0.5 to 129.32 NTU, whereas the "background" turbidity, in the absence of rainfall, is approximately 1.44 NTU. Unconsolidated particles are transported mainly during flood flow episodes in response to rainfall events. The blasting activities in the quarry may elevate the turbidity at SP-1 Spring by less than 2 NTU on an average basis. Its maximum response tends to occur 5–6 h after blasting and the impact lasts 22 h.

References

Abkemeier TJ, Stephenson RW (2003) Remediation of a Sinkhole Induced by Quarrying, in Sinkholes and the Engineering and Environmental Impacts of Karst, Proceedings of the Ninth Multidisciplinary Conference, Sept. 6–10, 2003, Huntsville, Alabama, Beck, BF (ed.) Geotechnical Special Publication No. 122, ASCE, pp 605–614

Akers JP, Jackson LE Jr (1977) Geology and ground water in western Santa cruz county, California, with particular emphasis on the Santa margarita sandstone: U.S. geological survey water resources investigation, 77–15 open file report

Alexander EC Jr, Quinlan JF (1992) Practical tracing of groundwater with emphasis on karst terranes, 2nd edn. Geological Society of America, Boulder CO, Short Course Manual

Aley T (1999) Groundwater tracing book. The Ozark Underground Laboratory, MO, p 33p

ASTM (American Society for Testing and Materials) (1998) ASTM D 6312-98, Standard guide for developing appropriate statistical approaches for ground-water detection monitoring programs. annual book of standards, West Conshohocken, PA. pp 1–14

Atteia O, Kozel R (1997) Particle size distributions in waters from a karstic aquifer: from particles to colloids. J Hydrol 201:102–119

Brabb EE (1989) Geologic map of santa cruz county, California: U.S. Geological survey miscellaneous investigation series maps I-1905

Bretz JH (1964) Vadose and phreatic features of limestone caves. J Geol 675–811

Clark JC (1981) Stratigraphy, paleontology, and geology of the central Santa cruz mountains. U.S. Geological Survey Professional Paper, California Coast Ranges

Crawford Hydrology Lab (2003) Laboratory procedure and quality assurance/quality control plan. Western Kentucky University, Bowling Green, KY

Drever JI (1997) The geochemistry of natural waters, 3rd edn. Prentice Hall, NJ, p 436

EPA (1989) Statistical analysis of ground-water monitoring data at RCRA facilities. interim final guidance, EPA/530-SW-89-026, Washington DC: office of solid waste, U.S. Environmental Protection Agency

EPA (1992) Statistical analysis of ground-water monitoring data at RCRA facilities: addendum to interim final guidance, in statistical training course for ground-water monitoring data analysis. EPA530-R-93-003

EPA (1996) Guidelines for wellhead and springhead protection area delineation in carbonate rocks: environmental protection agency, region 4, U. S. EPA 904-B-97–003. Prepared by Eckenfelder, Inc., Nashville, TN

EPA (Environmental Protection Agency) (2002) The QTRACER2 program for tracer-breakthrough curve analysis for tracer tests in karstic aquifers and other hydrologic systems: national center for environmental assessment—Washington Office, EPA/600/R-02/001

Even H, Carmi I, Magaritz M, Gerson R (1986) Timing the transport of water through the upper vadose zone in a karstic system above a cave in Israel. Earth Surf Proc Land 11:181–188

Field MS (2002) A review of some tracer-test design equations for tracer-mass estimation and sample-collection frequency. Environ Geol 43:867–881

Field MS, Wilhelm RG, Quinlan JF, Aley TJ (1995) An assessment of the potential adverse properties of fluorescent tracers dyes used for groundwater tracing: environmental monitoring and assessment. 38:75–96

Foose RM (1953) Ground-water behavior in the Hershey Valley, Pennsylvania. Geol Soc Am Bull 64:623–645

Fournie M, Massei N, Bakalowicz M, Dussart-Baptista L, Rodet J, Dupont JP (2007) Using turbidity dynamics and geochemical variability as a tool for under-standing the behavior and vulnerability of a karst aquifer. Hydrogeol J 15:689–704

Green JA, Pavlish JA, Leete JH, Alexander EC, Jr (2003) Quarrying impacts on groundwater flow paths. In: Beck BF (ed) Ninth Multidisciplinary Conference on Sinkholes and the Engineering and Environmental Impacts of Karst. Geotechnical Special Publication No. 122, pp 216–222

Gunn J (1986) A conceptual model for conduit flow dominated karst aquifer. In: Gunay G, Johnson AI (eds) Proceedings of karst water resources conference in Ankara, Turkey. International association of hydrological sciences. pp 587–596

Hubbard EF, Kilpatrick FA, Martens LA, Wilson JF Jr (1982) Measurement of the time of travel and dispersion in streams by dye tracing: U.S. Geological survey techniques of water-resources investigations, Book 3, Chapter A9, p 44

Käss W (1998) Preparation and realization of tracing tests. In: Käss W (ed) Tracing technique in geohydrology, A.A. Balkema, Rotterdam, Netherlands. p 315–39

Lolcama JL, Cohen HA, Tonkin MJ (1999) Deep karst conduits, flooding, and sinkholes: Lessons for the aggregates industry. Eng Geol 65(2–3):151–157

Mahler BJ, Lynch FL (1999) Muddy waters: temporal variation in sediment discharging from a karst spring. J Hydrol 214:165–178

McCarthy JF, Schevenell LS (1998) Processes controlling colloid composition in a fractured and karstic aquifer in eastern tennessee USA. J Hydrology 206:191–214

Mull, DS, Liebermann, TD, Smoot, JL, and Woosley, LH (1988) Application of dye-tracing techniques for determining solute-transport characteristics of ground water in karst terranes: U.S. Environmental protection agency, EPA 904/6-88-001. p 103

Ongley E (1996) Sediment measurements, Chapter 13. In: Bartran J, Balance R (eds) Water quality monitoring—A practical guide to design and implementation of freshwater quality studies and monitoring programs, UNEP/WHO

Phillips LR (1990) Depositional and structural controls on the distribution of tar sands in the Santa cruz mountains, California. In: Greene HG (eds) Geology and tectonics of central California coast region, San Francisco to monterey. Pacific Section AAPG Book GB67

Quinlan JF (1986) Water-tracing techniques in karst terranes: practical karst hydrogeology with emphasis on groundwater monitoring, a short course. National Water Well Association, Dublin, Ohio, p 25

Shevenell LS, McCarthy JF (2002) Effects of precipitation events on colloids in a karst aquifer. J Hydrology 255:50–68

Smart PL (1984) A review of the toxicity of twelve fluorescent dyes used for water tracing: national speleological society. Bulletin 46:21–33

Smart PL, Laidlaw IMS (1977) An evaluation in some fluorescent dyes used for water tracing. Water Resour Res 13:15–33

Sutter RD, Verhoeven R (2001) Simulation of sediment transport during flood events: laboratory work and field experiments. Hydrol Sci 46(4):559–610

Todd D (1963) Ground water hydrology of the bonnie doon quarry site, Santa cruz county. California. Prepared for Wisser and Cox, Berkeley, CA

Townsend MA, Macfarlane PA (2003) Seasonality effects on karst aquifers, flint hills region, central Kansas, Kansas geological survey. In: Beck BF (ed) Sinkholes and the engineering and environmental impacts of Karst, geotechnical special publication #122, ASCE. pp 277–286

Valdes D, Dupont J-P, Massei N, Laignei B, Rodet J (2005) Investigation of karst hydrodynamics and organization using autocorrelations and T-C curves. J Hydrol 329:432–443

White WB (1988) Geomorphology and hydrology of karst terrains. Oxford, New York, p 464p

White WB (1993) Analysis of karst aquifers. In: Alley WM (ed) Regional ground-water quality. Van Nostrand Reinhold, New York

Williams PW (1985) Subcutaneous hydrology and the development of dolin and cockpit karst. J Geomorphol 29:463–482

Wisser and Cox (1960) Groundwater conditions of quarry site, Santa cruz county. California, San Francisco, CA

Worthington SRH, Smart CC (2003) Empirical determination of tracer mass for sink to spring tests in karst. In: Beck BF (ed) Sinkholes and the engineering and environmental impacts of karst. ASCE geotechnical special publication No. 122

Chapter 10
Best Practices in Water Hazards Control in Coal Mines

10.1 General Guidance

These best practices are to strengthen the prevention and control of water in coal mines, prevent and reduce water hazards, and ensure the safety of coal mine workers. These practices comply with the "Safe Production Law", "Mine Safety Law", "Special Provisions of the State Council on Preventing Coal Mine Production Safety Accidents". These practices apply to coal mine enterprises (mines) and associated institutions. The work of water prevention and control shall adhere to the following principles:

- Predicting and forecasting.
- Investigating if safety is uncertain.
- Investigating prior to excavating/mining.
- Mitigating as needed prior to excavating/mining.
- Taking comprehensive measures for mitigation to minimize or eliminate water inrush risks.

1.1 The water hazards vary in coal mines. Qualified professionals shall be assigned to be responsible for water control in accordance with mine-specific situations. A water control team shall be assembled as appropriate and be equipped with specialized equipment for exploration and dewatering. Coal mine enterprises and mines with complex and extremely complicated hydrogeological conditions shall establish special water prevention and control institutions. Coal mining enterprises and mines shall educate and train employees on the prevention and control of water, ensure that employees have the necessary knowledge of water prevention, improve the skills of water prevention and prevention, and the ability to withstand floods.

1.2 Coal mining enterprises and mines shall establish and improve the responsibility system for water hazard prevention and control, the management approach for water hazard prevention and control techniques, the technical approach for

S. Dong et al., *Methods and Techniques for Preventing and Mitigating Water Hazards in Mines*, Professional Practice in Earth Sciences,
https://doi.org/10.1007/978-3-030-67059-7_10

predicting and controlling water disasters, and the proper measures for investigating and controlling hidden dangers of water hazards.

1.3 Coal mining enterprises and mines shall prepare their own long-term plans and annual plans for prevention and control of water and organize their implementation. Where the hydrogeological conditions within and outside the jurisdiction of a coal mining enterprise or mine are unclear, effective measures shall be taken to ascertain the water hazard. Mining activities are strictly prohibited until the water hazard situation is understood.

1.4 Upon identifying signs of water percolation, coal mines should immediately stop the mining operations in the area potentially threatened by water hazard, evacuate the workers to a safe place, take effective safety measures, and perform the root-cause-analysis of the water percolation.

1.5 Coal mining enterprises and mines shall encourage scientific research and technological innovations for prevention and control of mine, promote application of new technologies and equipment to preventing and controlling water, and improve scientific knowledge and skills of water-control professionals. Coal mining enterprises and mines with complicated hydrogeological conditions and extremely complex facilities shall be equipped with necessary equipment for water hazard emergency responses.

10.2 Classification and Basic Data Requirements of Mine Hydrogeology

10.2.1 Classification of Mine Hydrogeology

2.1.1 According to the distribution of aquifers and water bodies, mines and surrounding old air, or the distribution of water inrush or water inrush, mines affected by water damage and the difficulty of water prevention and control, mines The hydrogeological types are divided into four types: simple, medium, complex, and extremely complex (see Table 1.2).

2.1.2 The mine shall conduct studies on the hydrogeological conditions of the mining area, compile a report on the classification of hydrogeological types of the mine, and determine the hydrogeological type of the mine in the mining area. The report on the classification of mine hydrogeology is organized and approved by the chief engineer of the coal mining enterprise.

The report on the classification of mine hydrogeology should include the following main contents:

1. The location, mine layout and relationships with neighboring properties, geography and other natural factors;
2. Review of previous geological and hydrogeological investigations;
3. Hydrogeological conditions of the mine site and distributions and characteristics of aquifers and aquitards;

4. Analysis of mine water recharge sources, distribution of old mine pools within the mine site and its surrounding regions;
5. Analysis of the components of the mine water inflow, the locations of the main water inrush points, the amount of water inrush and the remedial measures;
6. Evaluating the extent to which the mining is impacted by water hazard and the difficulty of water prevention and control work;
7. Classification of mine hydrogeological types and suggestions for water prevention and control measures.

2.1.3 The hydrogeological type of mines shall be re-determined every three years. After a major water inrush accident, the mine shall re-evaluate the hydrogeological conditions and re-determine the hydrogeological type of the mine within one year. A major water inrush accident refers to a water inrush accident in which the amount of water inrush reaches 300 m^3/h or more for the first time or causes more than three deaths.

10.2.2 Fundamental Data of Mine Water Control

2.2.1 Mines shall prepare mine field geological reports, shaft construction designs and construction geological reports. The mine field geological report, shaft construction design and construction geological report shall have corresponding water prevention and control contents.

2.2.2 Mines shall prepare water control maps in accordance with the regulations and actual needs. The following are the minimum requirements:

Mine water recharge map

The mine water recharge map is a comprehensive drawing of the measured hydrogeological data of the mine. It is one of the main bases for analyzing the water recharge mechanism, predicting the water hazard and formulating water prevention measures. It is also an essential drawing for the prevention and control of mine water hazard. Generally, the plan of the mining project is used as the base map for compilation. The scale is 1:2,000–1:5,000. The main contents are as follows:

- All types of water inrush points should be numbered uniformly, including date of water inrush, water inflow, water level (water pressure), water temperature and other characteristics associated with the water inrush.
- Catchment area and amount of water in abandoned wells, abandoned shafts, goafs, and old adits.
- Underground waterproof gate, sluice wall, drainage hole, coal (rock) pillar, pump station, water storage tank, number of pumps and their capacity.
- Underground water conveyance route.
- Location of the underground water inflow observation station (point).
- Any other data that is important to the mine.

The mine water recharge map should be regularly updated with the progress of mining.

Hydrograph of mine water inflow and related factors

The dynamic curve of mine water inflow and various related factors is a map that comprehensively reflects the dynamic variation of mine water recharge and predicts the trend of mine water inrush. Each mine should select the following related factors according to the specific conditions:

- A plot of the relationship between mine water inflow and precipitation, and groundwater level.
- A plot of the relationship between the amount of water inflow from the mine and the length of the unit and the area of the unit.
- Graph of mine water inflow against surface water recharge or water level.
- Graph of mine water inflow against mining depth.

Comprehensive hydrogeological map

The comprehensive hydrogeological map of the mine is one of the drawings reflecting the hydrogeological conditions of the mine and is the main reference for the mine water prevention and control work. The integrated hydrogeological map is generally compiled on the basis of the topographic and geological maps of the mine field, with a scale between 1:2,000 and 1:10,000. The main contents are:

- The lateral distribution of the outcrop of the bedrock aquifer (including karst) and the aquifer at the bottom of the alluvium (sand, gravel, or sand raft).
- Surface water bodies, hydrological observation stations, wells, springs, and collapsed columns.
- Hydrogeological exploratory boreholes, monitoring wells, and pumping test results.
- Bedrock contours (applicable to concealed coal fields).
- Main roadway, recovery range and the underground water inrush point.
- Water level (pressure) contours of main aquifer.
- Locations and mining areas of old kiln and small-scale mine, and water inrush points.
- When conditions permit, divide into hydrogeological units vertically and hydrogeological zones laterally.

Comprehensive hydrogeological column

The comprehensive hydrogeological column of the mine is a drawing reflecting the combined relationship between the aquifer, the aquitard and the coal seam and the number of aquifers, thickness and water enrichment. Generally, the corresponding scale is used along with the comprehensive hydrogeological map of the mine. The main contents are:

- Aquifer lithology of, thickness, geologic time, and karst development.

- Hydrogeological test parameters of each aquifer.
- Type of water quality of the aquifer.

Hydrogeological cross-section

The hydrogeological cross-section of the mine mainly reflects the spatial relationship between the aquifer, the aquitard, the fold and fault structure and the coal seam. The main contents are:

- Aquifer lithology, thickness, burial depth, and karst development depth.
- Hydrogeological exploratory boreholes and their testing data.
- Observation wells and their observation data.
- Surface water bodies and their water levels.
- Location of main shaft.

Mine hydrogeological cross-sections are generally based on representative geological profiles along strike and dip.

Water level (pressure) contours of mine-recharge aquifer

The water level (pressure) contour map mainly reflects the flow field characteristics of groundwater. For hydrogeologically complex and extremely complex mines, the main aquifers (groups) should be consistently drawn with water level (pressure) contours to analyze the dynamics of mine dewatering. The scale is between 1:2,000 and 1:10,000. The main contents are:

- Aquifer, coal seam outcrop, and main faults.
- Location of elevations of hydrogeological exploratory boreholes, observation wells, monitoring wells, spring, and groundwater level (pressure) measurement.
- Locations of surface water bodies such as rivers, canals, mountain ponds, reservoirs, and subsided water reservoirs, ground elevations, and water surface elevations over the same period.
- Mine shaft location, mining scope, and primary highways and railways.
- Groundwater counts and groundwater flow direction.
- Isopleth of the aquitard thickness underlying the minable coal seam (when the main aquifer affected by mining is under the bottom of the minable coal seam).
- Location of water inrush and flow rate of water inrush.

Regional hydrogeological map

Regional hydrogeological maps are generally prepared after regional hydrogeological surveys and are based on the 1:10,000-1:100,000 regional geological map. It is essential to have a description to go with the drawing. Mines with complex and extremely complex hydrogeological conditions should prepare this map carefully. The main contents are:

- Divide of surface water system, watershed boundary and geomorphic unit.
- Main aquifer outcrops, loose layers and other thick lines.
- Groundwater natural discharge point and artificial exposure point.

- Karst feature inventory and structural fracture zone.
- Hydrogeological drilling and pumping test results.
- Groundwater and other water level contours and groundwater flow direction.
- Divide region based on groundwater recharge, runoff, and discharge.
- Divide region into different hydrogeological zones and hydrostratigraphic units.
- Comprehensive hydrogeological column and a regional hydrogeological cross-section with corresponding scales.

Karst map

In mining areas with karst, the karst map should be compiled based on the actual data of the investigation and exploration, which provides a reference for the study of the development and distribution of karst and the prevention and control of karst water in the mine.

The form of the karst map can be compiled into a karst distribution plan map, a karst cross-section, depending on the specific situation.

- The karst distribution plan map is based the comprehensive hydrogeological map of the mine. It can be used to show karst landform, sinkhole, depression, inlet and outlet of underground river, the natural discharge point of groundwater, man-made discharge point and the karst collapse area, divides of surface watersheds and groundwater.
- The karst cross-section is compiled based on the actual surveying and mapping data of the cave or the underground river.

2.2.3 Mine water prevention and control data should be carefully collected, catalogued, and preserved. Mine shall establish the following databases for prevention and control of water:

- Mine water inflow observation results.
- Meteorological data.
- Surface hydrological observation results.
- Water levels in boreholes, dynamic observation results of monitoring wells and springs, and river leakage measurement.
- Pumping test results.
- Mine water inrush points.
- Drilling logs for mine field geological exploration.
- Underground hydrogeological drilling logs.
- Water quality analysis results.
- Water quality measurements at the water source that is subject to the pollution.
- Inventory of water supply wells.
- Sealed boreholes.
- Relevant information on old mine pools in the mine and its neighboring mines.
- Sluice gate (wall) observation data.
- Information from other special projects.
- The database shall be updated every six months.

2.2.4 Newly constructed mines shall collect, catalogue, and analyze the relevant hydrogeological data of the mines during the construction of the shafts according to the relevant regulations for mine construction, and share the materials with the production institutions after the completion of the shaft construction.

The following main maps should be prepared for new mines:

- Hydrogeological observation data and analysis results.
- Water inrush point inventory, records and technical summary of water prevention and control, as well as grouting and water sealing records and related materials.
- Hydrogeological cross-section of the shafts and the main tunnels.
- Supplemental hydrogeological investigation for shaft construction.
- Hydrogeological report on shaft construction, which can be combined with the geological report on shaft construction.

2.2.5 Before a mine is closed, the mine shall prepare a mine closure report. The closure report should include the following main elements:

- Distribution of mined-out space at closure, analysis and evaluation of the possible water recharge source, pathway, water volume and water level.
- Impact of mine closure on the safety of adjacent active mines and the measures to be taken to prevent and control water.

The closure report (including the drawings and materials) shall be submitted to the local coal industry management department for the record.

2.2.6 The mine shall establish a hydrogeological information management system to achieve integration of mine hydrogeological data collection, drawing map, calculation and evaluation with prediction and forecast of mine water hazard.

10.3 Supplemental Hydrogeological Investigation and Exploration

10.3.1 Scope of Supplemental Hydrogeological Investigation

3.1.1 When the existing hydrogeological data of a mining area or mine cannot meet the needs of mine production and construction, a supplemental hydrogeological investigation shall be conducted to complete the data gaps. If the mining area or mine has not been hydrogeologically investigated or the hydrogeological data is inadequate, a supplemental hydrogeological investigation shall be carried out as well.

3.1.2 The scope of supplemental hydrogeological investigation shall encompass an entire groundwater system with independent recharge, runoff, and discharge conditions.

3.1.3 In addition to traditional methods, supplemental hydrogeological investigations may also adopt new technologies and methods such as remote sensing, global satellite positioning, and geographic information systems.

3.1.4 The supplemental hydrogeological investigation shall include the following:

- Data collection: Collect meteorological data such as precipitation, evaporation, temperature, air pressure, relative humidity, wind direction, wind speed and its monthly mean, minimum and maximum values. Collect data and results from previous investigations and research, dynamic observation data, exploration drilling log, water supply drilling and pumping test data. Geomorphology and geology: Collect data on mining-induced geomorphological changes such as collapses, landslides, artificial lakes, and various karst landforms in karst-developed mining areas. For the Quaternary unconsolidated overburden and bedrock outcrops, identify the age, lithology, thickness, water richness and groundwater recharge and discharge mechanisms, and differentiate aquifers from aquitards. Identify the shape, occurrence, nature, size, fractured zone (range, filling, degree of cementation, water conductivity) of any geological structure, and the presence or absence of springs. Perform preliminary analysis of their impact on mine.
- Surface water. Investigate and collect historical water levels, flows, accumulated water, maximum flood inundation range, silt content, and water quality of surface water bodies such as rivers, canals, lakes, waterlogged areas, mountain ponds and reservoirs in mining areas. Understand their hydraulic connections with the underlying aquifers. Detailed investigation should be carried out over the areas where surface water recharges groundwater, and the leakage should be monitored.
- Wells and springs. Investigate the location, elevation, depth, discharge formation, water inflow, water level, water quality, water temperature, presence or absence of gassing, type of gassing, gas concentration and source. Sketch the topographic and geological plan maps and cross-sections over springs.
- Abandoned mines: Investigate the location of the old kilns, abandoned or legendary mines and collect data on their mining history, water recharge and dewatering records, and the reasons for closure. Based on the terrain, circle the mined-out area and estimate the amount of water in mine pools.
- Producing mines. Investigate and study the water recharges, water recharge mechanisms, water inrush formation, water inrush point location and water inrush flow in mining areas, the dynamic change of mine water inflow and the relationship between mining level and mining area, and the records of previous water hazard data and water prevention measures and effectiveness.
- Neighboring mines. Investigate the location, extent, mining coal seam, water filling conditions, geological structure, coal mining method, coal mining volume, coal pillars and spatial relationship with adjacent mines, and observational or research data on water disasters in the past. Collect data and as-built maps on mining and tunneling.
- Surface karst. Investigate the features and distribution range of karst development. Detailed investigations of recharge and drainage channels that have a significant

impact on groundwater movement. Tracer tests and underground river mapping can be performed when necessary. Analyze the karst development mechanisms and groundwater runoff direction, delineate the recharge area, determine the infiltration in the recharge area, and estimate the groundwater runoff. Map karst collapses where karst collapses occur.

10.3.2 Surface Hydrogeological Reconnaissance

3.2.1 The hydrogeological reconnaissance for mining areas and mines shall include the following main contents:

- Conduct meteorological monitoring. Set up a meteorological monitoring station if no weather station is present within 30 km radius of the mining area (mine). Site selection and meteorological parameters shall meet the requirements of meteorological stations. If a weather station is present within 30 km radius of the mine, there is no need for a meteorological station; only a rain gauge will be sufficient.
- Conduct surface water monitoring. The surface water monitoring project has the same scope of work as the surface water survey. Under normal circumstances, surface water observations are conducted once a month. During rainy season or after heavy rain, the number of observations is increased according to the needs of the work.
- Conduct groundwater monitoring. Monitoring points should be placed in the following locations and formations:

  a. The main aquifer that has an impact on mine production and construction.
  b. Groundwater strong runoff zone (structural fracture zone) affecting mine water filling.
  c. An aquifer that may be in hydraulic connection with surface water.
  d. The area where the mine was firstly mined.
  e. A section where hydrogeological conditions may change during the mining process.
  f. A section where human activities may affect the water recharge of the mine.
  g. Near the main water inrush point, or a section with a threat of water inrush.
  h. Hydrogeological boundaries.

Existing boreholes, wells, and springs should be utilized as much as possible as the monitoring points. Observations include water level, water temperature and water quality. The flow should also be monitored at springs.

The monitoring points shall be uniformly numbered with fixed marks for consistent observations. Coordinates and elevations shall be surveyed, and plotted on the integrated hydrogeological map. The elevation of the monitoring point shall be re-surveyed once a year; if there is any change, it shall be re-surveyed immediately.

3.2.2 The mine shall carry out surface hydrogeological reconnaissance for one hydrological year before mining. In the process of mining, it is imperative to keep adequate monitoring frequencies. The monitoring frequency can be 7–10 days before the groundwater dynamics are understood. The frequency can increase to 1–3 times a month after understanding the groundwater dynamics. During rainy season or under an abnormal situation, the number of observations should be appropriately increased. The water quality monitoring is no less than two times a year, once in rainy reason and once in dry season.

When performing the reconnaissance work, technicians should follow the designed time and sequence, and complete the survey as soon as possible. Attentions are given to the continuity and accuracy of the observation. Two readings shall be taken for the borehole water level, and their difference should not be greater than 2 cm. The average shall be used. The measurement tool should be calibrated before use. Automatic water pressure transducers should be used to record and transmit data for mines that hydrogeologically complex and extremely complex.

10.3.3 Underground Hydrogeological Investigation

3.3.1 For newly constructed shafts, main gates and excavated roadways, hydrogeological observation and cataloging shall be carried out in time to develop hydrogeological sections. When the shaft and tunnels go through aquifers, notes that shall be taken of the aquifers include strike, dip, thickness, lithology, geologic structure, fissure or karst and filling materials. The location and elevation where water flow occurs, water flow rate and water temperature should be described as well. Water samples shall be collected and analyzed for quality.

When fractures are intercepted, their strike, length, width, number, shape, pinch-out, filling degree and fillings should be measured. Traces of groundwater activity should be documented. Rose diagrams should be prepared for the fissures. The fracture rate of the formation shall be measured over select sections using the following equation:

$$K_T = \frac{\sum lb}{A} \times 100\%$$

where

K_T fracture rate (%);
A measurement area (m^2); 1–2 m^2 for dense fractures; 4-10 m^2 for sparse fractures
l fracture length (m);
b fracture width (m)

When karst is encountered, characteristics of karst features such as shape, development, distribution, presence or absence of filling and filling components and water filling conditions, should be observed, and a sketch showing the karst features should be drawn.

When fault structures are encountered, fault displacement, strike, width of the fault zone should be measured. Filling composition of fracture zone, degree of cementation, and hydraulic conductivity should be observed.

When folds are encountered, its shape, strike, and associated fracture should be observed.

When collapse columns are encountered, data to be collected include lithology and strike of the strata inside and outside the collapse column, degree of karst development, and hydraulic conductivity. The development height of the collapse column should be determined. Factsheets, plan maps, cross-sections, and sketches should be prepared.

In case of a water inrush, details of the following observations shall be documented: time, geographic location, underground location, rock formation, lithology, thickness, inrush pattern, damage of surrounding rock. The water flow rate, water temperature, water quality and sand content should be measured. At the same time, the changes in water flow rate and water level in nearby springs and observation wells should be observed. The root cause analysis of the water inrush should be performed. The main water inrush points can be used to systematically observe the groundwater system. Factsheets should be prepared with plane maps and sketches.

Water inrushes that meet the following criteria should be reported to the local coal mine safety supervision agency and local government: flow rate of 300 m^3/h or greater in large and medium-sized coal mines; flow rate of 60 m^3/h or greater in small coal mines; or whenever a mining area or entire mine is flooded by water inrush.

Based on water inrush flow rate at water inrush point, water inrush can be divided into four levels:

- Small water inrush: $Q \leq 60$ m^3/h.
- Medium water inrush: 60 m^3/h < $Q \leq 600$ m^3/h.
- Large water inrush: 600 m^3/h < $Q \leq 1,800$ m^3/h.
- Extremely large water inrush point: $Q > 1,800$ m^3/h.

3.3.2 Mines shall strengthen observation of mine water inflow and monitoring of water quality. The mine should set up observation stations on water inflows at different shaft and mining level. The observation frequency is no less than three times per month. For fracture zones and collapse column with large water flow, independent observation stations should be set up and the observation frequency should be 1–3 times per month. The frequency for water quality analysis is no less than two times a year, once in the rainy season and once in the dry season. The observation frequency should increase appropriately when the amount of water inflow appears to be abnormal, water inrush occurs or in rainfall-sensitive mines during rainy season.

Flow at newly exposed water discharge points in the underground should be observed once a day until the water inflow becomes stable or the flow dynamics are well understood. Flow at new water inrush points should be observed once every 1–2 h before the cause of water inrush is determined. The observation interval can be extended appropriately afterwards. Water samples shall be collected and analysis for water quality. The flow can be observed according to the typical observation schedule once the water flow is stable.

Daily mine water flow is necessary when the mining areas are overlain by a surface water body or a water-rich aquifer; mining intercepts a structural fault zone that is connected to a water-rich aquifer, or mining is close to the old mine pools. The dynamic nature of the water flow must be fully understood. Below are the standards for water-rich aquifers according to specific water yield (q):

- Weak aquifer: $q \leq 0.1$ L/(s • m).
- Medium aquifer: 0.1 L/(s • m) $< q \leq$ 1.0 L/(s • m).
- Strong aquifer: 1.0 L/(s • m) $< q \leq$ 5.0 L/(s • m).
- Extremely strong aquifer: $q >$ 5.0 L/(s • m).

It should be noted that the specific water yield is based on a pumping well of 91 mm in diameter and 10 m drawdown. If the well diameter and drawdown are not consistent with the above parameter, the specific water yield should be converted for comparison. Conversion method is provided below:

- Plot water flow (Q) against drawdown (S) based on pumping tests.
- Determine their relationship using the least square method or graphic method.
- Use the Q and S relationship to calculate the water flow rate at the pumping wells when the drawdown is 10 m.
- Use the following equation to calculate the equivalent water flow rate at 91 mm diameter wells:

$$Q_{91} = Q_w \left(\frac{lg R_w - lg r_w}{lg R_{91} - lg r_{91}} \right)$$

where

Q_{91}, R_{91}, *and* r_{91}	water flow rate, radius of influence, and radius of a borehole assuming bore diameter of 91 mm;
Q_w, Rw, and r_w	water flow rate, radius of influence and radius of pumping well.

- Divide Q_{91} by 10 m to obtain the specific water yield.

For new vertical and inclined shafts, water flow shall be observed once at vertical advancement of 10 m. When a new aquifer is encountered, if the vertical advancement is less than 10 m, the water inflow should be measured on both top and bottom of the aquifer.

When mine water inflow measurement is conducted, attention should be paid to the continuity and precision of observation. Measurement methods include volumetric method, weir method, buoy method, flow meter method or other advanced water measurement methods. Measurement tools and meters should be periodically calibrated to reduce human error.

3.3.3 When dewatering occurs in the aquifer to reduce potentiometric pressure the water flow and water pressure should be measured 1–2 times per hour until both the inflow and pressure become stable. After that, the requirements of normal observation are to be followed. If the old mine pool is drained, it should be observed daily.

10.3.4 Conditions to Conduct Supplemental Hydrogeological Investigation

3.4.1 Supplemental hydrogeological investigation is required under one of the following circumstances:

- Hydrogeological investigation has not been conducted for the target coal seam.
- Original exploration work of the mine is insufficient, and the hydrogeological conditions have not been ascertained.
- Hydrogeological conditions are more complicated than the original exploration report after exposure of the coal seam formation by mining.
- Long-term mining has resulted in major changes of the hydrogeological conditions, and the original exploration report cannot meet the production requirements.
- There is need for mine expansion and deepening, mining of new coal measures (groups) or design of mine field expansion.
- The roof of the mine roadway is overlain by special geological formations or is underlain by strong water-filled aquifers. When the coal seam floor is under pressure water control measures must meet special requirements.
- There are construction needs whenever underground engineering projects pass through strong water-rich aquifers.

3.4.2 The supplemental hydrogeological investigation shall meet the corresponding data quality objectives to achieve water prevention and control.

When supplemental hydrogeological exploration is conducted the overall analysis and research of the regional groundwater system in which the exploration and mining area is located shall be performed. Hydrogeological reconnaissance is the scope of investigation outside the mine well field, whereas within the mine field the scope will include drilling exploration, hydrophysical survey, geophysical survey, and pumping or discharging tests.

Mine supplemental hydrogeological investigation should be based on the hydrogeological type and specific conditions of the mine. Traditional investigation methods and advanced techniques shall be used, and they include hydrogeological supplemental reconnaissance, geophysical exploration, hydrogeological drilling, pumping (discharge) test, water chemistry and isotope analysis, groundwater dynamic monitoring, and in situ testing.

Supplemental hydrogeological investigation work plan shall be prepared and reviewed by chief engineer of the coal mine enterprise prior to implementation. The work plan should specify the data collection design and data quality objectives. The existing conditions of the mine shall be used sufficiently with both surface and underground work elements.

After completion of the supplemental hydrogeological investigation, the investigation report shall be prepared and submitted in time for review and approval of the chief engineer of the coal mining enterprise.

10.3.5 Supplemental Hydrogeology Exploration on Surface

3.5.1 Boreholes used for hydrogeological investigations in coal mines shall be designed according to exploration specifications. The design parameters include borehole structure, inclination, coring, casing locations, annulus materials, final borehole diameter, termination formation, preliminary hydrological observation, pumping test, downhole logging, sampling, water-sealing quality, orifice device and requirement on civil survey.

The main quality indicators of borehole drilling shall meet the following requirements:

- For mines whose water hazards come from underlying aquifers, the supplemental hydrogeological drilling should not be terminated until the underlying aquifer is exposed.
- Downhole logging shall be performed in all exploration boreholes. When conditions permit, the following logging tools can be considered: heat pulse flowmeter, ultrasonic imaging, acoustic televiewer, and optical televiewer. Combination of these logging results with geologic logs and coring analysis provides essential parameters on aquifers and aquitards.
- The main aquifer or test section (screening interval) is preferably drilled with clean water. Where mud drilling is needed low particular content and high-quality mud can be used. However, change to mud drilling needs to be approved by geological department of the drilling institution with effective well development procedures in place.
- The borehole diameter is determined based on the purpose of drilling. Diameter of a pumping test borehole shall meet the design of pumping capacity and allow installation of pumping equipment. Diameter of an observation borehole shall meet the requirements of casing to seal hydraulic communications and allow water level measurement.
- The inclination for both pumping test borehole and observation boreholes shall meet the requirements on installation of the selected pumping equipment and water pressure transducer.
- Core drilling shall prescribe detailed description of rock cores. Basic coring rates are >70% for bedrock, >50% over fractured zone, >70% for clay, and >30% for sand and gravel layer. When use of downhole logging helps with identification of aquifers and aquitards core length can be appropriately reduced.
- Where hydraulic isolation is used for a testing or water-bearing segment the isolation effectiveness can be checked by such methods as water lifting, water injection

and downhole logging. The resting results are documented for the record. The isolation effort shall satisfy the effectiveness tests.

- In addition to long-term observation boreholes, the remaining boreholes are sealed with high-grade cement slurry. The quality of the seals is examined with sampling.
- Subsequent to drilling completion of the observation hole, borehole development is taken to ensure that the screening internal is open to the aquifer without siltation.

All hydrogeological boreholes should be accompanied with preliminary hydrogeological observations in accordance with relevant regulations and specifications. Boreholes without preliminary hydrogeological observations should be downgraded or not accepted for use.

All hydrogeological observation boreholes should be installed with appropriate well-head protection device and long-term observation marks. Effective measures shall be taken to protect them to ensure durability and convenient observation. Any boreholes that are damaged or silted shall be repaired or replaced in time.

3.5.2 The quality of pumping test in supplemental mine hydrogeological investigation shall meet the requirements of relevant national and industry standards.

At least three properly designed levels of drawdown should be conducted. Water level drawdown should reach the maximum of the pump equipment capacity if feasible. When the borehole water level is affected by to mining, one maximum drawdown test is allowed with extended observation duration. Unsteady flow calculation should be considered based on the prolonged monitoring data.

For hydrogeologically complex or extremely complex mines, underground water discharge test can be carried out if small-diameter pumping test is not sufficient to ascertain the hydrogeological and engineering (karst collapse) conditions. If underground conditions are not amenable to such tests, multiple well pumping tests with large-diameter wells should be carried out. Specific design and work plan shall be prepared for the multiple well pumping tests, which can be implemented after approval of the chief engineer of the coal mine enterprise.

Duration of the multiple well pumping tests with large-diameter wells shall be determined based on the trend of the water level change versus flow rate. Typically, it is no less than 10 days. Collaborative decision is made on the duration for scenarios in which the water level cannot be stabilized due to mining interference.

Tracer tests can be carried out in combination with pumping or discharging tests to investigate hydraulic connections between the mining-affected aquifer and other aquifers or surface water bodies.

Prior to pumping the water level (potentiometric pressure) and flow rate should be measured at test wells, observation boreholes, and hydrogeologically significant points on surface and in the underground. If necessary, an observation well can be installed to measure the central water level drawdown of the multiple well pumping tests.

3.5.3 Water injection test may be performed for rock permeability study to prevent water leakage into mine or because the water level of the aquifer is so deep that the pumping test is not feasible.

A work plan and design should be prepared for the water injection test. The test design includes start and end depths of the test interval; casing diameter and formation of installation, casing depth and water sealing method; water injection equipment, water injection test method, and the quality metrics of the water injection test.

The main technical indicators for water injection test construction shall meet the following requirements:

- Determine the test interval based on lithology and porosity and fracture depth in the bedrock. The testing interval shall be hydraulically isolated from other water-bearing formations.
- Thoroughly develop the injection well prior to water injection to ensure that the aquifer is free of siltation. Temperature of the existing borehole water and the temperature of the injected water shall be measured.
- Static water level prior to injection test and water level response at completion of the water injection shall be measured.

3.5.4 Geophysical survey design including survey line layout, number of receivers, duplicate error, and data processing, shall comply with the relevant national standards and industry standards.

Selection of geophysical survey method is based hydrogeological conditions of the survey area, the geophysical characteristics of the targeted geological body, and purposes of the survey. Multiple geophysical methods can be used for better results and more comprehensive investigation.

At completion of the geophysical survey, results should be submitted including report, raw data and processed map. Geophysical interpretations should be combined with geological exploration results. They can be used as a basis for mining design after mutual corroboration.

10.3.6 Hydrogeological Investigation in Underground

3.6.1 Hydrogeological investigation in underground shall follow the following rules:

- Hydrogeological investigations tools include underground geophysics, drilling, monitoring, and in situ testing.
- Integrate underground investigation with surface investigation.
- Reliable safety precautions are exercised in underground investigation to ensure the safe production of mines.

3.6.2 Underground hydrogeological exploration shall be conducted under the following conditions:

- It is difficult to understand the water-related conditions by using surface hydrogeological exploration, and it is necessary to carry out water discharge test or tracer test in the underground.

- Dewatering feasibility test is required where aquifer, flow-sand formation, or karst aquifer either overlies or underlies the coal seam.
- Surface access is limited on construction activities because of surface water body, topography or mining-induced subsidence.
- The depth of the borehole or the groundwater level is too deep to effectively conduct surface hydrogeological investigation.

3.6.3 Underground hydrogeological exploration shall meet the following requirements:

- Drilling design including specifications on technical requirements and safety measures cannot be implemented until it approved by the chief engineer of the mine.
- Drilling chamber shall be constructed and reinforced to ensure normal working conditions.
- The drilling rig shall be installed firmly. Pressure test shall be conducted after orifice casing in installed. Orifice safety valve shall be installed before construction to ensure capacity of safe control of water discharge. The resistance pressure of the safety valve is greater than the maximum water pressure. Orifice blowout preventer shall be installed before aquifer is intercepted.
- Carry out construction according to the design and strictly implement construction safety measures.
- Conduct tracer tests. Selected tracers should not cause water pollution.
- Any obsolete boreholes should be abandoned properly, and abandonment report is prepared and submitted to appropriate agency.

3.6.4 Water discharge test shall abide by the following principles:

- Prepare work plan and design of water discharge test. The design includes test method and anticipated water level drawdown and water release rate at each step. The amount of water released should not exceed the existing maximum dewatering capacity of the mine. In principle, the observation borehole that can be affected by the water discharge test should have measurable water level drop. The design should be reviewed and approved by the chief engineer of the coal mining enterprise.
- Conduct pre-test activities. The pre-test activities include assignment of personnel for various tasks, calibration of observation instruments and tools, and verification on dewatering capacity and conveyance lines.
- Collect background data prior to releasing of water. The background data includes synoptic measurements of water level, flow rate, water temperature and water quality at observation and water discharge points on surface and in underground.
- Determine the duration of the water discharge test according to the specific conditions. When the water flow and water level are difficult to stabilize, the test duration is generally no less than 10–15 days. Selection of the observation interval is based on need for unsteady flow calculations. Simultaneous observation should be made between the central water level or water pressure and water flow.

- Enter observation data into database promptly and plot graph of water flow against water level.
- Timely organize the data and submit a summary report at completion of the water release test.

3.6.5 When conventional hydrogeological exploration methods cannot evaluate mining safety for mines that are threatened by water hazard, dewatering or depressurizing feasibility test can be conducted by constructing stone gate that penetrates through multiple formations or sinking special shaft, depending to site-specific conditions.

Dewatering or depressurizing test should meet the following provisions:

- Specific construction design shall be prepared for the test. The design shall be reviewed and approved by the chief engineer of the coal mining enterprise.
- Predict the maximum water flow rate.
- Establish a conveyance system that is capable of drain the maximum water flow rate.
- Construct sluice gate at appropriate location.
- Perform advanced hydrogeological exploration, dewatering, and depressurization.
- Coordinate measurements of water level, water pressure and water flow on surface and in underground.

3.6.6 Geophysical surveys can be interpreted with reference to drilling data. Commonly used geophysical methods are direct current method (resistivity imaging method), audio electric penetration method, transient electromagnetic method, frequency domain electromagnetic sounding method, radio wave perspective method, ground penetrating radar method, shallow seismic, Rayleigh wave method, and trough wave seismic methods.

10.4 Mine Water Control Measures

10.4.1 Prevention of Water on Surface

4.1.1 It is essential to understand conditions within the mine and its vicinity including catchment and leakage of the surface water flow system, dewatering capacity and pertinent hydraulic engineering projects, and obstacles in local reservoir, hydropower dam, river levee, and river channel. Local precipitation and the maximum flood level data should be documented and carefully studied. A system including dewatering, water control and drainage should be established.

4.1.2 Elevation of mine shafts and industrial buildings shall be higher than the historical maximum flood level in the area.

Construction of mine shafts and industrial buildings in mountainous area, in addition to complying with the provisions of the first paragraph of this article, should avoid the areas where mudslides and landslides may occur.

Where elevation of mine shafts and industrial buildings is lower than the historical maximum flood level, dams, drainage ditches or other water control structures shall be constructed.

4.1.3 Safety precautions shall be taken when surface water body near the mine shaft or within subsidence area and beyond may flow into the underground.

It is strictly forbidden to mine water-resistance coal or rock pillars on coal seam outcrops.

Where water is easily accumulated on the surface drainage ditches should be constructed to drain water. Construction of drainage ditch should avoid outcrops and fractured and water-bearing rock formations. Low-lying areas that are not amenable to construct drainage ditches should be backfilled and compacted. If the low-lying area is too large to be backfilled, pump station or flood drainage system should be constructed to prevent the water in the low-lying area from seeping into the mine.

If mine is threatened by rivers and flash floods dams and flood discharge channels should be constructed to prevent intrusion of flood water into the mine.

Mine water that is discharged to the ground should be properly disposed of to avoid re-infiltration into the mine.

Leaking ditch (including irrigation ditch) and riverbed should be treated to reduce their permeability or diverted. Ground cracks and collapses should be backfilled in time. During backfilling, appropriate safety measures should be taken to prevent personnel from falling into the collapse pit.

In areas where there is a risk of landslides, when potentially threatens the safety of coal mines, measures should be taken to prevent occurrence of landslides.

4.1.4 It is strictly forbidden to pile up debris such as gangue, coal ash and garbage in areas where mountain torrents and river flows may occur. The debris can be washed to industrial sites and buildings or it may block rivers and ditches.

4.1.5 Wellhead covers of actively used boreholes shall be installed as required. Obsolete drill holes should be abandoned properly to prevent surface water or aquifer water from flowing into the mine. The orifice elevation of observing borehole, grouting hole, cable hole, and any borehole that connects with underground or aquifers should be higher than the maximum flood level in the area.

4.1.6 Retired or abandoned vertical shaft shall be backfilled or sealed. An alternative is to install a solid reinforced concrete cover that is larger than the cross section of the shaft. Fences shall be erected, and signs shall be posted.

Retired or abandoned inclined shaft should be filled and sealed. An alternative is to construct a brick or stone or concrete wall approximately 20 m below the wellhead. The space above the wall shall be backfilled with mud.

Retired or abandoned adit should be filled with clay at least 20 m into the adit with a wall built at entrance. Drainage ditch shall be constructed if surface water may affect the adit.

Engineering records and the drawings shall be documented and filed for backfilling of the abandoned vertical shaft, inclined shaft, and adit.

4.1.7 Mines shall establish an early warning mechanism for extreme weather events by interacting with meteorology, hydraulics, and flood control agencies. Coal mines should be knowledgeable of storm flood disasters that may endanger their safe production; pay close attention to forecast and early warning of severe or extreme weather; and timely grasp the water dynamics and take safety precautions. Timely communications between neighboring mines are vital. Abnormal conditions in one mine should be known immediately to other neighboring mines.

4.1.8 Dedicated personnel shall be assigned to be responsible for visual inspections of abandoned old kiln, ground collapse/subsidence, and mining-induced earth fissure as well as features that may pose threat to safe production of the mine such as reservoirs, lakes, rivers, conduits and embankments. Upon receiving the rainstorm forecast and warning, uninterrupted inspection should be carried out 24 h. Professional personnel should be sent to mines before and after severe or extreme rain events to monitor changes in the amount of water flowing into the mines.

4.1.9 Mines shall establish rules on emergency evacuation in response to mine-flooding storms so that underground workers can retreat in a timely manner. Criteria to initiate the emergency evacuation have to be clearly stated as well as responsible department, liaison personnel, and evacuation procedures. When observations suggest that storm is severe and may cause flooding of the mine, actions should be taken immediately to evacuate workers to a safe place. Production cannot be resumed until water hazard potential is completely eliminated.

4.1.10 Measures that are implemented to prevent severe storms from causing disasters in mines shall be inspected thoroughly prior to rainy reasons. Any identified deficiencies shall be corrected before rainy seasons start, and a responsible party shall be assigned for the corrective actions. Water control engineering work shall have specific designs. The project is considered completion only after acceptance by mine chief engineer.

10.4.2 Design of Water-Resistance Coal (Rock) Pillar

4.2.1 Water-resistance coal (rock) pillar is a safeguard between adjacent mines. Where the mine is demarcated by faults, water-resistance coal (rock) pillars should be on both sides of the fault.

4.2.2 Mine that are threatened by water hazard shall have water-resistance coal (rock) pillar if any of the following circumstances is present:

- Weathering zone of coal seam outcrop.
- Under surface water body or water-bearing alluvium, or adjacent to flooded area.
- Coal seams in contact with strongly transmissive faults or in contact with fractured zones or faults that are hydraulically connected to an aquifer.
- Old kiln and mine pool with large amounts of water.
- Water-bearing and transmissive collapsed columns, karst caves or underground rivers.

- Partitioning mining boundary.
- Observation wells, grouting holes and cable holes that are under protection.

4.2.3 Dimension of the water-resistance coal (rock) pillar shall be designed according to site-specific conditions including geological structure, hydrogeological conditions, coal seam occurrence conditions, physical and mechanical properties of parent rock, mining methods and rock movement mechanism. Water-resistance coal (rock) pillars shall be designed by a dedicated geological surveying team and shall be implemented only after being reviewed and approved by a committee organized and led by mine chief engineer.

Dimension of water-resistance coal (rock) pillar on coal seam outcrop

The following equations are used to calculate the dimension of water-resistance coal (rock) pillar.

- When the coal seam outcrops without cover or covered by slightly permeable clayey unconsolidated soil:

$$H_f = H_k + H_b$$

- When the coal seam outcrop is covered by high-yield unconsolidated aquifer (Fig. 10.1):

$$H_f = H_L + H_b$$

where

H_f Water-resistance coal (rock) pillar height (m);
H_k Height of collapse zone after mining (m);

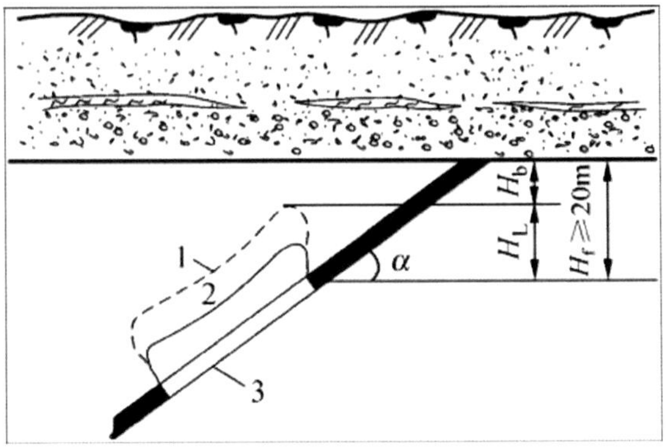

Fig. 10.1 Diagram of water-resistance coal (rock) pillar where coal seam outcrop is covered by water-rich unconsolidated aquifer

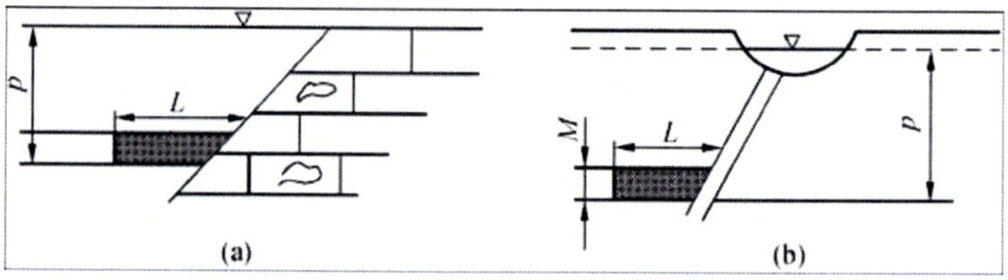

Fig. 10.2 Diagram of water-resistance coal (rock) pillar along transmissive fault

H_L Maximum height of mining-induced water-conducting fractured zone (m);
H_b Protective layer thickness (m);
α Coal dip angle, (°)

The value calculated from above equations shall not be less than 20 m. For the calculation of H_k and H_L in the equations, refer to the relevant provisions of the "Provisions for Coal Pillar Retention and Coal Mining in Buildings, Water, Railways and Main Tunnels".

Dimension of water-resistance coal (rock) pillar along transmissive fault

The following empirical equation is used to calculate dimension of water-resistance coal (rock) pillar along a transmissive fault (Fig. 10.2):

$$L = 0.5KM\sqrt{\frac{3p}{K_p}} \geq 20\,\text{m}$$

where

L Width of coal pillar (m);
K Safety factor, generally 2-5;
M Coal thickness or height (m);
P Water pressure or potentiometric pressure (MPa);
K_p Tensile strength of coal (MPa)

Dimension of coal (rock) pillar where coal seam is in contact with strong aquifer or water-conducting fault

When the coal seam is in contact with a strong aquifer or a water-conducting fault and is partially covered (Fig. 10.3), retention of the water-resistance coal (rock) pillar meets the following requirements:

- Where the aquifer top is higher than the upper limit of the highest water-conducting fracture zone, design of water-resistance coal (rock) pillar is shown in Fig. 10.3a and Fig. 10.3b. The calculation equation is:

$$L = L_1 + L_2 + L_3 = H_a \csc\theta + H_L \cot\theta + H_L \cot\delta$$

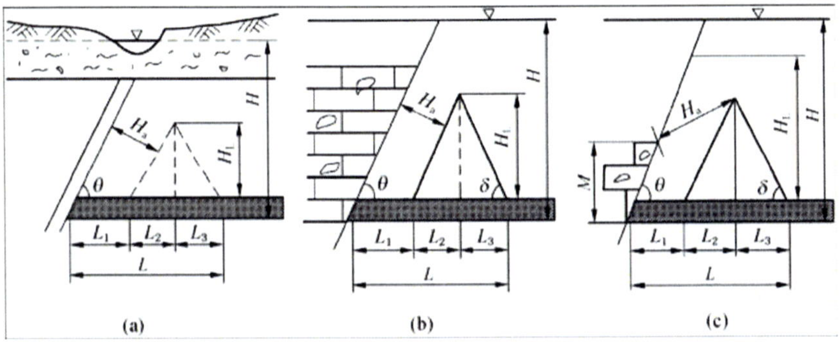

Fig. 10.3 Diagram of water-resistance coal (rock) pillar where coal seam is in contact with water-rich aquifer or water-conducting fault

- Where the upper limit of the highest water-conducting fracture zone is higher than the upper aquifer of the fault, design of water-resistance coal (rock) pillar is shown in Fig. 10.3c. The calculation equation is:

$$L = L_1 + L_2 + L_3 = H_a(\sin\delta - \cos\delta\cot\theta) + (H_a\cos\delta + M)(\cot\theta + \cot\delta) \geq 20\,\text{m}$$

where

L	Width of water-resistance coal (rock) pillar (m);
L_1, L_2, L_3	Segment width of water-resistance coal (rock) pillar (m);
H_L	Height of maximum water-conducting fracture zone (m);
θ	Inclination angle of fault (°);
δ	Angle of rock collapse (°);
M	Height between upper aquifer of the fault and bottom of coal seam (m);
H_a	Width of water-resistance coal (rock) pillar at fault (m)

The value of H_a should be determined according to the actual observation data, i.e. by summarizing the geological and hydrogeological data alone faults for both water inrush and safe mining scenarios. The basic data is water pressure (p) and retained thickness of water-resistance coal (rock) pillar (M). The ratio of water pressure and thickness is, $T_s = p/M$. The water inrush coefficient threshold can be estimated by plotting T_s as the horizontal axis and the buried depth H_0 as the vertical axis, as shown in Fig. 10.4.

H_a value can also be calculated as follows:

$$H_a = \frac{p}{T_s} + 10$$

Fig. 10.4 Relationship between T_s and H_0

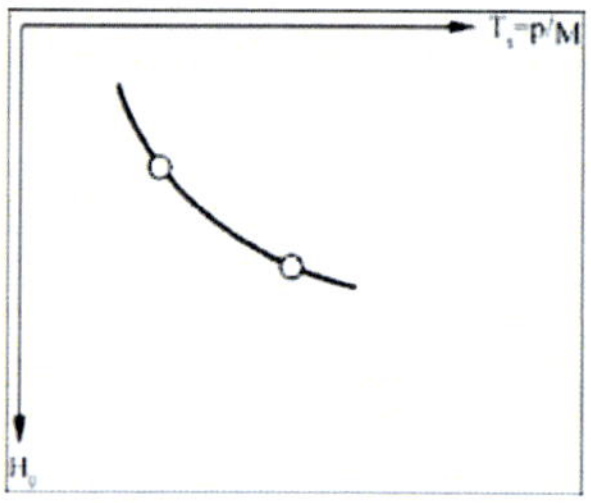

where

p Hydrostatic pressure exerted against water-resistance coal (rock) pillar (MP_a);
T_s Water inrush coefficient (MP_a/m);
10 Default value of protective zone thickness (m)

If there is no water inrush coefficient data available in the mine area, data from other mining areas can be used as reference with consideration of site-specific factors such as lithology, physical and mechanical properties of the aquifuge, roadway dimension, height of working face, coal mining method, and roof control method.

Dimension of coal (rock) pillar where coal seam overlies aquifer and water-conducting fault

Figure 10.5 shows the scenario in which coal seam overlies aquifer and water-conducting fault. Design of water-resistance coal (rock) pillar should take into account pressure in two aspects: potentiometric pressure of underlying aquifer that exerts on the formation below coal seam and potentiometric pressure within the fault in direction of the coal seam.

When the underlying potentiometric pressure is considered, the minimum distance (offset) between the coal seam floor and the fault plane should be larger than the calculated height (Ha) of the water-resistance coal pillar, and not less than 20 m. Below is the calculation equation:

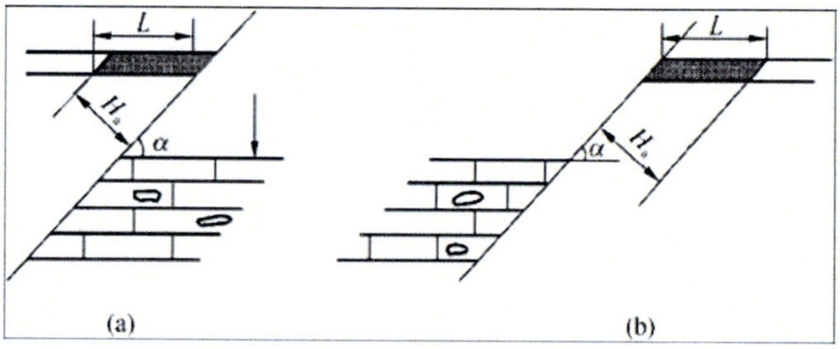

Fig. 10.5 Diagram of water-resistance coal (rock) pillar where coal seam overlies aquifer and water-conducting fault

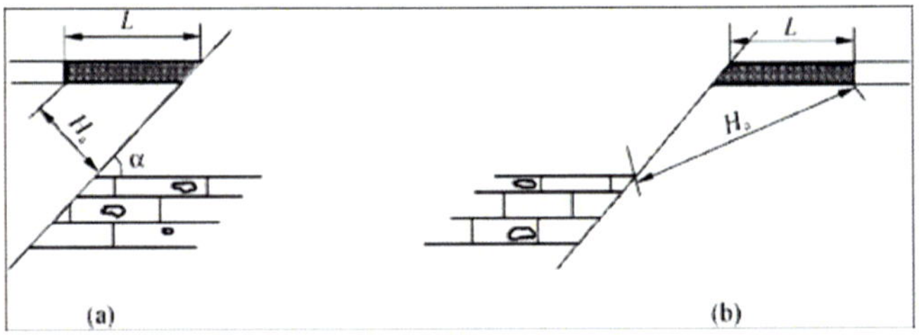

Fig. 10.6 Diagram of water-resistance coal (rock) pillar where the coal seam overlies aquifer and the fault does not conduct water

$$L = \frac{H_a}{\sin \alpha} \geq 20m$$

where

α Dip angle of fault (°);

Rest parameters are the same as before.

When the potentiometric pressure within the fault in direction of the coal seam is considered, calculation of the coal pillar follows the equation for Fig. 10.2.

The larger number from the two methods is taken as the width of water-resistance coal (rock) pillar; but it is still no less than 20 m.

If the fault does not conduct water (Fig. 10.6), size of the water-resistance coal (rock) pillar should ensure the minimum distance between intersection of the top surface of the aquifer and the fault plane to the floor of the coal seam, and the section perpendicular to the strike of the fault is greater than the height (Ha) of the coal pillar, but not less than 20 m.

Dimension of coal (rock) pillar where coal seam underlies mine pool or flooded area

Maintenance of anti-water-proof coal (rock) columns during mining in flooded areas or old kiln watersheds

- When roadway tunneling occurs under flooded area or underground mine pool, the distance between the roadway and the water body shall not be less than 10 times the height of the roadway.
- When mining is in the same coal seam as that in flooded area or mine pool and the boundary of the flooded area or the mine pool has been basically determined, the size of the water-resistance coal (rock) pillar should be calculated according to Fig. 10.2.
- When return mining occurs in the coal seam under the flooded area or under the mine pool, the size of the water-resistance coal (rock) pillar shall not be less than the sum of the maximum height of the water-conducting fracture zone and the height of the protection zone.

Water-resistance coal (rock) pillar for protection of surface water

Design of water-resistance coal (rock) pillar for protection of surface follows the "Procedures for Design of Water-Resistance Coal (Rock) Pillar and Mining under Buildings, Water Bodies, Railways and Major Tunnels".

Water-resistance coal (rock) pillar for protection of borehole

Design of water-resistance coal (rock) pillar is consistent with the method shown in Fig. 10.2 after the coordinates of the intersection between borehole and coal seam are determined. Calculation of the coordinates takes into account inclination of the borehole. If no inclination data is available, the calculation should error to the safety side.

Water-resistance coal (rock) pillar for adjacent mine

- Water-resistance coal (rock) pillar can be designed with the vertical method where mine hydrogeological type is from simple to medium, however, the total width should not be less than 40 m.
- Where mine hydrogeological type is complex or extremely complex, design of water-resistance coal (rock) pillar should be based on site-specific conditions such as coal seam occurrence, geological structure, hydrostatic pressure, movement angle of overlying formation, and height of water-conducting fracture zone.

  a. For multi-seam mining where the interval between the upper and lower layers of coal is smaller than the height of the water-conducting fracture zone induced by mining of the lower layer, the water-resistance coal (rock) pillar at boundary of the lower coal shall be based on rock movement angle in the uppermost layer of coal and the distance between coal seams, as shown in Fig. 10.7a.
  b. When the vertical distance between the upper and lower layers of coal is greater than the height of the water-conducting fracture zone after mining of the lower coal seam, the water-resistance coal (rock) pillar of the upper and lower coal seams may be designed separately (Fig. 10.7b).

The upper width of water-conducting fracture zone (L_y) can be calculated by the following equation:

$$L_y = \frac{H - H_L}{10} \times \frac{1}{T_s} \geq 20\,m$$

where

L_y upper limit width of water-conducting fracture zone (m);
H height of hydrostatic water column above coal seam floor (m);
H_L maximum height of water-conductive fracture zone (m);
T_s ratio of water pressure to rock pillar width, 1 as a default value.

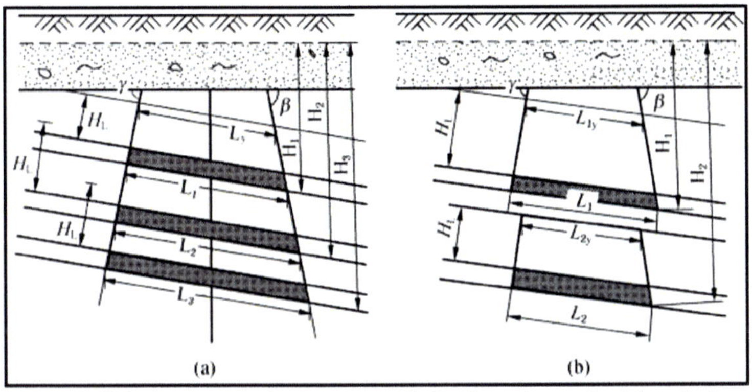

H_L—upper limit of water-conducting fracture zone; H_1, H_2, H_3—static water level above the floor of each coal seam; γ—moving angle of uphill rock formation; β–moving angle of downhill rock formation; the L_y, L_{1y}, L_{2y}—water-conducting fracture zone width; L_1—width of upper water-resistance coal pillar; L_2, L_3—width of lower water-resistance coal pillar

Fig. 10.7 Diagram of water-resistance coal (rock) pillar for multiple coal seams at boundary with other mines

Water-resistance coal (rock) pillar for adjacent mine at fault

For mines that are bounded by faults, water-resistance coal (rock) pillar at the boundary can be reserved with reference to method for faults. However, the coal seam characteristics of the other mine should be considered so that the coal (rock) pillar on the other side is not damaged. In addition to the method for fault, Fig. 10.8 can be used as a reference.

4.2.4 No change is allowed on water-resistance coal (rock) pillar once the design is finalized. Mining and excavation activities are strictly prohibited in any type of water-resistance coal (rock) pillar.

4.2.5 When mining of abandoned water-resistance coal pillar occurs under flooded area, the mine pool shall be completely drained. Operations are strictly prohibited where the mine pool is not drained.

4.2.6 In mines with a history of water inrush or mining with potentiometric pressure, isolation mining shall be implemented including segregation isolation and mining level isolation. Before partition, water-resistance coal (rock) pillar should be retained, and waterproof sluice gate should be constructed to control water flow, reduce disaster, and ensure continuous operation in the event of water inrush.

10.5 Dewatering System

4.3.1 Normal dewatering activities shall be guaranteed by equipping mines with pumps, conveyance pipes, power equipment, and water tanks that are compatible with the amount of water flowing into the mines.

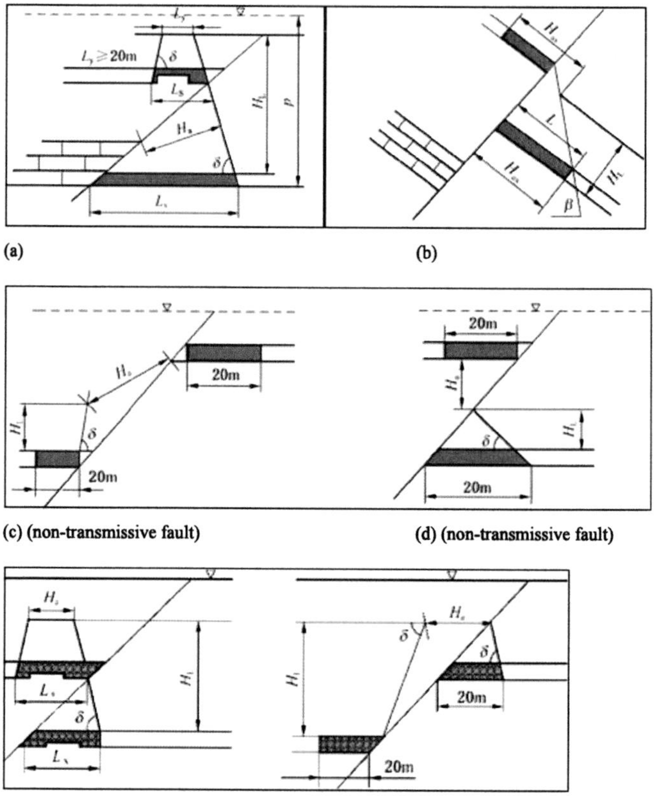

(a) (b)

(c) (non-transmissive fault) (d) (non-transmissive fault)

(e) (f)

L—coal pillar width; L_s, L_x—coal pillar width of upper and lower coal seams, respectively; L_y— upper limit rock pillar width at conducting water fracture zone; H_a, H_{as}, H_{ax}—required thickness of water-resistance rock pillar; H_L— upper limit of water-conducting fracture zone; P—hydrostatic pressure exerted on aquitard

Fig. 10.8 Diagram of water-resistance coal (rock) pillar for mines bounded by faults

4.3.2 Mine dewatering equipment shall meet requirements of mine drainage. There should be working water pumps and backup water pumps exclusive of water pumps being repaired. Capacity of the working water pumps should be sufficient to pump out within 20 h the water that is accumulated over 24 h in the mine under normal water inflow conditions (including filling water and other water). Capacity of the backup pumps should be no less than 70% of the capacity of the working pumps. The total capacity of the working and backup pumps shall be capable of pumping out within 20 h the maximum water inflow of the mine over 24 h. Capacity of the water pumps under repair should be no less than 25% of the capacity of the working pumps.

In addition to complying with the provisions of the first paragraph of this article, mines with complex or extremely complicated hydrogeological conditions may increase the drainage capacity of existing pumping or reserve space in the main pump station for installing additional pumps.

The conveyance pipes should have spare capacity. The capacity of the conveyance pipes should be able to discharge within 20 h the amount of water flowing into mine over 24 h under normal conditions. The total capacity of the working and backup conveyance pipes shall be capable of discharging within 20 h the maximum water inflow accumulated over 24 h in the mine.

Power equipment should be compatible with the total capabilities of working, backup and repair pumps. The power should be adequate to operate all pumps simultaneously.

Mines with risk of water inrush and flooding can be equipped with additional dewatering system in response to water disaster.

4.3.3 The main pump station of the mine shall have at least two emergency exits: one for the inclined tunnel leading to the shaft and at least 7 m higher than the pump station floor and the other connected to the parking lot at bottom of the shaft. A user-friendly door should be installed in the exit passage leading to the parking lot at bottom of the shaft. The door needs to be waterproof and fireproof. A reliable control valve should be installed within the tunnel between the pump station and the water tank.

4.3.4 Main water tank of a mine shall be accompanied with a standby water tank. When one water tank is under maintenance, another one can be used.

When normal water inflow is less than 1,000 m^3/h in mines that are newly constructed, under expansion, or mining on new level, effective capacity of the main water tanks should be able to accommodate the amount of water flowing into the mines over 8 h.

For mines with normal water inflow greater than1,000 m^3/h, the effective capacity of the main water tank can be calculated as follows:

$$V = 2(Q + 3,000)$$

where

V Effective capacity of the main water tank (m^3);
Q Normal water inflow per hour into mine (m^3)

Effective capacity of the water tank in a mining area should be able to accommodate 4 h normal water inflow in the mining area.

In mines that are characterized with large difference between the maximum water inflow and normal water inflow, the dewatering capacity and water tank capacity shall be designed by qualified institutions and reviewed and approved by the chief engineer of the coal mining enterprise.

A sieve should be placed at the entrance of the water tank. A sedimentation tank should also be constructed in mines that are being backfilled with water and sand, mined by hydraulics, and characterized with a lot of impurities in water. The empty storage capacity of the water tank should always be maintained at more than 50% of the total capacity.

4.3.5 Pumps, conveyance pipes, gate valves, power equipment and transmission wires shall be inspected and maintained regularly. They should be checked before rainy season each year. A joint dewatering exercise should be performed on all working pumps and backup pumps. Any identified problems shall be resolved timely.

Silt in water tank, sedimentation tank and drainage ditch shall be cleaned up in time, at least once before rainy season each year.

4.3.6 If adits are used to discharge mine water, the water passing capacity of the adit shall not be less than 1.2 times the maximum of historical inflow rates into the mine. Elevation of the drainage ditch or drainage tunnel shall be lower than that of the main transport tunnel.

4.3.7 During construction of new mines in hydrogeologically complex and extremely complex areas, submersible pumping pits shall be retained at bottom of the shaft. Similar submersible pumping pits should be contracted in old mines. At completion of a shaft, a temporary dewatering facility with a certain capacity should be constructed near the bottom of the shaft, which ensures safe construction of temporary substation and water tank.

4.3.8 In mines that are under construction, temporary dewatering facility with sufficient capacity shall be available in each construction area before establishment of the permanent dewatering system.

4.3.9 Extension of mining to lower level can take place only after water control and dewatering facilities are in place.

10.5.1 Sluice Gate and Sluice Wall

4.4.1 Sluice gate shall be constructed at parking lot near shaft bottom in mines with complicated hydrological and geological conditions. Alternatively, a submersible pump station is installed with a dewatering capacity no smaller than the maximum water inflow in addition to the normal dewatering system.

4.4.2 Sluice gate shall be constructed in the vicinity of a mining area where there is a risk of water inrush. If site-specific conditions do not allow for construction of the sluice gate, other strict water prevention and control measures shall be in place upon approval of an authority in charge of the coal mine enterprise.

4.4.3 Construction of sluice gate shall meet the following requirements:

- Design sluice gates by qualified institutions and use stereotype designs.
- Construction and quality of sluice gates meet design requirements. No leak is allowed in gate and gate chamber.
- Concrete linings of no less than 5 m thick are constructed on both ends of the sluice gate chamber. High-grade cements are used to reinforce gate chamber and sill with designed grouting pressure. Concrete is used to fill in empty spaces.

- Add 1 sieve door 15-25 m upstream from the sluice gate. Not parking and debris piling are allowed between the sluice gate and the sieve door. When water inrush occurs, close the sieve door first and then close the sluice gate. If water flows to the sluice gate in two directions, sieve doors are constructed on both sides.
- Sluice gate tracks, overhead power lines, and conveyor belts can be removed easily. The pressure resistance of various pipes passing through sluice gate and valves installed outside the gate is consistent with the design pressure of the sluice gate. Cables and pipes passing through sluice gate walls are tightly sealed with plugs and valves, and no water leakage is allowed.
- Sluice gate is equipped pressure transducers for observing water pressures and consists of drainpipes and drain valves.
- Verification and acceptance are performed at completion of sluice gate in accordance with design requirements. For newly constructed sluice gates, water injection test shall be carried out to test their resistance against water pressure. Test length inside the sluice gate shall not exceed 15 m, and the test pressure shall not be lower than the design water pressure. Keep a constant pressure for at least 24 h during the test. Safety measures should be in place during the pressure test.

4.4.4 Sluice gate shall be flexible and reliable. Closure tests shall be carried out twice a year, one before rainy season. Tools and spare parts for closing sluice gate shall be kept by a dedicated person and stored in a dedicated place. Borrow for other purposes is prohibited.

4.4.5 Underground sluice gate shall be designed by qualified institutions. Construction shall be done according to the design. Sluice date should not be put into use until it is verified and accepted in accordance with regulations.

4.4.6 When abandoned tunnels are sealed off, drainage holes shall be reserved in closed sluice gates located at bottom of the inclined tunnels. Observation frequency at the drainage holes is monthly and more frequent during rainy seasons.

10.5.2 Dewatering and Mining with Potentiometric Pressure

4.5.1 Dewatering is required where mining induced fractures connect to overlying water-rich aquifers.

The maximum height of mining induced collapse zone and fracture zone can be determined by in situ measurements and relevant equations presented in the "Provisions for Coal Pillar Retention and Coal Mining in Buildings, Water, Railways and Main Tunnels".

4.5.2 Dewatering may be required when the coal seam is gently inclined and is overlain by an unconsolidated and water-rich aquifer. Focused hydrogeological exploration or supplemental investigation shall be conducted to understand hydrogeological conditions. The results are used to determine areas to be dewatered and formulate a dewatering plan. Dewatering can be implemented upon approval by the chief engineer of the coal mine enterprise.

4.5.3 Dewatering and mining coal seams overlain by semi-consolidated or unconsolidated aquifers should comply with the following provisions:

- Identify distribution of quicksand layer and understand its phase transition and genetic type.
- Investigate water content and liquefaction threshold of the quicksand layer; estimate water inflow rate and dewatering feasibility; establish a dynamic observation network; monitor dewatering rate and radius of influence.
- Collect data for analysis of the following parameters: height of mining induced water-conducting fracture zone, water-sand separation method, angle of sand piling, minimum vertical distance of water-sand inrush at opening of tunnel, and safe distance on advanced dewatering.
- Conduct studies for prediction and treatment of ground subsidence or collapse caused by water and sand inrushes.

4.5.4 Advanced dewatering is necessary when mining induced fracture zone connects to a water-rich aquifer or water body. Focused hydrogeological exploration and testing shall be conducted to develop dewatering schemes, select dewatering type and method, and perform comprehensive evaluation of mining feasibility by dewatering and technical and economic practicability. The dewatering plan needs to be approved by the chief engineer of the coal mining enterprise.

4.5.5 Qualitative and quantitative analysis shall be carried out in the process of mine dewatering. The "three maps—two forecasts method" may be applied to predicting and assessing roof water hazard. If conditions allow, numerical modeling techniques should be used to simulate and predict changes in the height of mining induced fracture zone, amount of water to be pumped out, and dynamics of groundwater flow field.

4.5.6 Mining with potentiometric pressure can be a viable option when the aquitard between the confined aquifer and the mining coal seam can withstand the exerted potentiometric pressure; the aquitard is not easily destroyed by mining; and the water inrush risk from the underlying aquifer is minor. Mining with potentiometric pressure has to be accompanied with water control measures, which are approved by the chief engineer of the coal mining enterprise.

Below describe equations to calculate water inrush coefficient and required aquitard thickness for safe mining.

Equation to calculate the required thickness of aquitard for safe mining

$$t = \frac{L\left(\sqrt{r^2L^2 + 8K_pP} - \gamma L\right)}{4K_p}$$

where

t	Required aquitard thickness for safe mining (m);
L	Width of tunnel floor (m);
γ	Average specific gravity of aquitard (MN/m^3);

K_p Average tensile strength of aquitard (MPa);
P Potentiometric pressure exerted to aquitard (MPa)

This equation is applicable to tunneling. Excavation is unsafe when actual thickness of the aquitard is less than the calculated value.

Equation to calculate water inrush coefficient

$$T = \frac{P}{M}$$

where

T Water inrush coefficient (MPa/m);
P Potentiometric pressure exerted on bottom of aquitard (MPa);
M Thickness of aquitard (m).

This equation is more applicable to mining panels. Based on the nation-wide statistical data, water inrush coefficient is generally not greater than 0.06 MPa/m in areas with geological structures. Water inrush coefficient is not more than 0.1 MPa/m in areas without geological structures.

4.5.7 The following provisions shall be observed before mining when the aquitard between the confined aquifer and the mining coal seam cannot withstand the exerted potentiometric pressure:

1. Dewater and depressurize to decrease the potentiometric pressure of the confined aquifer to the allowable level for safe mining. Water control safety measures approved by the chief engineer of the coal mining enterprise need to be in place. To guide safe production, detailed analysis should be conducted to evaluate the safe hydraulic head suitable for the mine. Green mining and resource optimization should be considered that combine dewatering with water supply and eco-environment protection of the mining area.
2. Grout curtain may be installed in advance to cut off water recharge. Dewatering can then be carried out inside the curtain for safe mining. Pre-requirement of using grout curtain technique is the knowledge of primary recharge boundaries of the confined aquifer.
3. When mines are not amenable to either dewatering or installation of grout curtain, grouting can be conducted in the underlying aquitard to reinforce its strength or grouting in underlying aquifer to transform it into a weak aquifer. Implementation of this mining method requires preparation of a special design and trial mining. Trial mining has to be accompanied with adequate water control measures. Additional measures that shall be approved by the chief engineer of the coal mining enterprise need to be prepared to prevent flooding of the mine.

Below describes equations to calculate the safe hydraulic pressure:

Safe hydraulic pressure for tunneling

$$P = 2K_p \frac{t^2}{L^2} + \gamma t$$

where

- P Safe hydraulic pressure that aquitard can withstand (MPa);
- t Thickness of aquitard (m);
- L Width of tunnel (m);
- γ Average specific gravity of aquitard (MN/m^3);
- K_p Average tensile strength of aquitard (MPa)

Safe hydraulic pressure for coal mining panel

$$P = T_s M$$

where

- M Thickness of aquitard (m);
- P Safe hydraulic pressure (MPa);
- T_s Critical water inrush coefficient (MPa/m)

The T_s value should be determined according to actual data of the mining area. Under normal circumstances, 0.06 MPa/m is considered in areas with geological structures, whereas 0.1 MPa/m is considered in areas where aquitard is not affected by geological structures.

4.5.8 In mines where required data are available the "Vulnerability Index" method or "Five Maps—Two Coefficients" method can be used to evaluate water inrush risks for different region. Analog, analytical and numerical methods can be used to predict the maximum water inflow.

10.5.3 Grouting Techniques

4.6.1 Advanced grouting is required in shaft construction under the following provisions:

- When shaft is expected to sink through a thick fractured aquifer or multi-layered thin fractured aquifer, advanced grouting may be used on surface.
- Prior to development of grouting plan, pilot test boreholes shall be drilled to obtain preliminary hydrogeological data such as depth, thickness, lithology of aquifers.

Other groundwater data can be collected through groundwater monitoring, pumping (injecting) test, and water chemistry analysis.

- Grouting starts in competent rock formation below weathering zone and terminates at bottom of the aquifer underlying the shaft or at depth exceeding shaft bottom by 10–20 m.
- Grouting may occur directly in working face of the shaft if the aquifer is poor in water storage.

4.6.2 Sealing off water inrush points by grouting shall meet the following requirements:

- Determine location of the water inrush point; analyze geological structure near the water inrush point; identify shape of potentiometric pressure cone of depression, compare dynamic changes in monitoring well, borehole, and spring before and after the water inrush, and conduct tracer test if necessary.
- Investigate water sources such as aquifers that supply recharge to the water inrush point and their significance and characterize nature and size of water passageways that leads water sources to the water inrush point.
- Conduct tracer test and pressure (injection) test before and after grouting, respectively, to seal off the water inrush point and compare water inflow rates into the mine before and after grouting.
- Grouting working plan must be prepared to detail the objectives and activities. The plan must be approved by chief engineer of the coal mine enterprise before its implementation.

4.6.3 Installation of grout curtain requires a feasibility study.

Once installation of grout curtain is proposed as an alternative to mitigate water hazard, specific hydrogeological conditions need to be understood such as stratigraphic sequence, geological structure, boundary condition, impermeability at ends of the grout curtain that is in contact with either inactive fault or aquitard, seepage of groundwater to the mine, groundwater flow velocity, and flow direction.

Installation of grout curtain can proceed only after grouting plan is prepared and approved by chief engineer of the coal mine enterprise.

4.6.4 Advanced water exploration should be carried out when underground tunneling passes through water-conducting fault, fracture zone, and collapse columns, which are hydraulically connected to rivers, lakes, caves, and aquifers. If water is present in the front, construction is not allowed until advanced grouting is conducted to plug and reinforce the formation. If necessary, a sluice gate can be constructed in advance or other water prevention measures can be taken. Shafts or tunnels that pass-through aquifers shall be reinforced by grouting in accordance with requirements on water control.

4.6.5 No mining is allowed when faults, fractures or collapse columns are present in a working panel unless water-resistance coal (rock) pillar is designed and reserved per regulations. Alternatively, grouting can be used to seal off water-conducting passageways. Geophysical survey may be used to investigate hydrogeological conditions in working panel to be grouted. Grouting holes are placed based on the geophysical survey results. Geophysical survey and drilling can be used again to verify the effectiveness of grouting transformation.

4.6.6 A dedicated grouting team may be established in mines that are characterized with large water flow and high risk of water inrush. The team is solely responsible for grouting and water hazard control.

4.6.7 Obsolete local dewatering wells/holes shall be abandoned properly by grouting at completion of mining activities in each mining panel. Their positions shall be marked on relevant maps.

4.6.8 Before abandoned mine is inundated, geophysical, geochemical, and drilling explorations shall be used to detect damage of water-resistance coal (rock) pillars at boundaries of the mine and potential water break-in sections. Any potential hydraulic connections between the abandoned mine and adjacent production mines shall be mitigated by grouting to present water hazard from occurring in production mines.

10.6 Underground Water Exploration and Release

5.1 Mines that are affected by water hazard in mining panels need to comply with the principles: predicting and forecasting, exploring if safety is uncertain, exploring prior to excavating/mining, and mitigating as needed prior to excavating/mining. In addition, the following provisions shall be observed:

- Based on annual mining plan, hydrogeological data shall be analyzed at beginning of each year, identifying potential water hazards, and developing water hazard prediction tables and risk level maps.
- During mining, the forecast maps and tables are verified and updated monthly. Discovery of water hazard shall be notified of mine dispatching office, who issues water hazard announcement and inform those threatened by water hazard to evacuate to safe place.
- Annual and monthly water hazard prediction maps and tables shall be submitted timely to mine chief engineer and production safety department.

5.2 Mines with hydrogeologically complex or extremely complex conditions shall, when the hydrogeological conditions cannot be fully determined on surface techniques, shall adhere to the principle of exploring prior to excavating/mining. Artesian flow testing can be conducted in the underground.

5.3 Drilling, geophysical and geochemical methods should be used to investigate hydrogeological conditions in areas threatened by water hazard prior to roadway excavation. Hydrogeological analysis reports and proposed water hazard prevention measures shall be prepared submitted to mine chief engineer. The reports shall be reviewed and approved by those who are responsible for production, safety, geological survey, and regulation.

5.4 Drilling, geophysical, and geochemical methods should be used to characterize faults, collapse columns, and aquifer within mining panels prior to mining. Hydrogeological analysis reports and proposed water hazard prevention measures shall be prepared submitted to mine chief engineer. The reports shall be reviewed and approved by those who are responsible for production, safety, geological survey, and regulation. If faults, fractures or collapse columns are found to be hydraulically conductive, safety measures such as grouting reinforcement or retaining water-resistance coal (rock) pillar shall be adopted. Otherwise, no mining is allowed.

5.5 Groundwater investigation and artesian flow testing shall be conducted if mining work face encounters one of the following conditions:

- Approaching to abandoned mines with flooded mine pools or neighboring mines.
- Approaching to aquifers, water-conducting faults, underground rivers, caves, and water-conducting collapse columns.
- Breaking in water-resistance coal (rock) pillar for water release.
- Approaching to fractured zones of a fault that may be connected to river, lake, reservoir, retention basin, or supply well.
- Approaching to drilling holes that may discharge water.
- Approaching to areas with complicated hydrogeological conditions.
- Confined aquifers or water-bearing structures are within mining induced fracture zone or thickness of water-resistance coal (rock) pillar retained between aquifer and coal seam not clear.
- Close to grouting area with accumulated water.
- Close to areas where water inrush may occur.

Investigation plan should be prepared before exploring groundwater and plot on plane map for the mining area.

5.6 Groundwater detection design shall be prepared to discharge water in the underground. Exclusion zone should be determined for the exploration. Safety measures such as prevention of methane and other harmful gases shall be in place. The layout and lead distance of the exploration and drainage boreholes shall be determined based on height of potentiometric pressure head, thickness, and hardness of the coal (rock) layer. The work plan design is developed by geological surveying organization and approved by a committee organized by mine chief engineer.

5.7 Below are provisions that shall be followed in placement of boreholes for groundwater exploration and artesian flow discharge:

- When abandoned mine pools, collapse columns and boreholes are targets of exploration, boreholes are arranged in groups and are in fan-shape in both horizontal plane and vertical plane in front of the tunnel. Horizontal distance between boreholes is less than 3 m at their terminals. In thick coal seams, the distance is less than 1.5 m.
- When faults and karst features are targets of exploration, boreholes are arranged in from of tunnel and at formations below tunnel. The number of boreholes drilled into formation below tunnel should not be fewer than 2.

- No artesian flow discharge is allowed in coal seam that is connected to water-conducting fault, confined aquifer or collapse column and potentiometric pressure within these discontinuities is higher than 1 MPa. If it is absolutely necessary, flow discharge can take place after sluice gate is constructed, and groundwater can be investigated and discharged outside the sluice gate.
- Groundwater exploration in upper tunnel is typically conducted using dual tunneling method. One tunnel is used to discharge and collect artesian water, whereas the other is used to evacuate people in case of a need. A communication lane is excavated every 30–50 m between the dual lanes. Sluice gate is also necessary.

5.8 Specialized drilling rigs are used for underground groundwater exploration and artesian flow discharge. Drilling rigs used for coal exploration are strictly forbidden.

5.9 Before installing a drilling rig for groundwater exploration and artesian flow discharge, the following provisions shall be met:

- Reinforce roadway support near the boreholes and install solid bars and plates in front of working face.
- Clean roadway and dig drainage ditch. Dewatering facility should be installed with capacity compatible with the amount of water to be released when boreholes are located at the lower end of roadway.
- Install dedicated phone and phone line at or near the drilling location.
- Mark primary drilling locations by professional surveyors per design. Those who are responsible for exploration and release of artesian water must be onsite to determine orientation, inclination, depth, and number of boreholes.
- Install casing in advance if drilling occurs in areas with potentiometric pressure greater than 0.1 MPa. A safety valve is installed at the casing port, and the casing depth is specified in groundwater exploration and artesian water release design. Excavate evacuation chambers in advance and formulate exit strategy. Familiarize workers with evacuation routes.
- Drilling equipment shall be equipped with backflow pressure and blowout prevention device when potentiometric pressure in boreholes is greater than 1.5 MPa. Safety measures should be in place to prevent sudden bulging at pipe orifice and coal (rock) wall.

5.10 Borehole diameter shall not exceed 75 mm at termination unless they are used for discharging or sealing off water.

5.11 Advance distance of exploration borehole and length of water-seal casing shall comply with the following provisions:

- Lead drilling distance for abandoned mine pools is determined based on water pressure, thickness and strength of the coal (rock) layer and safety measures. The minimum horizontal distance shall not be less than 30 m, and the length of the water-seal casing shall not be less than 10 m;
- Table 10.1 presents the required lead distance and casing length when exploration and artesian water release occur in confined aquifer, fault and collapse column.

Table 10.1 Lead distance for drilling hydrogeological borehole and length of water-seal casing

Water pressure (MPa)	Lead distance for drilling (m)	Length of water-seal casing (m)
<1.0	>10	>5
1.0-2.0	>15	>10
2.0-3.0	>20	>15
>3.0	>25	>20

5.12 If water rush symptoms are observed during drilling, drilling should stop immediately without pulling out drilling rod. Water inrush symptoms include softening of coal or rock, collapse of side walls, water pressure increases, sudden increase of water flow, and resistance to drilling. Mine dispatching office should be informed immediately, and trained personnel be sent to monitor water situation. If situation is found to be serious, all personnel in area threatened by the water should be immediately evacuated to a safe place. Mitigation measures can follow to prevent water disaster.

5.13 Exploration and release of abandoned mine pools requires analysis and understanding of their spatial locations, accumulated water volume, and water pressure. Boreholes for exploration and water release should be drilled into the mine pools. Process of water discharge should be monitored. Amount of water discharged should be checked until the mine pools are drained. When borehole approaches abandoned mine pools, methane or other harmful gas may flow out. Gas inspectors or mine rescuers should be on duty at the scene to conduct air monitoring. If concentration of methane or other harmful gases exceeds their regulation limits, drilling should stop immediately. At the meantime, power supply should be cut off, and all personnel be evacuated. Mine dispatching office should be informed of actions and processes.

5.14 Amount of water should be estimated before water discharge occurs. Discharge flow should be controlled to prevent flooding based on pumping capacity of mine and capacity of water tank. Dedicated personnel should be arranged to monitor water discharge progress and document water pressure and flow in boreholes. Dedicated personnel take prompt actions in response to dramatic changes in water flow and report to mine dispatching office.

10.7 Coal Mining Below Water Bodies

6.1 Water-resistance coal (rock) pillars shall be designed and retained for coal mining under surface water bodies such as rivers, lakes, reservoirs, and oceans. For mining under unconsolidated aquifers, different types of water-resistance coal (rock) pillars (water-resistance pillar, sand-resistance pillar or anti-collapse pillar) should be set based on extent of deformation. For mining below bedrock aquifers or water-bearing fault zones, permeability of overlying formations and hydraulic connection between aquifers should be analyzed and evaluated before and after mining to decide better method of two alternatives: use of water-resistance coal (rock) pillar and dewatering.

6.2 Design of water-resistance coal (rock) pillars for coal mining under water bodies shall be based on hydrogeological and engineering geological conditions of the mine, mining method, mining height, and roof control method. Regulatory guidance is "Provisions for Coal Pillar Retention and Coal Mining in Buildings, Water, Railways and Main Tunnels". Feasibility study and mining design shall be prepared by qualified and accredited (Grade B or above) a coal mine design institution and submitted to provincial coal industry management department for approval. Mining processes must comply with the approved design requirements, in particular, mining area, mining height and dimension of water-resistance coal (rock) pillar.

6.3 Feasibility study shall be conducted if mining reveals that water-resistance coal pillars can be smaller because of changes in geological conditions. Trial mining can start once the feasibility study is approved by provincial coal industry management department.

6.4 Focused hydrogeological and engineering geological exploration shall be carried out on overlying strata of mining coal seam to reasonably determine size of water-resistance coal (rock) pillar.

Focused hydrogeological and engineering geological exploration includes the following:

- Identify hydrogeological structure of overlying strata related to coal mining, including thickness and distribution of aquifers and aquitards, water level, water quality and water-richness of aquifers, hydraulic connection between aquifers, recharge, runoff, and discharge conditions, water-richness and hydraulic conductivity of faults.
- Use drilling, geophysics, or other methods to ascertain undulations of bedrock surface and thickness of bedrock above the working face. Determine thickness and distribution of aquitard at bottom of unconsolidated layer when coal seams are overlain by unconsolidated aquifers.
- Characterize engineering properties of overlying strata through core logs and borehole geophysical logging. Determine engineering geological type and combinations and structural features of overlying rock layers. Collect rock and soil samples to analyze geotechnical parameters.

6.5 Design of water-resistance coal (rock) pillar for mining under water bodies shall be based on mining induced fracture angle in combination with extent of deformation.

"Provisions for Coal Pillar Retention and Coal Mining in Buildings, Water, Railways and Main Tunnels" is the general guidance to design size of water-resistance coal (rock) pillar and to calculate heights of collapse zone, fractured zone, and protective zone. Empirical data from mines with similar geological conditions can be used as reference in combination with model-based mechanical analysis and numerical simulation. Superposition analysis of natural water-conducting over mining induced fractures should be considered in overlying strata. For mining areas involved in mining under water bodies, in situ measurements shall be conducted to determine height and extent of mining induced collapse zone and fracture zone. Gradually accumulated in situ data can guide mining activities in the mine.

Thickness of protective layer used for gravity coal mining shall be determined based on analysis of overlying rock layer structure and lithology, roof collapse zone, water-conducting fractured zone, and mining experience. Where mining requires retaining of sand-resistance or anti-collapse coal (rock) pillars columns, seepage resistance tests shall be conducted to determine the critical hydraulic gradients in overlying soil layer and weathering zone so that neither water inrush nor sand inrush occurs.

6.6 The following requirements shall be observed in mining under water bodies:

- Adopt coal mining methods that effectively control mining height and range to prevent overlying strata of steeply inclined coal seams from sky lighting to surface bodies. If a high-angle fault is present in working panel, effective measures are taken to prevent the fault from directing water or causing collapse of overlying strata along the fault zone.
- Use layered longwall mining methods if coal seams that are under water are gently inclined. Minimize mining thickness of the first and second layers. Time lapse between different layer mining at the same position is no less than 4-6 months. The lapse time can be longer if overlying strata are hard. Reserve sand control and anti-collapse coal (rock) pillars. Use gravity mining method and start with trial mining before promoting.
- It is strictly forbidden to mine steeply inclined coal seams under water bodies.
- Use intermittent coal mining method for mining of coal seam groups. If safe mining cannot be guaranteed, coal pillar design shall be modified to increase sizes of the coal pillar.
- Adopt such mining methods as backfilling, strip mining or limiting mining thickness that effectively control development of mining induced fracture zone for the following scenarios: no apparent aquitards between surface water bodies and coal seams; strong water-rich aquifers overlying coal seams with no aquitard at bottom; shallow and thick coal seams overlain by mediocre aquifers; and predicted water-conducting fracture zone exceeding distance between water body and coal seam. Dewatering should be considered to ensure safe mining if overlying aquifers are amenable to be drained.

6.7 Strengthen monitoring efforts on water conditions and deformation at bottom formation of water body when conducting mining activities under water bodies. At completion of trial mining, a summary report shall be submitted with detailed analysis on mining induced fractures and deformation to overlying strata.

10.8 Water Prevention and Control in Open-Pit Mines

7.1 Open-pit coal mines shall formulate water prevention and dewatering plans and measures at beginning of each year. A comprehensive inspection of prevention and dewatering facilities should be conducted before rainy season. Berms, dams, and drainage ditches shall be constructed around buildings that are lower than local flood level.

7.2 Construction of water prevention and drainage facilities on surface and slope shall avoid areas with potential risk of landslide. Drainage ditches should be inspected and dredged frequently to prevent leakage, backflow, or overflow. Ditches should be constructed around the landslide area in the open pit to intercept water. Anti-seepage measures shall be taken where ditches pass through slopes with deformation and cracks.

7.3 When water storage occurs at bottom of an open pit for dewatering, pumping period shall comply with the following provisions:

- Not exceed 15 days when number of working faces that stop mining due to water storage is less than 1/3 of total number of working faces.
- Not exceed 7 days when number of working faces that stop mining due to water storage accounts for 1/3-1/2 of total working face.
- Not exceed 3 days when number of working faces that stop mining due to water storage is more than 1/2 of total working faces.
- Capacity of backup pumps shall not be less than 50% of capacity of active pumps when roadways are used for water storage and dewatering.

7.4 Advance dewatering shall be conducted to drain any water-bearing formations that are present and affect normal operation of mining. Dewatering should precede mining.

If earth fissures or collapses occur within cone of depression caused by dewatering, land use control measures shall be implemented to protect the area, install warning signs, and take safety measures. Pumping station partially located underground should be equipped with ventilation.

7.5 Groundwater level, potentiometric pressure and water flow should be monitored on slopes that are greatly affected by groundwater and dewatering activities. Data should be analyzed to determine influence of groundwater on slope stability and effectiveness of drainage and develop effective measures on groundwater management.

7.6 Groundwater dewatering shall be carried out if increase in groundwater level may cause landslides or slope instability.

10.9 Water Disaster Rescue

10.9.1 Emergency Plan and Implementation

8.1.1 Coal mines shall formulate emergency plans and on-site rescue plans in response to water disasters. The plans are based on types of water hazards and possible water disasters in the mine. Content of the emergency plan should be focused, scientific and operative. The rescue plan shall include specific measures for safe evacuation of personnel when unforeseen water hazards occur. The emergency plan shall be updated yearly. One rescue drill shall be conducted every year.

8.1.2 Mine management personnel and dispatching office personnel shall be familiar with the water disaster emergency plan and on-site rescue plan.

8.1.3 Mines shall establish emergency exits and stipulate evacuation routes to avoid inundation. Road signs on evacuation routes shall be clearly posted with reflective films. All employees should be familiar with the routes and can safely evacuate in response to water inrushes to avoid accidental fatalities.

8.1.4 Unmanned telemetry monitoring and control system shall be used when feasible in underground pumping stations. Dewatering system shall be inspected and maintained regularly to ensure normal operations. Water tank capacity should be kept at least 50%. Mines that are seriously threatened by water hazard should use unmanned telemetry monitoring of underground pumping stations and promote use of ground-controlled submersible pump systems.

8.1.5 First responders who discover water disasters on scene shall immediately inform mine dispatching office of location of the water inrush and conditions. Surrounding personnel should be notified to evacuate to a safe place or lift to surface.

8.1.6 Mine dispatching office shall immediately initiate mine water disaster emergency plan as soon receiving first responders' report, determining safe evacuation route for the personnel according to direction, location and water quantity of incoming water and notifying underground personnel in affected area to evacuate to a safe place or lift to surface. Those who are in charge of the mine and surrounding mines should be informed of the waster disaster.

8.1.7 Immediately prepare to close sluice gates when a water inrush occurs. However, the sluice gates can be closed only after it is confirmed that all personnel have been evacuated.

8.1.8 Adjust underground ventilation system timely based on degree of influence by flooding to avoid airflow disturbance and exceedance of harmful gases over regulatory limits.

8.1.9 Incorporate case studies of preventing coal mine accidents caused by storm floods into "Accident Emergency Rescue Plan" and "Disaster Prevention and Mitigation Plan". Materials, equipment and funds needed to prevent rainstorms and floods should be in place. Establish a professional rescue team or sign an agreement with a professional rescue team.

8.1.10 Strengthen communications with rescue and disaster relief agencies at all levels and master status of rescue technical equipment. Once a water disaster occurs, immediately initiate corresponding emergency plans, strive for assistance from society, and implement rescue.

8.1.11 Water disaster accidents shall be reported to relevant government departments in accordance with relevant regulations. Accident reports shall not be delayed; incomplete; false; or fraudulent.

10.9.2 Recovery of Flooded Mine by Dewatering

8.2.1 Before recovering a flooded well, an investigation report on water inrush should be prepared. The report should include the following main elements:

- (1) The process of water inrush, the location of water inrush, the time of water inrush, the form of water inrush, the analysis of water source, the change of flooding speed and water inflow.
- (2) The extent of water inrush and the estimated amount of water accumulated.
- (3) The amount of water in the drainage is expected. Find out the amount of water inflow in all parts of the well before flooding, calculate the maximum water inflow and stable water inrush at the water inrush point, estimate the amount of water inflow at different elevations, and design the displacement curve during the recovery process;
- (4) Provide dynamic data and curves related to hydrogeological points (holes, wells, springs) for analysis of water inrush, hydrogeological plan, section view, mine water filling map and water chemistry data.

8.2.2 When a mine is restored, it shall be provided with a special person to regularly measure the amount of water inflow and the elevation of the water surface, and make a record; observe the record of the top, the gang and the water in the well after recovery; observe the record The specific location of the water point, the amount of water inflow and the temperature of the water, etc., and make a sketch of the water inrush point; regularly observe the hydrogeological points such as the ground observation hole, well, spring, etc., and observe whether the ground has collapse or crack phenomenon.

8.2.3 Before the accumulation of water in the wellbore and downhill and the restoration of the flooded roadway, safety measures shall be taken to prevent the sudden emergence of harmful gases sealed by water. During the drainage process, the mine rescue team should check the air content on the water surface; find harmful gases and deal with them in time.

8.2.4 After the mine is restored, the drawings and materials of the two processes of flooding and restoration shall be comprehensively compiled, the cause of water inrush shall be determined, measures for avoiding repeated accidents shall be proposed, and the experience of hydrogeological work in drainage restoration shall be summarized. lesson.